Footprints of the Forest

T0262527

Biology and Resource Management in the Tropics Series
Michael J. Balick, Anthony B. Anderson,
and Kent H. Redford, Editors

Footprints

COLUMBIA UNIVERSITY PRESS
New York

of the Forest

Ka'apor Ethnobotany –
the Historical Ecology of
Plant Utilization by
an Amazonian People

William Balée

Biology and Resource Management in the Tropics Series
Michael J. Balick, Anthony B. Anderson, and Kent H. Redford, Editors

Alternatives to Deforestation: Steps Toward Sustainable Use of the Amazon Rain Forest,
edited by Anthony B. Anderson (1990)

Useful Palms of the World: A Synoptic Bibliography, edited by Michael J. Balick and
Hans T. Beck (1990)

The Subsidy from Nature: Palm Forests, Peasantry, and Development on an Amazon Frontier
by Anthony B. Anderson, Peter H. May, and Michael J. Balick (1991)

Contested Frontiers in Amazonia
by Marianne Schmink and Charles H. Wood (1992)

Conservation of Neotropical Forests
by Kent H. Redford and Christine Padoch (1992)

The African Leopard: Ecology and Behavior of a Solitary Felid
by Theodore N. Bailey (1994)

Columbia University Press
New York   Chichester, West Sussex
Copyright © 1994 Columbia University Press
All rights reserved

Library of Congress Cataloging-in-Publication Data

Balée, William L., 1954–
    Footprints of the forest: Ka'apor ethnobotany—the historical
ecology of plant utilization by an Amazonian people / William Balée.
        p.   cm.
    Includes bibliographical references and index.
    ISBN 978-0-231-07485-8 (pbk)
    1. Urubu Kaapor Indians—Ethnobotany.   2. Ethnobotany—Brazil.
3. Botany. Economic—Brazil.   4. Human ecology—Brazil.   I. Title.
F2520.1.U7B35   1993
681.6'1'09811—dc20   93–11044
                                                                  CIP
♾

Casebound editions of Columbia University Press books are printed on
permanent and durable acid-free paper.

Printed in the United States of America

To Conceição

Contents

Illustrations

Tables

Note on Orthography and Glosses for Indigenous Terms

For representation of Ka'apor speech sounds, I employ a standard, phonemicized orthography. For consonants, the following symbols are used: **p, t, k, kw** (labialized velar stop), **'** (glottal stop), **š** (alveopalatal fricative), **h, m, n, ŋ** (velar nasal), **ŋw** (labialized velar nasal), **w, r (ř)**, and **y**. The vowels are **i** (high front vowel), **e** (mid front vowel), **i** (high central vowel), **a, u, o** (mid back vowel) and nasalization (indicated with ˜). This orthography follows that of Kakumasu (1986:399–401) except for a few minor changes. Stress in Ka'apor generally falls on the final syllable and is here indicated only in exceptions to this pattern.

As with Ka'apor, stress in the Guajá, Tembé, and Wayãpi languages generally falls on the final syllable; only departures from this rule are marked. Stress in Asurini, however, tends to be on the penultimate syllable; here, too, only exceptions are indicated. Stress in Araweté is somewhat more irregular than in the other languages, but it is usually on the final syllable. I have therefore indicated stress in Araweté only when it is not on the final syllable.

I use the Bendor-Samuel (1966) phonemicization for Tembé and I recognize as phonemes the following: **p, t, k, kw, ', č** (alveolar or alveopalatal affricate), **z** (alveolar or alveopalatal affricate or palatal continuant), **h, m, n, ŋ, ŋw, w, r (ř), i, e, i, ə** (mid central vowel), **a, u**, and **o**.

There are no definitive phonemic statements for Araweté and Guajá, although some information on the Araweté appears in Viveiros de Castro (1986:145) and a tentative phonemicization of Guajá is given in Cunha (1987). I have chosen, accordingly, to use a liberal phonetic transcription for Araweté and Guajá. For Araweté, the following symbols are used: **p, t, k, ', b, d** (voiced interdental fricative), **c, č, h, m, n, ñ, w, r, (ř), w, i, ə** (mid central vowel), **e, i, i, a, u, o**, and nazalization. For Guajá I employ **p, pʰ** (aspirated), **t, tʰ** (aspirated), **k, kw, g, gw, h, m, n, č, r, (ř), ', y, i, e, i, ə, a, u, o**, and nasalization.

For Asurini words I employ a partial phonemic statement with some modifications (Balée and Moore 1991.260), using the following symbols: **p, t, č, k, ', j, g, m, n, ñ, ŋw, w, r, (ř), y, i, i, e, a, u**, and nasalization.

The Wayãpi phonemicization by Grenand (1980) is used, but the symbols have been standardized (as in Balée and Moore 1991:260). These symbols are: **p, t, k, ', s, m, n, ñ, ŋ, w, l, y, i, e, a, u, o, i**, and nasalization.

A simplified guide to the pronunciation of Ka'apor speech sounds is:
a   (*a* as in "father")
e   (*e* as in "pet")
i   (*ee* as in "sweet")
i   (high central unrounded vowel; no equivalent in standard American English; represented as **y** in Portuguese texts on língua geral and other Tupí-Guaraní languages)

o   (*ough* as in "sought")
u   (*oo* as in "root")
~   (signifies nasalization of any of these vowels)
h   (*h* as in "hat")
k   (*k* as in "key")
m   (*m* as in "mother")
n   (*n* as in "never")
ŋ   (*ng* as in "sing")
r   (*r* as in "very" in some dialects of British English)
s   (*s* as in "seek")
š   (*sh* as in "shoot")
t   (*t* as in "top")
w   (*w* as in "wonder")
y   (*y* as in "yen")
'   (glottal stop; the audible closing of the glottis heard just before *uh* in "uh-huh")

All terms from indigenous languages, such as Ka'apor, appear in italics in the text. Two basic glossing procedures have been used: equating the indigenous term with the scientific name of the referent species and making a morpheme-by-morpheme English translation. For example, a term such as *pinuwa'i*, which refers to the bacaba palm, may be glossed scientifically as *Oenocarpus distichus*. English glosses for Ka'apor and other indigenous terms follow distinctive procedures. Hyphens in indigenous names indicate morpheme boundaries. Morpheme boundaries are not, however, always indicated for Ka'apor words glossed in this book, as in *pinuwa'i* above (which consists of two morphemes). Hyphens in the English gloss reflect the indigenous morpheme boundaries, so that each term in a morpheme-by-morpheme gloss refers to a single morpheme in the indigenous term. Each morpheme-by-morpheme gloss is enclosed within single quotation marks. (In some cases, especially with regard to Ka'apor polymorphemic words that do not refer to specific plants, I have indicated morpheme boundaries in the native term but not in the English gloss. For example, *muyar-ha* is glossed simply as 'glue'—*ha* is an agentive suffix. In still other, similar cases, I have dispensed with quotation marks around English glosses to avoid distracting the reader from the text.)

The English gloss for each indigenous term occurs either immediately before or immediately after the indigenous name. If the English gloss precedes the indigenous name, the indigenous name is enclosed within parentheses. If the English gloss follows the indigenous name, either of two procedures may be used. First, if the gloss does not offer a translation to English of any literal morphemes ("simple opaque" terms [Berlin 1992:256] that are linguistically unanalyzable), the letter L is used to represent such morphemes and the gloss is not enclosed within parentheses. I have used this procedure especially in chapter 7 and appendix 7. On the other hand, if the indigenous term contains a literal morpheme and its gloss does offer a translation to English of that morpheme, the gloss is enclosed within parentheses.

For example, the term *pakuri'i* (referring to the tree *Platonia insignis*) is a literal plant term. It contains the literal plant morpheme *pakuri* and is nonpolysemous. The term for the same species in Portuguese (and presumably in English) is bacuri. Where emphasis on the literal nature of the entire term is desired (as in chapter 7), the indigenous term *pakuri'i* may be written and glossed as: *pakuri-'i* 'L-stem'. Where translation into English is a bigger priority than showing the linguistic structure of the literal indigenous term, it may be written and glossed as: *pakuri-'i* ('bacuri-stem').

David C. Oren, Helder Queiroz, and Lucila Manzatti assisted with glossing Ka'apor plant names that incorporated animal constituents. Aryon D. Rodrigues and Sidney Facundes supplied several glosses for Asurini terms. And Eduardo Viveiros de Castro supplied glosses for several Araweté terms. I translated all quotations which originally appeared in Portuguese or French.

Preface

This book is situated at the core of an Amazonian crisis, marked by the destruction of habitats and the consequent danger of massive species extinctions. The crisis also entails a tremendous loss of indigenous knowledge—much of it still unrecorded—about traditional ways of using and managing biological resources. These twin threats have prompted an ever more harried search for new economic plants that might supply humankind with food, fiber, medicine, and fuel. This crisis is coupled with a still poorly known history of Amazonia and its diverse peoples, habitats, and flora and fauna.

Footprints of the Forest is intended to document the historical and ecological foundations of the total relationship between an Amazonian people and the plants that not only surround them but also have to a large extent shaped (and been shaped by) their culture. As such, this book may also further understanding of the cultural and natural resource costs that attend a changing Amazonia. The people central to this account call themselves *Ka'apor*, a term that literally means "footprints of the forest" (from *Ka'a* 'forest' and *por*, the contracted form of *pipor* 'footprints'). A key concern here is how the Ka'apor, as individuals and as a culture, interact with the many plants of diverse vegetational zones of their homeland, including high forest as well as swamp forest, river edge, old fallow, young and old swiddens, and several transitional zones, both of a natural and anthropic sort.

The focus of the book is, therefore, Ka'apor ethnobotany, understood in a very broad sense. Both the study effort and the presentation of results reflect my adherence to a historical-ecological paradigm. The use, management, classification, and nomenclature of plants by the Ka'apor people of today did not appear precipitously; the boundaries of these domains were determined by historical antecedents and reigning ecological conditions taken together over many hundreds of years. These antecedents and formative conditions require explanation, which I attempt in these pages, if we are to grasp the fundamentals of Ka'apor ethnobotany. In addition, the historical and ecological substrate of Ka'apor ethnobotany parallels that of other societies in lowland South America. Where possible, therefore, I try to include here comparable information on the ethnobotanical systems of other indigenous, Amazonian societies.

This book is based on ten years of ethnological, historiographical, and botanical research among Tupí-Guaraní–speaking forest dwellers of eastern Amazonia, especially the Ka'apor but also the Araweté, Asurini, Guajá, and Tembé. Although my work surely has not exposed the full depth and breadth of the human-botanical relationship of my particular study group, I have aimed to present a treatise that may rank among the most exhaustive English-language accounts of the ethnobotany of any indigenous people in Amazonia. Sadly, little time remains for the kind of research I have been privileged to carry out.

The present account is intended to be comparable in scope to the ethnobotanical studies of Harold Conklin (1954), of Brent Berlin and colleagues (1974), of Pierre Grenand (1980), and of Janis Alcorn (1984a). I have learned much from the ideas and findings of these pioneers of modern ethnobotany, and the reader will find their influence evident here. This book nevertheless carries its own theoretical and substantive message. It appears at a time when its subject, namely, the forests and forest dwellers of extreme eastern Amazonia, lies vulnerable to seemingly unstoppable incursions by loggers, squatters, ranchers, and gold miners. *Footprints of the Forest* is today a book about an extant, culturally intact people. There are, however, about as many pages in this book as currently living men, women, and children of Ka'apor society. It may soon become an example of a kind of anthropological study no longer achievable and a record of a kind of human-ecological relationship that has disappeared.

Acknowledgments

For generous financial support of my long-term ethnobotanical research in Brazilian Amazonia, I am indebted to the Institute of Economic Botany of the New York Botanical Garden, the Edward John Noble Foundation, the Conselho Nacional de Desenvolvimento Tecnológico e Científico (CNPq), the Museu Paraense Emílio Goeldi, the Ford Foundation, the Jessie Smith Noyes Foundation, and the World Wildlife Fund. I am grateful to the Fundação Nacional do Índio and to the CNPq for authorizing my requests to carry out ethnobotanical fieldwork among the Ka'apor, Guajá, Asurini, Araweté, and Tembé peoples and for cooperating with this project in other ways. Very special thanks are also due the Center for Latin American Studies of the University of Florida, which generously gave me a visiting professorship, which allowed me to envision and begin the writing of this book during the spring of 1990. I wish to acknowledge the editors of L'Homme, who gave me permission to reproduce in chapter 6 portions of an article I wrote that appeared in that journal, vol. 33 (1993).

With regard to individuals, I would like to begin by thanking Jan Wassmer Stevenson, collections manager of the Institute of Economic Botany, for her expert processing of most plant vouchers herein cited. In addition, I am very grateful to Bruce Nelson, now of the Instituto Nacional de Pesquisas da Amazônia, for having arranged to ship many of my plant vouchers to the herbarium of the New York Botanical Garden over the course of five years. All plant collection numbers herein referred to, incidentally, are represented in the herbaria of the New York Botanical Garden or the Museu Paraense Emílio Goeldi.

I am also indebted to the professional botanists who identified many of my collections. They are W. R. Anderson, Lennart Andersson, D. Austin, V. Badillo, D. Baer, Rupert Barneby, Hans T. Beck, H. Bedell, C. C. Berg, Brian Boom, Sílvia Botta, R. Callejas, L. Constance, M. J. E. Coode, Thomas Croat, J. Cuatrecasas, Douglas C. Daly, Gerritt Davidse, L. J. Dorr, E. A. Christenson, Enrique Forero, Paul A. Fryxell, Al Gentry, Jim Grimes, Roy E. Halling, R. M. Harley, G. W. Harling, R. R. Haynes, W. H. A. Hekking, Andrew Henderson, P. Hiepko, Noel Holmgren, Helen Hopkins, R. Howard, Michael J. Huft, H. H. Iltis, Jacques Jangoux, C. Jeffrey, M. C. Johnston, Jackie Kallunki, M. L. Kawasaki, H. Kennedy, Tetsuo Koyama, Klaus Kubitzki, L. Landrum, A. J. M. Leeuwenberg, H. de Lima, Regina Lisboa, A. Lourteig, Jim Luteyn, H. Maas, P. J. M. Maas, Bassett Maguire, R. H. Maxwell, W. Meijger, A. M. W. Mennega, J. Mickel, John D. Mitchell, W. Morawetz, Scott A. Mori, T. Morley, C. Moura, Michael Nee, J. M. Pederson, Thomas D. Pennington, João Murça Pires, Timothy Plowman, Ghillean T. Prance, John Pruski, S. S. Renner, William Rodrigues, Pedro Acevedo Rodriguez, J. da Sacco, C. Sastre, J. Sauer, Ricardo Seco, L. E. Skog, L. B. Smith, J. C. Solomon, C. Stace, Berti Stähl, W. A. Stevens, C. Stirton, Wade W. Thomas, A. K. Van Setten, H. Van der Werff, Karl A. Vincent, Dieter Wasshausen, G. L. Webster, F. White, R. P. Wunderlin, J. J. Wurdack, E. Zardini, and Jim Zarucchi.

I also thank the local expert plant identifiers—Nelson de Araújo Rosa, Manuel Cordeiro, Carlos da Silva Rosário, and Benedito Ribeiro—with whom I worked on these collections in Belém. In addition, I am grateful to two woodsmen, Ferdinando Cardoso de Nascimento and Benedito Ribeiro, who were most helpful to me in the collection of plants and the execution of forest inventories.

My sincere gratitude goes to Ghillean T. Prance, who, as then director of the Institute of Economic Botany, gave me the initial opportunity to work full time on Amazonian ethnobotany; to Douglas C. Daly, for teaching me the rudiments of systematic botany; to David G. Campbell, for sharing with me during fieldwork in Acre a detailed methodology for forest inventory—much of which was subsequently used in the inventories described in this volume; to David C. Oren, for collecting and identifying many bird specimens from the Ka'apor habitat, thus helping me determine accurate glosses for Ka'apor plant lexemes that incorporate names of birds; to Lucila Manzatti and Helder Lima de Queiroz, for collecting and identifying mammalian and fish species in the Ka'apor habitat (which helped me ascertain glosses for plant lexemes), as well as for being good companions in the field. I am grateful to Pamela Van Rees for her assistance with the time allocation research reported in chapter 4. Thanks are also due to Bruce Albert, Dominique Buchillet, Elaine Elisabetsky, Pierre and Françoise Grenand, Denny Moore, Anna Roosevelt, and Eduardo Viveiros de Castro for helpful discussions related to various points raised in this book. I am also grateful to Michael Balick and two anonymous reviewers used by Columbia University Press to read the original manuscript.

I wish to thank sincerely the former directorship of the Museu Paraense Emílio Goeldi, especially Guilherme de La Penha and Celso Martins Pinto, for facilitating in countless ways my work at that institution and in the field. A special word of thanks is also due to my longtime friend and colleague from the Museu Goeldi, Adélia Engrácia de Oliveira Rodrigues, who kindly agreed to serve as my research sponsor when I began ethnobotanical research in earnest after 1984.

I thank my former secretary, Rosana Kerr, for transferring data from my early field notebooks to computer files. Thanks, too, to Paulo Cabral Filho, Altenir Pereira Sarmento, and Rafael Salomão of the Museu Goeldi for allowing me to use a program they had developed for analyzing forest inventory data. Harvey Bricker helped render the species/area curves in chapter 6. Thanks are extended to Ed Lugenbeel of Columbia University Press for encouragement and because he could not have recommended a more appropriate manuscript editor than Connie Barlow. She did splendid work in helping me rethink and clarify many aspects of the original manuscript.

My wife, Maria da Conceição Bezerra Balée (to whom this book is dedicated), and our children Nicholas and Isabel were understanding of my long field trips away from home; they also bore with grace the many hours I spent at the office analyzing data collected for the writing of this book. I am deeply grateful for this and for much more, and can only hope that the final product in some way justifies their sacrifices.

Finally, my sincere thanks to the Araweté, Asurini, Guajá, Tembé, and Ka'apor peoples. For the Ka'apor, in particular, I wish to acknowledge those individuals who not only tolerated my plant collecting and insistent questions, but who took an active interest in this project in every way. I am especially indebted to my principal informants, among them Timaparu, Pimenta, Hosi, Oriru (deceased), Tanuru (deceased), Lusiã, Ararihe, Meri, Mari, Šuã, Lusiãru, and Parasañru; also, I extend my gratitude to all the other Ka'apor, especially those from the villages of Gurupiuna and Urutawi, without whose consent this research would have been impossible, and without whose hospitality and friendship it would have been far more arduous.

ONE

Introduction

A few anthropologists still pigeonhole ethnobotany within the anthropological paradigm called ethnoscience (or cognitive anthropology). A few biologists continue to perceive ethnobotany (sometimes called economic botany or medical botany) as concerning lengthy enumerations of "useful" plants and active principles known to "native" peoples. Ethnobotany is probably best regarded as a field of biocultural inquiry, independent of any specific paradigm, yet rooted in a scientific epistemology.

As a field of inquiry, ethnobotany is located "at the interface of human needs and thought with nature" (Ford 1978:44). Fundamentally, the object of ethnobotany is to explain similarities and differences among societies in the use, management, classification, and nomenclature of plants. I proffer *historical ecology* as the soundest basis for such work. Historical ecology is more than a methodological improvement over cultural ecology, cognitive anthropology, and certain list-oriented approaches common to economic botany and medical botany. I have found, rather, that it is a powerful paradigm for comprehending interrelationships between Amazonian environments and associated indigenous societies, because it focuses on the interpenetration of culture and the environment rather than on the adaptation of human beings to the environment.

Historical Ecology as a Framework for Research

In cultural ecology, culture and environment become alienated subject and object; environmental "limiting factors" constrain the development of culture—the more "primitive' the culture, the more rigorous the environment. In cognitive anthropology, the nonhuman environment of plants and animals is taken to be principally a semantic domain of the lexicon, not an entity in constant material interaction with society itself. As for economic and medical botany, it is not often clear precisely who should benefit from the translation of indigenous ethnobotanical knowledge into new medical resources for the industrialized world. Historical ecology, in contrast, seeks a synthetic understanding of human/environment interactions within specific societal, biological, and regional contexts. In other words, the focus of historical ecology is a relationship, not an organism, species, society—not a "thing."

For example, the fact that many Amazonian hunter-gatherers, such as the Guajá, Aché, and Héta, heavily exploit the palms of old fallows abandoned by agricultural peoples can not easily be understood as cultural adaptation to a wholly natural habitat (Balée 1988b, 1989a, 1992a). These hunter-gatherers are engaged instead in a dynamic relationship with cultural as well as natural components of the environments they depend on for food, shelter, medicine, and fuel. They are, in a sense, not so much adapted to natural environments as to alien cultures—specifically, those cultures that establish swiddens in what later become patches of forest fallow.

An approach grounded in historical ecology has brought new insights to ethnobotanical research in Amazonia. Just as techno-environmental conditions can be inferred from indigenous Amazonian cultures, past and present, human activity can be inferred from the landscape, especially the vegetational changes wrought by prehistoric peoples throughout the basin (Anderson and Posey 1985, 1989; Balée 1987, 1988b, 1989a, 1992a; Balée and Campbell 1990; Denevan 1992; Irvine 1989; Moran 1990; Roosevelt 1991). A historical ecological approach has thus facilitated a dynamic synthesis of our understanding of culture and nature in Amazonia. More than other paradigms today, historical ecology befits the inimitable complexity of Amazonian culture and nature.

Summary of Chapters

Chapter 2 presents an overview of the botanical and ethnographic setting. The Ka'apor people, who are the principal actors in this book, are small-scale, village-level swidden horticulturalists of the extreme eastern Amazonian state of Maranhão. (See the map in this chapter of the pre-Amazonian region, which is the area immediately to the southeast of the lower Amazon River not actually drained by that river.) At a population density of only about .2 persons/km^2, the Ka'apor number some 520 individuals scattered in twelve autonomous settlements across 5305 km^2 (an area roughly the size of Trinidad and Tobago). The region receives between 2,000 and 2,500 mm of rainfall yearly with a fairly marked dry season between July and December. High forest (which is primary forest of the *terra firme*) predominates, but significant patches of the Ka'apor habitat may be classified as riverine forest, swamp forest, swiddens of various ages, and old fallows left behind by ancestors of the Ka'apor and refugee Afro-Brazilian slaves of the nineteenth century.

At least a thousand species of vascular plants are present in the Ka'apor habitat, of which I collected nearly eight hundred. In addition, I obtained more than five thousand plant voucher specimens from the larger region of lower Amazonia comprising the habitats of the Guajá, Tembé, Araweté, and Asurini peoples, as well as of the Ka'apor. Ka'apor utilization of this biologically rich habitat can be understood only in light of the regional history, which chapter 3 undertakes. Pre-Columbian forebears would have been associated with the chiefdoms extant along the lower Tocantins River in the late sixteenth and early seventeenth centuries. (See the map in chapter 3.) All these chiefdoms, as well as the Tupí-Guaraní chiefdoms of the Atlantic Coast, were fairly sedentary and relied heavily on agriculture (Balée 1984a, 1988a, 1989b). Linguistic evidence suggests that speakers of proto Tupí-Guaraní, the mother language of the Tupí-Guaraní family, practiced agriculture 1800 to 2000 years ago (Balée 1989b; Migliazza 1982; Rodrigues 1988).

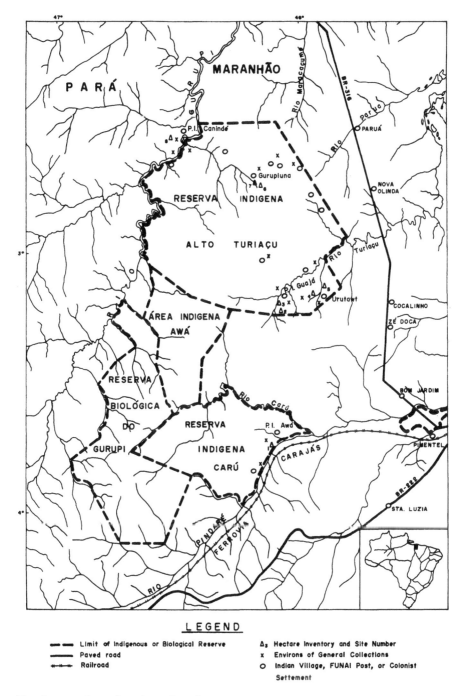

LEGEND

▬ ▬ ▬	Limit of Indigenous or Biological Reserve	
▬▬▬	Paved road	
╫─╫─╫	Railroad	

Δ₅ Hectare Inventory and Site Number
x Environs of General Collections
O Indian Village, FUNAI Post, or Colonist
Settement

Pre-Amazonia and region of study.

After the European conquest, the onslaught of colonial expansion, Jesuit missionary fervor, slave raids, debt peonage, and epidemics drove the aboriginal Ka'apor to
war with Luso-Brazilian society. Defeats c. 1825–1864 forced the Ka'apor to migrate
eastward, away from their berth in the Tocantins River basin. (See the map "Migration
of the Ka'apor" in chapter 3.) By the 1870s they had moved into the less populated
Maranhão region (Balée 1988a).

Although broad similarities exist, such as depopulation and colonial warfare, the
historical forces of contact affected indigenous societies of eastern Amazonia in different ways. Some groups, such as the Mundurucu (Murphy 1960), the Turiuara (Barbosa
Rodrigues 1875), and the Apinayé (Nimuendaju 1983), submitted to the military, serv-
ing as mercenaries in the subjugation of still-hostile tribes. Others adopted a nomadic,
nonhorticultural lifestyle, including the Avá-Canoeiro (Toral 1985) and the Guajá
(Balée 1988a, 1989b, 1990). Still others simply changed their agricultural habits; instead
of depending mostly on manioc (a slow-growing, cumbersome staple) they grew more
maize (a fast producing, easily transportable crop). The switch to maize brought an
overall reduction in the number of plant species cultivated by groups such as the
Araweté (Balée 1988a) and the Sirionó (Holmberg 1969; Stearman 1987). A few groups
resisted Luso-Brazilian domination fiercely within cul-de-sacs, and thus were eventually exterminated. An example is the Arawakan-speaking Aruã of Marajó Island
(Nimuendaju 1948a). Finally, many groups simply migrated into relatively uninhabited forests. Despite the difficulties, they were able to reestablish and maintain settled
villages. Most were dependent on the cultivation of bitter manioc. These were peoples
such as the Wayãpi (Grenand 1982), the Kagwahiv (Kracke 1978), the Tapirapé
(Wagley 1977), and the Ka'apor.

Even though the colonial juggernaut tended to depopulate all indigenous groups,
the diverse reactions of those that survived help explain many dissimilarities in the
modern comparative ethnobotany of Amazonia. Profound ecological differences, such
as that between horticultural production and full-time foraging, distinguish societies
of even the same language family that exploit the same habitat (e.g., the foraging Guajá
versus the horticultural Ka'apor). These ecological differences can only be understood
in a historical context because they are themselves largely historical artifacts. On the
other hand, the similarities in plant utilization among the various Amazonian peoples,
except those resources and techniques borrowed from Luso-Brazilians, most likely
trace their origins to pre-Columbian society.

Ka'apor ethnobotany is the prism through which I have tried to perceive and
understand human/plant interactions in Amazonia. Chapter 4 explores these interactions; I call them *activity contexts* to distinguish this investigatory approach from the
one-sided perspective implied by the term *plant uses*. Activity contexts involve the
Ka'apor and plants in hunting, fishing, swidden gardening, gathering, food preparation, manufacture and repair of material goods, and other activities. The usual
amounts of daily time devoted to each of these activities are recorded, along with
detailed descriptions of the activities themselves.

This contextual approach is a rather unconventional way of investigating and pre-
senting ethnobotany of the Amazonian region. For example, it markedly differs from
the ethnobotanical list-making of Vickers and Plowman (1984), Pinkley (1973), and

Davis and Yost (1983b). Those lists are, of course, excellent for summarizing data, but I have tried to take the next step in enhancing the uses and the ease of use of such data by keying plants to activities as well as activities to plants. I have tried, moreover, to convey the rich interweavings of human and plant that jointly produce both society and landscape. Chapter 4 (along with appendixes 6, 7, and 8) thus portrays the mosaic of Ka'apor ethnobotany in its day-to-day exuberance.

Amazonian ethnobotanical systems differ remarkably in their variable use of medicinal plants. Although medical anthropologists debate a distinction between magic and medicine in egalitarian societies (e.g., Augé 1985; Brown 1986), I have found the distinction useful in the specific case of the Ka'apor. In chapter 5, I base the distinction on the simple test of efficacy: medicine is that which, under stated conditions, may be expected to have therapeutic effects on people or their pets. This is not an attempt to reduce the totality of the Ka'apor category *puhan* 'medicine/magic' to an "ethnocentric binary concept" (Augé 1985)—particularly one drawn from the culture of the investigator. But it does recognize that any indigenous system has both an empirical sector and an opposed magical sector. Moreover, by making this distinction, I have been enabled to develop an understanding and a model of Ka'apor cognition of plant utility that would otherwise have been impossible.

Chapter 6 describes in detail the plants in the traditional habitat of the Ka'apor. It presents information that I obtained on site from 1985 through 1990, based on general surveys of vegetation conducted in partnership with indigenous informants and on statistically rigorous inventories of eight one-hectare plots (half old fallow and half high forest). I specifically calculated relative densities, frequencies, and dominance of tree and liana species and families, and, from these, derived ecological importance values for each. The "ground truth" cannot, in fact, be obtained in any other way; remote sensing cannot yet distinguish old fallow from high forest.

Chapter 6 also explores the differences in the breadth of plant utilization and intensity of forest management between societies, such as the Ka'apor, that are horticultural and semisedentary and those that are hunter-gatherers and trekkers. The Ka'apor cut and burn forest, often repeatedly in the same patch, hence modifying forest succession. Perhaps the most remarkable finding about indigenous forest management concerns an indigenous contribution to biodiversity. Old fallows, which are artifacts of human interference, appear to be insignificantly different from high forests in terms of plant diversity. And old fallows harbor certain species that occur nowhere else, suggesting a possible enhancement of regional biodiversity by indigenous agroforestry practices.

Pre-Columbian societies, with higher population densities, probably had an even greater effect on forest composition and structure than do modern indigenous horticultural groups (Balée and Campbell 1990; Smith 1980; Roosevelt 1989a). The impacts of today's indigenous peoples are nevertheless overwhelmed by other forces. Specifically, one must now distinguish those anthropic disturbances originating in modern industrial societies from those originating in egalitarian societies (such as the Ka'apor) or incipiently stratified societies (such as pre-Columbian chiefdoms) that rely (or relied) on renewable and local resources. The two kinds of anthropic influences have vastly different energetic potentials, meaning that their capacities to convert forests on a large scale to biologically more impoverished landscapes and to alter other basic environmental conditions, including climate, are fundamentally distinct.

It is plausible that Paleo-Indians of lowland South America, as elsewhere, may have contributed to the extinction of Pleistocene megafauna. But no evidence yet exists for extinction of species in the thousands of pre-Columbian years that followed the Pleistocene. The threat of extinction of Amazonian flora and fauna in modern times has its provenance principally in modern state societies. This is not to say that pre-Columbian horticultural peoples did not alter the environment in significant ways. Indeed they did, and with greater reach than today's indigenous peoples (Roosevelt 1989b).

Although indigenous peoples of the Amazon are organized today in sociopolitical structures different from those of their pre-Columbian forebears, similarities in languages, marriage customs, economic systems, and cosmologies suggest mostly pre-Columbian, neotropical origins, not so much influences from the Old World. Most of the crops in their swiddens and the species they exploit in forest and fallow are also of neotropical origin. Moreover, the extinct pre-Columbian chiefdoms and the extant indigenous foragers and farmers of lowland South America exemplify far less destructive ways of interacting with the environment, by any measure, than modern nation-states.

In chapter 6 I present an integrated approach to understanding Ka'apor forest management. I examine how horticultural activities affect not only plants but also game animals in the region. Plant and game management are linked in traditional Ka'apor ritual hunting and horticultural practices (Balée 1985; Balée and Gély 1989); some species are favored in their density and distribution at the expense (but not the extinction) of others. Specifically, the distribution and density of semidomesticated plants or "anthropophytes" (Pinkley 1973) cannot be understood without knowledge of both the natural and the human component.

Semidomesticates are, in a sense, an unconscious artifact of indigenous forest management. Although many of these species are useful in one way or another, it cannot be said that the Ka'apor are wittingly manipulating them for personal benefit, in contrast to recent findings among the Gê-speaking Kayapó of central Brazil (see Anderson and Posey 1989, Posey 1983; cf. Parker 1992). Rather, Ka'apor forest management is largely unconscious. The Ka'apor have no explicit "conservation ethic," yet their forest management practices have increased the ecological and perhaps even biological diversity of native Amazonian habitats.

Chapter 7 describes the plant nomenclature and classification used by the Ka'apor people. Linguistic evidence indicates that all modern Tupí-Guaraní languages are descended from a language spoken by a horticultural people about 1,800 to 2,000 years ago. Even contemporary foraging societies affiliated with the Tupí-Guaraní family, such as the Guajá, display linguistic and other relics of a horticultural past (Balée 1992a; Gomes 1988). Plant nomenclature in Ka'apor and other modern Tupí-Guaraní languages of eastern Amazonia has apparently been affected in patterned ways by this ancient cultural heritage.

The three life forms of Ka'apor botanical classification may be glossed as 'tree', 'vine', and 'herb'. Although many names for nondomesticated plants incorporate these terms, none of the traditional plant domesticates do so. In fact, traditional domesticates are not classified in any morphological way. Names for nondomesticated plants, moreover, may be based on analogy with names for traditionally domesticated plants, but not vice versa. For example, plant names which may be translated as 'sloth banana', 'macaw chile pepper', 'red brocket deer manioc', and 'white-lipped peccary peanut' refer to nondomesticated plants that are not, in our Linnaean system, actually types of banana, chile pepper, manioc, or peanut. These referents are, however, traditional domesticates of the Ka'apor. Names for traditional domesticates are also more stable and have higher rates of retention than names for other plants, including nondomesticates and semidomesticates. This becomes evident in the comparison of plant names between several different Tupí-Guaraní languages, including Ka'apor (Balée and Moore 1991; Berlin 1992:255–259). And this suggests that names for traditional domesticates have historical primacy over names for other plants.

Nondomesticated plants are, in addition, 'unaffiliated' with traditional domesticates in Ka'apor botanical classification. In partial contrast, Berlin et al. (1974) presented the basis of Tzeltal nomenclature as the classification of plants into such life forms as 'tree', 'herb', 'vine', and 'grass'. The few plants that were not so classified (including many domesticates) were considered to be 'unaffiliated' with this general classification. In Ka'apor ethnobotany, morphological criteria and factors relating to domestication are intertwined in the general classification.

Ka'apor plant classification is unlike Linnaean taxonomy in several important ways. First, Ka'apor plant classification makes no attempt to be universal, since it covers only plants with which the Ka'apor are already familiar. Second, although it does group some plants morphologically (into the life forms *mira* 'tree', *sipo* 'vine', and *ka'a* 'herb'), these are only nondomesticates and semidomesticates; traditional domesticates are in a real, cognitive sense placed in a different group. Third, the mode of transmission of Ka'apor botanical classification occurs completely within an oral tradition. Finally, in Ka'apor botanical classification, no codified rules exist that could be explicitly stated by informants, but unconscious patterns of nomenclature, shaped by an enduring legacy of plant management, can be deduced from the lexicon itself. These same patterns hold for the ethnobotanical systems of other Tupí-Guaraní peoples with whom I have worked.

It cannot be gainsaid that Ka'apor culture utterly depends on the maintenance of the environment in which it emerged. This dependence is not just a reflection of the people's dependence on nature, for the inhabited forests have felt the effects of human interference over many generations. Only recently has this external interference in the Ka'apor habitat become very intensive and ultimately destructive. The close connection of the Ka'apor with the forest is evident in their origin myth, which I summarize below.

> The high forest (*ka'a-te*) was the original habitat of the world. The culture hero, Ma'ir, who founded the Ka'apor people, beckoned the first generation of Ka'apor to come forth out of one of the hardest and most useful of forest trees, *tayi* (*Tabebuia* spp. in the bignonia family), which is used to make the bow (*irapar*). The ancestral Ka'apor emerged from these trees during a violent thunderstorm when the branches 'rubbed upon each other' (*mira kitik*). Next, Ma'ir one afternoon felled a tree that

has white heartwood: the kapok (*Ceiba pentandra* in the bombax family), which is called *wašiŋi*. He tossed it into a stream. By morning of the following day, white people (*kamarar*) had emerged.

Yes, the origin myth of the Ka'apor people today makes provision for a being that was unknown in pre-Columbian times. More importantly for our purposes, it is clear that humans, all kinds of humans, came originally from plants. The relation is not just a curiosity for those who may be interested in searching for structural parallels between plants and people. The material interdependence of the local universe of plants and the Ka'apor people in itself constitutes a fit topic for scientific inquiry. It is my contention, moreover, that neither can be understood except in relation to the other. This perspective provides the foundation for the final chapter, which suggests historical and ecological avenues for investigating the diversity of human/plant relationships in lowland South America.

TWO

The Botanical and Ethnographic Setting

The Ka'apor people of today, along with some of the Guajá and some of the Tembé, live within a federal indigenous reserve of 530,525 hectares. Called the Reserva Indígena Alto Turiaçu, it lies entirely within the state of Maranhão (see the map of pre-Amazonia in the previous chapter). The region is covered by dense forests (called *terra firme* by the Portuguese-speaking majority), by riverine vegetation, and by minor patches of swidden and young secondary growth. It is drained by the Gurupi, Turiaçu, and Maracaçumé rivers, all of which debouch directly into the Atlantic Ocean. To the south, the Pindaré River, inhabited along its upper left bank by Guajá and Guajajara peoples, also drains into the Atlantic. Although the rivers of the region do not therefore drain into the Amazon, the area shares many similarities (in terms of geology, climate, and biogeography) with Amazonia proper, which is narrowly defined as the basin of the Amazon River. This pre-Amazonian region nevertheless is distinctively Amazonian in both landscape and culture.

Geology and Climate

Very little systematic study has been carried out on the geology of the region. The evidence thus far available points to a fairly recent geological age for the major river basins. Except for the Gurupi basin which is pre-Silurian, the basins of the Turiaçu, Paruá, and Maracaçumé rivers are all Quaternary (Roeder 1967; Sombroek 1966:15). The chemical and nutrient compositions of the diverse waters of the region have not been well studied. The only solid research on this is that of Lucila Manzatti, who studied water composition on five minor tributaries of an artery (the Igarapé Gurupiuna) of the Gurupi. She found that although water color was black, the pH was relatively neutral. She concluded that the five streams in her study "can be considered to be excellent habitats for fishes" (Manzatti 1989).

Common and highly disturbed metamorphic deposits in the Gurupi basin consist principally of granite, schist, mica, and quartzite (Projeto RADAM 1973:13–15,27; Projeto Gurupi 1975:40; Moura 1936; Calogeras 1938:20). Elevation in the region of study is between 100 and 200 meters above sea level (Roeder 1967). Soils of the Gurupi, Turiaçu, and Pindaré basins are for the most part yellow podzols and oxisols of Precambrian origin (Projeto RADAM 1973:10–11). It is interesting that no *Terra Preta do Índio*

(Indian black earth, an anthrosol) has yet been encountered in the region, whereas this soil type is fairly common to the west, in the Tocantins and Xingu River basins. It is not clear whether edaphic factors or cultural ones (such as a possibly shorter and less intense indigenous occupation span for this region of Maranhão than for Pará) or both are responsible for this absence.

The entire region of the Gurupi and Pindaré basins, using the Koeppen classification, has a Humid Tropical climate (Aw); the dry season is of medium length and average precipitation in the driest month is less than 60 mm (MINTER 1984). Relative humidity is between 75 and 85 percent, and the average annual temperature is 26°C; the lowest average temperature recorded in any one year was 21°C and the highest average was 32°C (MINTER 1984; Projeto Gurupi 1975:20). Average annual rainfall in the Gurupi and Turiaçu regions is between 2000 and 2500 mm (MINTER 1984; Projeto Gurupi 1975:24; Nimer 1972:30). In the Pindaré basin, this drops to about 1500 mm annually (MINTER 1984). Whereas the average number of rain days per year in the Gurupi is 180, it is only 120 in the Pindaré. See figure 2.1 for a monthly comparison.

The Pre-Amazonian Forest

The region in Maranhão inhabited by the Ka'apor, Guajá, Tembé, and Guajajara is called pre-Amazonian forest (SUDAM 1976; Daly and Prance 1989:421) or Amazonian Maranhão (Fróis 1953). Its limits are roughly understood to be the Rio Gurupi on the west, Atlantic Ocean to the north, upper courses of the Rios Grajaú, Pindaré, and Gurupi on the south, and the left bank of the Rio Mearim on the east (Fróis 1953:99). The region

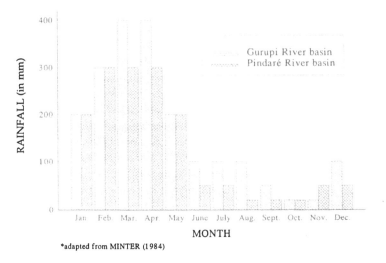

*adapted from MINTER (1984)

FIGURE 2.1 Rainfall in pre-Amazonia.

appears to represent an easterly "extension of" or "penetration by" the high forests that are drained by the Amazon River to the west (Fróis 1953; Ducke and Black 1953).

Depending on the source one uses, the pre-Amazonian forest may be further divided into areas of savanna, palm groves (especially nearly pure stands of the babaçu palm), and high forest that is called *hiléia* or *floresta densa* (Ducke and Black 1953; Projeto RADAM 1973:IV/7; Fróis 1953; Rizzini 1963). According to one source, the high forest lies between latitudes 3°18'S and 6°21'S and longitudes 44°35'E and 48°20'W (SUDAM 1976), its southern and easternmost limits being the upper Rio Gurupi and the left bank of the Rio Mearim. Ducke and Black (1953:6–7) were more conservative, indicating that the true high forest (*hiléia*) of Maranhão lies between the Rio Gurupi, the Rio Turiaçu, and the upper Rio Pindaré. According to Rizzini (1963), who agreed with Ducke and Black (1953), this region is part of the phytogeographic province called "Amazon forest." The region lies within the southeastern sector of the geological subprovince "Tertiary Plain," which runs east from the foot of the Andes. This southeastern sector includes the basins of the lower Tocantins, the area around Belém, lower Xingu, Gurupi, and upper Pindaré and Turiaçu (Rizzini 1963:45,51). The part of this sector lying in Maranhão corresponds well with the present habitat of the Ka'apor, Guajá, and Tembé, and I consider it to be definitive of the phytogeographic limits of the region under study.

Fróis (1953) used several key "Amazonian" tree species as indicators of the Amazonian nature of this region. These included maçaranduba (*Manilkara huberi*), maçarandubarana (*Manilkara paraensis*), a kind of rubber tree (*Hevea guianensis*), and the well-armed murumuru palm (*Astrocaryum murumuru*), all of which he collected (and I have subsequently collected) in the basins of the Rio Pindaré, Turiaçu, and Gurupi. Daly and Prance (1989:406) criticized Fróis's attempt to circumscribe the eastern limits of Amazonia in Maranhão by merely showing that a few typical "Amazonian" species occurred there. This is because no single Amazonia exists: "there are numerous genera and thousands of species endemic to Amazonia, but their distributions are discontinuous within the region" (Daly and Prance 1989:406). If Amazonia is composed of discrete phytogeographical provinces throughout, this should be no less true of its peripheral regions.

The pre-Amazonian region is not generally considered to have endemic plant species, although Oren (1988) has noted that the avifauna is essentially Amazonian, and Queiroz (1991) has registered species of monkeys not previously known to exist in the region. One of these species, *Cebus kaaporii*, is a new species to science (Queiroz 1992). Certain plant species with disjunct distributions elsewhere in Amazonia, such as the Brazil nut tree (*Bertholletia excelsa*), mahogany (*Swietenia macrophylla*), the riverine vine 'ituá' (*Gnetum*), and the 'mucajá' palm (*Acrocomia*) are simply absent from the *terra firme* forests of pre-Amazonian Maranhão (also see Rizzini 1963). On the other hand, I have subsequently collected in Maranhão a number of species in the Brazil nut (Lecythidaceae) and tropical rose (Chrysobalanaceae) families previously thought to be limited to the Belém refuge (a region considered to harbor endemic species, with eastern boundaries along the upper Gurupi River drainage; see Prance 1973, 1982). These new species include *Eschweilera piresi* Mori.

Overall, the pre-Amazonian *terra firme* forests appear to be as rich in plant species as any other region of eastern Amazonia (see below). Although there may be no endemic plant species restricted to pre-Amazonia, there are not as yet strong phyto-geographical reasons for segregating it from the Belém refuge.

Fróis reputedly made extensive collections in the regions of the Pindaré, Gurupi, and Turiaçu, but it is unlikely that he collected very much within the Ka'apor habitat. It is known that he saw some Guajá on the upper Carú River (a tributary of the left bank of the Pindaré) during a collecting expedition in 1943 (Nimuendaju 1948c:135), long before some of the Guajá entered into permanent contact with Brazilian society. Little record remains of any intensive botanical exploration in the region, however, perhaps because there were no intensive collections made during that volatile and dangerous time. Also, Fróis, despite his fame as the most knowledgeable botanist regarding the flora of Maranhão, never published a systematic and detailed treatment of the pre-Amazonian forest, nor has anyone else done so to date.

What remains of this forest today is threatened with conversion to pasture because of recent illegal invasions by ranchers and loggers. Nevertheless, it is an overstatement to contend that "the pre-Amazonian forest has been decimated before it could begin to be studied" (Daly and Prance 1989:421). The rest of this chapter seeks to evaluate sys-tematically the phytosociology of the pre-Amazonian forest occupied by the Ka'apor, Guajá, and Tembé Indians on the basis of extensive collections and inventories carried out between 1985 and 1990.

Vegetation of Rivers and Swamps

The first travelers into the region who were of Old World ancestry came by water. This was as true for the early Jesuits, who visited the mouth of the Gurupi and estab-lished a mission there in the early seventeenth century (Cleary 1990), as for much more modern travelers, such as the first ethnographers of the Ka'apor, namely, Lopes, Ribeiro, and Huxley (see especially Huxley's 1957 description of his long voyage from Bragança to Post Canindé). The only river that is navigable year round into Ka'apor territory is the Gurupi. The Maracaçumé is a mere stream by Amazonian standards, and the Turiaçu but little more. Not even a medium sized canoe, equipped with one of the ubiquitous Briggs and Stratton engines, is capable of penetrating either the upper Turiaçu or Maracaçumé at low water. There are no rapids on these rivers; the channels simply become too shallow and narrow.

Even though the Gurupi has four treacherous rapids to pass before reaching the limits of Ka'apor territory (see figures 2.2 and 2.3), these can be navigated rather easi-ly most of the year, even by boats with central, inboard diesel engines. Passage in the low-water season (September to December) is difficult, however, and dangerous regardless of one's craft. The Pindaré to the south, in Guajá territory, is also virtually non-navigable during the low-water season, although, in terms of length, it appears to be a respectable river on the map. The Pindaré has no rapids, but sandbars during dry months remind one of the Rio Moa in Acre. The navigable part of the river becomes so narrow and unpredictable as to location that one cannot easily maintain a boat in the channel; it is often necessary to disembark and push the vessel free from a sandbar.

The vegetation that casts shadows over the rivers of the pre-Amazonian forest is somewhat uniform across the major drainages, including the Pindaré. Trees common to the riparian vegetation in all four major basins, and which are familiar to travelers throughout Amazonia, include kapok (*Ceiba pentandra*), wild cashew (*Anacardium giganteum*), 'arapari' *Macrolobium acaciaefolium*, 'coumarurana' (*Taralea oppositifolia*), 'virola' (*Virola michelii*), 'anani' (*Symphonia globulifera*), 'imbaúba' (*Cecropia* spp.), and wild annatto (*Bixa orellana* and *Bixa arborea*). Common palms include açaí (*Euterpe oleracea*), marajá (*Bactris maraja*), jauari (*Astrocaryum javari*), and ubim (*Geonoma baculifera*).

Epiphytic philodendrons and bromeliads of Amazonia are also common in this part of pre-Amazonia (see table 2.1). Water-loving species, however, that thrive in inundated portions of the Amazon basin proper appear to be entirely absent. Four examples of missing species are giant water lilies (*Victoria amazonica*), *Hura crepitans*, *Piranhea trifoliolata*, and the beautiful chacouier (*Warszewiczia*). These absences are probably attributable to the tectonics that separated pre-Amazonian river basins from the Amazon itself.

A few minor phytogeographic differences, moreover, appear to distinguish even the Gurupi riverine vegetation from that of the Turiaçu. For example, I found along the Gurupi 'mungubarana' (*Pachira aquatica*) [a large tree of the bombax family], the vita-

FIGURE 2.2 Rapids on the Gurupi River just above Post Canindé.

min C–rich 'cámu-cámu' (*Myrciaria dubia*) [a small tree of the myrtle family], and *Montrichardia* sp. But despite extensive explorations, I never saw these plants along the Turiaçu, nor did the inhabitants report their presence.

Herbaceous vegetation is, by inspection, far more abundant on the Pindaré than on the river systems to the north. In fact, many aquatic plants commonly found in backwater swamps of the Pindaré are simply absent to the north. These include water hyacinths (*Eichhornia crassipes, Reussia rotundifolia*), 'water salad' (*Pistia stratiotes*), free-floating water ferns (*Salvinia auriculata*), and plants that anchor themselves in still, shallow water (*Ludwigia helminthorriza*). These plants are typical of river systems that debouch directly into the Amazon River. This biogeographical detail suggests that the Pindaré was connected to the Amazon basin proper (see above) for a longer period of time than were the Gurupi, Turiaçu, and Maracaçumé drainages (D. C. Oren, personal communication).

Further evidence supports this hypothesis. Although all four pre-Amazonian rivers lack water manatees and river dolphins that distinguish the Amazon, the Pindaré contains some Amazonian icthyofauna that the other three rivers lack. For example, electric eels, the white piranha, and stingrays of several species occur within the Pindaré and Amazon River basins, but none of these is present in the Gurupi, Turiaçu, and Maracaçumé basins—which seems to make bathers more relaxed there. Bird distribu-

FIGURE 2.3 Rapids on the Gurupi River near the mouth of the Igarapé Gurupiuna. The Gurupi is the only major river in the region that has rapids.

TABLE 2.1 *Common Species of Rivers, Streams, and Swamps in the Pre-Amazonian Forest*

Trees		Vines
Unonopsis rufescens (Gurupi only)	*Taralea oppositifolia*	*Machaerium ferox*
Miconia serrulata	*Virola michelii*	*Mabea pohliana*
Aniba citrifolia	*Coccoloba* sp. 2	*Hippocratea ovata*
Licania macrophylla	*Laetia procera*	*Masechites bicornulata*
Pachira aquatica (Gurupi only)	*Bellucia grossularioides*	*Dalbergia monetaria*
Eschweilera ovata	*Alibertia edulis*	*Derris* spp.
Myrciaria dubia (Gurupi only)	*Bixa orellana*	*Evodianthus funifer*
Diospyros artanthifolia	*Tovomita brevistaminea*	*Uncaria guianensis*
Myrcia sp. 1	*Tovomita brasiliensis*	*Philodendron* spp.
Crudia parivoa	*Inga* spp.	*Paullinia* spp.
Henrietta spruceana		
Phyllanthus martii	**Herbs and Epiphytes**	**Palms**
Vochysia inundata	*Guzmania lingulata*	
Macrolobium acaciaefolium	*Costus arabicus*	*Desmoncus polycanthus*
Ceiba pentandra	*Justicia pectoralis*	*Astrocaryum javari*
Anacardium giganteum	*Ischnosiphon petrolatus*	*Astrocaryum murumuru*
Anaxagorea dolichocarpa	*Ischnosiphon arouma*	*Geonoma baculifera*
Symphonia globulifera	*Ischnosiphon obliquus*	*Euterpe oleracea*
Croton matourensis	*Ananas nanas*	*Bactris maraja*
	Bromelia goeldiana	
	Limnocharis sp. 1	
	Sourobea guianensis	

tions are also slightly uneven between the Pindaré and pre-Amazonian basins to the north (D. C. Oren, personal communication).

In addition, evidence from the *terra firme* vegetation suggests a slight phytogeographical separation between the Pindaré basin, on the one hand, and the Gurupi, Turiaçu, and Maracaçumé basins on the other. For example, the leguminous trees *Cenostigma macrophyllum* and *Bauhinia acreana* are entirely absent, as far as my collections and surveys indicate, from the Turiaçu and Gurupi drainages; yet both are present in the *terra firme* of the Pindaré basin (see below). Both these species are also common in the Xingu River basin to the west (Balée and Campbell 1990).

Terra Firme Forests

These differences in fish and plants between the Pindaré, on the one hand, and the Gurupi, Turiaçu, and Maracaçumé on the other do not appear to be due to cultural factors; they more probably reflect ancient, geologically derived patterns of phytogeography that would have preceded the arrival of horticultural, indigenous peoples. Nevertheless, in the main, past horticultural events and processes are perceptible in the *terra firme* vegetation of pre-Amazonian Maranhão. I do not refer merely to the vast extent of cattle pastures and rice fields, to which any traveler along the BR-316 highway or any competent interpreter of recent satellite imagery may bear witness. Rather I refer specifically to what remains of the indigenous lands inhabited by the Ka'apor, Guajá, and Tembé peoples.

The most diverse forests of the region are, by far, those of the well-drained lands (*terra firme*). At least two kinds of *terra firme* forest occur in the region: old fallow (or simply fallow) and high forest. Fallow refers to sites that were once used for agriculture but which are now forested; the disturbance must have occurred between forty and more than a hundred years ago. High forest is not necessarily 'higher' than fallow; it is, rather, a forest of the *terra firme* which displays a primary character. It seems not to have been disturbed for agriculture within the last two hundred or three hundred years, if ever.

Satellite imagery to date does not distinguish fallow forest from high forest. The pre-Amazonian forests from the Gurupi in the north to the Pindaré in the south, regardless of their heterogeneity, are customarily mapped, moreover, as *floresta densa* ('dense forest'), a synonym for 'high' or 'primary' forest. (See, for example, Projeto RADAM 1973.) The ground truth, however, argues for a more refined interpretation of the forests in pre-Amazonian Maranhão, one that admits of substantial heterogeneity between stands of forest, depending on whether these were once used for horticulture, and regardless of the immediate, seemingly bleak future of these forests. Chapter 6 examines in detail the distinctions between fallow and high forest.

The Ka'apor as an Ethnographically Distinct People

The Ka'apor consider themselves to be a separate people. They are all *anam*, i.e., 'relatives', to each other. The term *anam* is, incidentally, polysemous; depending on context, it may mean 'another Ka'apor person', 'agnatic kinsman', or 'sister (female speaking)'. The people who call themselves Ka'apor are all interrelated through many generations of societal endogamy. They share a common language and distinctively similar mental representations about the seen and unseen worlds. They collectively demonstrate comprehensive systems of classification and utilization of living and nonliving things (e.g., Balée 1989b). Their economic activities affect the physical environment in ways that are distinguishable from other societies. Finally, they classify non-Ka'apor persons either as other indigenous peoples, whom they denote with specific epithets, or non-indigenous peoples, who are designated by the cover term *karaí*.

By these criteria, the Ka'apor do indeed appear to represent an analytical unit fit for ethnographic study. In other words, it is empirically possible to render a description of them, mutatis mutandis, as a linguistically and culturally bounded group. Amazonianist ethnographers of the 1950s—Francis Huxley and Darcy Ribeiro—referred to the Ka'apor as the 'Urubus'. 'Urubus' (meaning vultures in Portuguese) was bestowed on the Ka'apor by Brazilian society. In this book I use exclusively the people's self-designation: Ka'apor, or 'footprints of the forest'.

The Ka'apor speak a language of the Tupí-Guaraní family of languages. Other extant Tupí-Guaraní languages include Guajá, modern Guarani, Tembé, Wayãpi, Uru-eu-wau-wau, Asurini, Araweté, Kamayurá, Kagwahiv, Chiriguano, Sirionó, Guayaki (Aché), and Anambé. Extinct Tupí-Guaraní languages include Tupinambá and Archaic Guarani, which were spoken on the Atlantic Coast and in Paraguay during the sixteenth century. Tupí-Guaraní languages and cultures are among the most widespread in indigenous lowland South America; they have exerted important influences on modern Brazilian Portuguese and culture.

With a present population of about 520 persons, the Ka'apor occupy twelve villages in the basins of the Gurupi, Maracaçumé, and Turiaçu Rivers. Average village population in 1990 was 47 persons—up from 33 in 1982, when the population was somewhat lower and villages more numerous (Balée 1984b). Although their home range was somewhat more extensive until quite recently, Ka'apor villages today are all situated within the Reserva Indígena Alto Turiaçu (see chapter 3).

Familial Bonds, Headmen, and Marriage Practices

The Ka'apor lack classes of rich and poor, and professional and craft specialization. As to political organization, the Ka'apor do have headmen (*kapitã*) but a headman (almost always male) is not the village leader per se, since any village may possess more than one headman and because headmen lack genuine political authority. Headmen, rather, are informal spokesmen for clusters of related women, their parents, children, and in-married husbands. Upon marriage, husbands move away from their kinsmen to take up residence with their wives; daughters remain near their parents after marriage. Social relations in a typical village are centered on related mothers, daughters, and sisters. In a brilliant paper, Aguiar and Neves (1991) demonstrated

FIGURE 2.4 A Ka'apor village (Šoanĩ) in 1985.

genetic similarity among the adult female residents of three Ka'apor villages and genetic dissimilarity among the adult male residents of the same, which constitutes an independent measure of the ethnographically observed tendency toward uxorilocality. The men present, often with the exception of headmen, tend to come from elsewhere.

The primary function of a headman is to occasionally redistribute surplus goods and women among his people, among those who will and do constitute his residential unit. Although in Ka'apor theory, any male child may one day be eligible to become a headman, a tendency for patrilineal inheritance of this office exists for reasons that become clear below. A potential headman—the eldest son of a current headman—tends to remain in his village and residential unit after marriage. Unlike the sons of his father's sisters, the eldest son of a current headman has a wide array of possible spouses within the residential unit (marriage with a cross cousin—a daughter of his father's sister—is permitted). If the potential headman has several sisters he has even more reason to remain in the village of his birth.

The position, however amorphous, of a headman of a cluster of related households, derives from his having many daughters or sisters. The Ka'apor say that a man 'gives' (*mẽ'ẽ*) his sisters to other men. Headmen often take two wives, which is otherwise uncommon among Ka'apor men. Because of polygyny and the fact that half siblings are classified as full siblings, the oldest son of a headman is simply more likely than other men to have more sisters, which, in turn, establishes a wider basis for his jurisdiction.

On the other hand, an ordinary man, having few or no sisters, normally takes up residence with his wife's kin after marriage. His wife and her parents initially admit him into their household, under the same roof. He must then render brideservice by clearing forest and planting manioc and other crops for the use of his parents-in-law. He also supplies game meat to them. If the father-in-law is not too old for strenuous work, he and the son-in-law work together in the gardens and sometimes hunt together. After living under one roof with his wife and her parents for approximately one year, the young man usually constructs a house for himself and his wife adjacent to the house of his wife's parents. This first house of their own is often tiny and drab, since the man may expect little or no help from other men unless he has some kinship backing in the village—ordinarily this would occur in large villages only, where there are two or more out-marrying residential units.

A man living uxorilocally must prove himself to his wife's kin and to the others of the residential unit or village. His 'strangeness' erodes only after his wife gives birth. Men living under the same roof as their parents-in-law tend to be more reticent during group discussions than other, more established men. These discussions occur in the evenings in the house of an elder male. The uxorilocally wed man, in the first year or two of marriage, generally avoids eye contact with the other men, even if their talk centers only on sightings of animal spoor. One uxorilocally wed man I knew seemed to be asocial and withdrawn, but when I saw him during a naming ceremony at his home village, he was outspoken, he laughed loudly with other men, and seemed 'happy' (*huri riki*). It was more than just the occasion of the naming ceremony—which many people anticipate with anxiety because of possible quarrels during the ritual drinking—that induced this change in the man's behavior. He was on his own turf and temporarily free of the constraints imposed by living with one's in-laws.

Once a young man becomes a father, however, he speaks publicly more often in the residential unit of his wife's kin, without being addressed first. By this time, brothers-in-law cooperate more in hunting, household repairs, food sharing, and they may eventually sponsor each other's children in the naming ceremonies. The strongest cooperative ties between brothers-in-law are those between a headman and his sister's husbands, as the nature of the relationship between brothers-in-law is ultimately determined by the brother/sister dyad.

The headman finds himself at a crucial nexus of relationships with the other men in the village, since they tend to be married to his sisters or daughters. His brothers-in-law, therefore, are in some sense obligated to him. This is evident by the fact that a

FIGURE 2.5 Initiation rite of a Ka'apor girl just after her first menses. The girl's head was shaved minutes before; when her hair reaches shoulder length, custom allows her to marry. The man at lower right is making incisions on her legs with an agouti tooth.

headman's second wife is often his own sister's daughter (a permissible union generally in Ka'apor society), reciprocally given to him by his brother-in-law.

In many ways, the marriage bond is the most important relationship in Ka'apor
society. Marriages tend to be monogamous and remarkably stable. The divorce rate is
only about eight percent. Marriages are often arranged on a quid pro quo basis. A
father tends to approach his sister and her husband, asking them to give their daughter to his son in exchange for bringing their son into his own house to marry his daughter. Marriages of later-born children tend to be arranged by an elder brother.

A man and woman are recognized as husband and wife after their union has been sol-
emnized in a ceremony. On the morning of their wedding day, the groom or the groom's
father gives the bride's parents three yellow-footed tortoises (*Geochelone denticulata*),
which the groom himself has captured. Then the groom goes to the house of a headman
(*kapitã*), and sits in a hammock. Several people have by now gathered in the headman's
house. Next, the parents of the bride take her by each hand, leading her out of their house
and to her husband-to-be. The bride then sits in the hammock, to the left of the groom.
The groom gently clasps her neck in the crook of his left arm and places the palm of his
right hand on her head. The headman now faces the couple and exhorts the groom to
fetch yellow-footed tortoises (*yaši*) for his bride to eat during her menstrual periods,
warns him not to abuse her, and encourages the couple to be fecund (cf. Huxley
1957:158). Then in her parents' house the bride prepares the tortoises the groom sent her
that morning, along with soaking manioc flour in a calabash bowl, which she offers to

FIGURE 2.6 Hut of ritual seclusion. Ka'apor man stands outside the hut of ritual
seclusion (*kapi*) where his wife is soon to give birth.

him. They share this meal together. Then they depart to spend three or four days alone together in the forest in order to have sexual relations; they also hunt tortoises.

The gift of tortoises from husband to wife is very important in Ka'apor culture. The Ka'apor believe that a man should fetch yellow-footed tortoises (*yaši*) for his menstruating wife, for according to custom, she can partake of no other meat at these times (Huxley 1957:146,155; Ribeiro 1955; Balée 1985). The matrimonial bond, with reference to menstruation, is delicate and private. According to my informants, only a woman's husband can supply her with tortoise meat, which he must have captured himself. A man must also supply his wife with tortoise meat in the postpartum period. The dual injunction that a woman eat only tortoise meat and that her husband fetch it for her during ritually defined periods becomes a definitive attribute of the relationship between the sexes. It is one of the symbolic anchors of the integrity of Ka'apor marriage.

In the early period of marriage, a Ka'apor couple only gradually comes to focus their work on their partnership. At first, the husband tends to work more often with his father-in-law or brother-in-law, as the case may be, than with his wife, while she

FIGURE 2.7 Infant naming ceremony. Ka'por woman with her son in an infant carrying strap decorated with feather art, at an infant naming ceremony in the village of Urutawi in 1986.

tends to work more often with her mother. The couple's sexual activity, moreover, is generally confined to encounters in the forest by day on an improvised bed of açaí palm leaves, not in the house of her parents at night. This contrasts with older couples who live in their own houses and may have sex there. Once living independently, the man's chief obligation to his wife is to supply her with tortoise meat during her menstrual flow, as well as game meat and fish in general. The wife cooks, works in the swiddens, and processes manioc flour for her husband in turn.

Birth, Naming, and Family Life

With the first birth, by which time the couple already lives in a house of their own, the marital relationship deepens. Divorces after this point are rare. A woman gives birth (*ta'ĭ-u'ar-rahã*) inside a ritual room (*kapĭ*) of seclusion in her own house, sealed off by a door made from the woven leaves of bacaba palm (*Oenocarpus distichus*). Or her entire house may be sealed off this way as a ritual hut of seclusion. She gives birth in a kneeling position, supported by her mother, mother's sisters, and her own sisters. Her husband, alone among men, may also assist in the birthing process. To help speed delivery, these attendants rub on her belly leaves of any of several plants, or peeled skins of an ancient variety of bitter manioc called *mani'ĭ-se*.

After a woman has given birth to an obviously healthy child, she and her husband must remain inside the *kapĭ* for about fifteen days. The parents may bathe in lukewarm water fetched by their older children or members of another household, who are usually kin of the wife. During this period, both father and mother can eat only *yaši* tortoise meat and manioc flour. After the infant's umbilical scab falls off, the father may eat other foods, but not all. For about six months (until the child is named), he cannot eat any game except white-lipped peccary and tortoise, and the mother remains confined to a diet of tortoise meat and manioc flour.

Shortly after the umbilical scab falls off, the father designates someone (usually his wife's brother) to pierce the infant's ears. If the infant is male, his lower lip is also pierced, so that as an adult he can insert feather ornaments. After about six months, the couple's parenthood—and indeed, the infant's humanity—is publicly confirmed in a formal naming ceremony (Huxley 1957:186–187; Ribeiro and Ribeiro 1957:66–68). The naming ceremony, *ta'ĭ-hupĭ-rahã* 'when the child is lifted up' is the most public ceremony in Ka'apor society. It is the only ceremony, moreover, in which the many different feather pieces of Ka'apor art are displayed and worn. It takes place usually well after the end of the rainy season, to facilitate the journey of relatives of the couple who live in other settlements. It is timed to coincide with the day after a full moon.

About thirty days before the ceremony, the parents of infants to be named make tall, thirty-liter pots (*kamuší*) from clay, using ashes of the *karaipe* tree (*Licania* spp.) as pottery temper. Sixteen days before the ceremony, the villagers harvest from their swiddens a lot of bitter manioc (of certain varieties only), paring the tubers with knives. They place the tubers on storage racks in the houses of those who will name children. There the tubers collect mold for about ten days. Then they are placed on the copper griddle in the manioc shed and roasted for a few hours. The tubers meld into large gray chunks called *meyu-hu*. These are put into the tall pots, and small amounts of boiling water are poured over them.

After several days, the women whose children will be named walk to every house in the settlement, carrying calabash bowls (*kwi*) full of the semifermented syrup. Each of these women dips a small bowl into the larger one containing the syrup and offers a sip to the men in every house. About six days later, everyone washes their hammocks together at the stream, perhaps symbolizing a cleansing and renewal of village social life. The traditional soap used for washing hammocks is from the bark of a large tree of old swiddens called *šimo'i* (*Enterolobium* sp.).

On the following day, men and women—husbands and wives pair off—cut their hair to the length of the traditional tonsure. After the full moon rises, all adult participants begin to drink the now fermented manioc beer (*kawî*). The drinking, whether the beer is made from manioc, cashew, or banana, takes place at the full moon.

At dawn of the next day, all the women in the village assemble and walk slowly in a line (*yumupuku*) that leads to the house where the ceremony takes place. Each mother carries her infant in a cotton sling (or infant carrying strap). Many wear feather adornments. Meanwhile, the men recline in their hammocks, all of which now hang inside the house where the ceremony will take place—the house of one of the fathers whose infant will be named. When the women arrive, the mothers of infants to receive names sit on mats of bacaba palm (*Oenocarpus distichus*), which were woven for the occasion, and face east toward the rising sun. Then the *mai-aŋa* ('foster mother') takes the infant from its mother's arms and gives it to her own husband, the *pai-aŋa* ('foster father'). These individuals have been specifically designated by the infant's parents as sponsors. The role of sponsors is to give the infant one of its first names (see below), raise the infant should it become orphaned, and sometimes to later become its parents-in-laws. The *pai-aŋa* tends to be the brother of the infant's mother. In other words, the children of co-parents often stand as cross cousins, that is, potential marriage partners, to one another.

If the infant is male, his mother swaddles him in a red cloth, symbolizing headmanship for the purpose of this ceremony. If female, the mother wraps her in a white cloth. The ceremony is conducted in turns for each child. First, the *pai-aŋa*, with infant in arms, dances a few steps backward and forwards along an imaginary straight line. At the same time, he blows resolutely on a whistle made from the leg bone of a Harpy eagle and adorned with feathers. The whistle summons everyone's attention, for next the infant's name will be uttered for the first time publicly. Each infant receives two names, one bestowed by the sponsors and the other by the parents. The names, which refer to living and nonliving things, as well as culture heroes and qualities, are passed on bilaterally. They tend to skip one generation. The sponsor first announces the name he and his wife have chosen, then asks everyone, individually, to repeat the name—clearly a mnemonic device in this nonliterate society. Then the father, seated in his hammock, speaks aloud the name he and his wife have chosen, and again, everyone repeats it one at a time. Normally only one of the two names takes hold in everyday reference and address. (I was never able to determine the mechanism for this.)

If the child (regardless of sex) is first-born, the parents of the infant receive new names at this time as well. If the commonly used name of the child, for example, will be *Arar* 'Macaw', then his father will be addressed as *Arar-pai* or *Ara-ru* ('father of *Arar*') while his mother will be addressed as *Arar-mai* ('mother of *Arar*').

Teknonymy (the practice of naming adults for their children) in Ka'apor society serves as much to identify a man and woman with each other as it does to legitimize the link to their children. They take only the name of the first-born and retain that name even if the child dies. In the natural history of Ka'apor marriage, childbearing and child naming constitute a watershed, after which the economic, ritual, and psychoemotional bonds between a man and woman take on a more permanent quality.

The couple's union has been affirmed publicly and formally on three occasions by this time: in the wedding, in childbirth and the couvade (the ritual seclusion following birth), and in the naming ceremony. A division of labor has been established between them, and they have begun to raise their own family in their own house. The duties of brideservice, which a man fulfills for his father-in-law, no longer exist, although the couple tends to remain near the wife's kin. Care of the unweaned infant is the wife's responsibility and the two are almost always together.

Later, the young children of both sexes will play with other children their age in the village. Girls between the ages of about seven and ten, however, help their mothers get firewood. Boys of this age rarely get firewood. Older girls help their mothers in several chores, ranging from uprooting and carrying manioc tubers in baskets on their backs to preparing grills and cooking. But they do not work as much as married women. Young boys spend most of their time playing, but adolescent boys sometimes accompany their fathers hunting. They also fish with other boys with hook and line.

It is within the relatively restricted milieu of the nuclear family that the bulk of socialization takes place. Weaning and child rearing are responsibilities of the household, not the community. Mothers teach their daughters to weave and cook, while fathers teach their sons to hunt and swing an axe. And it is within the nuclear family that boys and girls learn the rituals of married life, which are the most basic expression of the social structure.

THREE
Ka'apor History

Origins

The Ka'apor, historically called the Urubus, have not always lived in the forests of northwestern Maranhão, where they are found today. They used to occupy a remote expanse of forest in southeastern Pará, much of which has recently become degraded pasture land. The Ka'apor speak of ancestors who hailed from the Rio Tocantins to the West (Huxley 1957:8; Arnaud 1978:6–7). Circumstantial evidence, including ethnobotanical evidence, indicates that an origin of Ka'apor society in the Tocantins basin is plausible. First, the name of the Rio Tocantins in the Ka'apor language is *i-takãšĩ* ('smoke river'), an allusion perhaps to the river's turbidity (even before the river was dammed in the late 1970s, its lower course was muddy because of tidal influences). The Portuguese name Tocantins, then, would have been a borrowing from a Tupí-Guaraní word for this vast river system, which in a sense drains directly into the Atlantic, not the Amazon (Sternberg 1976).

Second, the Ka'apor claim that they were at war with the *Karaya-pitaŋ*, who in the mid-1700s lived slightly below the confluence of the Araguaia and Tocantins Rivers (Noronha 1856:8; cf. Baena 1848:98). One scene in a Ka'apor myth, which recounts the power that a snakelike spirit named *Kuyamãye* has over swiddens, takes place in a swidden beneath an enormous Brazil nut tree. Brazil nut trees do not occur anywhere in the modern habitat of the Ka'apor, but the lower Tocantins, especially around the city of Marabá, was extremely rich in Brazil nut groves. Although the Ka'apor word for Brazil nut is *kãtãi'i* (an obvious borrowing from the Portuguese 'castanha'), the concept of a Brazil nut tree, even though the Indians have never seen one, is aboriginal and derived probably from the west, in the direction of the Tocantins River.

According to the missionary José Noronha (1856:8–9), at least eighteen indigenous societies occupied lands drained by the Tocantins River below the mouth of the Rio Araguaia in 1767. Noronha noted that most of these peoples spoke *Língua Geral* which would in effect have been one or more of several Tupí-Guaraní languages. Inhabitants of the left bank of the Tocantins he identified as *Turiuarus*, probably antecedents of the *Turiuara* (see Baena 1848:98), who are now extinct. He also identified the *Uayá*, probably antecedents of the modern Guajá. Both groups would later occupy lands east of the Tocantins and would separately become enemies of the Ka'apor. Noronha also mentioned the *Amanajóz*, probably the same as the Tupí-Guaraní–speaking *Amanajé*, who were still living near the Tocantins in 1775 (Nimuendaju and Métraux 1948:199) and would be seen near the Acará River in the early twentieth century by the English explorer Algot Lange (1914). In addition to Tupí-Guaraní–speaking peoples, Noronha also observed non–Tupi speakers on the right bank of the Tocantins, including the Apinayé, Timbira, and the *Acarajá-pitanga*, one of the acknowledged scourges of the Ka'apor ancestors (Noronha 1856:8).

Nimuendaju made the unsubstantiated claim that the Ka'apor and Turiuara were "local divisions of one people" (1948a:193). But his 1914 word list of Turiuara shows relatively few cognates with Ka'apor and it is clear that certain grammatical aspects of the two languages (such as negation) were different. Nimuendaju never visited the Ka'apor; he did visit the Turiuara, who were living in the Acará River basin to the north of the Ka'apor in the early twentieth century. His speculations on a possible Ka'apor/Turiuara connection may have been influenced by newspaper stories alleging such a connection that appeared in 1928 after overt hostilities between the Ka'apor and colonial settlers subsided (e.g., Anonymous 1928).

In fact, of the many Tupí-Guaraní groups of contemporary eastern Amazonia, few speak the same language or even what might be considered mutually intelligible languages. One exception is the linguistic affinity among the Parakanã, Suruí, and Assuri-

FIGURE 3.1 Brazil nut tree, *Bertholletia excelsa*, on a peasant homestead near the upper course of the Amazon River. Brazil nut trees are absent from the forests of pre-Amazonia.

ni-Trocará (all of whom are located in the Tocantins, with a few other Parakanã on the Rio Bom Jardim of the Xingu basin)—all of whom share linguistic affinity with the Tembé and Guajajara, whose language is known as Tenetehar. The Tembé and Guajajara share an identical language because of a very recent split of the mother population.

The branching history of the various groups speaking Parakanã has not yet been elucidated, but it is known that the Tembé separated from the Guajajara on the Pindaré River in Maranhão and migrated into the Gurupi, Guamá, and Capim River basins some time in the early 1850s (Gomes 1977:103; Wagley and Galvão 1949:4,n.2). From a historical linguistic point of view, this split is extremely recent. Although no other people speak a language mutually intelligible with Ka'apor, the Ka'apor language appears most closely related to Wayãpi, not to Turiuara nor even to the closest indigenous neighbors of the Ka'apor, namely, the Guajá, Tembé, and Guajajara. Linguistic affinity is, of course, not to be expected with the Wayãpi. Although the Wayãpi now live in northern Amazonia (in a region straddling the border of Brazil and French Guiana), they are rather recent migrants from the lower Xingu basin, where they were living in the early 1800s (Grenand 1982:94).

The Coming of the Portuguese

Linguistic, ethnobotanical, and historical evidence thus all suggest a westerly origin for Ka'apor society, probably between the Xingu and Tocantins Rivers. The Ka'apor left that region partly because of conflicts with the Portuguese.

The indigenous peoples of the Tocantins and elsewhere in eastern Amazonia were subjected in the seventeenth century to attempts to remove them from their lands and transport them to large mission settlements known as *aldeias* or *aldeiamentos*. These *aldeias* were, by law, under the jurisdiction of missionaries, but often they were influenced by civil and military authorities. Situated near Jesuit missions as well as large colonial towns, such as Belém, the *aldeias* served as sources of slave and corvée labor for Portuguese colonists, who arrived in 1616, expelling the Dutch settlers who had preceded them.

Between 1616 and 1621, the Portuguese colonists intermittently battled the "Topinambazes," a Tupí-Guaraní-speaking people who probably were organized into one or more native chiefdoms. The first major war of attrition and slave-taking against the Tupí-Guaraní peoples of Pará and Maranhão, however, took place in 1621. The supreme military commander of the captaincy, Bento Maciel Parente, slaughtered as many as 30,000 and captured an unrecorded number near Belém (Almeida 1874:18; Kiemen 1954:22). Most of the slaves taken were put to work in tobacco fields for Portuguese masters (Vieira 1925:308, cited in Hemming 1978:319). In 1649 and 1650, the governor of Pará deployed soldiers in slave-taking campaigns known as *entradas*. The captives were forced to till the governor's tobacco fields, to fish for him, and to work the salt flats (Kiemen 1954:22).

Entradas were outfitted in the northern towns of Belém and São Luís. From the south, inhabitants of the Tocantins basin faced *bandeiras*—rowdy bands of slave and gold seekers from São Paulo. Many Tupí-Guaraní groups had already fled from southern Brazil into the Tocantins basin (Berredo 1849:538–539). By 1673, despite moral appeals from Jesuit missionaries, *bandeirantes* had enslaved many of the indigenous inhabitants (Berredo 1849:539–544). The forest dwellers of the Tocantins thus found themselves trapped between three hostile fronts, to the north, south, and east. Some migrated west across the central Brazilian plateau (such as the Kagwahiv, who are today denizens of the Madeira basin; see Kracke 1978:7). Others were induced to accept "protection" from the Jesuits to the north and east.

In 1657 Antônio Vieira, the leading Jesuit in Pará (who would later become one of the most renowned clerics in all of the Portuguese empire), protested: "The injustices and tyrannies practiced on the Indians in these lands exceed by far those done in Africa. Within the space of forty years there were destroyed along this coast and in the interior more than two million Indians, and more than 500 Indian settlements as large as cities, and no punishments have been given for this" (Vieira 1925:468; quoted in Kiemen 1954:107–108).

In 1658 the Jesuit Manuel Nunes led 450 mercenary natives in canoes with an escort of an equal number of Portuguese soldiers up the Rio Tocantins. His mission was to pacify and neutralize the Inheiguaras, who were obstructing other peoples from visiting the Jesuit missions along the river. The Inheiguaras lived a many days' walk from the river, in remote forest. The expedition took 240 Inheiguaras as prisoners, who were declared slaves and divided among the soldiers (Betendorf 1909:112–116; Vieira 1925:554; Kiemen 1954:112).

In 1659 a group of Tupinambá still living on the upper Tocantins, perhaps having sought refuge there from the earlier slave raids and diseases prevailing on the Atlantic coast, migrated to Belém at the behest of some Jesuits. At the same time, about a thousand members of another Tupí-Guaraní group that had earlier fled the coast, the Potiguaras, also relocated to Belém (Betendorf 1909:113–114). The Potiguaras who chose not to flee are now a much reduced, deculturated people of Pernambuco in the northeast, where they were known since the 1500s (Balée 1984b). In 1660, two thousand more refugees of miscellaneous origins—some slaves (*escravos*) and some freemen (*livres*)—joined the *aldeias* near Belém and elsewhere in eastern Pará (Vieira 1925:555–556). Freemen, it must be noted, were not much better off than slaves. According to the meticulous Jesuit historian Matias Kiemen (1954:107), freemen tended to labor for the settlers ten months a year as constituents of a corvée labor force.

One can only speculate on what life was like in the Jesuit missions of the lower Amazon, but it surely would have involved partial acculturation for the refugees. They would have been exposed to church dogma, other peculiarities of colonial Portuguese culture, and the Portuguese language. The descendants of many of these people became assimilated to a Luso-Brazilian lifestyle, no doubt by marrying 'out'. The original village-based institutions would have withered in the pandemonium of peoples, cultures, and languages that characterized the *aldeias*.

Although far from ideal, the Jesuits nevertheless protected the original inhabitants to some degree from the state and its colonists. But this mediating role was eliminated in 1759 when the Marquis de Pombal, a Portuguese statesman who yearned to see his nation's empire grow, arranged for the expulsion of the Jesuits from the Amazon region. Paternalistic Jesuit fathers were replaced by army captains (*capitães*). A similar saga took place in Paraguay after the expulsion of the Jesuits in 1767 (Métraux 1948:79). Despite royal edicts prohibiting enslavement of indigenous peoples, many of the freemen of Pará who had settled near missions became indentured servants of the settlers after 1759. But equally large numbers fled the missions into the forest. Able-bodied men who remained were often conscripted for military service, which in those days largely entailed subjugation of hostile forest peoples.

With the disintegration of the *aldeia* system, thus, the cultural (and probably the linguistic) gap between the state and the forest Indians widened, and peaceful interchanges virtually ceased (Azevedo 1930:373–381). Sporadic contacts with missionized parties, nevertheless, both before and after the departure of the Jesuits, probably constituted a continuing source of Western influence on the forest peoples (cf. Watson 1952:49). Although groups such as the Ka'apor may have remained somewhat aloof from civilization, the political upheavals of this period probably had some effect on them, as historical linguistics and their own folklore indicate.

The earliest opportunities to study the Ka'apor language came only after 1928, when hostilities finally came to an end. Linguistic evidence suggests that Ka'apor society in 1928, which represented one of the most splendid indigenous enemies ever faced by Luso-Brazilians in the Amazon region, was hardly pristine and exclusively pre-Columbian. The language exhibits lexical borrowings, not only from Portuguese, but also from *Língua Geral*, a language derived from Tupinambá with much Portuguese influence.

Lexical borrowings from the Portuguese are very clear, for example, with regard to a few Ka'apor names designating nondomesticated plants. In the Amazon generally, linguistic borrowings for plants were in the reverse direction: from indigenous peoples to the Portuguese. Borrowings, however, often show complex influences in both directions. For example, the word for *Lacmellea* spp. (of the dogbane family) in Portuguese is *pau de colher*, i.e., spoon tree. Spoons and ladles are indeed obtained by the Ka'apor from the inner bark and heartwood of these high forest trees. The Ka'apor word for these native trees is *kuyer-'i* 'spoon-stem'. The word for spoon in Ka'apor (*kuyer*) was clearly borrowed from Portuguese *colher*. Another example of Portuguese influence that pre-dated the 1928 establishment of peaceful relations is the Ka'apor word for numerous species of *Protium*, a genus that with the exception of *Protium heptaphyllum* is found virtually only in high forest in the Ka'apor habitat. The name for this kind of tree is *kanei-'i* 'resin-stem', an unmistakable borrowing from Portuguese *candéia*, denoting 'wax candle' (Balée and Daly 1990). The resin of *kanei* is in fact burned by the Ka'apor to illuminate interiors. Finally, the preposed attributive in the word *kašima-'i* 'pipe-tree', which denotes several native trees in the spurge family genus of *Mabea*, came from Portuguese *cachimbo* (pipe), in turn derived from a West African language (Balée and Moore 1991). One use of the heartwood of this genus is in the manufacture of pipe stems.

A nonbotanical example of a pre-1928 borrowing from Portuguese is the vocative for 'father' in Ka'apor, which is *papai*. Referential terms remain indigenous; Portuguese for 'daddy' was recorded by Lopes (1934:168)—who put a question mark after it, it should be noted—shortly after the pacification of the Ka'apor in 1932. Since the

Ka'apor, even the eldest among them, never use an indigenous vocative for 'father', it may be assumed that *papai* entered the language long ago, prior to 1928.

All in all, a considerable body of linguistic evidence points circumstantially to sustained contact in the post-Columbian past between the antecedents of Ka'apor society and Portuguese speakers, perhaps including Jesuit missionaries. It comes as no surprise, therefore, that the Ka'apor language had a word for Catholic 'priest' (*pa'i*) in 1928—this was also one of the given names of the legendary young man who first 'pacified' (*mu-katu*) the *karaí* ('non-Indians'), according to the near unanimous voice of the living Ka'apor as well as the historical documents of the time. And there is even a 'priest-tree' (*pa'i-mïra*), that refers to *Sagotia racemosa* of the spurge family, one of the most common species in the high forests of pre-Amazonia.

Migration to the Acará and Capim Basins

The Ka'apor left the besieged Tocantins basin before they too were subjugated by the Portuguese. Sometime in the late eighteenth century the Ka'apor migrated northeastward into the upper Acará river basin, which is between the Tocantins and the Capim basins. There they appear to have settled for a time before moving on to their present homeland in the Gurupi river basin. In the mid–twentieth century, Darcy Ribeiro collected a genealogy from his principal informant, the headman Anakãpuku, who indicated that his most distant known ancestor was born in the Acará basin. On the basis of nine generations that Anakãpuku remarkably recalled, as well as 1200 named persons—with places of birth and death for each—this ancestor would have lived in the Acará basin in about 1790 (cf. Ribeiro 1951:372,384; 1956:38). Although none of my own informants could recall more than three ascending generations of their lineal kin (see chapter 5), Ribeiro's data are convincing (he kindly showed me the notebook in which Anakãpuku's testimony was recorded). Huxley also pointed out that one of his informants could recall ancestors from the fifth and sixth ascending generations (1957:265). One other piece of evidence indicates that the Ka'apor spent some time in the Acará basin. A myth concerning incest, menstruation, and the origin of the moon, which was current among caboclos of the Acará River in the early 1900s, bears striking resemblance to a Ka'apor myth (Oliveira 1951:29–30).

By the 1820s at least some of the Ka'apor had migrated still further east, living off lands drained by the Rio Capim. In the early years of occupying the Capim (called *ï-kãpï* or 'grass river'), because of the cattle pastures, the Ka'apor lived at peace with surrounding settlers, according to informants. That the Ka'apor at one time maintained harmonic relations with Luso-Brazilians—at least, relations that were short of outright hostility—was affirmed in newspaper reports at the time of their "pacification" (e.g., Anonymous 1928).

According to my informant Timapa-ru of the village of Urutawi, oral legend has it that many Ka'apor once lived in a settlement near a Brazilian town on the Capim River. One day a village headman named Tapõ'õ returned home from hunting only to find many *karaí* (non-Indians) as well as *Karaya-pïtaŋ* in his house. The visitors had been drinking. They planned to murder *Tapõ'õ*, since he and his people refused to live

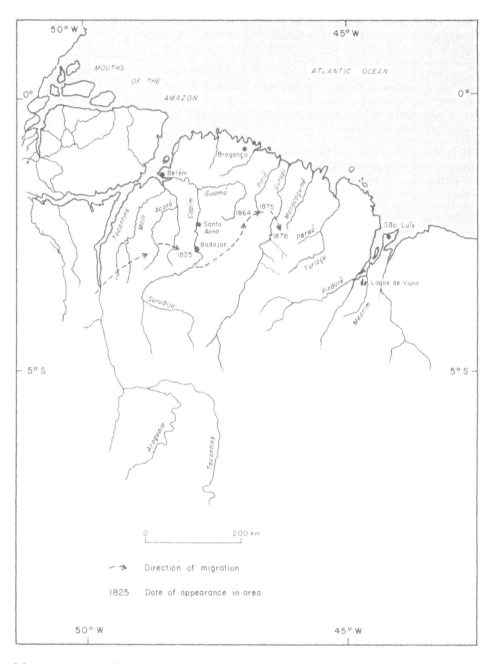

MIGRATION OF THE KA'APOR. Arrows show the routes of historically known migrations of the Ka'apor. Dates indicate approximate appearance in given regions.

in the town and work. The other Ka'apor of the settlement, including two wives of Tapŏ'ŏ , fled into the forest at the edge of a swidden, concealing themselves. Sensing danger, Tapŏ'ŏ leaped to the roof of his house. As the karaí and their accomplices fired at him with arrows and muskets, Tapŏ'ŏ played resounding notes on the long cere-monial flute (imi'a—made from Cecropia spp.) and danced at the same time, evading the arrows and lead shot. His people knew he was still alive, since they could hear the flute music. After several hours Tapŏ'ŏ shouted to his enemies from the roof, "Uruhu ne 'u ta" (Vultures will eat you). Then he descended, and with his ceremonial cudgel (tamaran) in hand, made from that most heavy of Amazonian woods, Zollernia paraen-sis, he dispatched them all to their graves. He joined his people at the swidden's edge, and with them fled to a new site, farther upstream. Thus began, according to Ka'apor lore, a period of unrelenting miaí 'hatred' between the Ka'apor and the karaí, which, although its chapter of open warfare closed, has perhaps never truly ceased.

There are few other accounts of events or legends during the time the Ka'apor resided in the Capim river basin. João Barbosa Rodrigues, one of Brazil's (and the world's) greatest naturalists of the nineteenth century, ascended the Capim River to its sources in 1874. He spoke with some of the caboclos, who informed him that the Ka'apor ('Urubús') had once raided the now defunct hamlet of Badajóz:

> Already in this century (182?), the river became the subject of assaults by the Urubú savages, who threatened the town. Francisco Nunes requested the support of the chief of the Turiuáras, Paraná-Mondo. They descended the river together, going to the governor. The governor ordered the rancorous Portuguese major, Manoel Manso Metello Manito Moreira to help them with militiamen under his command. Manoel Manito Moreira brought thirty soldiers, and with the Turiuáras, they attacked the Urubús, who today are on the Piriá River.
>
> (Barbosa Rodrigues 1875:33)

Barbosa Rodrigues's account is the only published reference I know that treats the Ka'apor prior to the 1860s. Some letters from Major Manoel Manso Metello Manito Moreira, however, survive in the public library of Belém. On the first of January 1825, Major Moreira, who was military commander of the Capim River, submitted a report to his superior officer in Belém, which I translate as: "Your excellency would do well to take account of the Indians.... Almost 400 Indians ... have stolen white daughters of citizens, as well as canoes, causing the greatest disorder possible on this river" (BPB 1825).

Major Moreira, in all of the legible correspondence he sent to Belém, never seems to have mentioned the Ka'apor (or Urubús, as they were pejoratively known) by name. But given Barbosa Rodrigues's account, it seems likely that the warriors who threat-ened the townspeople of the Capim River were antecedents of the modern Ka'apor. Other known inhabitants of the Capim River today, such as the Tembé, had not yet arrived from their berth on the Pindaré in Maranhão. The Turiuara evidently lived only on the lower Capim and Acará Rivers, and in any case already served the inter-ests of the state in attacking the Ka'apor. (This point makes Nimuendaju's [1948a] claim for a Turiuara/Ka'apor common origin, of course, dubious.)

Perhaps not all Ka'apor were hostile at this time. Some of the peoples of the Capim

River in the 1820s, for example, while living in the remote forests, maintained commercial ties with merchants in the towns, despite the hostility of the general populace. The business of extracting and selling raw materials from the forest, known as *drogas do mato*, focused mainly on copaíba oil—a resin that exudes from the bark of several species of *Copaifera*, a leguminous tree of the high forest. This resin is a proven cicatrizant and is used in Amazonia for several other ethnomedical purposes as well. Local merchants would have bought copaíba oil from forest dwellers, with the goal of selling it in Belém.

Copaíba trees grow dispersed in the forest and the incision made to draw out the resin appears not to kill the tree. Collecting the resin was generally an individual, rather than a cooperative, task—similar to rubber gathering (Gomes 1977:199). The merchants who traded with the Indians received little backing from subsistence farmers, slave-holding planters, and ranchers. Brazilian settlers who did not profit from the extraction business in the Capim river basin complained bitterly to government authorities about Afro-Brazilian slaves escaping along paths cut by Indians in search of copaíba trees (BPB 1822a, 1822b). The townspeople also objected to the paths being cut because, they asserted, game animals followed the trails away from the towns, causing a decline in game availability (BPB 1822b).

Only three merchants of the Capim traded goods for copaíba oil in the early 1820s (BPB 1822a). Unacculturated peoples evidently had little contact with other settlers. In the town of Ourém, situated on the Guamá River near the mouth of the Capim, the forest inhabitants inspired fear among would-be explorers and colonists. "The dwellers of this parish are greatly afraid of a large number of Indians, on the move in the headwaters, who could lay siege to the town" (BPB 1822b).

By 1829 the population of the Santa Anna parish (*freguesia*—an administrative unit) numbered 1,555 people, of whom 127 were classified as *índios* (BPB 1829). The parish seat was located at 2°6' South latitude on the right bank of the Capim (see the map). Badajóz, the hamlet which some Ka'apor raided in the 1920s, was some distance upstream from Santa Anna itself, but its population formed part of the parish of Santa Anna. The figure of 1,555 people reflects those living from the middle Capim to its lower reaches. But it did not include the forest dwellers who resisted adopting the ways of Luso-Brazilian town life. Many of the Ka'apor constituted one such group, although a few of them likely became town dwellers and would have been counted as *índios* living in the towns. Some of my informants said that, after hostilities began, some Ka'apor from the Capim did go to work for the *karaí* in the towns. While living in the towns, these Ka'apor continued to make 'featherwork' (*pu'ir*). But they were no longer regarded by their brethren as true Ka'apor. They had become 'false Ka'apor' (*Ka'apor-ran*). Those who remained in the forest soon considered them enemies.

Aside from the risk of falling victim to corvée and slave labor, the Ka'apor of the forest would have had another reason for minimizing contact with pioneers, once they perceived the devastating effects of smallpox. There were major epidemics of smallpox in Maranhão and Pará in 1695 (Betendorf 1909:585–595), in 1724 (Betendorf 1909:ix), and from 1743 through 1750 (Azevedo 1930:228, cited in Kiemen 1954:180). Between 1793 and 1800, another smallpox epidemic spread from Belém to outlying districts. The historian Artur Viana wrote that this epidemic "frightened and decimated the population, taking a horrifying toll on Indians and *mestiços* [i.e., creoles]" (1975:44). In 1819 a final epidemic struck, spreading again from Belém outward into the forests of Pará: "The smallpox took hold in Belém and in the country towns. . . . As to the number of afflicted and dead, no one knows" (Viana 1975:50).

This last epidemic coincided with the beginning of the war of the *cabanas*, or the *Cabanagem* (c. 1835–1836), a civil war that saw extraordinarily high casualties. This war pitted civil and military authorities against rebellious *cabanas*, who included those of mixed Indian and Portuguese descent, runaway Afro-Brazilian slaves, and acculturated Indians. The state eventually won this war, with the surrender in Pará of the leader of the *cabanas*, Eduardo Angelim, and his co-revolutionaries, among whom were Tupí-Guaraní–speaking peoples (Hurly 1936; Raiol 1970:979–980).

Toward the end of the war, after a devastating battle on the Acará River in which six hundred runaway Afro-Brazilian slaves in his group were captured and returned to bondage, Angelim sought refuge in the forests of the upper Capim River. Hurly wrote, "Eduardo and his wife and brothers, and few more, protected by the Indians, journeyed to the Capim River, ascending to its headwaters. Captain Mello [the same Mello cited above] had responsibility for pursuing him, and with other Indians, he followed Eduardo's trail, discovering the Indian house in which they took Eduardo's whole family prisoner" (1936:32; cf. Raiol 1970:978).

Did Angelim take refuge in a Ka'apor village? Given the location and the time described, as well as Ka'apor hostility toward governmental and military authority, this is possible. It was in these years, in any case, that the Ka'apor rejected civilization. Refusal to be subjugated accounts for why they survived into the present as a distinctive society.

Residency in the Gurupi Basin

By 1861 the Ka'apor had migrated into the Guamá and Piriá River basins, living "wild and hidden" (*bravios e occultos*) in the deep forest (Márques 1870:117). But by 1864 it was becoming increasingly difficult to hide from pioneers, who were encroaching upon formerly unexplored tracts of forest. Some of the Ka'apor took to raiding towns on and near the Guamá River, evidently to acquire iron tools. The authorities responded by searching out and attacking Ka'apor villages:

> The Indians called Urubús attacked [in 1864] various sites of the upper Guamá, stealing and murdering. In retribution, an expedition of 25 national guardsmen was organized under captain of the bush, José dos Santos Brandão. Upon penetrating the forest, they reached a settlement of these savages, where a bloody battle ensued. The superior arms of the invaders caused many casualties on the Indians' side, with some of the guardsmen dying as well. The savages fled, returning a little later with reinforcements to look for their dead. . . . Still in 1864 . . . a second expedition was organized against the Urubús, with 150 national guardsmen. . . . This force penetrated as far as the Indian villages, whose inhabitants were successively burning and abandoning them as the force approached, not permitting it to come

within reach of them. The expedition chased the savages into the headwaters of the Guamá and Gurupi. (Muniz 1925:154)

Most likely, the Ka'apor were also engaged in warfare with the Guajá and Tembé by this time. The foraging Guajá, who lived in the Gurupi basin in Pará in the 1860s, occasionally raided the swiddens of other forest dwellers and settlers for maize, sweet potatoes, and the like. Such raids entailed high risks; if the Ka'apor or Tembé caught them in their swiddens, they would "beat or kill them" (Araújo Brusque 1862:15; see also Nimuendaju 1948c:135).

Some Ka'apor settlements still remained in the Piriá basin in 1874 (Barbosa Rodrigues 1875:33). But by 1872, Ka'apor settlements had spread to the tributaries of the left bank of the middle Gurupi. The engineer Gustavo Dodt, who ascended and mapped the Gurupi in 1872, wrote of the Ka'apor:

> This tribe lives isolated and without relations with the civilized population. They enter into contact with the civilized population only in their raids, which extend to the banks of the Gurupi, which they also cross. . . . One must confess, however, that the cases in which they have fired some arrows over canoes on the river, or over the houses . . . are quite rare. (Dodt 1939:176)

Some Ka'apor were by 1872 at least crossing the Gurupi on occasion, probably during the dry season, for purposes of raiding and, no doubt, fishing and hunting. Francis Huxley (1957:103) speculated that the Ka'apor were nomadic foragers at the time they first crossed the Gurupi. But Huxley's probably ill-informed speculation does not agree with the observations of the engineer Dodt (1939:176), who was there at the time and concluded that the Ka'apor had swiddens and lived in settlements when they arrived in the Gurupi basin. In addition, Huxley's unsubstantiated statement that the Ka'apor had lived in Maranhão for more than one hundred years by the time of his research in 1953 is erroneous. David Cleary (1990:30) repeated this mistake, declaring without reference to any source that the Ka'apor were "still" in control of the upper Gurupi, Maracaçumé, and Turiaçu Rivers in Maranhão in the 1850s. Today it is clear that the first Ka'apor to cross the Gurupi River into Maranhão and remain there permanently did not do so until some time between 1873 and 1878.

Some of the Ka'apor initially may have swept beyond the Gurupi and into the Maracaçumé and Paruá River basins. In 1874 the chief of police for the province of Maranhão observed that a "horde of Indians attacked and robbed many houses in the Paruá district last June, the inhabitants of which have become terrorized." (APE 1874). In 1880 a district judge in the region wrote the provincial governor that: "The Chief of Police communicated to us that Dona Angelica Rosa de Nascimento . . . was killed by Indians in her own house in the Paruá district." (APE 1880). It is unlikely that the Indians invading the Paruá could have been other than Ka'apor, since the Tenetehara (Guajajara) in the nearby districts of Viana and Monção were already "civilized" by 1872 (Gomes 1977:92) and the Gamella and Timbira of the Paruá and middle Turiaçu Rivers, who may have once launched raids in the area, were but pacified remnants of formerly larger, more autonomous populations (Guimarães 1887:67; Marques 1870:491). As a distinctive society, the Gamella (as with the Tapajós), in fact, became extinct prior to their language becoming known, so that it is impossible to say to which aboriginal group and language family they originally belonged.

If by 1880 the Ka'apor were not in control of most of the Paruá River, they were soon after. Lieutenant-Captain José Guimarães, who navigated and surveyed the Paruá River in 1886, reported to the provincial Congress of Maranhão that:

> This river, which was once an agricultural center of manioc flour, corn, rice, coffee, and so on, is completely depopulated owing to the constant raids that the Urubu Indians launch, a tribe . . . that understands iron-working so well that they have bows, with arrow points of iron, use gunpowder, and comprehend the uses of paper currency. . . . This tribe, which . . . has at its continual disposal more than 3000 bows . . . has committed with impunity all crimes about which there is news from the Gurupi to the lakes of Viana. (Guimarães 1887:62–63)

Despite Guimarães' questionable estimate that the Ka'apor had 3000 or more warriors, he pointed out that their raiding parties were composed of only between 30 and 40 men (Guimarães 1887:63).

The Ka'apor who descended into the Paruá River basin in the late 1870s and early 1880s may have branched off from those who had already established a foothold between the Maracaçumé and Gurupi rivers around 1878 by destroying a settlement of runaway Afro-Brazilian slaves. Miguel Lisboa, who ascended the Gurupi River while conducting a survey of gold mines in 1895, wrote that the Ka'apor "employ the bow and arrow—the common weapon of all Indians of the Gurupi—as well as the club . . . with which they conduct violent hand-to-hand combat. Their arrows are equipped with a sharp iron point, admirably worked and polished" (Lisboa 1935:51–52). Although Ka'apor settlements became firmly established by the early 1880s in sparsely inhabited forest between the Gurupi and Turiaçu Rivers, the Ka'apor remained at war with several sociopolitical groups—among them, runaway Afro-Brazilian slaves, Kren-Yê Timbiras, Guajá, and the Guajajara and Tembé (Tenetehar).

Arquival and other sources attest that many settlements of fugitive Afro-Brazilian slaves existed in the Gurupi and Turiaçu basins since the early nineteenth century, long before the arrival of the Ka'apor in the region (cf. Cleary 1990). These settlements were called *mocambos*. Despite military raids against them, *mocambos* multiplied in the region until the "Golden Law" of 1888, which abolished slavery. Several primary documents report refugee Afro-Brazilian slaves living in the middle to upper Gurupi and Turiaçu basins, where they represented a challenge to the hegemony of the plantation class (APE 1839, 1840, 1854, 1874, 1878).

The fugitive slaves lived in remote forest, beset on their eastern boundary by rice- and cotton-growing peasants, as well as cattle ranchers and slave-holding planters. In the lower and middle Turiaçu basin, powerful landowners and the governmental-military axis looked upon fugitive slaves with alarm. In a book that argued for the transfer of the region between the Turiaçu and Gurupi rivers from the jurisdiction of Pará to that of Maranhão, which was accomplished in 1852, Cândido Mendes de Almeida wrote: "In addition to revolutionary invasions [i.e., the civil war known as the *Cabanagem*] the inhabitants of the Turiaçu have suffered continually from a scarcity of slaves, now fled, who live in large *quilombos* [a synonym for *mocambos*] united with others from the districts of Viana, Alcântara, and Guimarães of the Province of Maranhão" (Almeida 1851:29).

The *mocambos* formed a buffer between the farms and plantations to the east and the Ka'apor to the west. On both fronts, the fugitive slaves met violence. In 1878, for example, the Ka'apor sacked and burned a *mocambo* on a tributary of the Gurupi near the headwaters of the Maracaçumé River. This is near the site of the present village of Gurupiuna. It is now a *taper* 'fallow' dominated by bacuri (*Platonia insignis*) fruit trees (see chapter 6). The site destroyed by the Ka'apor, thus, was later occupied by them. Lisboa recorded testimony in 1895 from some former slaves in the Maracaçumé basin, who had witnessed the 1878 battle: "Primordially, they [the Ka'apor] were inhabitants of Pará. They crossed the Gurupi River, taking advantage of the dry season. They invaded the right bank, near the headwaters of the Maracaçumé, dislodging the *quilombos*. . . . The little we know about that tribe was transmitted to us by the Negroes of the *mocambo*" (1935:51–52; see also 1935:4).

In addition to raiding the settlements of fugitive slaves, the Ka'apor also attacked indigenous Amazonians of the Gurupi, Turiaçu, and Pindaré Rivers, who were all more or less mutually hostile. The foraging Guajá, for example, had long been one target of Ka'apor and Tembé raids (Dodt 1939:177), perhaps in retaliation for their pillaging of swiddens. In 1913 or shortly before, for example, a group of Ka'apor warriors massacred a Guajá camp (Nimuendaju 1948c:136). In fact, sporadic raids, some involving deaths, between the Ka'apor and Guajá continued until 1975, when FUNAI officials persuaded both groups to live at peace, in accord with the *Pax Brasiliana*, an unstated policy of the government to eliminate indigenous warfare.

Prior to 1913 and up to 1928, the Ka'apor also frequently attacked Guajajara settlements in the Pindaré River basin (Wagley and Galvão 1949:8–9; Gomes 1977:123). Along the Gurupi, the Ka'apor raided both Tembé and Brazilian settlements, primarily for steel tools (Hurly 1928, 1932a, 1932b). There, the Ka'apor also raided the Kren-Yê Timbira of the Gê language family, who were recent immigrants to the region from the savanna country to the south. In 1903, for example, some Ka'apor killed "many" Timbira who were collecting copaíba oil in the vicinity of the Igarapé Gurupiuna, a creek long occupied by the Ka'apor (Nimuendaju 1946:13–14). The Gê-speaking peoples of Maranhão, including the Kren-Yê and the Canela, are referred to derogatorily in the Ka'apor language by the name *Moi-yowar* 'snake-eaters', a term which incidentally shows a certain familiarity with one of the customs of these peoples.

Since the 1820s, the Ka'apor had raided farms, plantations, hamlets, and towns in Pará and Maranhão. Their chief objective was to obtain steel axes, machetes, and knives. Intrusions by gold prospectors and miners into the remote forests presented another potential source of steel implements for the Ka'apor. Gold mines were active in the hill country between the Gurupi and Maracaçumé basins since the early 1800s (Baena 1838:499; Moura 1936:7; Cleary 1990). In 1854, well before the Ka'apor had crossed the Gurupi River, the governor of Maranhão declared: "The report of the commission charged with exploring the gold territories of the Maracaçumé confirms the abundance and richness of mines" (Machado 1854:19; also see Marques 1870:302 and *Almanak do Maranhão* 1859:104).

The most important mining center in the Maracaçumé was Montes Aureos, which still exists today (Cleary 1990). At one time there were as many as three hundred workers in the mines of Montes Aureos, operated by the Mineração Maranhense company. But by 1862, there were only about a hundred (Marques 1870:302). An English company bought controlling interest in the mines at Montes Aureos in the 1860s but, like the earlier company, failed to turn a profit and eventually abandoned the enterprise (Marques 1870:302; Moura 1936:7).

Commercial interest in gold in the Maracaçumé and Gurupi regions diminished precipitously, but not because of the threat of Indian raids (cf. Cleary 1990). The gold was embedded in veins of quartz. Separating the gold from the quartz proved to be too costly in terms of wear and tear on sophisticated steel machinery (Moura 1936:62–63; Projeto RADAM 1973:24). The Ka'apor, upon arriving in the Gurupi region, took advantage of abundant stores of steel tools which the miners had left behind: "The iron and steel materials abandoned in the gold explorations around Montes Aureos by the English company, and the construction of a telegraph line through the backlands of Maranhão, furnished for many decades the raw material for the Urubus [i.e., Ka'apor] to make their terrible steel arrows" (Moura 1936:17–18; see also Lisboa 1935:2). In addition to projectile points, it is certain that the Ka'apor also used these raw materials for making axeheads, machetes, and knives.

The telegraph line mentioned above would eventually link Belém with São Luís. In 1885 an official of the state government in Pará wrote the Governor of Maranhão: "We ask you to expedite the necessary help to protect the engineer and the rest of the employees working on the construction of the telegraph line, now progressing in the headwaters of the Maracaçumé . . . who are threatened by Indian raids" (APE 1885). This threat clearly came from the Ka'apor. Despite these raids, the government continued work on the telegraph line, and in the 1880s the first permanent towns of the middle Turiaçu, Maracaçumé, and Gurupi Rivers came into being, some of which are still inhabited.

The threat of being ambushed by Indians did not deter some government planners and would-be profiteers who were becoming increasingly curious as to what the forests of the upper Turiaçu and Gurupi Rivers might conceal. In 1897 the Baron of Tromaí on the northern coast of Maranhão ascended the Turiaçu River up to the town of Alto Turi, an outpost of the telegraph line (Anonymous 1897:1). His objective was rubber. He returned to São Luís with a sample of more than one hundred kilograms of latex, harvested from 'rubber' trees on the banks of the Turiaçu (Domingues 1953:35). Although the true rubber tree, genus *Hevea* in the family Euphorbiaceae, occurs in the region and I have collected it, it is quite rare. More than likely, the rubber collected by the Baron of Tromaí came from *balata* trees of the far more common genus *Manilkara*, in the sapote family.

In any case, Ka'apor raids prevented this industry from assuming any importance in Maranhão: "The exploitation of this product in Maranhão has been until today encumbered by fighting between workers and savages in the region of the deep forest" (Fróes Abreu 1931:73). In 1912, during the worldwide rubber boom, one source recorded that: "The rubber collectors of the upper Turiaçu have retreated from that area, abandoning their labors at extracting rubber, in order to save their lives. They are threatened with being robbed, at any time, by the Urubu Indians" (Anonymous

1912a). By 1913, the year Amazonian rubber prices were peaking, there was only one small company involved in exporting rubber from the town of Alto Turi, on the eastern border of Ka'apor territory: "Unfortunately, the chief of that company, Sr. Luíz Antônio Alves, has exploited this area with great difficulty and at great danger, owing to the burning of his warehouses by the Urubus who infest that region" (J. P. Ribeiro 1913:41).

Attempts to "Pacify" the Ka'apor

In 1911 the Serviço de Proteção aos Índios (Indian Protection Service, henceforth SPI) first attempted to "pacify" the Ka'apor (D. Ribeiro 1970:177). Yet success in deterring the Ka'apor from raiding towns and farms would be delayed for seventeen more years. Between 1911 and 1928, hostile encounters were at the forefront of Ka'apor/karaí relations. As late as 1916, the upper Turiaçu and middle Gurupi were still hardly known to the colonists. According to Raimundo Lopes, a naturalist who would sixteen years later meet "pacified" Ka'apor Indians at the Canindé Indian Post: "The upper Turiaçu is a mystery. The denseness of the forests, the problems of navigation, and hostility of the Indians have impeded not only colonization, but mere exploration. . . . In the north, the great forest is cut by a road—a telegraph line. . . . Tragic incidents between savages and civilized people mark the history of the road. The town of Alto Turi, which appeared destined to become a nucleus of population in these wilds, has not existed for long" (Lopes 1916:184–185, 190). (Alto Turi still exists, as a small hamlet on the right bank of the Turiaçu, about five kilometers downriver from the BR-316 highway.)

Throughout the seventeen years of official attempts to pacify the Ka'apor, caboclos who extracted forest products, such as balata rubber, timber, and copaíba oil, those who farmed, and those who worked on the telegraph line, attacked Ka'apor settlements (Fróes Abreu 1931:217; Ribeiro 1970:179–180). One of my non-Indian, adult female informants from the town of Alto Turi recounted that in the 1920s several posses raided Ka'apor settlements, often killing women and children. One telegraph agent arranged raids against the Ka'apor of the upper Turiaçu and impaled the heads of his victims near the telegraph posts (Ribeiro 1970:177). Such terrorism as well as raids and counterraids impeded the efforts of SPI agents and no doubt forestalled the emergence of peacemakers among the Ka'apor.

In 1911, at a place about eight kilometers above Alto Turi on the right bank of the Turiaçu, SPI agent Pedro Dantas and his co-workers placed steel tools, cloth, and other gifts on the beach to 'attract' (atrair) the Ka'apor. But goodwill on the part of the government at this time was not reciprocated. Rather, a few Ka'apor warriors stalking the scene attacked the party, wounding one young telegraph worker, who got an arrow point embedded in his jaw (Anonymous 1912a). Mistrust of karaí gift-giving was, however, warranted. Shortly before this incident, some Ka'apor had been tricked into vulnerability in the very same region. A newspaper told the story of "the famous massacre of the Indians of the Alto Turi, planned and carried out by so-called 'civilized' men, who invited them to lunch, and then cowardly pounced on them, while the trusting Indians were peacefully eating" (Anonymous 1912b).

Meanwhile, Miguel Silva, who nearly thirty years later would become the SPI agent at Post Felipe Camarão, tried in 1911 and 1912 to lure the Ka'apor living near the Igarapé Jararaca, a tributary of the Gurupi; he was not successful (Anonymous 1912a). Between 1915 and 1917, SPI was short of funds, and agents in the field near the Ka'apor temporarily suspended their activities (Anonymous 1915:70; 1917:102). In 1918, after a respite of several years, Ka'apor warriors raided settlements in the Guamá river basin in Pará. What interested the Ka'apor in these raids were "utensils of iron . . . because they take advantage of these to prepare spears [sic]" (Anonymous 1920a). In 1920 some Ka'apor struck boldly into the territory of settlers on the outskirts of the city of Bragança, which is situated on Pará's Atlantic coast: "The Urubu Indians, who every September and October raid this county attacking and killing our farmers, already visited us this year. On the 14th of this month [October], the Urubus raided a house in the Benjamin Constant colony. . . . In this attack, they killed four people" (Anonymous 1920b).

September, October, and November constitute the peak of the dry season, when the Ka'apor especially require steel tools because this is when they fell the forest for crop space. Evidently they had been long accustomed to steel axes, machetes, and knives. No Ka'apor informants remember anyone using stone axes in the past. The stone axeheads one finds today in the habitat of the Ka'apor (admittedly not made by ancestors of the Ka'apor) are not even regarded by the Ka'apor as axeheads; they are *tupã-ra'ĩ* 'thunder-seeds' and are believed to be of divine origin. When the Ka'apor were at war, a warrior would heat a *tupã-ra'ĩ* in the fireplace and put it against his chest for courage; today, hunters apply the same to their dogs' chests to make them unafraid of jaguars (Balée 1988a).

A *terra firme* forest people, the Ka'apor in the early part of this century—like the Araweté in 1976 when they were first contacted—did not know how to swim. But at the peak of the dry season, one can easily ford the Gurupi River across the exposed rocks of its rapids. By 1919 or 1920, according to the lawyer Jorge Hurly, who had surveyed the headwaters of the Guamá River before arriving on the middle Gurupi, "The right bank of the upper Guamá . . . is at present uninhabited owing to frequent attacks by the bloody Urubus" (1928:32). Operating in small groups, the Ka'apor raided lands as far away as Viana and Bragança, attacking settlements two hundred kilometers or further distant from their adopted homeland in the Gurupi basin.

An interesting twist is the statement by some local peasants in 1928 that a white man had been directing Ka'apor raids. A director of Pará's Rural Health Service, who was visiting the upper and middle Gurupi, was told by local people that a white, blue-eyed, European polyglot named Jorge Cockrane Amir lived among the Ka'apor and commanded them in their raids (Anonymous 1928; Hurly 1928:34, 1932a:32; Ribeiro 1970:179). They claimed that Jorge Amir had business dealings with Guilherme Linde, a Swiss who mined for gold in the Gurupi and Guamá river basins. Linde's gold mining camps were not the targets of Ka'apor raids, but those of poor prospectors were, according to these reports. In any case, it seems most unlikely that any non-Indian could have commanded the Ka'apor (cf. Cleary 1990), given their long-standing belligerence toward the outside world. Linde, who claimed to know Amir personally, was said to have had a "fecund imagination" (Anonymous 1929a) and appears to have been nothing more than a small-time operator (cf. Cleary 1990). Once the Ka'apor finally entered into peaceful contact with SPI agents late in 1928, moreover, no white leaders emerged from their ranks. Nor do any modern Ka'apor, including the oldest among them, recall any such leader among their forebears.

Further calling to question the verity of statements made to that same director of rural health in 1928 is another claim recorded by him: "The Urubus pertain to the Turyrara [sic] tribe, and once were peaceful, inhabiting the Capim River, until the period of the *Cabanagem* [the civil war referred to above]" (Anonymous 1928). In light of linguistic differences, however, it is unlikely that the *immediate* cultural ancestors of the Turiuára and Ka'apor were the same people, even though it is true that the Ka'apor at an earlier time had inhabited the Capim basin.

By October 1928, the efforts of SPI agents to forge peace with the Ka'apor began to succeed. The Ka'apor themselves were ready for peace. Miguel Silva and Benedicto Araújo, with two other co-workers, set gifts of steel tools and cloth beneath a thatched shelter on the right bank of the Gurupi, directly across from the narrow islet called Canindé—once a crossing place of the Ka'apor, and then the site of the SPI post. A report of SPI Inspector Artur Bandeira stated:

> From the first days of October, the Urubu Indians withdrew gifts that we left under shelters. . . . On the 16th of October, an Indian appeared across from the Post, on the Maranhense side [of the Gurupi River]. Through his gestures, we understood him to be requesting axes, machetes, and clothes. . . . We prepared the launch to transport the requested objects. Raymundo Caetano [a Tembé Indian] and two other workers started to carry them to the Indian, who moved back somewhat from the river bank, when the launch approached. He signaled to Caetano to leave everything on the river bank, which Caetano did. Satisfied, he motioned for Caetano and his companions to withdraw, and when they were about thirty meters away . . . Indians from all corners of the forest came out toward the gifts.
>
> (Anonymous 1929a; see also Anonymous 1929b;
> cf. Ribeiro 1970:181–182).

By 15 December 1928, the Ka'apor and the SPI agents at Canindé had achieved a fragile peace, for on that day, ninety-four Ka'apor visited the post. Some of them remained for a few days (Ribeiro 1970:182). At about the same time, the Ka'apor of the Turiaçu and Paruá region also established a cease fire with the settlers of Alto Turi: several Ka'apor men approached the town, which was in a large open field. At the edge of the forest, the men turned the points of their arrows to the ground, symbolizing their friendly intent. They shouted *Katu Kamarar, Katu Kamarar!* ('Greetings Comrades, Greetings Comrades!'). *Yane katu apo* ('We are now peaceful'). A few of these Ka'apor men spent the night in the town, amid the *karaí*.

The 'time of rage' (*parahi-wa-rahã*) had ended. Although the Ka'apor were said to be "pacified," in Ka'apor lore, it was *Pa'i* ('priest'), one of their own, who 'pacified' (*mukatu*) the savage non-Indians. A new chapter of contact with Western society was just opening, one that would weigh more heavily on the Ka'apor than anything previous.

Peace and Decline, Post-1928

The year 1928 marked an irrevocable break with the past in Ka'apor history. It was one of those watershed events in post-Columbian Amazonia, like the arrival of the *bandeirantes*, establishment of Jesuit mission culture, expulsion of the Jesuits, and the bloody *Cabanagem*. First, the Ka'apor raiding complex abruptly ended, bringing relief to the settlers and many Ka'apor (the women especially were said to be, by one elderly female informant, *huri-riki* ['very happy'] with the end of hostilities). Second, the cessation of hostilities was perceived by many Ka'apor to bring new economic opportunities. Instead of risking their lives to obtain steel tools, the Ka'apor now received them free of charge.

But the pattern of pacifying the *índios bravos* ('wild Indians') has always been the same, regardless of the group, since the inception of SPI (see the insightful article by Gross 1982a). After several years of receiving gifts, the now *índios mansos* ('peaceful Indians') or even *índios aculturados* ('acculturated Indians') would be told that to receive, they must pay in labor, native crafts, game meat, agricultural produce, or anything else that had a price value. This continued to be the policy of FUNAI (the National Indian Foundation), heir of SPI. The agency has historically been committed to a program of integration (*integração*) for the remaining forest dwellers and their eventual assimilation into what it perceives to be the national society of Brazil. This policy envisions the ultimate incorporation of indigenous communities into the national economy, despite the fact that Indians, even under the most recent Constitution of 1988, remain legal wards of the federal government. (This policy changed somewhat when Sidney Possuelo became president of FUNAI in 1991. Under Possuelo's directive, FUNAI appeared to be genuinely engaged in fulfilling the Constitutional mandate for protecting indigenous rights and lands. Possuelo, however, may lack the requisite political and financial support from more powerful sectors of the government that would enable FUNAI to realize those aims.)

In 1928 the Ka'apor surely had no idea that accepting gifts from SPI agents would constitute the end of their sociopolitical autonomy. And they could not anticipate the massive depopulation they would suffer from introduced diseases. Finally, they could not then have predicted that activities of the Luso-Brazilian frontier, so long held in check by the uncompromising hostility of their ancestors, would expand into their eastern and southern border lands during the 1960s through the 1980s. A new and fearful order had indeed come into being, and the earlier inbuilt defenses of the indigenous society would be worn away.

In December 1928 there were two SPI posts on the Rio Gurupi, several days journey from most of the upstream settlements of the Ka'apor. By 1929 SPI had established a third post on the Maracaçumé River, but it was only manned intermittently because access was difficult during the dry season. It was permanently closed in the late 1940s. The SPI post near the mouth of the Igarapé Jararaca (Felipe Camarão) had been used since 1911 to deal with the already pacified Tembé and Timbira, as well as to attract the then-hostile Ka'apor. It closed in 1950, leaving only Post Canindé to represent the federal Indian agency in this remote region.

In 1929 Postos Felipe Camarão, Pedro Dantas, and Maracaçumé administered goods to an indeterminate number of Ka'apor. Relying on SPI records which now appear to be lost, John Duval Rice (1930:312) estimated the Ka'apor population at more than 5,000 in December 1928. In 1935 the engineer Pedro Moura (1936:19) indicated Ka'apor population to be about 3,000. One can infer that Darcy Ribeiro (1956:4–5) assumed Ka'apor population to be about 2,000 in 1928. Whatever the size of the population, the Ka'apor were clearly the most numerous Indians with whom SPI agents of the Gurupi basin were concerned.

In the early phase of contact, the principal function of SPI posts was to maintain peace between the Ka'apor and surrounding settlers. To accomplish this, the agents distributed goods, including steel tools, salt, cloth, tobacco, and fish hooks to those Ka'apor who visited the posts. At this time, the agents did not venture into the villages, which were no closer than about 25 kilometers from the posts. The first ethnographic memoir of the Ka'apor, by Raimundo Lopes (1934), was based on Lopes's visit of only a few days to Post Canindé in 1932. Evidently, going into the remote settlements was still considered unsafe.

No one would have guessed, moreover, that the frequent, peaceful, and lucrative visits of the Ka'apor Indians to the relative security of SPI Posts would wreak havoc on their population deep in the forest. But in April 1929, only five months after pacification, an unidentified respiratory infection swept through the Gurupi and Turiaçu River basins, killing many Ka'apor. It began when the SPI agent in charge of Post Canindé, Soeira Mesquita, seemingly anxious to prove to the world that the once war-like Ka'apor were now pacific, brought five Ka'apor men downstream with him to Vizeu, a port town at the mouth of the Gurupi (Ribeiro 1970:182; cf. Ribeiro 1956:4). Mission accomplished, Mesquita left the Ka'apor travelers with crew members of the launch, who were directed to return the Ka'apor passengers upstream to Post Canindé. On the return trip, the Ka'apor travelers caught a 'flu' (*gripe*). Two of them died in Itamoari, a town of Afro-Brazilian former slaves on the right bank of the Gurupi, which is still inhabited (Ribeiro 1956:4–5; 1977:182–183). The other three continued on to Post Canindé with the head of the launch's crew, Benedicto Araújo, who was second in command at Canindé. At about the same time, another party of about twenty-five Ka'apor visited the town of Itamoari, no doubt to trade or beg for tools. Several caught cold viruses and, upon returning to their villages, contaminated many people (Ribeiro 1970:182). Ribeiro (1956:4) estimated that about half the entire Ka'apor population died in this epidemic.

A Ka'apor named Oropok, well remembered by many of my informants, lost his two wives in this devastating epidemic. As if to retaliate against what he perceived to be sorcery, Oropok sought out and murdered Araújo, as well as the Timbira Indian who acted as Araújo's interpreter, shooting both with steel-pointed arrows (called *ita-takwar*). Oropok then fled to his village in the upper Turiaçu region. No one tried to stop him (Ribeiro 1956:5, 1970:183–184). The employees of the Post were true to the famous motto of SPI, coined by its founder, Colonel Cândido Rondon: "Die if you must, but never kill" (*Morrer, se preciso for, matar, nunca*). Oropok would never again visit any SPI post, according to my informants; he died of natural causes in a village in the upper Turiaçu. After the tragedies of 1929, the many Ka'apor deaths by epidemic and the murder of two of the Post's employees, several months would elapse before SPI could regain the confidence of the Ka'apor and vice versa (Ribeiro 1956:5).

By 1943 Ka'apor population had dwindled to 1,095. This figure came from an unusually complete census taken by SPI agents Miguel Silva and Elias Rosa Maia, who visited all known Ka'apor villages (Silva 1943, Silva and Rosa 1943). In 1949 an epidemic of measles—first contracted by a group of Ka'apor who had traveled to trade in Bragança—descended on the Gurupi and Turiaçu basins, killing 160 of an estimated Ka'apor population of 650, or one-quarter of the entire population (Ribeiro 1956:7). Darcy Ribeiro witnessed the epidemic, which occurred during his first visit to Ka'apor villages in the Gurupi. He wrote that prostrate mothers could barely feed themselves, much less nurse their helpless infants and that "Children abandoned on the ground were eating dirt—they burned themselves trying to keep the fires going." (1951:375).

After 1950 a series of reports were issued by the legendary SPI agent at Canindé, João Carvalho (called "Pinga-Fogo," i.e., 'drips fire' by friends and enemies alike). Carvalho's reports, along with others of SPI and its successor, FUNAI, demonstrated that respiratory infections continued to press Ka'apor population unremittingly downward. Between January 1954, when total Ka'apor population was estimated at 912, and June 1962, when total population was estimated at 822, there was a decline of 9.9 percent. Between 1962 and 1975, when population had dipped to 488, the decline was 40.6 percent (Carvalho 1954, 1962; Mariz 1975).

Following pacification in 1928, the most notorious epidemics occurred in 1929 and 1949. After 1949 the regional SPI headquarters in Belém received several urgent messages from the SPI agent at Canindé, which reveal profound shortcomings of SPI in handling medical emergencies. In January 1954 Carvalho described a malaria epidemic in the Gurupi: "The malaria is spreading as much in the Post as in the villages and we do not have enough medicine." In February Carvalho wrote, "The health situation continues without improvement, especially in the villages, where the infection is quite fertile, always taking victims, especially those of the Paruá [river basin]." This epidemic raged until April, killing at least fourteen Ka'apor and one Tembé. In November of the same year Carvalho wrote his superiors that "At the moment, a cold is going around, and we do not have enough medicine for the illness." (Carvalho 1954).

Four years later, in May, Carvalho declared, "A strong cold is passing through this region, including the villages. At present, the Post possesses few resources and the character of the cold is unknown. It has the following symptoms: fever with pain in the ribs and an excruciating headache. Some of the afflicted bleed through their noses and mouths. After two days, the exudate that comes forth from their chests is a true puss." Carvalho observed helplessly, "There are some Indians who only last two days." By its end in June, this epidemic had claimed thirteen Ka'apor lives (Carvalho 1958).

In each epidemic, the SPI agent complained of inadequate medical supplies and no knowledge of how to diagnose and treat these diseases. Ka'apor population decline, then, proceeded as if, in violation of its duty, no protection was offered by SPI. The history of SPI, in general, was marked by government failure to supply adequate funds, hence undermining the agency's ability to fulfill its legal mandate (Wagley 1977:283). FUNAI (National Indian Foundation), which replaced SPI in 1968 after a corruption scandal doomed SPI, had until the early 1980s more access to government funds (Wagley 1977:290; cf. Gross 1982a:5). But only from 1975 to 1984 were FUNAI personnel better equipped to meet medical emergencies. During this period, a FUNAI physician, nurse, and laboratory technician together made two visits per year to all Ka'apor settlements, treating any diseases they could diagnose. Between 1984 and 1990, however, the Gurupi region was once again largely ignored by the government. This was because of intertribal tensions between the Guajajara (who, at around ten thousand, are one of the most numerous native groups of Brazil) and the other Indians of Maranhão. Some Guajajara arranged for the appointment of one of their kinsmen as the FUNAI regional superintendent in São Luís. As a result—and widely acknowledged by older FUNAI employees—trained medical personnel for Ka'apor villages were lacking and FUNAI agents complained of a constant scarcity of medicines. Nevertheless, Ka'apor population evidenced a slight recovery during this time, rising from 494 in 1982 (Balée 1984b) to approximately 520 in 1990.

The Ka'apor were not the only group in the region to undergo major demographic decline in this century. The Tembé and Kren-Yê Timbira of the upper Gurupi, and the Guajá of the upper Turiaçu and Pindaré Rivers, have all been seriously depopulated. In 1872 Dodt (1939:172) estimated that there were 6,000 Tembé living on the Gurupi alone. In 1920 Hurly visited seventeen Tembé villages on the Gurupi River and counted a total population of only 1,091 (1932b:35–37). Little more than sixty years later, I took a census of the three remaining Tembé villages of the Gurupi and found a total population of only 111 people (Balée 1984b:61). It is probable that this decline owes more to mortality than emigration, since the only other Tembé settlements are located in the Guamá and Piriá basins and are today quite small (e.g., Anonymous 1980a). In addition, the Tembé of the Guamá and Piriá are not represented in the census figures of 1872 and 1920, which reflect only the Gurupi population. In little more than one hundred years, then, Tembé population has declined by at least 98 percent. Continual contact since the 1850s with Brazilian traders, known as *regatões*, who ascended the river in boats full of Western goods, has marked the recent history of the Tembé. The Tembé traditionally traded copaíba oil and other resins to the *regatões* in exchange for iron (and later steel) tools, shotguns, ammunition, and cloth. The *regatões* customarily bought goods from the Tembé at one-tenth the market value (Gomes 1977:198–200), which probably generated a relationship of debt peonage.

Both the Tembé and the Ka'apor appear to have experienced some of the same epidemics. The measles epidemic of 1949 that struck the Ka'apor, for example, also killed countless Tembé, according to Verônica, the headwoman of the village of Igarapé de Pedras and a 1981 informant of mine. She claimed that there were so many people prostrate in her village that they could not bury the numerous dead. Vultures descended and fed upon the corpses. This epidemic also affected the Kren-Yê Timbira.

The Kren-Yê, then only 100 to 150 strong, had migrated to the Gurupi from the lake region of the lower Mearim River in Maranhão between 1850 and the early 1860s. In the early 1880s they migrated to the Pindaré, but returned to the Gurupi in 1889. In 1915 there were 41 Kren-Yê living on the left bank of the Gurupi, near the mouth of the Igarapé Jararaca (Nimuendaju 1946:2,13,14). By 1943 only 28 were left (Silva 1943). And in 1981 there was only one surviving Kren-Yê who still knew the indigenous language—she died in 1983. She had two younger male kinsmen, now married to Brazilian women but who consider themselves Kren-Yê and who live within the jurisdiction of Post Canindé. Neither speaks Kren-Yê and both know more Ka'apor and Tembé language and lore than the Kren-Yê equivalents. In a way, demographic and cultural decline for the Kren-Yê between 1915 and 1990 was virtually complete.

The Guajá are spread across the upper reaches of two river basins: the Pindaré and the Turiaçu. Precise data on the Guajá population are as yet unavailable, since many members of this foraging people remain uncontacted in the upper Caru river basin (a tributary of the Pindaré River). FUNAI employees who first saw the Guajá of the upper Turiaçu in 1975 estimated the population then to be 100. Many subsequently died in a 'flu' (gripe) epidemic, purportedly introduced by an ill member of the contact team. The population as of 1990 of the Turiaçu Guajá was only 36, which represents, therefore, a decline of 64 percent. Taking the known populations of the two FUNAI Posts, P. I. Alto Turiaçu and P. I. Awá, and drawing estimates of the uncontacted population based on camp remains, I estimated the total Guajá population to number 226 in 1988 in the Turiaçu and Pindaré river basins combined (Balée 1988b).

For the Guajá, especially the uncontacted ones who will probably be contacted between 1995 and 2000 at the latest, and given the pace of destruction of their habitat by ranchers, loggers, and landless peasants, this unhappy chronicle of disease and death is probably not yet over. They may go the way of the culturally and linguistically extinct Kren-Yê Timbira. Or, if lucky, the Guajá may come to be like the now-expanding Ka'apor and Tembé populations—many of whom are now becoming, however, what the Ka'apor call yumupara, i.e., of mixed ancestry, making a specific indigenous ethnic identity difficult to ascertain.

The principal threat to the well-being of the forest dwellers of the pre-Amazonian forest of Maranhão, including the Ka'apor, Guajá, and Tembé, is not merely disease, but far more generally the invasion and destruction of their forest habitat. After about 1959, the eastern frontier of Ka'apor territory began to change radically. The state government of Maranhão opened a dirt road that ultimately connected Belém with Teresina, Piauí by 1960 (Anonymous 1972:30). This road could not support motorized traffic, and was almost totally useless in the rainy season. Most travel and commerce between the ports of São Luís and Belém at this time continued to be by ship.

This dirt road was, however, the forerunner of a paved road, the BR-316, which was an integral part of a governmental plan to alter the transportation networks, ecology, economy, and society of northern Maranhão. In 1962 officials of SUDENE (Agency for the Development of the Northeast), which is a federal agency, formulated an "escape-valve" plan to colonize the upper Turiaçu with landless peasants from the arid states of Ceará, Bahia, and Pernambuco (Anonymous 1970). This plan was called Projeto de Colonização do Alto Turi ('Colonization Project of the upper Turiaçu') (Anonymous 1980c). The federal government designated for colonization 939,000 hectares between the Pindaré and the Gurupi valleys. This region was mostly tropical forest and it contained five Ka'apor settlements in the Paruá and Turiaçu basins (IBGE 1981:xiii; Anonymous 1975). While SUDENE maintained nominal title to the land, it offered rights of usufruct and agricultural credit to incoming colonists. Each resettled family could farm an area of fifty hectares and would receive relatively lenient terms on a few head of cattle, rice seeds, and tools. The objective was to expand the agricultural frontier of Maranhão and at the same time absorb part of the flood of emigrants who were abandoning the Northeast at a time of increasingly prolonged droughts (Anonymous 1975, 1980c). At no time did the agency officials appear to consider the effects this plan might have on the mostly monolingual Ka'apor and Guajá peoples.

Between June 1964 and October 1970, SUDENE had resettled approximately 6,000 people as *colonos* in the upper Turiaçu (Anonymous 1970). Along with the legal settlers came, however, 58,646 *posseiros* ('squatters') living on small plots of land on isolated homesteads and with no ties to SUDENE (IBGE 1981:5). During the 1972 dry season, the Brazilian army's paving crew for the BR-316 had reached the Paruá River. They cleared a right of way of about forty-five meters on either side of the old dirt road and raised the roadbed to one and one-half meters above ground level. In doing so, they destroyed some swiddens and swidden fallows of the Ka'apor. The inhabitants of three Ka'apor villages east of the highway moved about twenty kilometers westward into the remaining forest in 1974, which was the year the highway opened to motorized traffic (Andreazza 1974:202).

The new highway and commercial bus travel led to a 111 percent increase in population in the region between 1970 and 1980, with 94 percent of the increase taking place on or near the highway (IBGE 1981:xviii). In 1980 population had reached 115,994 on government-owned lands just to the east of the Ka'apor (IBGE 1981:5). The population density of the settlers in 1980 was about 4 persons/km². In contrast, Ka'apor population density was only about .2 persons/km². By 1982 all Ka'apor settlements of the Paruá River basin had been abandoned—the Indians had migrated to a new settlement in the upper Maracaçumé basin called *šimo-rena* ('fish poison–place'), which was within the reservation that had been demarcated in 1978 and 'approved' (*homologada*) by presidential decree in 1982 (see the map in chapter 2). By 1985 the Paruá River basin, the richly forested homeland of the Ka'apor since the 1870s, had been completely deforested by settlers and abandoned by the Ka'apor.

In 1972 a new government agency succeeded SUDENE in the region. COLONE (Companhia de Colonização do Nordeste, or Company for Colonization of the Northeast) in 1980 had operating capital of about 120 million US dollars, financed by the state of Maranhão, SUDENE, the Bank of London, and the World Bank (Anonymous 1980b). COLONE has encouraged the *colonos* to plant cash crops new to the region, such as black pepper (*Pimenta nigrum*), which require intensive cultivation and use of fertilizers and fungicides. The agency has helped to form agricultural cooperatives among the *colonos*, has continued to offer them credit, and makes available rural extension programs, limited hospital care, and educational facilities.

The largest town in the region immediately east of the Ka'apor homeland is Zé Doca, which hosts the regional headquarters of COLONE (the agency's main office is in São Luís). Zé Doca has a population of about 12,000. It grew quickly since its inception in about 1960. Zé Doca has had electricity only since 1979. By 1990 there were many general stores, haberdasheries, and nationally franchised appliance stores, plus hotels, bars, and single family dwellings. There are even two restaurants, two jewelry stores, a small movie theater, and several small supermarkets. An agricultural hub of northern Maranhão, Zé Doca has two bus stops and two truck stops, at which traffic is constant virtually twenty-four hours a day. Yet in the early 1950s, Zé Doca was nothing but forest and hunting grounds of the Ka'apor. The town takes its name from that of the first homesteader, who arrived about 1958. The man Zé Doca, probably unaware that his name would carry on, abandoned his homestead around 1963 for an obscure destination in the Northeast.

A dilapidated bus travels from July to January over a mostly dirt and gravel road from Zé Doca to Igarapé Grande, a town thirty kilometers to the west. For the rest of the year, the stretch of road beginning five kilometers east of Igarapé Grande becomes impossible to travel by motor vehicle. One continues either on foot or by mule. A large creek, for which the town is named, flows one kilometer east of Igarapé Grande, making a slow descent to the Turiaçu. One must cross it by canoe. Although this town of about 1,000 is situated on land nominally owned by COLONE, none of its residents are *colonos* and COLONE has begun no project in this region. Igarapé Grande has one hotel, four general stores, and about two hundred single-family dwellings where mostly poor squatters and landless agricultural laborers live. There is no electricity. Most of the residents make regular trips to Zé Doca for provisions, since prices (regardless of sporadic government controls) are higher in towns such as this that are not along the paved highway.

To the west of Igarapé Grande, along narrow, twisting trails, lie the homesteads and fields of numerous squatters. In this landscape only a few tall, charred and leafless trees—which were too hard and thick to fell before the burning of the fields—testify to a once imposing forest. Most of the secondary growth is very young and weedy, suggesting short fallows and heavy pressure on the land. Beyond this, twelve kilometers from Igarapé Grande, the forest and homeland of the Ka'apor begin at the Reserva Indígena Alto Turiaçu. At this juncture, along the line of demarcation, the forest legacy of the Ka'apor meets in an uneasy truce with the open pastures and monocultural rice swiddens of the newly arrived settlers.

In 1989 several dozen families of *colonos* had illegally occupied about four hundred hectares of the eastern part of the reserve (PIB 1991a). A far more serious threat to the integrity of Ka'apor lands, however, was a massive invasion, involving several hundred families of squatters as well as more powerful ranchers and loggers, which was taking place across the southern and western limits of the reserve (Anonymous 1992; PIB 1991a). If this invasion and the accompanying forest destruction are not soon brought to a halt, it is unlikely that research of the kind I have conducted in the region will be possible for much longer.

FOUR

Activity Contexts of Plants and People

Upon arriving in a Ka'apor village after a long walk through the forest, a visitor is usually greeted with the squealing of young children, shouting of adolescents, barking of dogs, squawking of parrots, and the more subdued, respectful welcomes extended by adults. Some adults are, of course, away from the village temporarily, engaged in hunting, fishing, gardening, collecting fruits, gathering material for firewood, house construction, or other artifacts, bathing, and so on. Those who have remained in the village are likely processing manioc flour, repairing a leaky roof, carving a new bow, eating manioc meal, nursing an infant, or perhaps just relaxing in their hammocks. In general, a visitor is struck by how much of the day each Ka'apor villager spends working with plants and plant products from the immediate environment. The central objective of this chapter, then, is to examine how the Ka'apor employ hundreds of different species of plants in their routine activities. Appendixes 6, 7, and 8 present detailed lists of activity contexts of the people and plants in the Ka'apor habitat.

The apportionment of time to miscellaneous work activities, food preparation and consumption, child care, hygiene, and leisure constitutes an important marker of any society. In 1981 and 1982, we conducted more than 2,000 randomly scheduled observations in two Ka'apor villages in order to learn how adults apportion and spend their time (Balée 1984b). The method we employed, known variously as random spot-checking, instantaneous scan sampling, or simply time allocation, involves randomizing and quantifying observations of behavior in order to obtain statistically sound patterns of how people use their time. This approach has become a popular tool for ethnographic study in South America (e.g., Johnson 1976, 1978; Gross et al. 1979; Hames 1980; Hames and Vickers 1982).

Ka'apor men and women are constantly using plants, in one way or another, in their daily activities. Although the time allocation data interpreted here cover many of these activities and uses, certain uses of plants were not recorded during the time allocation study. These unrecorded uses include application of medicinal or magical plants, uses of plants in ritual, and some other miscellaneous uses (which are described in chapter 5).

I call these human interactions with plants *activity contexts* rather than uses, in order to highlight the reciprocal nature of many of the activities. In Ka'apor society, humans and plants have complex relationships in which humans affect and nurture, as well as use, plants and entire plant communities. The most significant activity contexts of plants and people, in terms of time allocation data, concern subsistence (which includes hunting, fishing, gardening, plant food gathering, food preparation), manufacture and repair (of tools, weapons, baskets, garments, decorative objects), child

TABLE 4.1 Time Allocation by Married Adults, Villages of Soanĩ
and Urutawi (1981-82)

ACTIVITY	MEN[1](n=26)	WOMEN[1](n=28)
SUBSISTENCE:	588 (54.6%)	512 (45%)
Hunting[2]	208 (19.3%)	73 (6.4%)
Fishing	72 (6.7%)	56 (4.9%)
Gardening[2]	212 (19.7%)	174 (15.3%)
Gathering (plants)	31 (2.9%)	28 (2.5%)
Food preparation[2]	65 (6.0%)	181 (15.9%)
MANUFACTURE AND REPAIR	67 (6.2%)	96 (8.3%)
Tools and Weapons[2]	42 (3.9%)	01 (0.0%)
Basket Weaving	10 (0.9%)	01 (0.0%)
Cotton Weaving	0 (0.0%)	23 (2.1%)
Sewing[2]	02 (0.2%)	29 (2.5%)
DecorativeObjects[2]	13 (1.2%)	42 (3.7%)
CHILDCARE[2]	03 (0.3%)	75 (6.6%)
EATING[2]	43 (3.9%)	75 (6.6%)
HYGIENE	55 (5.1%)	70 (6.1%)
LEISURE	237 (21.9%)	284 (24.9%)
OTHER	85 (7.9%)	27 (2.5%)
TOTAL OBSERVATIONS	1078 (100%)	1139 (100%)

1 Numbers in left part of columns represent the number of observations for each activity; numbers in parentheses are percentages of the total observations.
2 The differences between married men and women are significant at the .01 probability level on Student's t-test.

care, and eating (see table 4.1). Although it would be inaccurate to claim that separate domains of men's and women's plants exist, there are significant differences in the ways in which men and women apportion their time; as a result, one observes significant differences in the ways in which men and women handle and work with plants. These gender-related differences are evident especially in hunting, gardening, food preparation, manufacture and repair of tools and weapons, eating, and child care (see table 4.1).

Swidden Gardening

Swidden gardening consumes more time of both men and women than any other activity except leisure. On average, married men and women both spend about one-fifth of their daily time engaged in swidden gardening. Once a site for a new swidden has been selected (for details of the selection process, see chapter 6), the male household heads individually or communally cut brush along what will become the limits of the swidden. This boundary cutting is called *maɲa-ha* ('measuring'). Young swiddens (*kupiša*) and old swiddens (*taperer*) harbor the most important energy sources in the Ka'apor diet, especially varieties of bitter manioc (*mani'i*).

Having selected and delimited a site, the laborious yet methodical procedure for converting it into a major food producing unit of land begins. First there is the cutting (with machetes, *kise*) of underbrush, treelets, seedlings, and vines less than about seven centimeters in diameter. This phase is called *ka'a-wiro* (Port. *brocar*); both men and women as well as their older unmarried children work at this activity. *Ka'a-wiro* begins normally in about mid-June. This is the season, according to the Ka'apor, when old leaves of many tree species, especially *Tabebuia* (the pau d'arco tree, from which the Ka'apor obtain their bow wood) become dessicated and ready to fall (*mira-ro-upa-ta-rahã*). *Ka'a-wiro* takes place from June to about late August, during the season of 'falling tree leaves' (*mira-ro-kukwi-rahã*). From about late August through mid-October, only married men and older, unmarried boys continue the work, using steel axes (*yir*) to fell the bigger trees; this is called *ka'a-mono* 'forest-felling'. Very hard trees of enormous girth (such as *Tabebuia* and *Zollernia*, which are found only in old fallow) are often simply girdled, since it is believed that an attempt to cut them down might break one's axehead. Other large trees are deliberately felled in such a way as to bring down others. This procedure is called *amu-mira'a-pu'ar-oho* ('one tree knocks another down') or *mira-mu-pen* ('tree breaking'). In general, two to five clustered trees are first weakened by axe; the biggest is then felled so as to 'break' (*mu-pen*) the rest as well.

Felling on the swidden is finished by about late October or early November, with the onset of the season of 'emerging new leaves' (*mira-piahu-uhem-rahã*). Also at this time, many high forest trees, such as *parawa'i* (*Eschweilera* spp.), 'come into flower' (*mira-putir*). The newly felled swidden is left to 'dry out' (*širik*). By the end of November and early December, when the fallen trunks, leaves, and branches have become crisp, men 'burn the swidden' (*kupiša-ukwai*). They burn it with torches of bark from *parawa'i* trees (*Eschweilera* spp.), which have been stripped from desiccated, fallen trees. Burning begins only after the sun has dissipated the morning mist. To ensure a good burn, the male household head to whom the fruits of this swidden will pertain may blow on an ox horn to summon the ancestral spirit of the wind (*iwitu-ramũi*). Although the ox horn is an introduction from regional settlers, the spiritual ideas attached to the wind and its significance toward ensuring a proper burn are traditional.

Swidden burning is always a group effort; several men from different households help each other in work groups for the burning of each other's swiddens. They space themselves at one end of the swidden. With matches they ignite their shredded bundles of bark and kindle leaves and twigs, moving in a zig-zag formation across the swidden, to assure an even burning. As the flames rise high into the sky and the heat becomes intense, one can often hear loud popping sounds—these are the exploding shells and armor of tortoises and armadillos. Although unusual, sometimes a swidden fire spreads into the surrounding forest, burning mostly the underbrush. This peripheral region is then known as *ka'a-kai* ('burned forest') until the underbrush regenerates. *Ka'a-kai* may also occur as a result of lightning strikes and thus is not merely an anthropic formation.

After burning the swidden, the people await the first heavy rain (*aman-uhu-rahã*). The time to plant is signalled by, among other forest events, the loud nocturnal croaking of an edible frog called *marywa* (*Leptodactylus* sp.), which usually is heard around mid-December. The adult villagers, mostly men although there appears to be no rule separating the sexes in this activity, then begin 'breaking earth' (*iwi-murywai*) with their short-handled steel hoes (*purure*). Next comes the planting of manioc cuttings from a *taperer*, an old swidden in either the same or a vacated village (see chapter 6 for more detail).

The Ka'apor recognize two basic types of planting: *yitïm* and *omor*. These terms are polysemous. For example, *yitïm* means not only 'planting, digging' but also 'burial', as of humans. *Omor* refers to 'scattering' of 'seeds' (*ha'ï*). It also means the tossing and throwing of any objects in general. Transplanting requires seedlings or suckers (the latter, as in the case of bananas and plantains); both seedlings and suckers are called *ta'ïr*. These are stocked in old swiddens and old villages (*taperer*) and then transplanted to dooryard gardens (*kar*). There, they join the domesticated, bromeliaceous fiber plant called *kïrawa* (*Neoglaziovia variegata*), lemon grass, chile peppers, coconuts, purge nut plants, and arrow cane. A brief description of planting and harvesting methods, by species, follows. The rich terminology is indicative of a society directly dependent on plants in everyday life (also see chapter 7).

Yitïm 'planting' of *ha'ï* 'seeds' is appropriate for maize, rice, beans, squash, bottle gourds, peanuts, passion fruit (gathered from high forest), watermelon, chile peppers, lemon grass, citrus fruits, mango, papaya, soursop, guava, bath sponge, cacao, cupuaçu (gathered from forest), hog plum, *Inga* spp. (gathered from forest), cotton, Indian shot (*Canna indica*), *Ormosia* (gathered from forest), Job's tears, castor bean tree, *Chenopodium ambrosioides*, annatto, and *api'a* (*Guazuma ulmifolia*, gathered from old swidden). *Yitïm* 'planting' of a *hapo* 'tuber' or 'part of tuber' is appropriate for yams, sweet potatoes, cocoyam, *pipiriwa* (*Cyperus corymbosus*), ginger root, and turmeric. *Yitïm* 'planting' of a *hãkã* 'branch' or 'cutting' occurs with *Chenopodium*

FIGURE 4.1 Swidden fire. Ka'apor man burns his young swidden (*kupiša*). He carries a torch made from a bundle of *parawa'i* inner barks (*Eschweilera* spp.).

ambrosioides, calabash tree, and manioc (bitter and sweet varieties). *Yitim* 'planting' of *ta'ir* 'seedling' or 'sucker' is appropriate for *arapawak* (*Ocimum microanthum*), *kirawa*, lemon grass, chile peppers, bananas and plantains, *mikur-ka'a* (*Petiveria alliacea*), purge nuts, and arrow cane. *Omor* 'scattering' of *ha'i* 'seeds' occurs with tobacco, the fish poisons *Tephrosia sinapou* and *Clibadium sylvestre*, and certain palms (such as inajá and bacaba).

Harvesting methods are described with a rich terminology as well. Dehiscence is *pororok* 'exploding'; this term is also used in association with the word for thunder. Dehiscent edible fruits are called *pororok* at harvest time. Nondehiscent fruits when ripe are called *tiarö*; unripe edible fruits of all kinds are called *yakir*. Ripe, nondehiscent fruits (as with leaves of all kinds) are called *kukwi*, those that 'fall to the ground'. Ripe tubers, as with manioc, are called *tiha* 'big', whereas immature tubers are called *ta'i* 'small'. Domesticated species are harvested in six linguistically encoded ways. Many are simply *po'ok* 'collected on the ground'. Others are *yo'ok* 'pulled up from the ground' or 'pulled off the tree'. Rice and the shoots of *u'i-wa* (arrow cane) and sugarcane are 'cut' (*monok*). The medicinal branches of *piyã* (purge nuts) are 'broken' (*mu-pen*). Cultivated cashew trees are 'shaken' (*mukatak* or *muŋwi*) so that the fruits will fall to the ground.

Many similarities and a few minor differences are noted between the activities of men and women with regard to planting and harvesting plants. Both men and women plant bitter and sweet manioc. Cuttings of bitter and sweet varieties, which average forty-five centimeters in length, are kept separate; the Ka'apor recognize the differences by leaf color. The leaves of sweet manioc are believed to have reddish dorsal sides, whereas bitter manioc leaves regardless of variety are said to have green dorsal sides. Many other specific criteria are used to distinguish between subvarieties of bitter and sweet manioc.

After manioc has been planted, by around the end of January, the men and women together plant cocoyam, squash, bottle gourd, watermelon, guava, annatto, turmeric, and the fish poison *Tephrosia sinapou*. Maize, too, is planted—but only by men. Planting of these crops takes place sporadically until the beginning of the dry season in late June. Come June, one full year after the swidden was begun, both men and women plant yams, sweet potatoes, *kirawa*, beans, papaya, peanuts, bananas, arrow cane, and bath sponge. At this time, too, a few other crops are planted, but strictly by women: cotton, Indian shot (*Canna indica*), Job's tears, and *pipiriwa* (*Cyperus corymbosus*)—these are all used exclusively either in textiles or for body ornamentation.

In addition to planting and harvesting domesticated plants of the swidden, villagers also spend time cultivating plants in dooryard gardens and in fallow. But the amount of time devoted by both sexes to dooryard and fallow gardening is far less than that spent in swiddens.

House Building

Establishment of large communal swiddens that are destined to become new villages (see chapter 6) brings an attendant need for house building. House building consumes less than one percent of the time a Ka'apor adult spends working. In addition to sleeping, the Ka'apor spend much time in their houses and in the village at large, which itself began as a swidden and later represents, in terms of vegetation, the juncture of many dooryard gardens (Balée and Gély 1989). Ka'apor houses conform to an architectural pattern involving a rectangular floor plan with a pitched roof. This was probably introduced from neo-Brazilians hundreds of years ago, since aboriginal houses tended to be conical.

The basic components are house posts, which are inserted into deep holes at the corners of a raised clay (*iwi-tuyuk*) floor, beams, rafters, and roofing thatch. The timbers used for house posts are from a very select group of trees, more so than rafters and beams. Only ten species are used for posts. Most preferred is *yowoii* (*Minquartia guianensis*), a fairly rare tree, notable for its sulcate, straight trunk and tremendous resistance to termites and rot (Milliken et al. 1992:39). This species is so important as a house post that it is tabooed as a firewood source in Ka'apor culture. Other important hardwoods used for house posts include *yawi'i* (*Xylopia nitida*) and *yeyu'iran* (*Diplotropis purpurea*). Finally, all seven members of the genus *Eschweilera* in the region also make suitable house posts, probably in part because of large amounts of silica grains typically found in the rays of the secondary xylem (Ter Welle 1976:115). The silica appears to inhibit *kupi'i* 'termites' (G.T. Prance, personal communication, 1986).

FIGURE 4.2 Ka'apor man and his house in the village of Gurupiuna. The sulcate, central post behind the man is from a trunk of the prized, termite-resistant *yowoii* tree, *Minquartia guianensis*.

In contrast to house posts, which are the most permanent part of any house, the selection of beams and rafters comes from a corpus of forty-eight botanical species. The most important include *araraka̱'i* (*Aspidosperma* spp.), *ayu'i* (*Ocotea* spp.), *akušitiṟiwa'i* (*Pouteria* spp.), *tata'iran'i* (*Duguetia riparia, Guatteria* spp.), *mukuku'i* (*Licania* spp.), and *parawa'i* (*Eschweilera*). Like *Eschweilera* spp. noted above, and for the same reason, *Licania* also appears to be resistant to termites. These two genera are therefore preferred for beams and rafters.

House posts, beams, and rafters are harvested from forest plots destined for agriculture. In other words, the structural supports of houses are, in a sense, built from the future swidden. Clearing of crop space, in the case of new and large communal swiddens, is to a certain extent synonymous with clearing space for houses and felling trees for the construction of them.

Although only men place the posts and beams of a new house, mostly (but not exclusively) women and children prepare the weave of the roofing thatch. The raw material for thatch is *owi* (*Geonoma baculifera*) palm leaves. This is a caespitose, short palm of seasonally flooded forests and swamps. The leaves of *owi* are woven onto slats of açaí palm (*Euterpe oleracea*) over the beams of the house. These are fastened together with the *sipo-te* (*Heteropsis*) vine, which is now also a commercial item for wicker furniture. Some Ka'apor sell this vine to FUNAI. *Owi* is rapidly depleted near settlement sites (see chapter 6). Since roofing thatch must be changed every four or five years because of insect infestation and wear and tear under heavy rainfall, large older villages (such as Gurupiuna) may substitute for this thatch the leaves of *inaya'i* (the inajá palm, *Maximiliana maripa*). A few houses also include walls made from slats of açaí.

Some families, especially in frontier villages, put 'fences' (*kurar*) around the dooryard garden of their house (although swiddens per se are never fenced). This is to keep neighbors' livestock out. These fences are made from *Diospyros, Eschweilera*, and *Lecythis*, which is also good wood for rafters and beams. *Cecropia*, although of low quality as a wood, is so abundant around villages it is sometimes used for temporary fence repair. Benches in the house are made from açaí with legs of *Eschweilera*.

Hunting

Men spend about 2.3 hours of their daylight time hunting, whereas women, who sometimes help their husbands hunt and even hunt without men on occasion, spend about 46 minutes of their daily time in this activity (table 4.1). The tools of hunting consist of the bow and arrow, in addition to guns, as well as implements also used in the swidden, such as machetes and steel hoes. Ka'apor men and older teenagers carve their bows, usually about 1.3 meters in length, from principally two species, *tayi* (*Tabebuia impetiginosa*, called 'pau d'arco roxo' in Portuguese) and the moraceous *mirapitaŋ* (*Brosimum rubescens*, called 'muirapitanga' in Portuguese). Men fell these trees with a steel axe and chop out the heartwood for bow-making. The bowstring, which is knotted to the narrow, whittled ends of the bow, is made from rope fiber of the domesticated bromeliad, *kirawa* (*Neoglaziovia variegata*). All bows are equipped in the center with a grip made from strands of either of two species of *Desmoncus*, a spiny, climbing palm (and close relative of East Asian rattan palms) whose vine bast exhibits great 'tensile strength' (*hayïk*). They wrap the strands transversely around the bow's mid-section.

By far most arrow shafts are made from *u'ïwa* (*Gynerium sagittatum*), known as arrow cane in English and 'flecha' in Portuguese, which is planted by cuttings. I have often seen this species growing spontaneously on side channels of the upper Amazon River. *Gynerium sagittatum* does not occur spontaneously on the floodplains of rivers in the Ka'apor habitat; it is planted, moreover, also by the Amahuaca of eastern Peru (Carneiro 1970) and by the Waimiri Atroari of northern Brazilian Amazonia (Milliken et al. 1992:27). Recently, some Ka'apor men have begun also making arrows from the petioles of a stemless palm, *Bactris humilis*, which grows in the high forest. This is the traditional source of arrow shafts for the Guajá (see chapter 8), from whom it has been introduced.

Ka'apor arrow feathering is sewed and cemented, meaning that arrow feathers are sewn to the butt of the arrow and are flush with the shaft. Feathers are usually from the curassow *Mitu mitu*. The thread is cemented (i.e., glued) to further bind the feathers (cf. Lopes 1934:142–143). The only plant the Ka'apor use in feather binding is *irati-'i* 'wax-stem' (*Symphonia globulifera*); the Waimiri Atroari also use it as an adhesive (Milliken et al. 1992:28–29). Lopes (1934:142) noted the use of this species in feather-work and arrow feathering of the then quite recently contacted Ka'apor in 1932. The yellow resin, which becomes black upon oxidizing, serves as 'glue' (*muyar-ha*). Ka'apor men fashion chunks of this hardened, charcoal-black resin into a cylindrical block, about 15 cm long and 5 cm in diameter. Every adult male keeps a block or two of this resin in a storage space in the roof of his house, usually wrapped in a maranta-ceous leaf (arrow root family) to keep it clean.

Thirteen species of plants are used for arrow points. The length of the arrows depends upon the points used. The Ka'apor employ lanceolate bamboo points (*Guadua glomerata*) 31 cm long, called *takwar*. A number of points that are serrate on both edges and about 28 cm long are made from *Bauhinia viridiflorens*, *Brosimum rubescens*, aff. *Myrcia*, *Talisia micrantha*, *Calyptranthes* or *Marlierea*, and *Eugenia* sp. 8. The most common of these serrate points is made from *mira-pitaŋ* 'tree-red' or ('redwood') [*Brosimum rubescens*]. A blunt point, which consists of two cylindrical pieces of wood, each about five cm long and bound together in a cross, is made from wood of the vio-laceous *Rinorea* spp., which are small forest trees; these points are used only for killing birds, especially birds whose feathers are used in feather art. The steel pointed arrow, *ita-takwar*, used for killing large game and human enemies of the past, is composite. Mostly three small myrtaceous trees (*Eugenia* sp. 5, *Eugenia* sp. 8, and aff. *Myrcia*) of the high forest are employed in making the joint of the composite arrow. Wood from a species of Clusiaceae (*Tovomita brasiliensis*) found in swamp forest may also be used as the arrow joint. The joint is fitted into a carved socket at the end of the arrow shaft and to this a steel point is attached. In addition, bamboo and steel points are cemented to the joint by means of a round, hollowed-out tucumã palm seed (*Astrocaryum vulgare*), a species which only grows in old fallow, into which is pasted black, sticky resin from the *irati* tree (*Symphonia globulifera*). Because of the tucumã seed, which constitutes the joint between the shaft and the point, these arrows in flight produce a characteristic whistling sound, which in the past terrorized surrounding settlers and indigenous enemies, such as the Guajá, Guajajara, and Tembé (Lisboa 1935:52; Ribeiro 1955). Arrows range in length from 127 cm (for steel pointed ones) to 183 cm (for serrate and blunt pointed ones).

The use of steel points by modern indigenous peoples of the Amazon is fairly common. The Ka'apor acquired knowledge of steel before they migrated into Maranhão. In 1872 Dodt had noted that some of their arrows were equipped with an iron point (1939:177). Subsequent sources confirmed this observation. Iron points were in the past made as steel points are made today. A man heats a worn machete directly in the coals of a fire for several hours. Then, with a steel or stone mallet, he pounds the softened metal against a hard stone. He then cuts free a small strip with a machete, and hammers it into the shape of a willow leaf, using heat again when necessary to make the material more malleable.

Most hunters living in settlements near the border of the reservation own shotguns. These shotguns are usually dilapidated breach loaders with a single barrel. A few men also own .22 rifles. The possession of firearms neither reflects nor confers status differences among men. In the deep forest, most hunters own only bows and arrows. All hunters, even those near the reservation's border, own bows and arrows, regardless of whether they also own a firearm, since lead and gunpowder are frequently scarce.

In addition to these tools, most hunters also own two or three dogs. This practice was known to occur prior to the 1928 peacemaking (Lopes 1934:150). Dogs are of limited value, however, in hunting (Ribeiro 1955) and are best at tracking tortoises and already wounded game. Ka'apor dogs tend to have short lives and small litters, and they are perpetually undernourished. They usually tire before a long hunt is over. Dogs are not usually taken on hunts of white-lipped peccary, for fear they would be killed.

Hunters usually travel alone or with their wives. Sometimes a man hunts with his father-in-law or brother-in-law, especially when tracking the white-lipped peccary, which travels in bands of more than a hundred individuals. In fact, a man who spots a herd of white-lipped peccary typically runs back to the village to alert other potential hunters. All then return quickly to the site. White-lipped peccaries are often found grazing in a *Bactris maraja* palm swamp, an açaí palm grove, or in a spot in the high forest that hosts edible herbaceous plants such as *Heliconia* or the tubers of various members of the arrow root family, which are also edible to people. The men then quietly align themselves in a U-shape around the herd. Once everyone is in position, they begin shooting, and the herd stampedes through the open part of the U. Mature individuals are killed, butchered, roasted, and eaten. Live young that linger are frequently taken back to the settlement for rearing. It is said that the distress cries (*pukai*) of the juveniles in the settlement will attract the herd to return nearby, which may provide opportunity for future kills.

Hunting strategy depends on the season, the location of game, and the species being stalked. In the early part of the dry season, from mid-June to September, game tend to congregate about relatively few resources. Birds, deer, rodents, peccaries, and tapir, for example, exploit the fruit of a few trees such as the açaí palm, andiroba (*Carapa guianensis*), and copaíba (*Copaifera* spp.), which are found in swamps and seasonally inundated forest. In the case of some plants, such as açaí palm, bacaba palm, and copal trees (*Hymenaea* spp.), hunters normally collect the fruit for themselves and inevitably scout around for and notice spoor. In the rainy season, numerous trees flower and fruit, such as enormous cashew trees (*Anacardium* spp.), bacuri (*Platonia insignis*), maçaranduba (*Manilkara huberi*), and piquiá (*Caryocar villosum*). All these attract game animals as well as people. As Redford et al. (1992) pointed out, the animals do not appear to be in much competition with the human population for these resources, since the Ka'apor prefer game meat and fat to fruit.

Hunting game near the village, such as in the swidden, shades into horticulture itself (Balée and Gély 1989). It is noteworthy that men, putatively going to fell trees in their new fields, also carry with them a bow and two or three arrows, or a loaded gun. The weapons are not primarily for protection from jaguars, of which adult men are not afraid anyway; they are carried just in case an opportunity arises to take game. Women, who often accompany their husbands to the swidden, are frequently indispensable in a hunt. A man's wife may drive a small game animal, such as a paca, an armadillo, or an agouti, toward her armed husband, waiting in a forward position. If the animal takes cover in a burrow, the wife often exerts much of the effort in extracting it.

When women's productive activities are not curtailed by menstrual or postpartum taboos, they may hunt in small groups, especially if their husbands are away hunting in the deep forest and game is sighted near the village. One afternoon in the village of Šoanǐ in 1981, a small boy spotted an agouti in a fallow about half a kilometer from the village. He ran back to the village shouting "*Akuši! Akuši!*" His mother Waraširan ('West Indian gherkin'), her brother's wife Meri, Waraširan's daughter Itašǐ ('white rock'), with her own infant daughter in a sling, the boy, and I as observer took off toward the fallow in haste, followed by one scrawny dog. The dog located the agouti and gave chase. The agouti scampered through the tangled underbrush until finding a burrow, which it entered. Waraširan and Meri dug seven holes with their long-handled steel hoes (*tasir*) into the agouti's snakelike tunnel, which was about eight meters long. Itašǐ sat on the buttressed roots of a *Cecropia* tree, where she nursed her infant and watched the older women. The diggers inspected each hole by running a long strand of liana down its length to the next hole, constantly pushing away the dog, which kept trying to squeeze into the tunnel. In each hole, after inspection, they inserted a branch large enough to prevent the retreat of the animal. They decided where to dig each succeeding hole by putting their ears to the ground, listening for barely perceptible scratching sounds. Meri finally caught the agouti by its hind legs in the last hole and pulled it out with a forceful tug. While Meri held the hapless creature by its hind legs, Waraširan picked up her machete, clutched the agouti by its nape in her free hand, and cracked its skull with the flat side of the machete. The entire episode lasted fifty minutes.

Men commonly pursue small game, especially tortoises, alone or less frequently with their wives. Hunters are profoundly knowledgeable about the food habits of game. They seek out particular game species near nondomesticated as well as domesticated plants that are known to be attractive. Red brocket deer, which are nocturnal feeders and known to eat copaíba seeds, for example, will sometimes be hunted at night during the fruiting season of this tree, which occurs from the middle to the end of the dry season. One evening in the village of Šoanī in August 1981, a few hours before the full moon that night would be directly overhead, I observed a young man and his father-in-law set off for a copaíba tree about thirty-five minutes walk from the village. The young man had seen deer tracks and distinctively chewed copaíba seeds there a few days before, and the hunting party now returned with the aid of one of my flashlights. About an hour later, the village was aroused with the distant blast of a shotgun. The two men returned to the village at midnight, but with no red brocket deer. The young man had only wounded a deer, which fled. Early the next day, the two men and the older man's wife searched for the wounded deer with a dog. After following a trail of blood, they found it taking refuge in an açaí grove (*wasaí-tɨ*) about twenty minutes walk through forest from the scene of the shooting. Indigenous knowledge of plant/animal interactions, therefore, is seen in this instance to greatly enhance hunting success.

Knowledge and exploitation of game animal feeding habits by the Ka'apor represent less a use than an aspect of the decoded "activity signature" (Hunn 1982) of a plant. Ka'apor knowledge of the relationships between flora and fauna—in particular, knowledge of the faunal species that consume parts of given plants—is extremely diverse. I determined that the Ka'apor recognize 170 species of nondomesticated plants as being edible to three or more game animals. (For details, see the plants identified with ANF 'Animal Food' codes in appendix 8.)

It is difficult for field investigators to confirm whether the knowledge of plant/animal interactions expressed by informants is an accurate rendering of what actually occurs in nature. If a large enough sample of animals' stomach contents confirm indigenous knowledge with respect to the feeding habits of many game species, then it can be extrapolated to suggest a high level of accuracy in general—even absent material evidence. Such knowledge is, of course, testable in principle, if not easily in practice. For example, if informants claim that red brocket deer eat leaves of wild manioc (*Manihot quinquepartita* and *M. leptophylla*), which are indeed called *arapuha-mani'i* 'red brocket deer–manioc', it is simply a matter of opening up red brocket deer stomachs and finding, eventually, wild manioc leaves there to prove the assertion. It is, in fact, the case that the Ka'apor say that red brocket deer eat manioc and wild manioc leaves and that, thus far, Helder Queiroz (1990) has confirmed the existence of manioc leaves in stomach content analyses of red brocket deer. Most important, data I acquired from interviews and observations and the interpretations I made have been tested by Queiroz (1990); on the basis of a large sample of game animals, he found this work to be extremely accurate.

Jacques Huber is widely credited with 'discovering' that agoutis disperse Brazil nuts (e.g., Kubitzki 1985). Their scatter hoarding habits, combined with small brains, supposedly lead them to "forget" some of their caches of Brazil nuts, with the consequence that one or more of the seeds in these forgotten caches may germinate and grow into a Brazil nut tree. It may be inaccurate, however, to credit Huber with discovery of a relationship already known to many generations of indigenous peoples. Huber probably learned about the association either from caboclos or from Tembé Indians in Pará, with whom he worked. Huber made a contribution to science, but the identical knowledge, if not in the codified form of field notes, manuscripts, and so forth, was already present in the heads and language of his informants. This is not to discredit Huber, merely to suggest that indigenous knowledge, especially with regard to ecological relationships, may represent a genuine shortcut to theory generation in tropical ecology (Posey 1986) and should be acknowledged by scientists—not categorically (and ethnocentrically) dismissed.

Knowledge about the food habits of game animals tends to be inconsistently, or incidentally, reported in the economic botanical literature. It is all the more interesting to note that much of the data extant on the topic comes precisely from ethnographic and economic botanical reports on the comestibility of species to one or more game animals (see Redford et al. 1992). Knowledge of the habits of game animals and of the fruiting periods of trees that attract them is not coincidentally related to hunting success. If the knowledge is accurate, it probably reduces what would otherwise be a higher proportion of chance in hunting. This is especially so in the case of 'waiting' (harõ) for game, which is practiced only where flashlights with good batteries are available. The tree selected for harõ is typically not only in fertile condition, with its usually big-seeded fruits littering the ground, but tracks of highly desirable nocturnal game species (such as brocket deer and paca) are often nearby, confirmed by unmistakable teeth marks on the fallen fruits themselves.

Darcy Ribeiro had distinguished two periods in the consumption of protein foods among the Ka'apor. One was a period of "plenty" from December to March and the other was one of "penury" from May to August, "when they depend almost exclusively on cultivated produce for food, suffering a true scarcity of foods of animal origin." (Ribeiro 1955, also 1956:16). My data showed, however, that in the period of June, July, and August the per capita daily consumption of game meat in a Ka'apor village of twenty-seven people was 51.5 grams, which is more than adequate and better than that ingested by many other peoples of the world (Balée 1985). If fish consumption is added to the total, the Ka'apor are consuming protein at a very high level by any standard. Queiroz (1990) has found, similarly, a more than adequate per capita protein intake in both dry and wet seasons at the very large village of Gurupiuna as well— which now has ninety-eight residents, the largest village in Ka'apor history. It is likely that knowledge of game animal feeding habits as well as anthropic enhancement of the catchment area contribute to this success.

Fishing

Fishing is considered a relatively unproductive activity in the high Amazonian forest, since waters are typically acidic, contain few nutrients, and are shaded (Smith 1981:5). Nevertheless, Ka'apor men spend an average of 48 minutes and women and average of 35 minutes of the daylight hours fishing. The difference between these figures, moreover, is insignificant; there is no real gender bias to fishing. Points are used by both men and women for 88 percent of the total time spent fishing, while hook-and-line fishing is practiced only 12 percent of the time. I observed no net fishing, although this is practiced by the Tembé. It is unlikely that fishing poles were used in aboriginal times, due to the probable absence of fishhooks, but the Ka'apor now show a highly specific knowledge of the tree species with the proper combinations of elasticity, tapering length, and diameter for highly flexible fishing poles.

Poles are crafted from the saplings of four annonaceous species of plants, three of which are in the genus *Duguetia* and grow in the understory of the high forest. (It is interesting that the Waimiri Atroari also use a species of *Duguetia* for precisely this purpose—see Milliken et al. 1992:31.) The fourth species of annonaceous plant is *Unonopsis rufescens*, which grows along river banks. All four species used for fishing poles are denoted by the folk genus name *pina-'i* 'fishhook-stem'. Although the term *pinar* 'fishhook' appears to be aboriginal, its modern referent probably is not. In other words, *pinar* may have meant something else in pre-Columbian times.

Fishhooks (*pinar*) and line (*pina-ham*) are obtained through trade. The bait is usually worms, but two species of plants serve an intermediate role. Fruits of the rubiaceous forest herb *Psychotria colorata* and seeds of *Ricinus comunis*, a treelet of dooryard gardens, are used for baiting hooks and catching small fish: characins called *pirapiši*. Fishermen cut chunks of the characins as bait for hooking larger, more desirable fish.

Unlike fishhooks, fish poisons are most probably traditional. These include two cultivated piscicides (*Clibadium sylvestre* and *Tephrosia sinapou*) as well as at least four species of nondomesticated fish poisons, which are *Derris utilis*, *Serjania* aff. *lethalis*, and two species of a tall, solitary palm of the high forest, *Syagrus* spp. Although I never saw the stems of *Syagrus* spp. being used for fish poisoning, the Kayapó Indians of Pará state use it in the same way as stated by the Ka'apor, that is, by pounding the stem underwater with a club (Anderson and Posey 1985). *Derris* is the fish poison of choice for large stream meanders, since *Clibadium* and *Tephrosia* are useful only on tiny creeks. *Serjania* (*kururu-šimo* ['toad-timbó']) is also useful in larger, fast-moving streams. This sapindaceous vine is the model *šimo* for the Guajá Indians, while the *Derris* species they call *kururu-šimo* ('toad-timbó') is an example of false cognacy.

All forms of fishing by poison entail cooperation. Either men or women may initiate an expedition by gathering *šimo* (*Derris utilis*) in villages, or by harvesting a shrub of the swidden, *kanami* (*Clibadium sylvestre*) which reaches a height of two meters. (Although the Ka'apor do not use *Derris amazonica* as a fish poison, with informants claiming that it is completely ineffective, the Chácobo of Bolivia and the Waimiri Atroari do employ it as a fish poison—see Boom 1987; Milliken et al. 1992:30.) I observed Meri initiate a fishing expedition with *kanami* at Šoanī in the mid–dry season of 1981. First she went to the swidden which she and her husband worked. She stripped off the green leaves of several *kanami* plants, which ultimately kills them. Upon returning to the village with the leaves packed in her carrying basket, she dug a hole with her steel hoe, about 33 cm deep. Then she placed the *kanami* leaves in the hole and pounded them with the end of a long pole. After reducing the leaves to a mash, she scooped them up and put them into a basket woven from fresh açaí palm leaves. She and several other women and children then walked about ten minutes upstream from Šoanī, along the winding creek called *Mirití-rena*. Meri dipped the basket into a pool of muddy brown water at a bend in the shallow creek. Then she lifted the basket out of the water, while everyone watched the green liquid filter into the stream. She repeated this several times during the next half hour, while greenish billows of *kanami* spread downstream in the sluggish current. About ten minutes after she had begun, some stunned fish, the gill muscles of which had been paralyzed by ichthyothereol in the *kanami* leaves, rose clumsily to the surface, where they became visible. Tiny characin fish of the genera *Tetragonopterus* and *Moenkhausia* (called *pirapišī*)—died immediately and were taken. To kill the larger stunned fish, including cichlids, *Leporinus* (*waraku*), and eels, one boy gigged them with a 2 m shaft, from which protruded three 15 cm prongs of steel. Another woman sliced off their heads with her machete, and quickly grasped the fish corpses with her hands, before they slipped down into the muddy bottom.

The small fish were captured by hand and killed by pinching their heads. As fewer and fewer fish came to the surface, Meri and her party moved downstream. Because of the green discoloration amid the russet water, one could easily follow the course of the poison. As the other women and children stood downstream in thigh-deep water, Meri periodically dipped her basket into the water, squeezing the mash of leaves inside with her hands. Upon reaching a wide bend in the creek, Meri brought forth from beneath her skirt a piece of the liana *šimo* (*Derris utilis*) as well as roots of *Tephrosia* (*šimo-'i* ['little timbó']), both of which she proceeded to pound with a club against a fallen tree that now bridged the creek. When one end of the liana was sufficiently shredded, resembling a skein of flax, she dipped it into the water. She squeezed out a milky sap that soon stunned the fish below. When that was finished, she pounded the roots of *Tephrosia* in like manner, as more poisonous exudate filtered downstream. The other women and children spaced themselves at short intervals, up to thirty meters downstream and out of sight. They covered three-fourths of a kilometer in all, wading through the small, meandering stream. After two and a half hours, they had caught twenty fish weighing about 250 g each and numerous small fish, such as *pirapišī*.

Derris is the fish poison of choice for large stream meanders, since *Clibadium* and *Tephrosia* are effective only on small creeks. Because he never saw it in a noncultivated state and because the plant is not known to flower, Ducke (1949:197–198) concluded that *Derris utilis* must be an aboriginal domesticate. Yet the Ka'apor today harvest it in areas of high forest and swamp forest. They too claim that it does not flower, nor have I collected it in a fertile condition. Where it has been cut and removed, according to the Ka'apor, it does not regenerate. Indeed, of nondomesticated plant resources, *Derris* appears to be one of the first to be depleted near Ka'apor settlements (Balée and Gély 1989). According to informants in the village of Gurupiuna, none is to be found any longer between that village and the Gurupi River. In the village of Urutawi, in the Turiaçu basin, the nearest stand of *Derris utilis* was located three and a half hours walking time from the village in 1985. This suggests the possibility that the remaining stands of *Derris utilis*, a very large liana, were planted before the arrival of the Ka'apor, perhaps by fugitive Afro-Brazilian slaves in the nineteenth century or even earlier by unknown horticultural Indians. It appears to be the only traditional plant species in the Ka'apor habitat that is threatened with local extinction by harvesting practices that have become 'traditional' in the adopted homeland of the Ka'apor (see chapter 6). Both the domesticated fish poisons—*Clibadium* and *Tephrosia*—are herbs, while the nondomesticated fish poisons are either lianas (*Derris* and *Serjania*) or a tall palm (*Syagrus*).

The harvestable portion per plant is much larger for the nondomesticates. For fish poisons as a group, the active principles are diverse. *Derris* contains rotenone, *Serjania* contains saponins probably responsible for its toxicity, the cultivated *Clibadium* contains ichthyothereol, and *Tephrosia* contains both rotenone and toxic tephrosin (Moretti and Grenand 1982). These poisons act in different ways on fish nervous systems and appear to affect different species in different ways.

Gathering

Married women spend an average of 18 minutes per day and married men spend an average of 21 minutes per day gathering edible nondomesticated plants and plant parts. The difference between the amount of time men and women allocate to this activity is insignificant. Gathering is usually incidental to hunting, fishing, or swidden horticulture. People returning from a distant swidden, for example, may find that a tree is bearing fruit and stop to collect the ripe ones. Although I have not calculated the nutritional significance of nondomesticates in the Ka'apor diet, some of these are potentially very important sources of vitamins, minerals, and energy; others may be important as snack foods or during times of starvation, should there be some village-wide calamity with swidden production.

Scientists have been voicing growing concern for the conservation of the genetic resources of domesticated plant foods, especially threatened landraces known principally to Third World farmers (see Oldfield and Alcorn 1987; Plucknett et al. 1983; Shulman 1986). In addition to the general problem of losing tropical forest species to deforestation (Myers 1988; Raven 1988), the loss of proven utilitarian species and landraces looms as a potentially costly threat to humankind's survival. With notable exceptions (e.g., Myers 1983; Prance et al. 1987), little attention has been given to the identification and conservation of nondomesticated plant foods, such as edible species traditionally gathered in tropical forests. In fact, at least in Amazonia, many edible nondomesticated plants are simply unknown to a public wider than small-scale indigenous societies (Cavalcante 1976:11).

Recently, the Food and Agricultural Organization of the United Nations published a list of 74 important "food and fruit-bearing forest species" of Latin America (FAO 1986). Yet 12 (16 percent) of these species are not "forest" species at all, but well-known domesticates, such as papaya, cashew, pineapple, and soursop that grow almost exclusively in dooryard gardens, swiddens, and other areas quite recently perturbed by agricultural peoples. The state of knowledge of edible nondomesticated plant foods in Amazonia remains rudimentary and far from comprehensive. The new economic plants and endangered germplasms that economic botanists and ethnobotanists are seeking to identify in the Third World (Alcorn 1984b:392) should encompass nondomesticates (NAS 1975:1) as well as landraces of domesticates.

Although one could show *ex post facto* that cultivated plants tend to be more economically important than nondomesticates to indigenous peoples, it cannot be assumed that the process of plant domestication in the lowlands of South America was complete by the time of Columbus. Neo-European colonization probably precluded the further development of many pre-Columbian agroecosystems in lowland South America. This can be inferred from several formerly agricultural groups who became foragers after 1500 (see chapter 8). Amazonian Indians possibly were experimenting with some species, including what are today called "wild" food plants, and using traditional tools of domestication prior to this disruption.

The Ka'apor recognize as edible 179 nondomesticated plant species. Several of these species are widely known as edible Amazonian plants (some are even sold commercially in cities), while many more are not known as edible beyond Ka'apor culture. Although one may reasonably argue that any plant the Ka'apor regularly ingest is "edible" for human beings in general, one cannot conclude, ceteris paribus, that all plants the Ka'apor regard as inedible are biologically toxic. The Ka'apor, like most Amazonian peoples (e.g., Kensinger and Kracke 1981), proscribe or regard as inedible some species of flora and fauna that are otherwise edible, according to the criteria of other Amazonian cultures.

For example, the Ka'apor do not eat the seeds (nor any other part) of the rubber tree (*Hevea*), which elsewhere in Amazonia local peoples eat roasted (e.g., Schultes 1977). In addition to eschewing all otherwise edible parts of given species, the Ka'apor avoid some parts while consuming others of the same species. They do not, for example, eat manioc leaves, which are widely consumed by lower Amazonian city dwellers in a dish known as *maniçoba* (Albuquerque and Ramos Cardoso 1980:75). Yet manioc tubers are of paramount calorific importance in the Ka'apor diet. (The Waimiri Atroari, who also depend on bitter manioc in their diet, likewise eschew the leaves— Milliken et al. 1992:25.) In spite of these and other Ka'apor avoidances and although clearly no Amazonian society exploits all edible plant foods of its indigenous habitat (cf. Meggers 1971:101, Meggers et al. 1988:281), the number of nondomesticated plants regarded as edible by the Ka'apor is to my knowledge the highest yet reported for an Amazonian people.

Many species edible to the Ka'apor are little known or completely unknown as such in the economic botanical literature. This is the case with several trees in the tropical rose family (Chrysobalanaceae), such as *Exellodendron*, *Couepia*, and *Licania*. Mesocarps of these genera are eaten by the Ka'apor, but they are not classified as edible by Prance (1972:134,145,199,208–214), whose attention to food uses in his monograph of the neotropical Chrysobalanaceae is otherwise and generally reliable. The annonaceous *Duguetia riparia* and *Guatteria elongata* offer edible pulps and/or mesocarps to the Ka'apor, but this use has not been reported elsewhere (cf. LeCointe 1947:75). The edibility of the annonaceous fusaia, araticum, and *Rollinia exsucca*, all of which the Ka'apor regard as comestible, is also known from other sources (e.g., Roosmalen, 1985:6,14,16). The fruits of *Perebea guianensis* (mulberry family) are edible to the Ka'apor and to the Wayãpi Indians (Grenand 1980:268), but the edibility of this species is not mentioned in any other sources where its general uses are described (e.g., Corrêa 1984:(2)155; LeCointe 1947:134). The cupiúba tree, *Goupia glabra* (Goupiaceae), has edible berries for the Ka'apor and Wayãpi (Grenand 1980:245–246), and at one time an edible oil was evidently extracted from this species (Corrêa 1984:(2)483), but no part is mentioned as edible by other authors (LeCointe 1947:134; Roosmalen 1985:132).

On the other hand, many species considered edible by one or more societies are never eaten by the Ka'apor. This is the case, for example, with Job's tears (*Coix lacry-ma-jobi*), the seeds of which are reported to be consumed in Asia (Smith 1976:308). The Ka'apor do not regard as edible any parts of the following species (all of which occur in the Ka'apor habitat): *Alibertia edulis, Canna indica, Casearia decandra, Casearia sylvestris, Ceiba pentandra, Guarea guidonia, Hibiscus rosa-sinensis, Licania apetala, Luffa cylindrica, Sida cordifolia,* and *Urena lobata.* According to Kunkel, however, one or more parts of all these species are consumed by one or more peoples around the globe (1984:16,71,78,82,174,183,210,217,337,373).

In spite of the relative ease in identifying food plants, as opposed to efficacious (in the economic botanical sense) medicinal plants (see chapter 5), ethnobotanical studies of Amazonian societies tend to list far fewer species of edible nondomesticated plants than I report here for the Ka'apor. For example, in their list of 224 useful species of the Siona-Secoya of Ecuador, Vickers and Plowman (1984) described only 28 nondomesti-cated species, by my count, as being edible to these Indians. In a list of 58 economic species of the Suruí of southwestern Amazonia (Rondônia state, Brazil), Carlos Coim-bra (1985:39–50) mentioned only 29 nondomesticated food species, by my count. From a corpus of 183 collected, identified, and useful plant species, Glenboski (1983:81–84) noted a mere 11 nondomesticated food species for the linguistically isolated Tukuna of Amazonas, Colombia—almost certainly an underestimate. Pinkley (1973) reported only 24 nondomesticated plant foods for the Kofán of Ecuador. Davis and Yost esti-mated that they collected 80 percent of the useful plant species of the Waorani of Ecuador; of these, only 44 were listed as "wild" food species (1983:162,167). By extrap-olation from the Davis and Yost study, one might conclude that the Waorani would recognize only 55 nondomesticated species as edible. For Indians who obtain "most of their protein and virtually all their minerals and vitamins from the forest" (Davis and Yost 1983:162), this would indeed seem to be a small number of edible nondomesti-cates, compared to that of the Ka'apor—who derive most of their food energy from horticulture, not the forest.

Other than here, the most comprehensive list of edible nondomesticates for an Amazonian people is to be found in Pierre Grenand's pioneering ethnobiological study (1980) of the Tupí-Guaraní–speaking Wayãpi of French Guiana. One can infer that the total number of edible nondomesticates, based on a collection of at least 552 species of trees, lianas, herbs, grasses, and palms, is 129 for the Wayãpi (Grenand 1980:219–303). This figure is still less than the 179 I have recorded for the Ka'apor, but it is comparable because the identified species corpus for the Ka'apor (768) is propor-tionally larger. Thus, 23 percent of the species collected by Grenand are edible to the Wayãpi, and intriguingly 23 percent of the species I collected are edible to the Ka'apor.

The evident similarity between Ka'apor and Wayãpi regarding the edibility of non-domesticated plant species may seem not too surprising, since both the Ka'apor and Wayãpi speak very closely related languages of the Tupí-Guaraní family, occupy like habitats, and probably share similar cultural criteria of edibility. But several species common to the habitats of both groups and regarded as edible by one are not described as such for the other. For example, *Helicostylis* (mulberry family), *Bellucia* (melastome family), *Bactris maraja* (palm family), and *Solanum stramonifolium* (tomato family) all have edible parts to the Ka'apor. Names and uses of these same species are also supplied for the Wayãpi by Grenand (1980:230,235,248,269,297), yet these species are either explicitly considered to be inedible or their potential edibility is unstated by the Wayãpi.

Independent data on the edibility of all these fruits beyond Ka'apor culture are available, if sparse. For example, Roosmalen (1985:308,310) described fruits of the genus *Helicostylis* (including *Helicostylis tomentosa*) as being "edible, sweet-tasting" and, likewise, fruits of the genus *Pourouma* (including *P. minor* of the mulberry family) as being "fleshy, edible, sweet-tasting" (cf. Milliken et al. 1992:93). (One must give Roosmalen credit for bravery.) The fruits of *Bellucia grossularioides*, regarded as edible only to game animals by some authors (e.g., Silva et al. 1977:96) or whose uses are said by others to be unknown (e.g., Denevan and Treacy 1988:22), have been elsewhere reported to be edible for human beings (e.g., Cavalcante 1976:8; Corrêa 1984:[2]140; LeCointe 1947:45; Roosmalen 1985:275). *Bactris maraja*, a heavily armed, caespitose palm of inundated forests, is known to possess edible mesocarps (e.g., Pesce 1985:51). And fruits of the Amazonian *Solanum stramonifolium* are known to be (or to have been) an ingredient in sauces in Hindu cuisine (Corrêa 1984:(4)385). This species is even cultivated for its edible fruits by the Siona of Ecuador (Vickers and Plowman 1984:31–32).

Fruits that are deemed edible by the Wayãpi but not by the Ka'apor include *Casearia pitumba* and *Conceveiba guianensis*. The edibility of *Conceveiba guianensis* caruncles is also known outside Wayãpi culture (e.g., Roosmalen 1985:117). Sleumer, the modern botanical monographer of neotropical Flacourtiaceae, wrote that "no proper edible fruits or seeds are known in this group [i.e., Flacourtiaceae]" (Sleumer 1980:6). But Roosmalen (1985:129) observed that *Casearia pitumba* exhibits a "sweet-tasting, edible aril," knowledge of which the Wayãpi also share. Davis and Yost (1983:168,198), in addition, listed two edible species of *Casearia* for the Waorani (see also Kunkel 1984:78 on the general edibility of these species).

If variability in the criteria of plant edibility exists between different groups of the same language family, it is not surprising that one finds variability between groups of contrasting language families. The Siona and Secoya, for example, do not collect (and presumably do not eat) the pulps of *Inga marginata* (Vickers and Plowman 1984:17), which are nevertheless edible (LeCointe 1947:219) and which are collected, according to my own observations, by the Ka'apor and Asurini. The Waorani eat the fruits of *Cecropia sciadophylla* and *Heteropsis* sp. (Davis and Yost 1983:167,177,182–183), both of which are present in the habitats of the Ka'apor, Asurini, and Wayãpi, but are nevertheless inedible to these people.

Any estimation of the number of edible nondomesticates to a given Amazonian society, such as that of the Ka'apor, is hence likely to be much lower than the potential number of such food plants in the region, were the criteria of edibility based on actual parameters of biological toxicity, not highly variable cultural criteria. (See chapter 5 for a discussion of the possible causes of such cultural variability.) Nevertheless, in this nascent area of research, indigenous knowledge of plant edibility is still the best source for data on toxicity and can provide the basis for discovery and possible domestication of new food species.

The Ka'apor classify nondomesticated species into two primary categories based on edibility: (1) *awa-mi'u* 'human-food' and (2) *awa-mi'u-'im* 'human-food–not.' They recognize several secondary categories of edibility, such as *u'u-katu-awa* ('delicious, high-ly edible'), *u'u-we-awa* ('somewhat edible, slightly desirable'), and *ta'ɨ-ta-mi'u* ('children's food', which may also be considered 'starvation food' in relation to adults). Some edible fruits are restricted to segments of the population defined by age or sex. For example, the sweet pulp of 'wild' cacao (*Theobroma speciosum*) is consumed only by males and postmenopausal women; if other females were to eat this cacao, it is said that they would *yaɨ-hu* 'menstruate excessively'. Likewise, men 'occasionally eat' (*u'u-we*) the fruits of *Guatteria elongata* and *Duguetia riparia* (both in the soursop family). These fruits are proscribed for women because, again, it is said they would *yaɨ-hu* if they ate them. The men's concern about *Yaɨ-hu*, by the way, seems unusually great, compared with our own cultural standards. It is associated with a concept of female pollution, which is widespread in lowland South America.

Many fruits that are small (less than 2 cm in diameter) and sugary or starchy, such as *Protium* spp., *Rollinia exsucca*, *Mendoncia hoffmannseggiana*, are described as *ta'ɨ-ta-mi'u* 'children's food'. Adults may also eat such fruits, but they tend not to as a matter of course—fortunately, I can report no cases of chronic hunger during my study. On the other hand, children are ritually prohibited from eating certain "adult" foods. For example, small children are not permitted to eat the kernels of the monkey pot tree *yapukai'i* (*Lecythis pisonis*—Brazil nut family), otherwise they would become 'lazy' (*yɨtɨm-'ɨm*). All forest fruits are prohibited to individuals in certain ritual states. This includes men and women in the couvade (*nino-ha*), menstruating women (*yaɨ-rahã*), girls undergoing initiation to womanhood (*yaɨ-ramõ*), and sponsors of infant-naming ceremonies (Balée 1984b).

The principal fruits of Ka'apor subsistence, i.e., those which are described as *u'u-katu-awa*, include wild cashew (*Anacardium giganteum*), açaí palm fruits (*Euterpe oleracea*), bacaba fruits (*Oenocarpus distichus*), piquiá (*Caryocar villosum*), bacuri (*Platonia insignis*), bacurizinho (*Rheedia* spp.), sapucaia (*Lecythis pisonis*), ingá-titica (*Inga alba*), ingá-açu (*Inga capitata*), abiu cutite (*Pouteria macrophylla*), and cacao and its relatives (*Theobroma* spp.) [cf. Ribeiro 1955:147]. The procurement of these fruits is often a principal activity of adults. It appears that almost all Ka'apor fruit gathering time (at averages of 18 minutes per day for adult males and 21 minutes per day for adult females) was spent in search of just these eleven kinds of economically important fruits. Although this is a small amount of time compared to that devoted to hunting and horticulture, fruit gathering occurs as a subsidiary to some other activity, such as hunting (this is more common than the reverse). For example, although a hunter's principal activity may be coded as stalking a particular game animal, he often does so near a tree whose ripe fruits might attract the prey. Once pursuit ends, the hunter is likely to harvest fallen fruits in addition to the carcass of his prey.

Even though probably about 80 percent of Ka'apor energy requirements are satisfied by horticulture, forest fruits clearly make some, as yet unmeasured, contribution to the Ka'apor diet in terms of calories, vitamins, minerals, protein, and fats. Phytochemists, regrettably, have studied the nutritional values of only a very small minority of the edible parts of Ka'apor food plants.

Almost all the foods listed in appendixes 7 and 8 (under the headings PFS and SFS, as coded in appendix 6) are edible raw. Only three species must be cooked or roasted to render them edible. These are piquiá (*Caryocar villosum*), *Sacoglottis amazonica*, and *Sacoglottis guianensis*. Edible parts of the palms *Euterpe*, *Maximiliana*, *Oenocarpus*, and *Orbignya* are consumed either raw or cooked, but preferably (to the people themselves) cooked. Only two nondomesticated species (*Anacardium giganteum* and *Anacardium parvifolium*), both congeners of cashew, are fermented before consumption as beer at infant naming ceremonies. To make these brews 'potent' (*iro*, which literally means "bitter"), the Ka'apor add heartwood from one of several tree species. These are *kupa-pa'i* (*Pouteria* spp.), breu manga (*Tetragastris altissima*), and *Caryomene foveolata*. The efficacy of these additional ingredients in contributing to the mind-altering effects of Ka'apor 'brew' (*kawī*) is unclear, but it is interesting to note that one of these plants (*Caryomene*) occurs in the moonseed family (Menispermaceae), which is very rich in alkaloids.

By far, most of the 179 edible nondomesticates have the edible part somewhere in the fruit—usually mesocarps and arils. Five species supply edible, sweet latex. Interestingly, four of these five species are in the dogbane family (Apocynaceae): *Ambelania acida*, *Lacmellea aculeata*, and *Parahancornia* spp.—which contains numerous toxic

FIGURE 4.3 Detoxifying manioc meal. Ka'apor woman using a sleeve press (*tapeši*) to detoxify manioc meal prior to sifting and toasting.

exemplars. The other is maçaranduba (*Manilkara huberi*) in the sapote family. These latices are normally eaten directly from the tree and usually only by hunters—they are
snack foods. The Tembé Indians put latices from these species in their *chibé* (farinha of
manioc soaked in water) to sweeten it. Finally, five species, all large lianas of the high
forest, supply potable water. All these species are in the dillenia family and are generically called *tiriri-sipo*. Normally, only hunters in the dry season, far from any flowing
water, drink from these vines. They sever a liana with a machete and turn the hanging
end, from which water gushes, toward their mouths.

Food Preparation

Once people return to the settlement with food, be this the product of hunting, fishing, gardening, or gathering, they normally must process it before eating it. Married
women allocate more of their productive effort (an average of 1.9 hours per day) to
preparing food than to any other activity. Although men spend significantly less time
than women preparing food, it is noteworthy that they spend as much as an average
of 43 minutes per day, helping their wives with this domestic chore. Food preparation
includes fetching water and firewood, processing manioc, cleaning game meat and
fish, and cooking. Although fetching water and firewood are also useful for bringing
in drinking water and fueling a fire for warmth at night, these activities primarily support food preparation.

Married men seldom fetch water or gather firewood, except when their wives are
menstruating or in the ritual postpartum period. As the sun is going down, it is common to hear a man anxiously remind his wife, *Pitun tate* ('It's nearly evening', which
is another way of saying, 'It's time for you to get firewood'). On the other hand, men
sometimes help their wives with all stages of food processing, even if they rarely do
such work alone. Women clean most of the game and fish that come into their households, as well as boil and roast the food.

The Ka'apor dedicate most of their food preparation time to the processing of bitter
manioc (*mani'i* for the plant, *mani'ok* in reference to the tubers). Most of the bitter manioc is processed into farinha (*u'i*), with a smaller proportion becoming manioc bread
(*meyu*) or tapioca (*tupi'a*)—which in the past served as war farinha. Upon harvesting,
bitter manioc tubers are placed (*mani'ok omor*) in a bend of a shallow creek for 'soaking' (*mani'ok-i-pe-hi*). After about three days of soaking, a woman, either with her husband or with other females (especially real and/or classificatory daughters and sisters,
who tend to live in the same settlement because of uxorilocality), gathers up the
soaked manioc tubers. After soaking, the skin becomes so loose that the tubers can be
peeled by hand. The Ka'apor 'peel' (*mani'ok pirok*) the tubers at the creek's edge, leaving behind the brown skins, and place the white pulp in their carrying baskets
(*panaku*). They return to the village with these heavy loads on their backs.

In the manioc shed (*paratu-rok*), they dump the pulp into a wooden trough, called
mira-kwar 'wood-hole'. This trough is made from the wood of one of three folk species,
kupi'i'i (*Goupia glabra*), *tarapa'i* (*Hymenaea courbaril* and *H. reticulata*), and *kiki'i* (*Newtonia* spp. and *Pithecellobium comunis*). Mortars (*ayu'a*) for the grinding of other plant

foods are preferably from *iwise'i* (*Laetia procera*), *tareka'i* (*Bagassa guianensis*) and *akaú'i* (*Helicostylis tomentosa*). Once the pulp has been poured into the trough, it is 'pounded' (*sosok*) with wooden pestles. Pestles (*sosok-ha*) are made from the heartwood of *mukuku'i* (denoting three species of *Licania*), *wapini'i* (*Licania canescens* and *L. kunthiana*), and *mira-pitaŋ* (*Brosimum rubescens*). They remove by hand the long and coarse central 'fibers' (*šurer*) of the tubers. Then they pack this pulp into the cylindrical manioc press (*tapeši*), which has loops at both ends. The top loop is placed around either an extended ridgepole of the manioc shed or a separate pole attached to a vertical pole especially for this purpose. A wooden bar is placed through the lower loop. Sometimes a woman sits on this bar, to expedite the process of removing prussic acid and other moisture. Often, however, she places another, heavier wooden log transversely over the bar. With either method, it takes about fifteen minutes to remove enough prussic acid and other moisture from the manioc mass to ensure edibility. Tapioca (*tupi'a*) is made from the starchy white residue that coalesces in a pot beneath the liquids extracted in this way. It is washed free of its impurities by soaking in successive changes of

FIGURE 4.4 Sifting and toasting manioc flour. Ka'apor couple toasting manioc flour on a ceramic griddle (*iwi-paratu*).

clean water. Then it is dried, crumbled, and eaten like farinha. Like popcorn, tapioca is light in weight; for this reason it was preferred as war farinha over the heavier *u'i* (the manioc flour itself).

Next they unpack the press, dumping the now semi-dry mass of manioc (*mani'ok maha*) over a square sieve. They sift out more fibers and inedible pieces. The relatively dry meal, then, is emptied from the sieve onto the copper griddle (some villages still make clay griddles), which rest on the furnace in the manioc shed, a process called *paratu-pe-omor*. Finally, the toasting (*u'i-karãi*) begins. A man or woman moves the meal back and forth over the griddle with a rake (*u'i-yïwïrï-ha*) made from very hard wood of the sapindaceous *Cupania* or *Matayba* or from *Diospyros* in the Ebenaceae family. Within about twenty minutes, the 'manioc meal is almost dry' (*u'i-širik-tate*). After about forty minutes of toasting the meal in this way, the coarse manioc flour (*u'i*), which is the caloric staple of the Ka'apor diet, is ready. The flour is stored in baskets lined and covered with marantaceous leaves—one of the preferred leaves for food storage is *ka'a-ro-hãtã* 'herb-leaf-strong' (*Ischnosiphon* sp. 2). They hang these baskets from beams in the house to protect the flour from dogs, chickens, and other potential pests. The supply of manioc flour usually lasts the household for nine days. After six days, the couple returns to the swidden to harvest more and to begin the several-day process anew. The flour is most commonly presented for eating in a calabash bowl (made of *kwi*, *Crescentia cujete*), although bowls are also obtained occasionally from the

FIGURE 4.5 Asurini woman toasting manioc flour on a clay griddle.

large rounded fruits of the semidomesticated *Posadea sphaerocarpa*, which occurs in the *yanama*, or intergarden 'messy' zone near houses. The already toasted manioc flour is most commonly consumed plain, in a kind of mash form (*u'i-tikwar*), after it has soaked in water for about ten minutes.

With sweet manioc (*makaser* for most varieties), women scrape off the skin with knives and boil the tubers. The skin of *maniaka*, a very large tuber, is also pared with knives, but the tubers themselves are grated over a tin grater. Formerly, the 'grater' (*iwise*) was made from the spiny pneumatophoric roots of the palm *paši'i* (*Socratea exorrhiza*). The grated mass is boiled in a large pot, resulting in a truly sweet sauce that is eaten with *makaser*. Sweet potatoes and yams are boiled with the skin unpeeled, but skins are discarded during eating. Most fruits are eaten raw, although piquiá, açaí, and bacaba are all boiled. Bananas, mashed and then mixed and boiled with coarse manioc flour, becomes a child's gruel.

When hunters bring fresh game into the village, women usually take charge of cleaning it, unless the hunters have already done so at the kill site, to reduce the burden they must bear on the way home. Hunter or wife first guts the animal, throwing

FIGURE 4.6 Guajá family preparing manioc tubers. Guajá of the FUNAI Post P.I. Awá clean and peel tubers alongside a creek, in the first stages of manioc flourmaking. Manioc cultivation is a fairly new activity for them; an undetermined number of Guajá still maintain an exclusively hunting-and-gathering lifestyle in the pre-Amazonian forest.

aside the intestines, stomach, spleen, lungs, and heart for the dogs to fight over. The wife typically places the animal next to the fire for about thirty minutes, to singe off the hair, after which she skins and quarters it.

With large game, the hunter keeps the spinal column (*šape-kaŋwer*) and its tenderloin for himself and his family, while his wife and children distribute the quarters first to her kin, then to the rest of the community. They cut the meat of most game animals into small pieces, inserting skewers (*sepetu*) made from *pïwa'i* branches (small high forest trees, *Rinorea* spp. in the violet family) into each piece. Also used for skewers are *Ouratea* and *Hirtella racemosa*, small trees that are also used for making grills. They set the skewer at about a 60–70 degree angle over the fire and roast the meat slowly, turning it occasionally. They consume the meat only when they perceive no speck of blood (cf. Huxley 1957:85), usually about three hours after roasting begins. If the game is relatively large, women lash a grill (*mu-ka'ë-ha*) together. The grill is a frame standing about 40 cm off the ground, above the fire. Its legs and grill slats are made from green branches of select species easily found in nearby swiddens and fallows, such as *Hirtella racemosa* (*mukuku'ïwi*), the meliaceous *Trichilia quadrijuga*, and two fabaceous

FIGURE 4.7 Ka'apor man inspecting a banana plant (*pako*) in a young swidden.

species, *Andira* sp. 1 and *Diplotropis purpurea*. These species are selected in part because when green they do not easily catch fire. Meat is roasted slowly on the grill, during half a day or more. They almost always place the liver on the grill first, then the rest of the cuts. Most people prefer the liver (*ipi'a*), which is totally consumed on the first day, to all other cuts, regardless of the animal. The roasted meat, if in sufficient amount, may last for three or four days. But if demand is low and the animal large, people may still be eating its meat seven days after the animal was killed. During this period, the meat is slowly smoked over the fire. If the meat is from that of the gray brocket deer, called *maha* (*Mazama goazoubira*), it is boiled first for ritual reasons.

Fire and firewood are thus essential for most Ka'apor food preparation—whether manioc or meat. Although most Ka'apor today use matches, virtually every household still keeps a fire drill and hearth on hand. Both the drill and hearth are made from the same materials, usually wood of the treelike liana in the soursop family, *Guatteria scandens*, called, fittingly, *tata-'i* 'fire-stem'. They also use wood from both cultivated and noncultivated annatto (*Bixa orellana*), as do the Araweté (Balée 1989b:20). The technique for making fire is by rotary friction (see Cooper 1949:283). Tinder, wood shavings, is placed near the socketed and slotted hearth. As meal from the hearth is heated by friction, it ignites the tinder. A man twirls the drill downward with both hands in rapid movements. Upon reaching the end of the drill, he quickly begins again from the top downward, repeatedly, until the sparks from the wood meal ignite the tinder. I have seen Ka'apor men obtain a flame in about one minute using this method.

It is interesting that the Guajá claim no knowledge of how to make fire in the recent past—they kept fires going with slow burning species used as torches, such as *Sagotia racemosa*. The Guajá kept these torches burning by rubbing on the burning end appreciable quantities of the already oxidized and hardened sap from maçaranduba (*Manilkara huberi*). For the Ka'apor, smoldering fires are kept alive and intensified with a fire fan in which young leaves of the bacaba palm (*Oenocarpus distichus*) are woven in a very simple lattice type of weave.

The Ka'apor distinguish two special types of firewood (*yape'a*), namely, those which are good for the manioc furnace (*u'ikaraiha*) and those which are useful for roasting meat, either on the grill or the skewer (*musepetuha*). Choice firewoods for manioc toasting are *kanei'i* (*Protium* spp.), *ayu'iran* (Lauraceae family), *mirawawak* (*Sagotia racemosa*), *pa'imira* (*Dodecastigma integrifolium*), *yasisipope* (the monkey ladder liana, *Bauhinia* spp.), *sipo-piran* (a large heterogeneous group of lianas from several different families), *yawi'i* (*Xylopia* spp.), *parawa'i* (*Eschweilera* spp.), *iwiri'i* (*Lecythis*), *araruhukãtãi'i* (*Eschweilera obversa*), *iratawa'i* (*Pouteria* spp.), and all species of *Inga*. The Waimiri Atroari also prefer species of *Protium* and *Eschweilera* for manioc toasting (Milliken et al. 1992:41). It is probably no coincidence that the species used by the Guajá to maintain a fire, *Sagotia racemosa* and *Dodecastigma integrifolium* (called collectively *mirĩka-i* 'woman-stem'), are among the preferred slow-burning species for manioc toasting of the Ka'apor.

Species definitely not useful as *u'ikaraiha* include the lauraceous *ayu'i* (*Ocotea* spp.), because it 'stinks' (*piher*) when burned and is considered to be very hard, not producing a good flame—they say *heni-'im* 'flame-not'. Other fast-burning species are also avoided as firewood for the manioc furnace, including *yapuriwa'i* (*Pouteria egregia*), which the Ka'apor say would burn the toasting manioc meal, although wood of this species is useful for roasting meat. Also useful for roasting meat are *wapini'i* (*Licania canescens*), *mukuku'i* (*Licania* spp.), *mu'i* (*Bellucia grossularioides*), and *piwa'i* (*Rinorea* spp.). Generally speaking, most firewood comes from swiddens and fallows, specifically from tree trunks that had been felled during swidden clearance.

Women and children distribute food, raw or cooked, to other households. If a game kill is large, such as deer, peccary, or tapir, and the village is small, everyone receives a share. With small game, such as caviomorph rodents, monkeys, birds, as well as fish, families normally part with little or none of their catch. Since most game is small, food distribution, as with food preparation and other productive activities, tends to be a household affair.

Manufacture and Repair of Material Goods

Married women allocate less time to subsistence production than do married men, but they spend more time than men engaging in tasks related to the manufacture and repair of material goods. This activity category involves the construction and upkeep of tools and weapons, basket weaving, thread making and weaving, sewing, and the production of decorative objects. Married women spend an average of 60 minutes per day, performing some of these activities, while married men spend an average of 45 minutes per day, doing the others.

Only women weave hammocks and infant carrying slings and they carry out most work involving needle and thread, especially when confined to the house because of menstruation. Women spin cotton into thread on the spindle (*i'im*), the axis being made of *mirapitaŋ* (*Brosimum rubescens*) and the disk from wood of *irikiwa'i* (*Manilkara huberi*) or *irari* (*Cedrela odorata*) in the mahogany family. Women weave on rectangular simple looms set in place by their husbands. Only two species are used in constructing the four-part loom. These are the palm *Syagrus* and *Sterculia pruriens* of the cacao family. The preferred diameter for the loom posts used is about 8–10 cm. The diameter is the same from one end of the post to the other. The surface of the posts is extremely smooth. In the case of *Sterculia pruriens*, the outer bast is stripped off. The loom is held together at the four corners by vines and/or fibrous barks. Ka'apor women twine warp and weft elements on this loom with a bobbin made from the wood of *Tabebuia impetiginosa* or *Brosimum rubescens*. They do not use a weaving sword. In addition to hammocks and infant carrying slings, they also twined their own cotton skirts in the past. Today, however, Ka'apor women obtain manufactured cloth and make Mother Hubbard–type dresses. (See figure 4.3, for example.) Most of the hammocks and all of the slings are still made by Ka'apor women themselves, but factory-made hammocks are increasingly seen. The Ka'apor method of weaving hammocks is similar to the Guajá method, except the Guajá are less choosy with respect to the woods used in the frame, and the weaving material is sword leaf fibers of the semidomesticated tucumã palm (*Astrocaryum vulgare*), not cotton (Balée 1988b; see also chapter 8).

Ka'apor men make and repair tools. This includes manufacturing bows and arrows of all kinds as well as carving axe and hoe handles and other woodwork, such as the rectangular feather chest (*patawa*)—the native word for which has expanded in reference to include 'storage container'. This chest is made only from the fragrant heartwood of *Cedrela fissilis* (*irari*), a member of the mahogany family. The feather chest consists of two parts, the box itself and a grooved lid. It is approximately 40 cm long by 20 cm wide. A man and his wife's feather art are kept inside this chest and are retrieved only for the infant naming ceremony. The lid is tied around the box with twine from *kirawa* (*Neoglaziovia variegata*) fibers. The craft of making feather chests appears to have changed little in recent years. A Ka'apor feather chest pertains to the ethnographic collection of the Museology Department of the Goeldi Museum, having been collected after a settlers' raid on a Ka'apor settlement near the Turiaçu River in 1908, according to the label. The style, material, and size of this exemplar are identical to those of the feather chests still made today by Ka'apor men.

Axe and hoe handles are made from three species of the genus *Pouteria* (Sapotaceae), denoted by the terms *kupapa'i* and *wiririmi'u'i*. These woods are selected not only for elasticity, but also evidently because they are fairly light. The men also use wood from the less common *tamari-mira* (*Diospyros* spp.) for axe handles. All woodwork is sanded with the asperous leaves of *Pourouma guianensis* spp. *guianensis*. The name of this species, *ka'ame'i*, is very close to Portuguese caimbé, which on Marajó Island, according to herbarium labels at the Goeldi Museum, refers to the dillenciaceous *Curatella americana*, a common species of Amazonian savannas, which also possesses quite asperous, thick leaves.

Work in Plant Fibers

All garments are made from cotton, a traditional cultigen of the Ka'apor. Lopes (1934:13) wrote, "I was told that the Urubus possess the same agricultural complex of manioc, banana, and other cultigens, to say nothing of cotton, which is so important in their artifacts." In eastern Amazonia, only groups lacking horticulture or those who are drifting between a full-time foraging lifestyle and some link to horticultural society use products other than cotton for garments and hammocks. A group of seven Araweté Indians who had become separated for a long time from their kinsmen, for example, lost cotton as a crop and were making their hammocks from the soft bast fibers of *Couratari* sp. (see chapter 8). For the Ka'apor, all hammocks, infant carrying straps, and items of clothing are from cultivated cotton. This is in contrast to the foraging Guajá, who, for example, make their hammocks, carrying straps, and women's skirts from the sword leaf fibers of the *Astrocaryum vulgare* palm.

Men are mostly responsible for working objects with fiber plants other than cotton. The Ka'apor use nondomesticated fiber plants for only a few purposes, but the large number of species so used is impressive. A fundamental distinction made in economic botany is also implicit in the Ka'apor classification of useful fibers. This distinction is between fibrous barks and entire stems. Entire stems are used in rope making, especially from the traditionally cultivated bromeliad, *Neoglaziovia variegata* (*kirawa*). In the absence of this species, the Ka'apor make rope of (inferior) quality from malvaceous and tiliaceous herbs that, although not planted by the Ka'apor, are frequently cultivated by other Amazonian peoples for commercial purposes. These are *Sida*, *Urena*, and *Triumfetta*. The bark of only two trees (*Cochlospermum orinocense* and *Rollinia exsucca*) are occasionally used for rope, the first being rare and the second a fairly common tree of old swiddens, fallows, and dooryard gardens. Although the Ka'apor use rope only for hanging their hammocks and making bowstrings, the above species, with the exception of the trees, also contain soft bast fibers used for spinning and weaving sacks and bags in the lower Amazon—generically, these are called "malva" in the region.

Urena lobata (*mira-kirawa*), known in English as aramina fiber, is not planted per se by the Ka'apor and is a fairly recent introduction, circa 1930s. The Ka'apor of the Turiaçu and Maracaçumé basins employ it for bowstring only in the absence of *Neoglaziovia variegata*, which is considered superior fiber. After cutting *Urena* with a machete, the plants are left in the sun to dry. The defoliated stems are then retted for three days in a trough of water. After retting, the stems are stripped of coarse material and dried, at which point very fine fibers become accessible for spinning into rope of varying thicknesses. Almost the entire commercial production of this species in Brazil is limited to the lower Amazon in Pará state (Dempsey 1975:351; Abreu and Juca 1968). The Tembé Indians and some Ka'apor of the Gurupi basin harvest *Urena* in the dry season, selling it to middlemen who then ship it to Belém. The species is so widespread in the tropical world that its origin is unknown, but lowland South America appears to be excluded as a possibility (cf. Dempsey 1975:372). Ridley (1930) suggested Africa as the origin for *Urena*, and added that since it is not found on uninhabited oceanic tropical islands, humans must have been mainly responsible for its distribution (also see Purseglove 1968 2:374). The Ka'apor tend to weed *Urena*, *Sida*, and *Triumfetta* from their swiddens, claiming that it *uhem-uhu* ('grows too much') there. Purseglove (1968 2:374) wrote that *Urena* "can be a troublesome weed . . . and is declared a noxious weed in Fiji." Hence, it is useful on one hand, as a substitute for *Neoglaziovia variegata*, but should be seen as a pest on the other, especially considering that it tends to exhaust the soil, having the highest nutrient uptake of all other annual bast fiber crops, including flax, jute, hemp, and roselle (Dempsey 1975:376).

Because of evidently similar ecological adaptations, *Urena* often grows next to and intertwined with (almost as if they were one organism) two species of *Triumfetta*, a genus native to tropical Africa (Purseglove 1968 2:613), and two species of *Sida*. The similarity of habit, in other words, appears to override morphological differences. All these species are covered by the term *mira-kirawa*, although the two species of *Sida* are distinguished by the postposed attributive meaning false, *ran*, hence, *mira-kirawa-ran*. The uses of these species, nevertheless, are identical to those of *mira-kirawa*. These fiber "crops" (although not 'planted'—*yitim* or *ha'ï-omor*—by the Ka'apor, they are cultivated for commercial purposes elsewhere in the lower Amazon region), share a property with tree barks used for lashing material, vines used in tying house parts together, and material used in basketry, namely, *hayïk* 'tensile strength'.

The word *iwir* covers fibrous barks used for lashing material. The Ka'apor use forty species of trees that have *iwïr*. These plants occur in ten botanical families, the most important of which are Lecythidaceae (thirteen species), Annonaceae (six species), Boraginaceae (nine species), and Tiliaceae (six species). Bombacaceae, Sterculiaceae, Caesalpiniaceae, Celastraceae, Cochlospermaceae, and Ulmaceae are each represented by one species useful in making *iwïr*.

Unlike many peoples of the upper Amazon, the Ka'apor and other lower Amazonian Tupí-Guaraní peoples did not obtain bark fibers for tapa cloth, a canvas-like fabric widely used for clothing and, in the case of the Ticuna Indians, as canvas for ritual paintings. Perhaps because of the great length of *Ficus* fibers (used in tapa cloth, in addition to other members of the mulberry family), these are not used for lashing material. The underlying similarity of barks classified as *iwïr* by the Ka'apor is stratified phloem (Roth 1981:185). In other words, *iwïr* may be as accurately glossed 'stratified phloem' as 'lashing material'. The annonaceous and boraginaceous species of fibrous barks are quite similar in that their fibers have a zigzag pattern in the outer bark (Roth 1981:162). Perhaps for this reason, in addition to their utility as lashing material, the folk genus *ape'i* covers five species of *Cordia*, *Rollinia exsucca*, and is also a synonym for *Apeiba tibourbou*, even those these three genera are otherwise quite morphologically distinct and each is from a different botanical family. Another feature of several barks used in lashing material by the Ka'apor is that the hard bast (tissues of the secondary phloem that contrast with soft bast), if present, occurs only in the form of fibers (Roth 1981).

Iwïr is used in the following ways: as straps for carrying home entire or butchered carcasses of game from the hunt, and as straps for baskets. The potential for rope making from many of these taxonomic groups is well known. For example, although the Ka'apor use only one sterculiaceous species (*Sterculia pruriens*) for lashing material, some sixty genera and seven hundred species of this family may have been used for fiber by indigenous people worldwide (Medina 1959:701). *Sterculia pruriens* is known for its bark's capacity to make good rope (Medina 1959:717). The Waimiri Atroari also employ this species in preparing lashing material (Milliken et al. 1992:32). *Apeiba tibourbou* was at least formerly used in Panama and northern Brazil for making strong rope (Medina 1959:586), while *Couratari guianensis* was used also for strong rope and cigar paper in northern Brazil (it is today used as cigar paper by the Ka'apor—see below). Medina (1959:734) noted that lecythidaceous species were of little commercial value as fiber products, but nevertheless observed that many of these species, including *Eschweilera*, offer fiber for sacks and bags (Medina 1959:736).

Ka'apor men exclusively weave baskets used for carrying, processing, and storing manioc and manioc tubers. As with other sleeve presses of the South American tropical lowlands, the Ka'apor manioc press (*tapeši*) is twill woven, usually from green but dried splints of the marantaceous *warumã* (*Ischnosiphon arouma*). Occasionally, split bast from a very elastic vine, *Cyclanthus funifer*, which grows in swamp forest, is used as raw material. The *tapeši* is approximately 2.5 meters long and 10 cm in diameter (see figure 4.3). Ability at making this is considered so much a sign of prestige for adult males, that magical-catalytic procedures are undertaken to ensure that boys will learn to make the manioc press (discussed in chapter 5).

Young men often make their manioc presses together, usually following the lead of an older man, considered to be an expert at the task. Skill at basketry generally is put to use only after marriage. Men also make the manioc sieve, which consists of splints of *Ischnosiphon arouma* arranged in a lattice-style weave attached to a square frame. The cylindrical braces composing the frame are from *u'ɨ-tɨma* 'arrow-leg' (aff. *Myrcia*), which is also used in arrow points and as the midshaft in the composite arrow. This wood is very hard and straight.

Aside from the sleeve press and the sieve, there are several other kinds of Ka'apor baskets used in and out of the household. First, there is the rucksack (*wasaikã*), which is improvised in the forest. Both men and women may make this. This disposable rucksack is used for transporting home açaí fruits, other fruits, and game meat, both butchered and not. They make the rucksack by knotting fresh açaí leaves together. Shoulder straps from *iwɨr* 'lashing material' are then attached and the rucksack is carried on one's back. (The Ka'apor do not use tumplines in their basketry.) The most sophisticated basket is the *panaku*, which is a carrying basket on a wooden frame. Women and sometimes men use this in knapsack fashion for carrying manioc flour on long journeys, for carrying harvested manioc tubers to either the creek for soaking and peeling or home, and for carrying manioc cuttings for planting. It is in the shape of a slipper. Most often, this basket is made from *Ischnosiphon arouma*, but several other species of *Ischnosiphon* are also acceptable (although not for the sleeve press and sieve) as well as *Cyclanthus funifer*. The lashing material used for the shoulder straps is virtually only from what is considered to be the best source of durable stratified phloem, namely, *yašiamɨr* (*Lecythis idatimon*), called "caçador" in local Portuguese. The frame is made from straight branches of *u'ɨtɨma* (aff. *Myrcia*), which is also used for the frame of manioc sieves.

Another woven item is the floor mat (*pinuwaro*) which is made, as the name implies, exclusively from leaves of the *Oenocarpus distichus* palm (*pinuwa'ɨ*), the bacaba palm from which the fire fan is also made. The pinnae of a single leaf are braided together into a mat which is the site of birthing (as with the Araweté). In addition, at the infant naming ceremony, a man makes such a mat for his wife to sit on, while holding their infant during the ritual. The floor mat is a temporary item and is only encountered in households either with parturient women or with women who have unnamed children.

Decorative Objects and Body Ornamentation

Both men and women make decorative objects, although wives spend more time at this than their husbands. Although the most noted handicrafts of the Ka'apor are their various works of feather art, they make a number of necklaces, bracelets, waist adornments, and rings solely from plant parts. Fourteen species, including five domesticates, are used. The cultivated species include Job's tears (*Coix lachryma-jobi*), Indian shot (*Canna indica*), and bottle gourds (*Lagenaria siceraria*). The round white seeds of Job's tears and the small black seeds of *Canna indica* are used in men's armbands and women's waistbands. Women and children puncture small holes in these seeds on a wooden bowdrill (made from either *Tabebuia impetiginosa* or *Brosimum rubescens*, both of which are used in making the hunting bow itself) with a sharp steel bit, made by the men. Women string the seeds, using either cotton thread or nylon fishing line, for necklaces and waistbands. Sometimes they tie brightly colored macaw feathers onto these objects as well as onto the armbands made for their husbands. Occasionally, tiny gourds of *Lagenaria siceraria* are attached to the women's skirt adornment. Necklaces, made from the red or red-and-black seeds of *Adenanthera pavonina*, *Andira retusa*, *Ormosia coccinea* (all called generically *pu'i-piraη* 'bead-red') and black seeds from *Colubrina* sp. 1 and *Colubrina glandulosa*, are also produced for both men and women. These are also sold to FUNAI. Rings with points are made from the enormous hard seeds of *Astrocaryum murumuru* (*muru'i*), which is rare in this habitat, and from *Astrocaryum vulgare* (*tukumā'i*), which is known only from old fallows.

Ka'apor parents routinely fashion necklaces, armbands, and legbands for their infants and small children. In contrast to everyday adult ornaments, many of the objects found on children's ornaments possess properties that protect the child from evil divinities; shielding the child from colds and other illnesses; for catalyzing a long life; and, in the case of boys, for making good hunters of them. For example, pungent bark of the cultivated *Petiveria alliacea* (*mɨkur-ka'a* 'opossum-herb') is wrapped in a tiny cloth to ward off the evil divinity *āyaη*. Hardened resins from several species of *Trattinickia* (*kɨrɨhu'i*) in the bursera family are wrapped in cloth and attached to the bead necklaces of children to prevent them from becoming 'ill' (*ahɨ-ma'e*). (See chapter 5 for more details on magical and medical uses of plants.)

Men make a comb (*kɨwa*) for smoothing feathers as well as human hair from two plants. The spine is fashioned from the hard petiole of the palm *inaya'i* (*Maximiliana maripa*); the teeth come from the wood of *Brosimum rubescens* (*mɨrapɨtaη*), a wood that is ubiquitous in Ka'apor handicrafts and tools. The comb is decorated with feathers for use at the infant naming ceremony. Makeshift combs in the forest are available from the spiny fruits of *āyaηkɨwa'i* trees (Port. 'pente de macaco'). The Ka'apor term covers about four species of the genus *Apeiba* in the linden family.

Women more than men are involved in preparing plant dyes for the human body as well as for objects. The dye complex of the Ka'apor is well developed in terms of number of species, although it includes no mineral substances for fixing or mordanting the dyes (Buhler 1948). Men and women paint their faces with annatto dye, which occurs in the pods of this treelet around the seeds. The design normally has a horizontal line below the lower lip, a diagonal line on each cheek leading to the lower line, and a short horizontal line on the nose bridge. No ritual significance normally applies to facial paint, although the female sponsor in an infant naming ceremony paints the pattern slowly and methodically onto everyone's face. Facial paint, rather, is principally a marker of Ka'apor group identity, as with their distinctive featherwork and other personal adornments.

The yellow pigment from *Vismia*, a treelet of old swiddens, may be used for facial paint, but only in the absence of annatto. Although uncultivated annatto in swamp forest supplies red dye, it is not used for body painting (Balée 1989b:20). The purple pigment in the pulp of *Renealmia alpinia* fruits, a zingiberaceous shrub of old swiddens, fallows, and high forest, is also used in face painting, albeit rarely. Genipapo dye, from *Genipa americana*, is in quite limited use today. Apparently, it was commonly used in the 1930s (see Lopes 1934). The Ka'apor paint the faces of the dead with black dyes from tannin-rich *mukuku'i* (denoting *Licania heteromorpha, L. latifolia,* and *L. glabriflora*). These species are also used in painting *Crescentia* bowls; only the inside of the bowl is painted. The dye is obtained from the inner bark of these majestic high forest trees. The bark is stripped off vertically on one side only with a machete. This piece is cut into manageable strips of about 1 m in length by 10 cm in width. A woman then pounds the bark with a steel mallet for several hours until the bark becomes a mash. She then puts the mash in a kettle of boiling water. The residue is a thick reddish liquid, which upon cooling is applied to bowls with a cloth. The dye turns black on the bowl. Bark from the melastomataceous tree *Bellucia grossularioides* (*mu'i*) is used in the same way. The inner barks of *Inga alba* and *I. brevialata* (*iŋašiši'i*, called ingatitica in Portuguese) are also crushed and boiled. But the residue from the bark of these species is applied with charcoal—never alone—for dyeing bowls a deep black.

The Ka'apor use three species for dyeing arrows. The black glue from *Symphonia globulifera* resin is burned off onto the butt of arrows. Arrow points are dyed deep red with the residue of the crushed seeds of crabwood (*Carapa guianensis*). Arrow points may also be dyed deep purple with pigment of the berries of *Phytolacca rivinoides*, an old swidden shrub.

Featherwork is the artistic activity for which the Ka'apor are most renowned, thanks to Darcy and Berta Ribeiro, whose book *Arte Plumária dos Índios Kaapor* (1957; also see Gerber 1991), praised Ka'apor featherwork for chromatic brilliance and an abiding concern for the smallest details. The Ribeiros rightly classified Ka'apor featherwork as a kind of jewelry. The knowledge of how to make feather ear ornaments, bracelets, combs, lip ornaments, and diadems persists, even in the villages with most contact with the outside world. The full range of Ka'apor featherwork is still seen in the deep forest settlements, although only at the naming ceremonies of children. Most men and women from time to time adorn themselves with small ear ornaments from the feather of cotingas; a yellow tail feather of the crested oropendola is habitually worn in the perforation of the lower lip of men; and bracelets of scarlet macaw down are worn by both women and men. Some women, even young women, work beautiful belts of cotton thread adorned with yellow, red, and black feathers of macaws and toucans, which are worn by men. The only plant used in featherwork, with the exception of the feather comb whose plant components are discussed above, is *Manilkara huberi* (*irikiwa'i*), the maçaranduba tree of the sapote family. The oxidized black sap of this species is the only 'glue' (*muyar-ha*) in Ka'apor featherwork (Ribeiro and Ribeiro 1957).

Although it is the prerogative of both sexes, women spend more time working on decorative objects than men (27 minutes versus 9 minutes of the daylight time) because they are periodically confined to the village and prevented from engaging in most other productive activities by menstrual taboos. During menstruation, women can work only on decorative objects, yarn spinning, weaving hammocks and infant carrying straps, sewing, and cleaning the house.

Child Care

In Ka'apor society, women are the caretakers of young children. An average adult woman spends only about 48 minutes a day exclusively and actively in child care activities. But it should be noted that women with grown children are included in this estimate, as well as those with infants to care for. Then too, after about the age of five, children become rather autonomous, spending much time in their own play groups away from adults or (at least the girls) in helping their mother with chores. More importantly, a young mother is able to manage her infant or young children in ways that allow her to be active in other ways. Thus, during many of the hours I observed and recorded Ka'apor women engaged in a particular activity, they were simultane-

FIGURE 4.8 *Bagassa guianensis* of the mulberry family, which the Ka'apor call *tareka'i*, is an enormous tree found in both high forest and fallow. Many Ka'apor believe that eating its fruit causes baldness. On the other hand, informants avow that brocket deer, tapir, and yellow-footed tortoises feed on its fallen fruits.

ously caring for their children. I tended to record the particular activity as primary, whereas child care was treated as an adjunct activity.

Shortly after birth, and until about the age of three, an infant is constantly with its mother. The infant spends most of its time in the carrying strap, which is a woven band of cotton, looped around the mother's shoulder and waist. Young mothers perform few demanding chores for about six months after childbirth. After this, a mother may accompany her husband on occasion into the forest, always carrying the child with her. She may go to the swidden to harvest manioc, but her load of tubers is usually light, since her infant remains in its strap in front of her.

Whenever the infant cries, the mother pulls her breast out of her dress and nurses the infant until the crying stops and the infant is satisfied. Infants nurse often, and mothers show no signs of displeasure in giving them all they want. The infant sleeps with mother in her hammock until the age of four or five. Children are weaned at about the age of three. If another child is born before this time, however, the first child is removed abruptly from the breast, and with no small trauma. I did not learn of any bitter herbs used by mothers to discourage their children from nursing. A mother almost never attempts, however, to nurse two infants at once. On the other hand, because of birth spacing and frequent miscarriages, such conflicts rarely arise.

After it is weaned, the child is no longer transported in the infant carrying strap. But it remains close to its mother, following her about. When a young child must eliminate, the mother takes it with her into the forest near one of the refuse sites— these are never located in active swiddens. A mother also takes her young child to the stream to bathe. Children younger than about four or five do not go into the forest and swiddens about the settlement alone, although they may accompany older children there.

Shortly before a child is weaned, she is given, in addition to mother's milk, soft foods such as manioc or tapioca gruel and mashed bananas. By the age of two and a half, the child may eat roast fat and liver from the hunt, as well as boiled yams, sweet potatoes, and sweet manioc. Later the child begins to eat u'itikwar (flour of bitter manioc soaked in water), the main food of adults. At about the age of five, a child tends to eat most foods not avoided by adults. There are, however, food restrictions that apply to all children younger than about the age of fifteen. Several fruits are strictly prohibited to them, such as that of tareka'i (Bagassa guianensis—see figure 4.8). Many restrictions that apply also to menstruating women and parents in the couvade also apply to children.

Although infants spend more time with their mothers before they are weaned, their fathers maintain an active interest in their development. Fathers will hold their children by the hands trying to help them learn to walk. They will sometimes play with their otherwise nursing infants in their own hammocks, with the mother looking on. When a boy is about four years old, his father makes for him a miniature bow and arrow. The materials used for toy bows and arrows are more varied than those used for the adult equivalents. For the toy bow, a father may use not only tayi (Tabebuia impetiginosa) and mɨrapɨtaŋ (Brosimum rubescens), he may also use pɨwa'i (Rinorea) and the petiole of the inajá palm, inaya'i. The shaft of the arrow may be either from arrow cane (*u'iwa*) or from *marato'i* (*Schefflera morototoni*). Arrow points tend to be the same as those used by adults, including *tayi, u'itima* (aff. *Myrcia* sp.1), *wapini'i* (*Licania* spp.), *mirapitaŋ*, with the exception of *mukuku'i*. Children also more often use points of *piwa'i* (the small *Rinorea* trees of the violet family) than do adults, for the purpose of killing birds without ruining the feathers. For a daughter, a father often captures a small bird, monkey, or tortoise for her to raise.

By the time a boy is about five or six, he spends much more time playing with other children. Boys shoot their small arrows at makeshift targets, such as large limes, mangos, and papayas. They also shoot at moving targets, such as rats and lizards. Parents warn their young sons not to kill too many small animals. One six-year-old in the village of Urutawi who had killed several rats fell ill with a malarial fever. His parents claimed than an evil divinity (*āyaŋ*) of one of the dead rats had entered his body.

Boys tend to play much more than girls. From about the age of eight, while her brothers are off on their own fishing or shooting at targets, a girl accompanies her mother and father to the swidden, begins carrying heavy baskets full of manioc tubers, and learns how to prepare manioc flour in the manioc shed. Girls also wash clothes, help clean the house, and do some of the cooking. Their mothers teach them how to weave hammocks and infant carrying straps, and how to make decorative objects and dresses. Meanwhile, fathers teach their sons how to shoot, use an axe, and make a basket. Parents are very tolerant of their offspring. Mothers may scold their children for playing too close to the fire or for making too much noise in their play

FIGURE 4.9 Ka'apor friends relaxing in the forest near the village of Gurupiuna.

groups. But it is decidedly rare for a mother or father to take physical action against their children.

In general, child care takes up relatively little time for adults except for nursing mothers. This is because children become relatively autonomous, spending more time with other children, after about the age of five, with boys having somewhat more autonomy than girls. The hopes and dreams of Ka'apor parents lie with their children. This becomes evident to an outsider very quickly. One of the first questions they ask people who visit them is "Do you have children?" If answered in the affirmative, this is followed rapidly by "How many?" and "What are their names?" and "Boy or girl?"

Eating

Married women preside in the rearing of infants; they also spend more time eating than men, snacking often during the day. I have no evidence to show which gender may be better nourished, however. One of the most striking markers of the relative independence of the nuclear family is the fact that the members of each household tend to eat together as an isolated group, not breaking bread with members of other households. There are no specific mealtimes, however, except shortly after daybreak. As the morning sun rises, the women of each household prepare a hot manioc gruel with roast or boiled meat from the day before for themselves and their families.

With spoons and ladles the women dish out the food into black-dyed calabash bowls. Spoons for eating snacks and ladles for cooking are obtained by carving wood from two species in the dogbane family (both of which have edible latex and fruits) called by the folk generic term of *kuyer-'i* 'spoon-stem.' The word *kuyer* 'spoon' is a direct borrowing from Portuguese *colher* and appears to be fairly old in the Ka'apor language. It is likely that Indians already made spoons in pre-Columbian times from these species but called them by a different name or names.

The Ka'apor drink water with their meal from large bottle gourds (*kawasu-te*), cultivated in the swidden. These bottle gourds do not leave the village because hunting and gathering parties do not bother to carry water; water is always obtainable in the forest itself, either from running creeks or from water vines.

Hygiene

Ka'apor men spend about 37 minutes and women about 44 minutes a day keeping themselves and their households clean. There is thus no significant difference between the amount of time the genders spend on hygiene, which is a category of composite data. In fact, men tend to bathe more frequently than women because their work—especially in the swiddens—is dirtier than women's work. Women, on the other hand, spend more time cleaning house, delousing themselves, their husbands and their children, laundering clothes and hammocks, and washing pots and utensils. It should be pointed out, however, that these tasks are not exclusively done by women. Sometimes their husbands help them. For example, men do most of the clearing of asteraceous and grassy weeds from around the house, which become tall and cumbersome in the rainy season. (I have coded "yardwork" here as a kind of "housework.")

Each family usually has a place on the meandering creek near the settlement where its members habitually bathe. If the settlement is large, two or three families may use the same bathing spot, but they almost never go to bathe together. Members of different households, except for children, seldom bathe together even if of the same sex. The Ka'apor normally bathe in the late afternoon during the temperate rainy season, but in the dry season, when work is dirty and days quite hot in open clearings, men tend to bathe three times a day. Another aspect of personal hygiene is delousing. Members of the same family pick lice from each other's hair, and seldom does one observe members of different households delousing one another, except among children.

Women do almost all the house sweeping in the early morning. They sweep away chicken and dog droppings as well as dust from the clay floor. The brooms are usually made from the rachis of açaí. Usually, the rachis is bound with a vine at its midpoint to make it more handy. Although it is uncommon, a few households use a broom with a handle. This broom is from *Scoparia dulcis* and the handle is normally made from one of the same species used in making the manioc flour rake, i.e., *Cupania*, *Matayba*, and *Diospyros*.

Women do most of the laundry, which, as with bathing and the occasional cleaning of pots and utensils, takes place at the family's spot on the creek. Soap was aboriginally obtained from the leaves of *Carica papaya*, the leaves of *Gouania pyrifolia* (a spontaneous, rhamnaceous vine of young and old swiddens), and the inner barks of *Enterolobium* sp. and *Parkia paraensis*. I have seen these materials used and can confirm that all produce suds. These aboriginal soaps were used mainly for hammock cleaning in the absence of detergent soap. Also, the body and hair can be cleaned with these soaps, according to informants. Finally, Ka'apor adults possess plant deodorants and scents. Body odor can be controlled by rubbing resins of *Protium* spp. (*kaneiape'i*) in the armpits—only men use this (Balée and Daly 1990). Women only perfume themselves with the fragrant *Cyperus corymbosus* (*pipiriwa*) of the dooryard garden. They attach a sprig of this herb to their skirts. They also use perfume from the tonka bean (*Dipteryx odorata*), the seeds of which are put in small sacks of cloth and attached to necklaces and other adornments.

Leisure and Other Activities

Aside from leisure, which occupies about 2.6 hours of the men's daily time and 2.9 hours of the women's daily time, the other activities not discussed here but which engage people have minimal representation in the time allocation data. In other words, many other activities, in particular, ritual, medicinal, and magical practices involving plants, fall into the category of 'other' ways of using one's time. Any one of these activities, taken alone, would not constitute 1 percent of the daylight time allocation of either men or women. But these activities are so important with regard to comprehending how Ka'apor adults "know" the plants that surround them and the environment they inhabit that a separate chapter has been reserved for discussing them.

FIVE

Medicine, Magic, and Poison

I think forgetting is very healthy. There's a lot of stuff in
our heads that we don't need.
> —Rajan Mahadevan, former memory champion,
> Guinness Book of World Records

Oral Traditions and the Limits of Knowledge

A kind of "mental economy" (a term suggested to me by Eugene Hunn) surely is operating in Ka'apor plant knowledge. Because oral transmission is the only route of Ka'apor plant knowledge, the dimensions of this knowledge are limited by the capacity for human memory to store it in a retrievable, communicable, and replicable manner. About five percent of Ka'apor youths and a smaller percentage of Ka'apor adults are today somewhat literate. But even this small incursion of literacy is very recent—owing only to the efforts of James and Kay Kakumasu, linguists of the Summer Institute of Linguistics who worked from the 1960s into the 1980s, mostly in the village of Gurupiuna.

In attempting to comprehend the cognitive basis of Ka'apor ethnobotany, an essential distinction to be made is not between "modern" Western thought (connoting science, rationalism, empiricism) and indigenous "traditional" thought (connoting magic, irrationality, faith), but between modes of communicating ideas in the two kinds of society. I agree with Lévi-Strauss (1966:65) that the form of the mind is constant among cultures; only its contents differ. The so-called Grand Dichotomy (e.g., Lloyd 1979:239), which posits fundamental differences in mental capacities, is an artifact of the presuppositions of an earlier era in anthropological studies.

The mind's contents, in terms of the kinds of information to which it has access, however, are influenced by certain cultural practices, chief among them literacy. As Jack Goody (1977:43) cogently explained:

> Members of oral (i.e., 'traditional') societies find it difficult to develop a line of skeptical thinking about, say, the nature of matter or man's relationship to God simply because a continuing critical tradition can hardly exist when skeptical thoughts are *not* written down, *not* made available for men to contemplate in privacy as well as to hear in performance.

Because Ka'apor plant knowledge has not been developed in a literate context, it cannot be scrutinized systematically by the actors themselves, as could be the same knowledge in, say, a botanical garden. An opportunity for skepticism with regard to plant knowledge has not arisen in Ka'apor society. Oral transmission tends to make knowledge increasingly anonymous (Goody 1977:27 et passim; Horton 1982:218,251) and also unstable over the generations.

In oral traditions, according to Goody (1977:27), the "individual signature is always getting rubbed out in the process of generative transmission." This is not the case with literate traditions, which, in principle, can maintain the acquired knowledge originating with specific, authentic individuals for indefinite periods of time and in more or less unaltered form, should they so choose. According to Michel de Certeau (1988:216), "To writing, which invades space and capitalizes on time, is opposed speech, which neither travels very far nor preserves much of anything. In its first aspect speech never leaves the place of its production. In other words, *the signifier cannot be detached from the individual or collective body*" (emphasis in original).

Likewise, in the words of Walter Ong (1967:33), "For oral-aural man, utterance remains always of a piece with his life situation. It is never remote." Ka'apor plant knowledge, among the people themselves, is communicated *only* by means of spoken words. Ka'apor plant knowledge, to the actors themselves, therefore, cannot be visualized in its totality, as can the written word (including this book) within a literate tradition. Hence, Ka'apor ethnobotany cannot be decontextualized, systematically tested, corrected, and reworked (Goody 1977), at least in its indigenous context. Although increments in rationalistic knowledge about Amazonian nature may occur from time to time in Ka'apor culture, these are inherently labile because of an oral mode of communication and transmission.

Lévi-Strauss (1966:154) observed that there seems to be a "sort of threshold corresponding to the capacity of memory and power of definition of ethnozoologies or ethnobotanies reliant on oral tradition." (Also see Berlin 1992:96–101.) I would add here that not only does memory place upper limits on the number of lexical entries about plant species and their names that can be made in the brain (cf. Sperber and Wilson 1986:86), but the fact that Ka'apor society exists in a nonliterate tradition influences processes whereby *rationalistic* knowledge is acquired about ecological associations, active biological principles in nature, and modes of plant reproduction, distribution, and behavior.

I am not distinguishing here, incidentally, between any kind of "primitive" or "prelogical" mentality and the modern mind, as did the early Lévy-Bruhl (1923), but rather between the *capacities* of empiricism and rationalism in nonliterate versus modern societies. One can also argue, of course, that Ka'apor society lacks scientific instruments, such as microscopes, to detect the smaller relations of nature, and hence a fundamental distinction should be drawn at the level of scientific tools. But the development of optics and the like occurred only in literate societies and arguably depended on literacy.

As Goody suggested (1977:90), writing facilitated a greater knowledge of the natural world, because visual inspection of written words enables one to order and reorder materials. Whereas Linnaean botany recognizes and gives a unique name to each of some 30,000 identified plant species in Amazonia alone, for example, no indigenous language in the region would have more than 1,000 distinct names for this flora. In fact, it is likely that there is an upper limit of about 500 generic names (Berlin

TABLE 5.1 'Medicine' Words in Several Tupí-Guaraní Languages

Language	Word for 'medicine'	Source
Araweté	tóhã	Balée, field notes
Asurini	puhã	" "
Guajá	pohã	" "
Tembé	puhã	" "
Ka'apor	puhan	" "
Língua Geral	posaŋa	Platzmann 1896:145; Ruiz de Montoya 1976:312; BarbosaRodrigues, 1894:77; Barbosa 1951:129
Modern Guaraní	poã	Dooley 1982:149

1992:96–99). The Ka'apor certainly do possess rationalistic knowledge about their habitat, but this knowledge is less than the total which could materialize in a scientific tradition depending on a literate mode of communication.

The recent discussion about the value of indigenous knowledge and 'intellectual property rights' of indigenous peoples should take account of the in-built limitations of this knowledge. These limitations reflect an exclusively oral mode of transmission and a variety of ancillary conditions: lack of a state, classes, craft and professional specializations, money, tribute, monumental architecture, and so on.

Absence of a written tradition thus precluded Ka'apor culture (or any other indigenous culture of the South American lowlands) from acquiring a completely rationalistic knowledge of their immediate physical environs. In other words, the absence of literacy as a traditional mode of preserving knowledge about the local habitat coupled with the intrinsic complexity of this habitat precludes an indigenous true science.

Distinguishing Magic and Medicine

Much of current ethnobotanical concern focuses, and rightly so, on medicinal plants used in indigenous Amazonian cultures. As these cultures are eroded and the habitats with which they are associated succumb to invading cultures, many truly useful biological principles found in plants and employed in indigenous therapies may be forever lost to the world. This emphasis on medicinal plants, however, has too often been mired in a conundrum regarding the exact meaning of 'medicine' in indigenous societies. It seems that until the semantics of indigenous therapeutics, indeed, of indigenous plant "knowledge" more generally, have been agreed upon, the utility of listing the medicinal plants of this or that tribe will continue to remain quite limited for the ethnopharmacological sciences.

The indigenous concept most often glossed as 'medicine' in English covers an extremely wide semantic range in Ka'apor. The word 'medicine' (*puhan*) itself has several cognates in other Tupí-Guaraní languages (see Table 5.1). No evidence suggests that this term was borrowed among these languages. Rather, the word for 'medicine' in these languages, along with an essentially similar concept, was inherited from proto–Tupí-Guaraní which was spoken some 1,800 to 2,000 years ago (Migliazza 1982;

A. D. Rodrigues, personal communication). The word *puhan* and its cognates in other languages are frequently translated into Portuguese as *'remédio'* (remedy) or *medicamento* (medicine) by modern lexicographers (see, for example, Dooley 1982:149; Kakumasu 1988:232). But the Western medical sense of 'remedy' is far more restricted than that of *puhan*. Barbosa (1951:129) captured the sense of the Língua Geral word *posaŋa* very well by listing as its referents: *'remédio'* (remedy), *'mezinha'* (magic), *feitiço* (fetish or charm), and *veneno* (poison). All these meanings appear to be conjoined in the word *posaŋa* as well as in its cognates, such as *puhan*.

Puhan is a polysemous expression, denoting various referents which one might wish analytically to divide into at least two categories: 'medicine' and 'magic'. I confine this discussion only to botanical referents of these categories, which are in any case far more numerous and salient in everyday life than are animals and minerals. 'Medicine' or a 'medicinal plant' refers to a plant which is used in the treatment of a human disease in a falsifiable way, "falsifiable" being defined in the Popperian sense (Popper 1959). A magical plant is used deliberately to effect a change in a being or object from one state to another; it is perceived as a catalyst, but its function and efficacy remain untestable. In other words, the therapeutical mode of use of a *medicinal* plant is stated by informants in such a way that it can be rigorously tested; the catalytic functions of *magical* plants, on the other hand, are not subjectable to criteria of scientific falsification.

To divide the concept *puhan* into subcategories of seeming rationality and nonrationality does not imply necessarily a false dichotomizing of Ka'apor magico-medical knowledge. The testable-nontestable distinction I offer is provisional and designed for explicating Ka'apor plant knowledge. I am not proposing, as Augé (1985) has claimed in relation to the work of Ackerknect (1946), to reduce this totality to an "ethnocentric dualist conception," whereby any indigenous system exhibits an empirical-rational sector and an opposed magical sector. In fact, Ackerknect's conception did not apply only to indigenous societies (see below). I am also against setting up a dichotomy of Ka'apor indigenous thought and Western thought as unbridgeable, incommensurably different domains, as did the early Lévy-Bruhl. In his *Primitive Mentality* (1923), Lévy-Bruhl argued that non-Western peoples could not distinguish between subject and object, could not form abstractions, and that they perceived nature in a mystical law of participation, a conflation of representations and nature itself (see Wallace 1966:9).

It may be the case that dualistic notions such as rational-nonrational, testable-untestable ultimately derive from a list-making mentality which is itself derivative from writing (Goody 1977). I, however, favor viewing logical and nonlogical mentalities as coexisting universally, as did the later Lévy-Bruhl (see Tambiah 1990:84). Ackerknecht wrote, "Our own rational scientific medicine is not entirely free from supernaturalistic tendencies and notions" (1946:467). Stanley Tambiah (1990), who painstakingly traced the development of the concepts "magic," "science," and "religion," showed that the distinctions between these domains are artifacts of Western history. It was seventeenth century Protestant thought that divided the concepts of "science" and "religion," whereby scientific activity and Puritan doctrine were complementary, but not magic and religion nor magic and science (Tambiah 1990:31).

One of the most lasting (and polemical) contributions to the notion of "magic" as a distinctive semantic domain came from Edward Tylor and James Frazer, Victorian armchair scholars who were affiliated with the great evolutionary schemes of the nineteenth century. Tylor envisioned magic, science, and religion in an evolutionary order. Occult practices that continued to exist in modern Europe he deemed to be "survivals" of a more barbarous past (see Tambiah 1990:45). Frazer attempted to ground magical practices in two complementary "principles of thought": the law of similarity, which stipulates that like produces like, and the law of contact or contagion, wherein things that have been in contact continue to act on each other after contact has been discontinued (Frazer 1963:12). To Frazer, magic paved the way for science, since magic and science rest on the assumption that order is the "underlying principle of all things" (Frazer 1963:825). Lévi-Strauss likewise considered magic as "science yet to be born" (1966:11).

Nevertheless, Lévi-Strauss also rigorously separated magic (associated with a "science of the concrete") from science ("science of the abstract") in a series of binary distinctions, which have been criticized by Goody (1977:8) as artifacts of a list-making mentality associated only with literate cultures. The philosopher Wittgenstein criticized Frazer's characterization of 'savage concepts' and 'magic' because of their having been taken out of context. Foreshadowing the rise of philosophical relativism, Wittgenstein argued that "the truth of certain empirical propositions belongs to our frame of reference" (quoted in Tambiah 1990:64). Taylor (1982) has further posited that nonrationality might be defined as logical inconsistency in the context of cultural articulations.

According to these relativist views, typical magical practices of a given society are not irrational, since they are not inconsistent in the context of cultural standards. Although someone who practiced magic in our society may be seen as irrational, since the dominant culture does not articulate standards for such activity, magic as a general concept cannot be considered irrational given the diversity of experience in different cultures. In spite of the strength of this relativist argument, nevertheless, the concept of magic as nonrational and unfalsifiable is still too useful for Western observers to ignore, especially those who wish to understand the kinds of phenomena under discussion here (see M. Brown 1986:20 for the same point).

Although Popper has been criticized, notably by Kuhn, for what Popper claims to be the *sui generis* form of scientific propositions (Kuhn 1977; Tambiah 1990; cf. Horton 1967), it is no less true that some kinds of magical, catalytic procedures and presumable functions defy definition within a rationalistic framework. They are not only invisible to the naked eye, but to any degree of technical investigation and falsification. Simply stated, some effects of *puhan* are plausibly stated and observable, while others are not, at least according to indigenous formulation.

A Mental Economy in Ka'apor Ethnobotany

My proposed division of rational and nonrational in Ka'apor culture is not, therefore, fruit of an a priori bias concerning egalitarian societies, since all human societies, regardless of the presence of science in some of them, share nonrational beliefs. Representational beliefs, in other words, are universal (Sperber 1985). In addition, all societies require some kind of mental economy to manipulate strategically the natural environment, including pathological elements. Without such manipulation, there cannot be survival. Western societies possess means of storing information, through books, institutions, computer hardware and software, as well as through craft and professional specializations, such that mental economies are diffused through various sectors and individuals of the population in unequal ways. Societies such as that of the Ka'apor, the so-called egalitarian ones, lack the baggage of the nation-state, including artificial intelligence (I consider writing a kind of artificial intelligence). Hence, they require that each adult individual possess a mental economy fundamentally similar to that of the other members of the society—with exceptions sometimes based on gender and sometimes enhanced by special shamanic practices in otherwise egalitarian societies.

This mental economy covers not only knowledge about nature, such as ecological relations between species of flora and fauna, but also shared notions of kinship, marriage, edible versus inedible foods, cosmology, ritual acts, handicrafts, and so on—in short, that which one customarily calls 'cultural' in its broadest sense. Magic from this perspective may be a possible foundation for future increments in the corpus of rationalistic knowledge (Horton 1967). But there is no guarantee that magical knowledge will become rationalistic, for it is not customarily written or recorded on storage devices external to one or more human brains and certain spoken languages.

The defining power of literacy may be used to account for some salient cognitive differences between Ka'apor culture and Western culture(s). This account is, of course, predicated on other fundamental differences, such as intensive agriculture, class stratification, industrialization, and the rise of modern state society (see Gellner 1988). But however useful as an analytic tool literacy may be for discerning unlike mental economies of indigenous societies and the West, it obviously cannot purport to unravel enigmatic differences among the diverse ethnobotanical and ethnomedical systems of native Amazonia.

One observes in indigenous populations of the lowlands of South America a problem in the replicability of ethnomedical knowledge. Since Amazonian peoples do not have a written tradition, this knowledge is perhaps unstable over time. In a recent paper, Davis and Yost (1983a) pointed out that Waorani Indians of the Amazonian forest of Ecuador have only thirty-five medicinal plants used against common diseases such as fungal infections, snakebite, dental problems, fever, botflies, and burns. They argued that the recently contacted Waorani would not have needed more medicines, considering that only among indigenous groups with longer histories of contact does one observe developed pharmacopoeias, a response to the Conquest and its epidemiological ills. That is, the natural state of pre-Columbian Indians would have demanded few remedies for treating few diseases.

I take issue with the conclusion drawn by Davis and Yost. The Ka'apor, in contrast to the Waorani, use 112 plant species for medicinal ends, as defined above. The Ka'apor recognize 37 different diseases of which only two, colds and measles, are clearly introduced (see the MED entries in appendix 8). The great majority of medicinal plant species thus are used to treat human diseases of pre-Columbian origin. These diseases include stomachache, vomiting, diarrhea, fungus, worms, botfly, cuts, burns, boils, toothache, earache, eye lacerations, snakebite, pregnancy, labor pains, postpartum hemorrhage, tiredness, and rheumatism, all of which are treated within the ambit of traditional Ka'apor medicine—that is, according to testable principles and formulas. (I have included pregnancy in this list because unwanted pregnancy may be regarded, like disease, as a condition in need of a cure; see, for example, Hern 1975.)

Other than medicines used by the Ka'apor to treat the above-listed diseases, there are also 'magical' plants within the category of *puhan*. Here I include those plants used, among other objectives, to win love, secure luck in the hunt, gain protection against evil divinities, ensure the abilities of speaking and walking in children, restore sanity (after known taboo violations), cause death of enemies, impart courage in war, and guarantee a long life. The principles and formulas in their use are not testable by sci-

FIGURE 5.1 *Ryania speciosa*, a denizen of many Amazonian forests, is one of the most poisonous plant species known. Source: H. O. Sleumer, 1980, *Flacourtiaceae, Flora Neotropica* (vol 22). Copyright 1980 by the New York Botanical Garden. Reprinted with permission.

ence. It is interesting to observe that only 51 species are used in magic, in comparison to 112 in medicine (some of which are used to treat more than one disease)—although all fall within the single domain *puhan*. That is, the rational part of *puhan* exceeds significantly, in its number of botanical referents, the nonrational part (even though a few species have both medicinal and magical uses). Together, the total number of medicinal and magical aspects of plant use in Ka'apor culture is impressive, considering the precariousness of transmission of experiential knowledge in societies without writing.

Amazonia (and the Ka'apor habitat in particular) comprises an extremely diverse biological environment; hence, it would be impracticable if not impossible to register faithfully this diversity with proper names and concepts in indigenous cultures, although in literate societies this problem, theoretically, need not exist. An example in Ka'apor culture concerns the lack of explicit recognition of many poisonous plants, especially internal poisons. Internal poisons in the Ka'apor habitat are listed in table 5.3. These include milkweed (*Asclepias curassavica*) which contains a cardiac glycoside; purge nuts (*Jatropha* spp.) which contain a protein molecule of high toxicity called curcin; castor bean which contains ricine, also a protein molecule of high toxicity and related to curcin; yellow allamanda (*Allamanda carthartica*), a gastric poison whose active principle is unknown; lantana (*Lantana camara*) which contains a poisonous crystalline material known as lantanine; cotton (*Gossypium barbadense*) which has an unknown active principle; and 'mata-cachorro' (*Ryania speciosa*) which contains ryanodine, an alkaloid that is one of the most violent gastric poisons known. These plants have certain nonfood uses for the Ka'apor, and each has a folk name. For example, informants recognize that castor beans can be used as fish bait and purge nuts can be deployed in the treatment of contact dermatitis (see appendixes 7 and 8). One informant indicated that 'mata-cachorro' (*Ryania speciosa*) may be used as firewood. But if it is true that the wood smoke may have been responsible for past accidental deaths elsewhere in Amazonia (Monachino 1949, cited in Milliken et al. 1992:41), it is highly unlikely that any Ka'apor persons have, in recent memory, used it as fuel. I also did not see it employed in this way. In any case, I recently sent a letter to local FUNAI authorities requesting them to alert the Ka'apor people to the potentially lethal consequences of using this rare species as fuel or food.

Despite their documented toxicity, it is surprising that none of these species is expressly considered poisonous by the Ka'apor. The frequent Ka'apor reply to a query about a plant's edibility, *U'u 'im awa* 'One doesn't eat it', does not often imply knowledge of any possible internal toxicity.

The Ka'apor do not, however, lack a conception of poisons. They call poison *ahi*, which also means 'pain'. (The term *puhan* in certain contexts also means poison; see below.) In my view, *ahi*, like *puhan*, has both a rational and a nonrational domain. Consider: it is said that one can kill an enemy by burying leaves of a species of a fern (*Lomariopsis japurensis*) in his house. In fact, this plant contains cyanogenic substances that are highly toxic, according to a recent laboratory assay (Klaus Kubitzki, personal communication, 1986). For the plant to be considered lethal in a Western medical sense, it would have to be ingested. It may be the case that this fern was mistakenly ingested by someone hundreds of years ago. But it was not transmitted as an internal poison down to the present generation of Ka'apor because of (a) the rarity with which one would eat a food not on a preferred mental list of species and (b) the precarious-ness of oral transmission.

Unexplained deaths are attributed to sorcerers who 'know the medicines' (*puhan ukwa katu*). These 'medicines' are not always from plants. For example, bushmaster (*Lachesis*) fangs are believed to be eternally poisonous and if one were to step on such a fang, even detached from the snake itself, it is believed that the effect would be the same as a bite from a living bushmaster. A sorcerer is presumably capable of putting a bushmaster fang on a trail so that his intended victim might step on it. But would-be poisoners do not tend to use plants as though these were activated by internally poi-sonous properties.

Inedible foods (i.e., those that are universally avoided) are in an unstressed catego-ry (cf. Lévi-Strauss 1966); in contrast, it is the edible foods that are stressed. The major-ity of 'poisonous' plants (*ahi*) sting with urticating hairs (as with stinging nettles and the *yanu-mira* 'spider-tree', *Eugenia* spp.), burn or cause itching upon combustion (as with *Caryocar*), or are believed to cause boils, chills, headache, bad luck, loss of hair, and undesirable weight loss if touched or ingested (see table 5.3). Only tubers of unprocessed bitter manioc and five other species of plants (the active biological prin-ciples of which, if any, are unknown) are indeed internal poisons in the Western med-ical sense.

A failure by the Ka'apor to note the presence of internal poisons in plants that are scientifically known to be poisonous is not, in principle, related to possible mistaken identification of the relevant organisms themselves. For example, *Ryania speciosa* (Fla-courtiaceae; called *mata-cachorro* in Portuguese), although very rare in the Ka'apor habitat, was correctly identified by Ka'apor informants. One can say the plant was "correctly" identified not only because the majority of informants gave as its name *parani-ran-'i*, but also because a close relative of *Ryania*, and which is far more common in the Ka'apor habitat, *Laetia procera* (Flacourtiaceae), is called *parani'i*. *Parani-ran-'i*, the name for the extremely poisonous *Ryania speciosa*, can be translated as '*parani*-similar-stem'—or, depending on perspective, '*parani*-false-stem'. (Because *parani* is a literal plant term, moreover, another acceptable gloss would be 'L-false-stem' or 'L-similar-stem' (see chapter 7 and appendix 9). In other words, *parani-ran-'i* is a plant name mod-eled on another plant name, the referent of which is a close biological relative (see chapter 7 for similar examples). Although its poisonous nature does not comprise any Ka'apor knowledge of this plant, the Ka'apor do, in a sense, "know" *Ryania speciosa*; they just do not know that it is deadly poisonous. Hence my attempt to alert them to this fact.

Mistaken identification did not occur with the other plants known to Western sci-ence to be internal poisons, such as milkweed, pokeweed, castor bean, and purge nuts, which are in any case familiar plants of the dooryard garden or secondary swidden. It is true that internal poisons such as castor bean and purge nuts were recently intro-duced. One might thus argue that the Ka'apor have not had enough time to evolve a rational conception with regard to their toxicity. But this does not explain why the Ka'apor already have developed uses for these exotic plants (they are cultivated, in fact, which implies at least some direct knowledge of the plant.) Recent introduction does not explain either why only certain indigenous species, such as the raw form of

TABLE 5.2 *Dangerous or Avoided Plants in Ka'apor Culture*

Ka'apor	Scientific Name	Relevant part(s)	Reason for Avoidance
inayakami	*Markleya dahlgreniana*	all	unstated
tareka'i	*Bagassa guianensis*	fruit	causes botfly infections or baldness if eaten; prohibited food for children
yanumira	*Myrciaria pyriifolia; Eugenia omissa*	bark	causes contact dermatitis or blindness if touched
anĭ-ran	*Dracontium* sp. 1	all	causes blindness if touched
ka'ayuwar	*Solanum rugosum*	all	causes contact dermatitis or blindness if touched
tukumã'i	*Astrocaryum vulgare*	fruit	causes baldness if eaten
yowoi	*Minquartia guianensis*	wood	causes death if used as firewood
yaŋaramira	*Solanum leucocarpon*	all	causes insanity if touched
wamesipo	*Philodendron* spp. all; *Monstera* spp.	all	causes contact dermatitis
arakãsĭ'aran'i	*Chrysophyllum sparsiflorum*	bark	causes contact dermatitis if touched
purakeka'a	*Laportea aestuans*	all	causes contact dermatitis if touched
piki'a'i	*Caryocar villosum*	wood	causes contact dermatitis if used as firewood
piki'aran'i	*Caryocar glabrum*	wood	causes contact dermatitis if used as firewood
taši'i	*Tachigali myrmecophila*	wood	combustion results in noxious odors
mahasipo	Connaraceae sp. 2	wood	internal poison
yahisipo	*Dioclea reflexa*	all	internal poison
aŋwayarmira	*Pseudyma frutescens*	wood	internal poison
arariãkã'i	Rubiaceae sp. 2	wood	causes headache if burned
warumãširik	*Ischnosiphon graciles*	stem	causes undesirable weight loss if eaten
u'iwaran	*Imperata brasiliensis*	root	noxious for manioc plants
irati'i; iratiãtã'i	*Symphonia globulifera*	all	unstated; children are prohibited from touching it
tamaran'i	*Zollernia paraensis*	all	causes lameness in women or children who touch it
irakahuka'a	*Xiphidium caeruleum*	all	causes boils if touched

cultivated bitter manioc, and nondomesticates like *Solanum leucocarpon*, Connaraceae sp. 2, *Dioclea reflexa*, and *Pseudyma frutescens*, are taken to be toxic for consumption (table 5.2), when clearly many nondomesticates harboring internal poisons occur in the habitat of the Ka'apor.

I suggest that the reason for this lack of knowledge is to be found in a mental economy appropriate to a society lacking a written tradition. I submit further that it is simply unnecessary to remember the properties of plants whose parts contain gastric poisons when there is an evidently sufficient number of plants known to be edible and having no known lethal effects. The only exceptions (as indeed noted above for the Ka'apor) would be poisonous plants of the dooryard garden or swidden, which might pose dangers to young children if not recognized as poisonous by adults.

To borrow the terminology of Sperber and Wilson (1986), it is "irrelevant" to

TABLE 5.3 *Internal Poisons Not Recognized By The Ka'apor*

Species	Ka'apor Name	Toxic Part	Sources
Allamanda cathartica	**sipo-tuwir**	fruit	Kingsbury 1964:263; Albuquerque 1980:36; Hoehne 1939:239-240
Asclepias curassavica	**ira-kiwa-ran**	all	Kingsbury 1964:270; Lewis and Lewis 1977:52; Albuquerque 1980:74; Hoehne 1939:244
Gossypium barbadense	**maneyu**	seed	Kingsbury 1964:175; Hoehne 1939:192
Jatropha spp.	**piyã**	seed	Kingsbury 1964:191; Lewis and Lewis 1977:20; Albuquerque 1980:76-78; Hoehne 1939:175-176
Lantana camara	**kanami-ran**	fruits	Kingsbury 1964:296; Lewis and Lewis 1977:56; Albquerque 1980:24
Phytolacca americana	**ka'a-riru**	all	Kingsbury 1964:225; Lewis and Lewis 1977:90; Albquerque 1980:28; Hoehne 1939:113
Ricinus comunis	**kahapa**	seed	Scarpa and Gaerci 1982:120
Ryania speciosa	**parani-ran-'i**	all	Lewis and Lewis 1977:35; Monachino 1949

know about internally poisonous properties of plants if the edible ones occur in a sufficient quantity. By the same reasoning, it is irrelevant to know about potentially hallucinogenic plants in the environment if the society makes no use of such plants. For example, although species of caapi (known in Brazil as *Santo Daime* and in Peru as ayahuasca), including *Tetrapterys styloptera* and *Banisteriopsis pubipetala* (see Purseglove 1968 (2):637), both of which the Ka'apor call *sipo-piraŋ* 'vine-red', are used as hallucinogens by many peoples of the Northwest Amazon, these plants are not used for any mind-altering purpose by the Ka'apor. Ka'apor informants did not claim that these species possess mind-altering properties, although they do believe that other plants do, such as marijuana, which they do not use. Likewise, although at least four hallucinogenic species of *Virola* trees occur in the habitat of the Waimiri Atroari, these people do not use them for any such purpose (Milliken et al. 1992:43–44).

According to Sperber and Wilson (1986:86–91), the human deductive device has access only to elimination rules. Edible and inedible foods in Ka'apor thought can be conceived of as encyclopedic entries in memory; one need only know what objects are (edible or not) rather than what they mean (whether they contain internally poisonous principles, for example). The human deductive device contributes to the economy of thought: "Mental processes, like all biological processes, involve a certain effort, a certain expenditure of energy. The processing effort involved in achieving contextual effects is . . . to be taken in account in assessing degrees of relevance. . . . Other things being equal, the greater the processing effort, the lower the relevance" (Sperber and Wilson 1986:124).

In terms of Ka'apor plant knowledge, the processing effort involved in determining a plant's inedibility would reduce its relevance. In other words, the inedibility of a particular plant is a relatively irrelevant bit of information. It would also be an uneconomical one. The Ka'apor simply lack the *habit* of eating plants or any plant parts that are not already on a prescribed list of edible foods. According to Gregory Bateson (1972:257), who wrote a good deal about habit formation and the economy of thought: "In the field of mental process, we are very familiar with this sort of economics, and in fact a major and necessary saving is achieved by the familiar process of habit formation. We may, in the first instance, solve a given problem by trial and error; but when similar problems recur later, we tend to deal with them more and more economically by taking them out of the range of stochastic operation and handing over the solutions to a deeper and less flexible mechanism, which we call 'habit'."

In Ka'apor culture, to simplify further, all one needs to know is what are the edible plants—the others are eliminated by definition and no explanation of the whys and wherefores is indicated by the principle of relevance and by the basic economy of thought and habit. The irrelevance of internal toxicity of plants which are not prescribed (and therefore relevant) foods appears to be general in lowland South America. This shared mental economy helps explain myriad differences in food edibility among related societies. Differences of environment, moreover, appear to have little to do with dietary differences among Amazonian indigenous peoples (Milton 1991). For example, the Asurini of the Xingu traditionally did not eat hog plum, *Spondias mombin*, and the Parakanã did not eat açaí, *Euterpe oleracea*, or bacaba, *Oenocarpus distichus* (Magalhães 1985:35), although other indigenous (and nonindigenous) groups esteemed these fruits in their diets, including the Ka'apor, Guajá, Tembé, and Araweté. A mental economy that generated and stored knowledge of prescribed foods need not

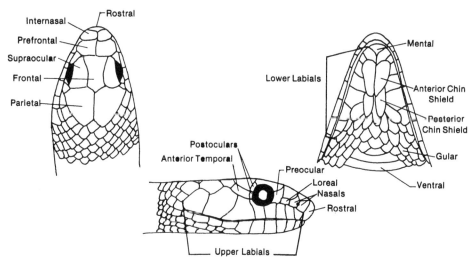

FIGURE 5.2 A harmless colubrid snake. Source: Marcos Freiberg, 1982, *Snakes of South America*. Copyright 1982 by T.F.H. Publications. Reprinted with permission.

store rational knowledge about toxic principles. The reverse procedure, i.e., that all that is not inedible is edible would seem to be more wasteful of mental energy in a non-literate society, since there are probably far more universally inedible plants than edible ones in Amazonia. In this sense, edible plants are "stressed" and inedible plants are "unstressed," which is another way of designating the comparative irrelevance of the category "inedible."

One can argue that this kind of mental economy with respect to plant use has a parallel with the widely noted phenomenon of "genealogical amnesia" in the social structure of lowland South American Indian societies. Frequently in Amazonia, informants cannot remember the names of their grandparents, much less the names of their great-grandparents (see Murphy 1979). For example, Wagley (1977:97) wrote that: "The Tapirapé [a Tupí-Guaraní people of Central Brazil] . . . had short genealogical memories. In recording genealogies of individuals no one could remember the personal names of relatives further back than their grandparents, and even then it was difficult to be certain whether the individual referred to was a biological grandparent or a classificatory one." Likewise, with regard to the Wayãpi in the state of Amapá, near the border with French Guiana, Alan Campbell (1989:35) recently noted that "people genuinely forget names and relationships." Regardless of the specific causes of genealogical amnesia, for the purposes of correct marriage (avoiding thereby the category of prohibited kinsmen, such as parallel cousins) it is simply not necessary to calculate degrees of relationship and keep genealogies. One already knows with whom one can marry, either because of affiliation with an exogamous unilineal descent group that maintains one or more alliances with other such groups or through the kinship terminology itself, which explicitly indicates potential spouses. The first mechanism is commonly noted, for example, in the Northwest Amazon (see Jackson 1983); while the latter applies to the Ka'apor.

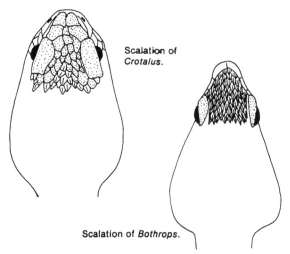

Scalation of
Crotalus.

Scalation of Bothrops.

FIGURE 5.3 Two pit vipers. Source: Marcos Freiberg, 1982, *Snakes of South America*. Copyright 1982 by T.F.H. Publications. Reprinted with permission.

For example, Timapa-ru once told me that he knew he could marry a certain woman because he had heard his mother call the woman's father by the term for 'brother' (*mu*). In other words, he "knew" she was a cross cousin (*reko-har* 'sex-place'), whether classificatory or not, without having to know the genealogical links that connected them. She was his mother's brother's daughter and it was enough to know that his mother called her father 'brother'. (In anthropological jargon, cross cousins are children of siblings of the opposite sex.) The Ka'apor state a preference for marrying individuals who are, in fact, cross cousins, even if there is never a need to calculate and "remember" the actual genealogical links in the relationship. The same pattern, moreover, is seen with the Trio and Pemon of Guiana (Rivière 1984:50) as well as with the Nambikwara of Mato Grosso (David Price, personal communication, 1984). In lowland South America, a preference for marriage with one's cross cousin is extremely common.

As for the ethnobotanical domain, if someone of a hypothetical society of the remote past died because of having ingested a toxic plant, and, if because of this, that plant was then considered to be internally poisonous by that society, it would likewise not be necessary to store this information, when the prescribed edible foods were limited to nontoxic plants, that is, when the mental economy of the society promoted a positive rule in relation to food (as it does in relation to marriage). To draw upon another analogy, in societies that do not eat and in general do not utilize snakes, it is not of immediate importance to know how to distinguish the poisonous from the nonpoisonous ones. According to Peter Roe (1982:12), the Shipibo Indians of the Rio Ucayali in Peru "kill a harmless snake with the same exaggerated panic as any American." For the Ka'apor, not all snakes are considered to be *ahi* 'poisonous'. An exception, however, is the *cobra-cipó* (as it is called in Portuguese). The *cobra-cipó* is a harmless colubrid snake, which the Ka'apor call *yararak-pu'i* 'fer de lance–thin.' This snake is considered to be extremely venomous, as are all other folk species of *yararak*, which denotes all pit vipers of the genus *Bothrops* as well. The *cobra-cipó* or *yararak-pu'i* virtually never bites, hence there could be no memory of attacks by this snake. Yet the Ka'apor routinely kill it and with notable anxiety.

Unlike the *yararak-pu'i*, the snake called *panuũ*, another colubrid, is indeed aggressive. It bites. But it is not considered to be poisonous. In fact, the bite of the *panuũ* is used to ward off future bites by poisonous snakes. The *panuũ* thus is regarded as 'snakebite-medicine' (*moi-puhan*). When captured live, parents of young children torment the snake into biting each child on the legs. It is believed that if bitten by this snake, the truly poisonous ones, such as bushmasters and fer-de-lances (including the quite poisonous and far too common *Bothrops atrox*), will not strike one later in life.

All snakes that are poisonous (and indeed some, like the *panuũ* which are not) are regarded by the Ka'apor as poisonous. But not all snakes are regarded as poisonous. In contrast, for the Shipibo, it is clear that to avoid snakebite, it is enough to "know" that *all* snakes are poisonous. It economizes mental energy, is a relevant assumption, and as such has a greater chance of being passed on to the next generation, given the inherent precariousness of oral transmission.

But not all instances of erring on the side of caution can clearly be attributed to the survival benefits of mental economies. I recall the day I learned the Ka'apor word for gecko, which is any of several species of lizard with very webby feet. I was with two Ka'apor companions collecting plants in the high forest near the village of Gurupiuna that day. As we walked down a trail, suddenly one of them shouted, with marked anxiety, "*tarakupe! tarakupe!*" The other, who had a machete, immediately came running toward a tree next to the man who had just shouted. He quickly cut in half a beautiful red, green, and black gecko which had been clinging to the tree. I asked him why he killed it. He replied that a 'gecko' (*tarakupe*) is extremely poisonous; the sting of its tail is said to kill a man. I got interested in geckos after that and learned, to my surprise, that no poisonous geckos are known to exist anywhere.

To return to ethnobotany, it is obvious that plants which compose a category of 'edible plants' could be consumed by humanity in general were it not for ethnocentric concepts about food. These concepts are the nonrational components of a mental economy that contribute to the desuetude of otherwise edible resources. That is, it is enough to consider something not edible (rather than toxic) to dissuade use of a particular item as a food source.

It is less obvious, however, that medicinal plants, as here defined, would have direct effects, according to the folk formula of their use, on the human body. This is because most have yet to be rigorously tested by science. Their active principles have yet to be identified and clinical trials have not yet been conducted. Absence of evidence is not, however, evidence of absence. At least the possibility exists that any or even all medicinal plants of the Ka'apor could be shown scientifically to have therapeutic effects on the body, since the formulas for their use are potentially observable and testable by Western science. Traditional uses of medicinal plants presumably have survived for centuries, suggesting further evidence of possible adaptive functions. This longevity is in spite of the fact that rational knowledge is always in a precarious state, owing to a mental economy that is able to detect and retain only some of the active principles in such a biologically complex environment.

Plants Used in Magic

It is a commonplace in the literature of economic botany to refer to indigenous medicinal and magical plants as conforming to the "doctrine of signatures" (see, for example, Milliken et al. 1992:36). The fifteenth and sixteenth century notion of a doctrine of signatures was held by herbalists to mean that the name of a plant was an essential aspect of it and that its "signature" corresponded to some visible quality which was of presumed importance in human affairs. The basic idea was that like begets like (Vogel 1970:33). The correct applications of medicinal plants were concealed in the external form of the plant itself (Tippo and Stern 1977:47). Moreover, the doctrine of signatures intimated a theory about the nonarbitrariness of language, since the name of a plant was inextricably related to its use (Focault 1973, cited in Tambiah 1990:21,163).

Hence, in some sixteenth century herbals, hemorrhoids (or piles) could be cured with the roots of pilewort; pulmonary diseases could be treated by the spotty leaves of lungwort; and bloodroot, with its reddish sap, could be deployed against blood diseases (Tippo and Stern 1977:47). Likewise, kidney beans could cure kidney infections, walnuts (because of their resemblance to brains) were good for migraines, and the pantaloon-like flower Dutchman's breeches (*Dicentra*) was therapeutic on venereal diseases (Klein 1987:306). As for poisons, the English mystic Nicholas Culpepper stated that "the greatest sign" of poisons is "signature" (Culpepper 1813:9 [orig. 1680]).

Modern economic botanists have suggested that the doctrine of signatures is universal, since similar patterns of "resemblance" have been observed in ancient Asia, classical Greece, medieval Europe, and pre-Columbian America (Klein 1987:306; Tippo and Stern 1977:280). For example, the Hopi were described as "specialists" in the doctrine of signatures: they rubbed hot ashes on burned skin, treated falling hair with the hairlike fibers of clematis and cowinia, and placed a twisted piece of wood next to a person undergoing convulsions (Stone 1962:30). Although most economic botanists tend to regard remedies conforming to the doctrine of signatures as nonefficacious, Tippo and Stern (1977:235–236,280) noted that snakeroot (*Rauwolfia serpentina*) of the dogbane family, whose roots resemble a snake, and which Central American Indians used as an antidote against the toxins of fer-de-lances and coral snakes, does contain the alkaloid reserpine, a psychotropic drug now used in treatment of schizophrenia.

Lévi-Strauss (1966:16) argued for the rationality of "signatures" not because of supposed inherent relationships between a name and its referent, but because the inferred relationship reflects thinking about a *connection*:

> It seems probable . . . that species possessing some remarkable characteristics, say, of shape, colour or smell give the observer what might be called a 'right pending disproof' to postulate that these visible characteristics are the sign of equally singular, but concealed properties. To treat the relation between the two as itself sensible . . . is of more value provisionally than indifference to any connection.

The doctrine of signatures is, in fact, merely a specifically named example of sympathetic magic. For Lévi-Strauss, Frazer, and others, it is a first step toward science: order and explanation pervade the seen and unseen worlds. It is only doctrinaire to the extent that it was codified and given permanent form in writing by alchemists and herbalists during the Middle Ages. And it has remained a distinctive concept in economic botany, although it lacks generality as an explanatory tool.

In several examples of plant fetishism discussed in the next section, one perceives a supposedly inherent relationship between a name and the thing it describes, a characteristic feature of the doctrine of signatures. On the other hand, as for plants used as game fetishes, for example, it is important to keep in mind that not all bear a resemblance, real or imagined, to the designated animal (cf. Berlin 1992:259). Although I discuss Ka'apor plant nomenclature in chapter 7, it is important now, at least, to point out one of the negative rules: a plant's name is not related, *in principle*, to its activity signature. Given that most plants have more than one "use," it is unlikely that plant names could follow any rule (or "doctrine") whereby the name directly indicated all the diverse uses (whether medical, magical, or otherwise).

A considerable number of Ka'apor magical plants, therefore, do not conform to the doctrine of signatures. This is because a plant can be named for any of several properties. For any given plant named in the Ka'apor language, its medicinal or magical uses (as *puhan*) may constitute but one dimension of a very complex activity signature.

Ka'apor amulets, charms, and talismans correspond fairly well to the old (i.e., pre-Freudian) notion of fetish. Fetishism, according to E. B. Tylor, pertains to "the doctrine of spirits embodied in, or attached to, or conveying influence through, certain material objects" (1958:230). The term *fetish* in English, incidentally, was borrowed from Portuguese. Portuguese sailors in the fifteenth century called the amulets and talismans of West African peoples with whom they came into contact *feitiços*. The Portuguese term itself derived from Latin *factitius*, meaning 'magically artful' (Tylor 1958:229; also Marett 1957:201). Although Marett (1957:202) suggested that the term applied only to "a limited class of magico-religious objects in Africa" and should therefore be expunged as a technical term from ethnology, it is clear that a similar idea is at work in Ka'apor ethnobotany. The term *fetish*, therefore, will be used to describe the magical plants employed by the Ka'apor for catalytic functions. All of these functions are classified, moreover, under the rubric *puhan* 'medicine.'

Love fetishes. The leaves of two folk species, *pirapišíka'a* and *yukunaka'a* (referring, respectively, to the aromatic herbs *Justicia* spp. and *Ardisia guianensis*) are used by young, generally unmarried men to attract female sexual partners. These fetishes, both of which contain the essential oils known as coumarins, are described by the generic term, *inamohar-puhan* 'girlfriend-medicine'. It is believed that by concealing the dried, crushed leaves of these aromatic species on one's person, women will become *ka'u* ('horny' in this usage) for the bearer. Male informants specifically requested that I not reveal this use to any Ka'apor woman. On the other hand, when I asked female informants what was the use of these same plants, which I showed them without revealing any "secret," they unabashedly replied *a'e inamohar-puhan* 'it is girlfriend-medicine'. This episode supplies some evidence for a genuine lack of secret knowledge about plants confined to one of the sexes. This lack of gender-confined plant knowledge is not surprising, given that the Ka'apor have no men's houses, sacred flutes, trumpet cults, secret societies, or other evidence, in their social organization, for a formal dichotomy in knowledge between the sexes.

Hunting fetishes. Fetishes for killing game are in comon use and are applied in several ways. Some of these are called 'dog-medicine' (*yawa-puhan*), conferring, when properly applied, it is believed, a special ability to hunting dogs for finding given species of game. These medicines work by contagion. In most cases, some external property of the plant resembles a feature of the game animal in question. Odor is one of these. Several are believed to 'smell' (*piher*) like the game animal against which they are active. One of the highly aromatic folk species used to attract girlfriends is also believed to be active on game animals, *pirapi šíka'a*. It is combined with the dried body of *wira-puru* (*Hylophilus ochraceiceps*, the brown-headed greenlet of the vireo family), and then wrapped in a marantaceous leaf and concealed in the hunter's pocket. It is said that this *so'o-puhan* 'game-medicine' will cause an agouti to stop in its tracks, attract red brocket deer, and lure howler monkeys to descend from the treetops to within shooting range of the hunter. Use of the brown-headed greenlet is clearly a sympathetic device, since this species is a creator of mixed flocks. Its song, according to informants, attract several other species. When I have seen it, it was surrounded by other birds, in search of insects.

A dog's necklace may contain wood wrapped in a cloth of *makahi-mira* 'collared peccary–tree', which denotes three species in the annona family (*Duguetia yeshidah*, *Duguetia* sp. 2, and *Ephedranthus pisocarpus*). It is said that the wood of these species 'smells' (*piher*) like the hide of collared peccary and that, when attached to a dog's collar, helps the dog to track collared peccaries. Interestingly, the Waimiri Atroari encourage their dogs to sniff bark shavings of a species of *Duguetia* in order to be more effective hunters (Milliken et al. 1992:35–36). Wood from *maha-mira* 'gray brocket deer–tree', which refers to three species of the genus *Ocotea* in the laurel family (*Ocotea guianensis*, *Ocotea opifera*, and *Ocotea glomerata*), is said to smell like the hide of *maha* 'gray brocket deer' (*Mazama guazoubira*). Either rubbing the wood on a dog's nose or putting a piece of the wood on its collar is said to help the dog track this important, yet elusive and uncommon game animal. Finally, flowers of *Psychotria poeppigiana* (*tapi'i-kanami* 'tapir–*Clibadium* fish poison') are wrapped in a piece of cloth and affixed to a dog's collar so that it may more easily find the enormous, highly desirable, and decidedly uncommon tapir. Only one plant species is useful specifically for hunting agoutis. This is *Celtis iguanea*, a rare treelet of old fallow called *akuši-ka'a* 'agouti-herb'. In this case, the hunter 'rubs' (*kitǐk*) the leaves on his head.

It is not surprising that there is a fetish for capturing one of the most important game species in Ka'apor ritual and culture, the yellow-footed tortoise (*Geochelone denticulata*) (Balée 1985). This fetish is *yaši-mira* 'tortoise-tree' (*Coccoloba* sp. 2). A seed from this species on the necklace of a boy is believed to help guarantee that upon marriage, when he will be required by the norms of society to capture tortoises for his menstruating or postpartum wife, he will become a good tortoise hunter.

Fetishes used for hunting one of the largest (and most common) game animals, red brocket deer, *arapuha* (*Mazama americana*), are disproportionately represented. *Coccoloba* sp. 1 (*arapuha-sipo* 'red brocket deer–vine'), a liana whose wood is said to smell like the hide of red brocket deer, is used in hunting this species—the hunter puts a piece of the wood in his pocket. Several celastraceous lianas (*Cheiloclinium cognatum*, *Salacia insignis*, and aff. *Cheiloclinium* or *Salacia*), called generically *maka-wa'ẽ-sipo* 'capuchin monkey-fruit–vine', are also useful for dispatching red brocket deer. The leaves are chewed by the hunter to straighten his aim. A hunter may also bathe in the leaves of *ipe-ka'a* (*Psychotria ulviformis*), an herb of the madder family, to kill more easily red brocket deer.

Prevention of incubus. In order to prevent nightmares, which are specifically defined as the appearance of the incubus *ãyaŋ*, one bathes one's head in leaves of the introduced herb, *Chenopodium ambroisiodes* (*motoroi*). Several species are deployed against waking visions of the *ãyaŋ*, the occurrence of which is believed to kill the viewer. These are *kirihu'i'pitaŋ* (referring to several species of the cashew family, *Astronium* cf. *obliquum*, *A. lecointei*, *Thyrsodium* spp.), *mirapirerhẽ'ẽ'i* (*Pradosia* of the sapote family), and the cultivated *mikurka'a* (*Petiveria alliacea*). In addition, a number of plant species are used in protecting children, who are considered to be especially vulnerable, from visions of the incubus. *Caladium picturatum* (*ka'aropinī*), a terrestrial high forest herb, is transplanted to the dooryard garden for this purpose (Balée and Gély 1989:139,142); *Petiveria alliacea* is also planted in the dooryard garden for its apotropaic effects. Also to prevent visions of *āyaŋ*, children's necklaces often contain dried resin of *kirihu'i* (*Trattinickia burserifolia*, *T. rhoifolia*, and *T.* sp. 1 in the bursera family) wrapped in cloth.

Fetishes for catalyzing speech in infants. In a society where, until recently, about 2 percent of the population is congenitally deaf, parents tend to be concerned about whether their newborns will speak. Evidently, infant deafness was attributable to widespread congenital syphilis, which was discovered in 1978 and has since been treated (Gomes 1988:94). Two folk species are used to catalyze speech in infants. These are *ka'a-ro-tiapu* 'herb-leaf–loud' (referring to *Lindackeria latifolia* and *L.* sp. 1) and *ka'a-ro-yĕ'ĕ-hātā* 'herb-leaf–speech-hard' (*Moutabea* sp. 2). The leaves of these species are rubbed on the infant's lips while the parent recites an invocation (the precise form of which I did not record, although I heard it—it resembles poetry) referring to the child's future speech.

Fetish for children to walk. The seed of *Astrocaryum* sp. 1 (*yu'i-pihun*) is placed on the infant's necklace so that he or she will walk at the appropriate age.

Fetish for children to sleep. The spiny *Acacia multipinnata*, a vine of secondary growth, is hung above the infant's hammock to soothe it to sleep.

'Medicine' for insanity. 'Insanity' (*ka'u-ha*) in the Ka'apor nosological system derives from taboo violations, specifically if and when the parents of a newborn eat some forbidden food during the couvade or when a young woman in some way violates the taboos that surround the first menses and subsequent menstruations. Several 'medicines', called generically *ka'uha-puhan* 'insanity-medicine', are deployed to restore a person's normal mental state. These include *ka'uwapusan* (*Siparuna amazonica*), the leaves of which are rubbed on one's head. Also, the leaves of the fern *wariruwai* (*Lomariopsis japurensis*) may be tied over the eyes for this purpose. The leaves of *Phytolacca rivinoides* (*ka'ariru*) are tied about the head to cure insanity. In a curious example of a plant which is seen to be inherently dangerous as a cause of insanity (see table 5.2), *Solanum leucocarpon* (*āyaŋaramira*) is also considered to be a cure for the same by some informants.

Homicide fetishes. Fetishes used for murder of personal enemies (called *anam-puhan*, which at one level of translation could mean 'relative-medicine'—see chapter 2—but which is best glossed as 'other Ka'apor-medicine') are few in number. Three folk species, *anī-ran* (*Dracontium* sp. 1), *apo'i* (*Clusia* sp. 1), and *wariruwai* (*Lomariopsis*) are believed effective if planted in the clay floor of the victim's house. At least two of these, *Dracontium* and *Lomariopsis*, have internally toxic principles. *Lomariopsis* in particular has been shown by recent laboratory assays to contain cyanogenic substances which are highly toxic (Klaus Kubitzki, personal communication, 1986). It is possible that in the past someone accidentally ate this plant and died. But oral transmission over many generations would have likely distorted its true, internally toxic principle. The principle of contagion is seen to work not merely in an internal sense, of course, but in terms of contact with his possessions—in this case, his house. Finally, aside from unprocessed bitter manioc, the wood of *Pseudyma frutescens* (*aɲwayarmɨra*), is one of the only internal poisons explicitly recognized by the Ka'apor. Its active principles, if any, are unknown. It is said that a sorcerer may put wood shavings of this species in his victim's food (especially the dish of manioc soaked in water called *u'itikwar*).

Fetish for catalyzing plant growth. In spite of the importance of horticulture in the Ka'apor economy, only one plant fetish is here involved. It is believed that the brilliant red, elongated inflorescence of *Norantea guianensis* (*araruhuwairimo*), which is an epiphyte, will help the growth of cotton plants if it is made to grow above a swidden.

'Medicine' to Prevent Rheumatism. Both preventive measures and curative catalysts may be called *puhan.* Tying the vine *Cyclanthus funifer* onto one's leg not only is used to cure a case of rheumatism (*tɨma-risã* 'leg-cold') but is also said to prevent it.

'Medicine' to prevent excessive menstruation. A decoction of the leaves of either of two annonaceous tree species called *pinahu'ɨ* (*Duguetia surinamensis* and *Unonopsis rufescens*), if taken orally by a woman, is said to prevent her from menstruating excessively.

Catalysts for manioc press-making ability. It is said that a boy should rub leaves of *tapeši'ɨ* (*Cephalia* sp. 1) or leaves of *Psychotria colorata* (*tapi'ika'a*), both of which are in the madder family, between his hands so that upon maturity he will be able to fashion well the manioc press (*tapeši*), an exclusive task of men.

Growth fetish. Tapping the hollow stem of *mususɨpo* (*Styzophyllum riparium*), a vine, on a child's legs is believed to stimulate physical growth.

Catalyst for bravery. Rubbing black paint made from the fruits of *Genipa americana* (*yenipa'ɨ*) onto one's face not only contributes to one's invisibility; it is believed to impart courage in war.

Longevity catalyst. A child's necklace normally contains a dried seed of *puru'ãrimo* (*Philodendron* sp. 1) to guarantee longevity.

Cold preventative. It is believed that bathing in water suffused with rhizomes of the domesticated and highly aromatic *Cyperus corymbosus* (*pipirɨwa*) will prevent one from catching cold.

Fever preventative. It is believed that seeds of *pikahuămpŭirimo* (*Mendoncia hoff-mannseggiana* and *Mendoncia* sp. 1) on a child's necklace will prevent 'fever' (*hakuha*).

Plants Used in Medicine

Plants in Ka'apor medicine are used in the treatment of 37 recognized diseases or disease complexes. Most plant medicines consist of decoctions of leaves or bark. They are either taken internally or 'rubbed' (*kitřk*) directly on the body.

Headache. Headache (*ăkă-ahř*) can be treated by four folk species of 'herbs' (*ka'a*): *teyu-pitřm* 'skink-tobacco', *mřkur-ka'a* 'opossum herb', *purake-ka'a* 'stingray-herb' [stinging nettles], and *řra-hu-ka'a* 'bird-big-herb'. With all these species, the leaves are rubbed on the head to alleviate pain.

Fever. Fever (*haku-ha*) is treated by seven folk genera, namely, *teyupitřm*, *ka'apiši'u*, *tapiša*, *ka'uwapusan*, *mřkurka'a*, and *purakeka'a*. Some of the fever remedies are identical in application (and even in species) to the headache remedies. The exact source of relief from one of the medicinal plants, *taši'ř* (*Tachigali myrmecophila*), is actually from an insect species that exists in a state of mutualism with the plant itself: the painful stinging of the ants, *taši* (*Pseudomyrmaex* spp.), which inhabit the petioles and leaves of the tree *taši'ř*, is believed to reduce body temperature. The Ka'apor cut a fresh branch and rub it over the chest and back of the febrile patient; the ants, meanwhile, scramble angrily out of their abodes in the branch, stinging mercilessly.

Cold symptoms. General cold symptoms (*katar*) are treated with three folk species, *sřpo-pirap* (which is phylogenetically very heterogeneous), *kurenami*, and *marakatai* (ginger root). Teas are prepared from these. It has been suggested that ginger root contains bactericidal principles—especially against *Staphylococcus*, *Salmonella*, *Mycobacterium*, and *Trichophyton* (Sousa et al. 1991:309–310).

Decongestant. For relief, the congested individual snorts the ground bark shavings of *tapi'irăkwăipe'ř*, a tree of the elm family, or snorts resin of one of two species of the bursera family, which are called *kaneiape'ř*.

Sore throat. For treating sore throat (*yurukwa-ahř*), four folk species are used: *tuk-wămřra*, *marakatai*, *kuyer'řpuku*, and *kuyer'řpu'a*. In the case of the tree of the nutmeg family, *tukwămřra*, the patient drinks the red sap. With ginger root (*marakatai*), a tea is made from the root. In the case of the two species from the dogbane family, *kuyer'řpuku* (*Ambelania acida*) and *kuyer'řpu'a* (*Lacmellea aculeata*), known in Portuguese as *pau de col-her*, that is, the 'spoon tree', a patient should drink the sweet white sap.

Stomachache. To provide remedies for stomachache (*pusu-ahř*) or diarrhea, the Ka'apor prepare teas from the pounded and 'ground barks' (*pirer-tikwer*) of several tree species, including *teremumira* (*Anaxagorea dolichocarpa*), *kurupiši'i* (*Croton* spp.), *kirihu'i* (*Trattinickia* spp.), *taši'i* (*Tachigali myrmecophila*), *taši'iran'i* (*Sclerolobium* sp. 20), *parawa'i* (*Eschweilera* spp.), and the liana *yašisipope* (*Bauhinia* spp.). *Chenopodium ambrosioides*, the introduced herb called *motoroi*, is used for its putative effect of relieving stomachache; the liquid derived from the pulverized stem is sipped. In addition, a tea made from *Chenopodium*'s leaves is used for diarrhea. Although Ka'apor informants do not claim that *Chenopodium* is a vermifuge, the plant does contain ascaridol, an extremely effective toxin against intestinal worms (Sousa et al. 1991:382–383). Finally, the water of a malpighiaceous liana, *Stigmaphyllon hypoleucum*, is drunk for stomachache. This is taken directly from the liana and drunk immediately without additives.

Kidney stones. A tea from the shrub *ita-mira* 'kidney stone-tree' is made from the leaves of *Phyllanthus urinaria* (which is not a tree, but a small herb) for treating kidney stones (*ita*) and kidney infections in general. This species and several of its conspecifics are widely used in Brazil and Amazonia by rural people for the same purposes. Recent research suggests that the anti-spasmodic effect of certain (as yet unidentified) substances in *Phyllanthus* may be responsible for a genuinely remedial effect with regard to kidney stones; these species also appear to be effective against viral hepatitis (Sousa et al. 1991:377–378).

Vermifuge. The Ka'apor know traditionally only two folk species with vermicidal properties: *karawata* (several species of bromeliad) and *yeta'i* (*Hymenaea parvifolia*), one of the copal trees. The seeds of *karawata* were in the past eaten for killing worms. In the case of *Hymenaea parvifolia*, its burned resin is taken orally with water.

Skin eruption. This is one of the largest categories of the Ka'apor nosological system, not in terms of ethnotaxonomic types of diseases, but in terms of the number of therapeutic folk taxa and biological species. Skin eruptions of all kinds are generically called *pere*, which is synonymous with *kuru*, from which Portuguese derived *pereba* and *coruba*. The Ka'apor use 19 species in several different ways to treat *pere*. The most common treatment consists of applying crushed inner bark and water on the eruption. *Sipo-piraŋ* (*Pristimera, Salacia, Banisteriopsis, Tetrapterys,* and *Memora*), *eremeni* (*Ampelozizyphus amazonicus*—the bark of which is ground and boiled before application), and *pere-pusan-'i* 'skin eruption–remedy-stem' (*Simaba* aff. *cavalcantei* and *S. cedron*) are applied in this way. Latex from *irakiwaran* (*Asclepias curassavica*) or *kuyer'i* (*Ambelania* and *Lacmellea*) in the dogbane family is also topically applied. Root shavings of the aromatic *mira-wawak* (*Sagotia racemosa*); fresh crushed leaves of *yaŋwate-ka'a* (*Selaginella*); several composites called *teyupitim* (*Conyza, Porophyllum,* and *Pterocaulon*); burned leaf of *tatu-ruwai* (Polygonaceae sp. 1); and stem water of *kupi'ipu'ã* (*Euphorbia*) and the orchid *a'ihupako* (*Trigonidium acuminatum*) are all useful, according to informants, in the treatment of skin eruptions.

Cuts and burns. Droplets of stem sap from purge nut (*Jatropha* spp.) are used to cure minor cuts, abrasions, and burns.

Cicatrizant. Remedies to aid in the rapid closing and healing of deep cut wounds, which are fairly common among a people who are constantly using axes, machetes, and knives in sundry chores, are very important in Ka'apor society. Many such remedies are in frequent use, regardless of new Western remedies. Foremost is the remedy made from the inner bark of all species of *Eschweilera* (*matá-matá* in Portuguese), generically called *parawa-'i* 'mealy parrot–stem'. The bark is stripped off the tree and wrapped around the wound, like a bandage. I have seen flowing blood from a deep machete cut wound on a young Ka'apor man stop very quickly after use of one of these species. It is described as a *huwi-hu-puhan* 'blood flow–copious–remedy.' Also used in the same way is bark of the related *Gustavia augusta* (*mitüpusu'i*). Resins from copaíba trees (*kupa'i*), *Copaifera duckei* and *C. reticulata*, are also applied directly to wounds as cicatrizants. Shavings from the petiole of the inajá palm (*Maximiliana maripa*) are also applied directly to wounds.

Boils and pimples. Latex from *kaŋwaruhu-mira* 'paca-tree', referring to *Tabernaemontana* and *Agonandra*, is rubbed on boils or pimples (*yaŋoyar*). Irritating leaves of *Solanum rugosum* are also applied on boils and so are bark shavings from *Pseudyma frutescens* (believed to be poisonous if eaten—see table 5.2).

Moles. The Ka'apor find moles unsightly, and claim that latex of *Asclepias curassavica*, when properly applied, will cause them to go away.

Itching. Itching (*yuwar*) can have a wide range of causes and therapies, many of which were identified in the section on skin eruptions. But two species are especially useful for alleviating the irritation of jock itch and athlete's foot. These are root shavings of the aromatic *mirawawak* (*Sagotia racemosa*) and bark shavings of *teyuka'a* (*Galipea trifoliolata*), which are rubbed directly on the irritated spot.

Mouth sores. Sores at the corners of the mouth, which in Portuguese are 'boqueira', are called *yuru-pere* 'mouth–skin eruption'. The treatment of choice for this condition, which usually affects children, is topical application of the reddish sap of *Virola* spp. (*urukiwa'i* and *tukwāmi'u'i*). Also, sap from *Licania* spp. (*mukuku'i*) of the tropical rose family may be used. (It is not clear whether this condition may have a nutritional basis.)

Botfly. The cure for botfly infections is smothering the infected area with tobacco.

Measles. Although measles is technically incurable, the Ka'apor use two species, claiming that, when rubbed on the body, these reduce the severity of the infection. One species is turmeric (called *tawa* 'yellow'); the yellow pigment to be rubbed on the body is drawn from the rhizomes. Turmeric does have many bactericidal properties similar to those of its close relative ginger root (see Sousa et al. 1991:309–310); for this reason, it has been included here as a medical not a magical treatment of a disease. Crushed and boiled bark of the *eremeni* vine (*Ampelozizyphus* in the Rhamnaceae) is also used to treat measles topically, although its active principle, if any, remains unknown.

Toothache. Toothache is a common malady among the Ka'apor. Six forest species are used in controlling toothache; these act mainly as painkillers and do not appear to fight decay. The drug of choice consists of the rhizomes of *yamɨr* (*Piper ottonoides* and *Piper piscatorum*), a small forest herb. I have tried this remedy. The effect is sharp stimulation and tingling of the mouth and tongue for several minutes after initial tasting. Resin of *kanei'i'tuwɨr* (Port. *breu branco*), referring to *Protium giganteum*, *P. pallidum*, and *P. spruceanum*, are also used for this purpose, being rubbed directly on the tooth and gum. Crushed leaves of the citrus family member, *Zanthoxylum juniperina*, a tree of old fallows, may also be applied topically.

Earache. The Ka'apor ethnomedical system has evolved three folk species used in the treatment of earache, which affects children especially. These are oils, topically applied, from *kupa'i* (*Copaifera duckei* and *C. reticulata*) and *kumaru'i* (*Dipteryx odorata*). In fact, the use of *Dipteryx odorata* oil (or tonka bean oil) is widely used in Brazil for earache; its active principle is coumarin (Sousa et al. 1991:250). The ground inflorescence of *tatu-ruwai* (Polygonaceae sp. 1) may also be placed in the ear.

Sore eyes. Sore eyes are treated by applying resins or saps from five species directly to the eye. These medicinal species include four vines and one tree. The vines are *iraí-sipo* (*Schubertia grandiflora*), *kawasu-ran* (*Cayaponia* sp. 1 and *Gurania eriantha*), and *parawa-sipo* (*Uncaria guianensis*). The tree species is *yeta'i* (*Hymenaea parvifolia*), a copal tree.

Snakebite. Although modern medical procedures for treating snakebite victims are based on the use of horse serum, it is scientifically plausible that several plants contain active substances which work either to antagonize certain properties of snake venoms or to effect some immunological changes in the victim (Mors 1991). Plant remedies for snakebite used by the Ka'apor come from two species, a tree of the Brazil nut family, *Gustavia augusta* (*mitũpusu'i*) and a tree of the fava bean family, *Poecilanthe effusa* (*moi-mɨra* 'snake-tree'). Both species are used in the same ways for treating the bites of pit vipers, such as bushmasters (*Lachesis*) and fer-de-lances (*Bothrops* spp.). A decoction of the inner bark is taken orally over the course of several days; wet bark shavings may also be rubbed on the site of the wound. Informants claim that such therapy does reduce the tissue damage occasioned by such a bite. It is interesting that the Guajá also use *moimɨra* for snakebite; they call it, however, by a completely unrelated name, *arakowã'i*. This raises the possibility that the Ka'apor and Guajá independently discovered the uses of this plant. It is certainly not a case of doctrine of signatures, since the plant bears no likeness to a snake. The name *moimɨra* is nonarbitrary only insofar as it designates a tree whose use is treating snakebite; it is perfectly arbitrary as far as "resemblance" is concerned.

Pregnancy prevention and termination. Unwanted pregnancy is not unique to societies undergoing the demographic transition. The preferred Ka'apor family size, however, follows no explicit rule. Women appear to decide whether they want to maintain or terminate a pregnancy according to highly individualistic criteria. Two plant species are used either to prevent conception or as abortifacients. These are *Abuta grandifolia*—which is widely used as an abortifacient in Amazonia and appears to contain alkaloids (Rocha et al. 1968, cited in Milliken et al. 1992:89)—and *Symphonia globulifera*. A woman drinks a decoction of the woody *Abuta* or she drinks sap from *Symphonia* to abort. It is claimed that *Symphonia* sap may be taken orally over a course of three days to prevent menstruation (*yai'ïm puhan*) and thereby extend the infertile phase of the menstrual cycle.

Birthing aids. Three folk species are used to alleviate labor pains or to speed delivery of a woman already in labor. These are *yakare-ka'a* (referring to the ferns *Ctenitis pretensa* and *Pteridium aquilinium*), *mani'ï-se* (a very old cultigen of bitter manioc), and *ape'i'tuwir* (referring to four species of *Cordia*). In the case of *yakareka'a* and *ape'i'tuwir*, the leaves are rubbed (*kitïk*) on the parturient woman's belly. As for *mani'ïse*, the skin of the tubers is rubbed on the belly for speedy delivery.

Postpartum hemorrhage. Two species are used to prevent or stop postpartum hemorrhage. A decoction from the leaves of *mira-ki'ï* (*Myrciaria tenella* in the guava family) is taken orally by a mother shortly after delivery. A decoction of the roots of the sturdy vine *Cyclantus funifer* (*sïpo-hu*) is drunk over the course of three days after birth by both father and mother so that the mother will not hemorrhage.

General tonic. Plants used as tonics (i.e., therapies for general well-being and strength) are called *he'õ-ha-puhan* 'weary-agentive–remedy'. Five folk species are used as tonics, *yašisïpope* (*Bauhinia* spp.), *kururu'iran'ï* (*Drypetes variabilis*), *parani'i* (*Laetia procera*), *wari-mira* (*Clarisia racemosa*), and *tiriri-sïpo* (several water vine species of the dillenia family). As tonics, a root decoction of *yašisïpope* is taken orally; a bark decoction of *kururu'iran'i* or *parani'i* is taken orally; root shavings of *warimira* are taken orally; stem water from *tirirïsïpo* is drunk.

Strength tonic. A decoction of root shavings from the kapok tree, *wašiŋi* (*Ceiba pentandra*), is taken orally to strengthen a person weakened by illness.

Dog tonic. Tonics for dogs are saps from *Himatanthus sucuuba* (*kanaú'i*) and purge nut (*Jatropha curcas*), *piyã'i*, which are added to the dog's food. Bark shavings from the ant tree, *Tachigali myrmecophila*, may be also added to a dog's food.

Dog mange. Three folk species are used to treat dog mange. These are *kaŋwaruhumira* (*Tabernaemontana* and *Agonandra*), *aŋwayarmira* (*Pseudyma frutescens*), and *kanaú'i* (*Himatanthus sucuuba*). Latex from *kaŋwaruhumira* and *kanaú'i* and a decoction of bark shavings from *Pseudyma* are topically applied. It is interesting that *Pseudyma* is believed to be a gastric poison (see "homicide fetishes").

Insect repellent. Oil from *Carapa guianensis* (*yaniro'i*) is applied to prevent insect bite, especially the bites of blackflies (*Simulium* spp.), known in Ka'apor as *mariwĩ* and in Portuguese as *pium*.

Excessive menstrual discharge. The Ka'apor believe that excessive menstrual discharge (*yai-hu*) is an illness. To prevent it, a woman may take orally a decoction of *Abuta grandifolia* (which, as noted above, may also in appropriate quantities prevent or terminate pregnancy, according to informants), or she may ingest droplets of resin from *Hymenaea parvifolia* (*yeta'i*).

Regulating urine color. It is believed that drinking a decoction of the ground rhizomes of turmeric will make the urine less yellow.

Rheumatism. Rheumatism, *tima-manõ* 'leg-numb', is treated with the still hot, boiled leaves of *mani'i'se*, a variety of bitter manioc, which are applied directly to the affected area.

Bloody urine. It is believed that drinking a decoction of the bark of *yašimukuku'i* (Annonaceae sp. 1, Flacourtiaceae sp. 1, and *Neoptychocarpus apodanthus*) can cure 'bloody urine' (*sirukwa*).

Historical Ecology and Ethnomedical Knowledge

The traditional knowledge of medicinal plants in Ka'apor is much greater than that of the Waorani of Ecuador (Davis and Yost 1983a, 1983b). This is probably not because the Ka'apor may have had more contact with non-Indians and the diseases they brought with them from the Old World than did the Waorani. Other aspects of the Waorani ethnobotanical system, as reported by Davis and Yost, also indicate a certain truncation of the potential repertoire of useful plant species, in comparison with the Ka'apor and other, more settled indigenous peoples of Amazonia, such as the Wayãpi. For example, while the Ka'apor recognize 179 and the Wayãpi recognize 129 nondomesticated species of plants as being edible (see chapter 4 and appendix 8), the Waorani recognize only 55 nondomesticated species—this is from a people for whom "most of their protein and virtually all of their minerals and vitamins" come from the forest (Davis and Yost 1983b:162; also Davis and Yost 1983a:275). The forest inhabited by the Waorani, moreover, is one of the botanically most diverse in Amazonia if not the world (Gentry 1988).

It has been shown that a relatively small percentage of the medicines of the Ka'apor are used to resolve post-Columbian diseases. Most are used, rather, in therapy of diseases native to Amazonia. It seems likely that, with regard to the Waorani, one is witnessing not a pristine system of indigenous medicine or even indigenous ethnobotany in its totality, but rather a regression of knowledge due to the Conquest and its many aftershocks (see chapter 8).

First contact with diseases of the Old World, of course, tends to bring major population losses for indigenous groups. The Waorani supposedly were free of post-Columbian diseases until quite recently (Davis and Yost 1983a; Larrick et al. 1979). But their population density—660 people distributed over 20,000 km², or .033 persons per km² (see Davis and Yost 1983a:275)—is approximately one-sixth of that of the average indigenous population of modern upland Amazonia, at .2 persons/km² (e.g., Denevan 1976; Hames 1983). It is also now clear that smallpox epidemics as well as other alien epidemics ravaged Ecuadorian Amazonia beginning in the early 1600s (Newson 1992:104–107).

It is likely that the Waorani are (or recently were) a trekking people (Yost 1981a), by which I mean a society that spends six or more months per year away from a central settlement site. It is also plausible that agriculture did not simply come late to the Waorani (cf. Yost 1981a), but rather that they regressed, at some time in the past, from a more sedentary lifestyle—perhaps because of the pressures, direct and indirect, of contact. Yost (1981b:682) describes the traditional settlement pattern as "semipermanent sedentarism," pointing out that the Waorani trek between gardens throughout the year, not unlike the trekking Hotí of Venezuela (as discussed in chapter 8). That the Waorani have only 19 domesticates or semi-domesticates (Davis and Yost 1983b:164) further suggests a trekking lifestyle, one that may have succeeded a more sedentary agricultural society. Finally, the Waorani appear to have grown no bitter manioc in the recent past (cf. Davis and Yost 1983b:192–193). Their principal crop is sweet manioc which, unlike bitter manioc (Balée 1992a) can be successfully cultivated by a trekking society.

Regardless of the generative reasons, the profoundly low demographic density of the Waorani reduces the breadth of interpersonal contacts upon which lexical entries about the biological world would depend. In other words, it can be argued that differences of historical ecology lead to differences of mental economy among lowland South American cultures, a point I address in more detail in chapter 8.

For now it is sufficient to point out that, in addition to the trekking Waorani, many small surviving groups of hunter-gatherers such as the Guajá, and groups that are marginally sedentary such as the Araweté, appear to possess truncated inventories of medicinal as well as other useful plants, in comparison with the sedentary Ka'apor. The Guajá and Araweté also probably regressed from more sedentary and more agricultural lifestyles. Key evidence is that they both speak natural languages of the Tupí-Guaraní family, the mother language of which, proto–Tupí-Guaraní, was associated with an agroforestry complex that controlled many domesticated plant species (Balée 1989b, 1992a).

The Conquest, respiratory and other diseases from the Old World, and subsequent severe depopulation thus pushed some sedentary groups into the more nomadic lifestyles of hunter-gatherers or trekkers—or put an end to them altogether as unique societies. Although seriously depopulated after 1928, the Ka'apor did not lose their agricultural lifestyle; a significant part of their traditional knowledge about medicinal plants has apparently survived. Such was not the fate of the presumable plant knowledge of many other groups, probably including most seminomadic trekking societies like the Waorani.

In other words, the relatively small number of medicinal plants in Waorani ethnobotany most likely does not owe to the putative existence of but a few diseases in pre-Columbian times, or to the people's own failure to make use of their environs. Rather, a more likely cause is that diseases in the postcontact period reduced significantly their population, bringing about a more and more nomadic lifestyle, incapable of preserving all ancestral knowledge of medicinal plants.

The Ka'apor, in contrast, by maintaining a sedentary lifestyle, have been able to retain a comparatively large repertoire of plants to treat or prevent diseases that existed in the New World prior to 1500. This persistence of ancient knowledge is remarkable in light of the limitations that an oral tradition places on the transmission of rationalistic knowledge about nature.

Indigenous Forest Management

Although not by conscious design and intent, traditional Ka'apor society may be much closer to a state of ecological equilibrium than disequilibrium with the environment. By ecological equilibrium I mean that Ka'apor society exhibits a small population density, basically a subsistence (or domestic) economy, economic self-sufficiency through use of local organic and inorganic resources, a technology whose principal sources of energy are solar, fire, and human muscle; virtually all Ka'apor are in constant contact with the living, nonhuman environment, and they (at this time) have a minimal impact on this environment—they do not, for example, convert the forest to cattle pasture (see Bennett 1976; Sponsel 1992).

In contrast to the Ka'apor, societies in disequilibrium with their environments typically manifest high population densities, market economies, dependence on and importation of resources from distant peoples and lands, technologies that depend mostly on nonrenewable resources (such as fossil fuels), relatively few individuals in daily contact with the environment, and massive impacts on biological and ecological diversity in various habitats (Bennett 1976). In a sense, the contrast between societies basically in or out of equilibrium with Amazonian environments may be understood in terms of fundamental differences between stateless (egalitarian) peoples and those incorporated into a nation-state society (Balée 1992b; Sponsel 1992).

My purpose here is to show that the Ka'apor effectively manage their environment and that they have done so throughout their history. For purposes here, I define management rather stringently. *Management is the human manipulation of inorganic and organic components of the environment that brings about a net environmental diversity greater than that of so-called pristine conditions, with no human presence.* My definition is thus similar to Raymond Hames' (1987:93) definition of resource management: "activities that enhance the environment of game animals with the effect of increasing their abundance." By not mentioning plants, however, the Hames definition of management is narrower than the one I use.

"Management" thus transcends the usual binary distinction between "preservation" and "degradation" of Amazonian environments. First, unlike preservation, which implies human noninterference with other species, management involves direct and indirect human interference in species' populations, distribution, and behavior. In a managed ecosystem, some species may become locally extinct as a result of human interference, even though the overall effect of such interference may be a net increase in the ecological and biological diversity of the particular locale or region. Management also differs from degradation or pollution (see Rambo 1985), since it does not perforce defile the pre-existing, nonmanaged environment. Indigenous resource management may, in fact, enhance living conditions for many species, including the human one, even if such enhancement is unintended.

Neither Preservation nor Degradation

Some confusion appears to have arisen as to whether indigenous Amazonian societies tend to manage—that is, effectively increase the local or regional diversity of—biological resources of their immediate environment (e.g., Denevan and Padoch 1988; Moran 1990; Posey and Balée 1989). Redford (1991:46), for example, has made the well-intentioned argument that evidence for past Amazonian societies suggests deleterious effects on virgin forests and plant and animal species therein: "These people behaved as humans do now: they did whatever they had to feed themselves and their families" (see Sponsel 1992 for an insightful critique). In general, Redford (1991) avowed that Amazonian Indians are no different from Westerners in their desire for commodity goods, that they will be eventually assimilated into Western society, and, in fact, that they represent a threat to the maintenance of biological and ecological diversity in Amazonia. Specifically in relation to modern indigenous peoples, Redford (1991:47) declared, "They have the same capacities, desires, and, perhaps, needs to overexploit their environment as did our European ancestors." Redford's views with regard to indigenous, nonstate societies of Amazonia echo Rambo's (1985) views on the Semang, an egalitarian people of the Malay Peninsula. Unlike Redford's article, Rambo's treatise seems explicitly biased in places. For example, he writes (1985:43), "As jack-of-all-trades opportunistic foragers, the Semang are not particularly skillful at any specific activity"; also, "The Semang themselves . . . are not systematic thinkers" (1985:47). These observations, incidentally, are largely based on Rambo's (1985:6) total of only one month and fourteen days of fieldwork, spread over three visits between 1975 and 1978, among the Semang. In a perceptive critique, Laderman (1987:704) points out that equating the environmental impact of the Semang with that of modern civilization is "to encourage the very greed that Rambo has elevated to an existential necessity of social life."

Like Redford, Rambo implicitly invoked the assumed unity of humankind to argue that indigenous peoples, no different from Westerners, degrade natural environments. He (1985:1, 79–80) even equated "environmental change" effected by all human societies with "pollution." Both Redford (1991) and Rambo (1985) independently suggested that environmental changes induced by human beings, regardless of societal type, represent the nemesis of nature.

This notion is perhaps comprehensible in the modern context of increasing threats to the survival of nonhuman species. Myers (1988) predicted that unless forceful conservation measures are implemented soon, more than 17,000 or about one half of the 34,200 endemic plant species in "hot spots" of tropical biodiversity will be extinct by the year 2000. This predicted extinction would entail more than 13 percent of all tropical forest plant species worldwide. Myers's estimate for the number of animal (mostly insect) species immediately facing extinction in these areas is 350,000. This coming extinction 'spasm' will be like nothing seen on earth since the mass extinctions of plants and animals in the late Cretaceous, some 65 million years ago. This extinction spasm is, of course, a direct result of certain human activities. A recent report to the National Science Board declared that the extinction spasm of our time "has been caused by a single species" (NSB 1989:1). The human species itself, regardless of profound societal differences within it, is frequently seen to be the juggernaut of modern extinctions on earth.

Perhaps one ought to consider establishing more biological reserves with few or no people in them (e.g., Redford and Stearman 1989). If it is true that all stateless societies of Amazonia will sooner or later be integrated into the wider society, however, one questions whether a monocultural state would be a better custodian of forested ecosystems than indigenous, stateless peoples of our time. After all, state societies, not indigenous stateless ones, are in the first instance responsible for these continuing ecological crises.

It is evident that much of the commercial hunting in Amazonian forests (as of caiman for their skins) is unsustainable—including that conducted by modern forest dwellers (Redford 1992; Redford and Stearman 1989; Stearman and Redford 1992). The same can be said for commercial logging by indigenous groups, including the Guajajara (Balée 1990) and Kayapó (Turner 1992). These harmful activities implicate the state-level economic and political conquest in parts of the Amazonian hinterland—not traditional resource management practices by the forest dwellers. These forest dwellers have partly become, in effect, constituents of a society fundamentally in disequilibrium with the environment. Placing blame with the Indians is tantamount to blaming the victim. The frontier moved toward them, not they toward it. The indigenous groups and individuals now involved in some aspects of unsustainable exploitation of natural resources did not autocthonously evolve this way of life. Such exploitation is, moreover, nowhere near characteristic of all modern forest dwellers in Amazonia (Milton 1991).

Existing Amazonian forests have survived in spite of indigenous perturbations. What is more, the presence of luxuriant canopies rich in plant species implies that faunal populations also retained their health, because the vegetation depends on animals for pollination and seed dispersal (Redford 1992). Today's lush forests are thus evidence of highly sustainable resource management by forest dwellers of the past— whose populations and consequent environmental effects would have been far greater in pre-Columbian times than they are today (Balée 1989; Piperno 1990; Roosevelt 1991; Denevan 1992).

To make a conceptual distinction between state and nonstate patterns of natural resource exploitation is not, therefore, tantamount to elevating (or demoting) the peoples of nonstate societies to the condition of Noble Savage. Indigenous management has had a cumulative and noticeable impact on biotic and abiotic environmental resources, such as the atmosphere, water, soils, and biodiversity of flora and fauna. But this impact is quantitatively (and, in the case of biodiversity, qualitatively) distinct from modern state societies. For example, Amazonian Indians probably through the centuries have released only minute amounts of carbon dioxide and other greenhouse gases into the atmosphere by the burning of their swidden fields, in contrast to modern states, whether of the First or Third Worlds. Carbon dioxide in the earth's atmosphere rose from 315 parts per million to about 340 parts per million only within the last generation (NRC 1983:1; cf. Tirpak 1990:41). Indigenous Amazonian societies, however, reached their climax, in terms of use of fire energy and forest burning, long before this, that is, prior to the Conquest.

Effects on Rivers and Soils

Although the Ka'apor and many other Amazonian societies use plant poisons to augment the efficiency with which they capture fish, these poisons in the indigenous context do not result in irreversible damage to aquatic wildlife. First, these poisons rapidly dissolve. They do not usually appear to have effects beyond about one kilometer downstream on small creeks. Larger rivers, such as the Gurupi, would dissolve these poisons so rapidly that the Indians do not attempt to use poisons there. Second, no evidence exists for local extinctions of aquatic wildlife in rivers of the Ka'apor habitat, regardless of the deployment of indigenous fishing technology. Ka'apor informants claim that, on average, within three to four days after a fish poisoning expedition, the same stretch of stream originally affected regains its fish stocks.

The Ka'apor do not, as a rule, fell forest along creek margins, if only because the most significant plant domesticates (in terms of area planted and calorific production) in Ka'apor swiddens, such as manioc, sweet potatoes, and yams, cannot tolerate waterlogging (cf. Chernela 1982). This avoidance of streamside agriculture is important in the blackwater river systems the Ka'apor inhabit, since many fish, especially characins relied on for food and for bait, in turn depend on fruiting floodplain trees (Goulding 1980; Gottsberger 1978).

The Ka'apor never dispose of their bodily wastes in the nearby river or creek, but rather in given locales of either high forest or old swiddens, devoid of edible domesticates. They do, however, discard manioc skins in creeks and along creek margins, upon 'peeling' (pirok) the tubers that have 'soaked' (mani'ok-i-pe-hĩ) for three days in the creek. Rather than lead to eutrophication (which often occurs in truly polluted or excessively fertile streams of many modern state societies), this practice may enhance, in fact, certain desirable aquatic life forms. The Cocamilla Indians of the Peruvian Amazon, for example, dispose of all their garbage in the nearby lake, upon which they rely heavily for fish. According to Stocks (1987:116), the garbage supplements the diets of the particular kinds of fish that are harvested in the high-water period when fishing is otherwise relatively inefficient. In general, indigenous utilization of Amazonian rivers has not brought about eutrophication, species depletion, or streamside deforestation—unlike the effects of state-level society. It should also be mentioned that no indigenous analogy for the mercury poisoning of rivers during the modern Amazon goldrush has ever existed.

Rambo (1985:48–50 et passim) argued that swidden gardening and other agricultural activities of indigenous peoples may cause soil erosion and local air pollution by excessive production of dust, soot, and smoke. While these effects are at least partly obvious, it can also be shown that indigenous agroforestry practices have had the long term effect of enhancing the productivity (or productive potential) of many arable lands in Amazonia. The most noteworthy example of this enhancement effect concerns *Terra Preta do Índio* ('Indian black earth'), an anthrosol which is arguably the most fertile soil type in Amazonia (Smith 1980). This soil horizon, which is charcoal black in color, is associated with prehistoric indigenous societies. For reasons apparently related to a lack of long term intensive occupation in the remote past, the Ka'apor habitat does not display any *Terra Preta do Índio*. Yet arable soils are variable in fertility, and this partly owes to Ka'apor management practices.

Fallows that range between 40 and 200 years since the last planting normally exhibit a layer of about 20 cm or more of what the Ka'apor refer to as *iwi-pihun* ('black earth'). Although this soil type is not technically *Terra Preta do Índio*, and is, in fact, somewhat less fertile than the soils of true black-earth sites I have surveyed in the Xingu River basin (Balée 1989a, 1989c), it is more fertile than the soils of the primary *terra firme* forest of the Ka'apor habitat. Soils in Ka'apor fallows tend to be richer in carbon, nitrogen, phosphorus, and exchangeable bases (in particular, magnesium levels appear to be very high) than the soils in nearby high forest sites. Expressly because of prior indigenous occupation, the soils of fallows are probably more arable than those of high forest, even if not by design.

Many of the indicator trees of the fallow appear to contribute to increased soil fertility. For example, the babaçu palm (*Orbignya phalerata*) seems to recycle soil nutrients and improve soil structure. Upon death, a deep pocket of loose, decaying organics is left in the soil where the palm once stood—an artifact of babaçu's crytogeal germination. The pocket soon collapses, in effect turning the soil (Anderson et al. 1991:71). A caesalpiniaceous tree species common to old fallows, *Dialium guianense*, (Ducke 1949:112; Huber 1909:162) is also a soil enhancer. This pan-tropical, nitrogen-fixing legume is used by farmers in Nigeria "to improve the nutrient status of the fallow" (Whitmore 1992:136). *Dialium guianense* is rare in high forest, but it is the sixteenth most ecologically important species of fallows in the Ka'apor habitat. Some human activities in old fallows may directly enhance soil. Firewood burning and disposal of organic wastes are common activities in the old fallows that surround villages and active swiddens. In spite of prior occupation by forest dwellers, then, whatever erosion may have occurred in Amazonian environments did not result in an overall loss of soil fertility.

Effects on Wildlife

As for indigenous effects on biodiversity, some observers claim that indigenous hunting inexorably leads to declines of, and in some cases local extinctions of, edible game species (Hames 1980; Hames and Vickers 1982; Redford and Robinson 1987; Redford and Stearman 1989; Vickers 1980). This perspective perhaps gains support from explanations concerning late Pleistocene extinctions of megafauna. South America lost more mammalian genera than any other continent in the late Pleistocene, for which human hunting was probably at least in part responsible. Human hunters probably contributed directly to the extinction of enormous, slow-reproducing and low-density species such as gomphotheres, giant ground sloths, and toxodonts, which would have been easy to track and kill (Martin 1984:374,375). These species became extinct around 10,000 to 8,000 years ago.

No evidence exists, however, for extinctions in Amazonia, either of plants or animals, *after* the development of settled village life. Settled village life in Amazonia was certainly associated with the domestication and intensive management of plants. This process of sedentism may have begun at least 5,300 years ago, as suggested by the age of the oldest known phytoliths of maize—a plant that does not grow wild—at Lake Ayauch in Amazonian Ecuador (Bush et al. 1989; Piperno 1990). While no evidence has yet been presented for Holocene extinctions of fauna and flora by Amazonian peoples, one can argue that, on the other hand, indigenous agricultural lifestyles may have actually increased the abundance of certain desirable species.

Ritually regulated hunting behavior by the Ka'apor Indians, for example, rather than leading to declines in game species, appears to increase the densities of some of these near human settlements (Balée 1985; Queiroz and Balée, in prep.). Ka'apor hunting behavior thus conforms well to Raymond Hames' (1987:93) definition of resource management: "activities that enhance the environment of game animals with the effect of increasing their abundance." Ka'apor game management must be regarded within the complex ritual system. Instead of temporal or activity-related proscriptions on important game animals, such as deer, tapirs, and peccaries, which could be conservative of large game (McDonald 1977; Ross 1978), the Ka'apor ritual system prescribes hunters to capture yellow-footed tortoises (*Geochelone denticulata*), which are small game, for their menstruating and postpartum women. Over the long term, compliance with this food prescription appears to focus hunting increasingly far from settlements and swiddens (Balée 1985; Queiroz and Balée, in prep.). This is because tortoises are rapidly hunted out while women continue to menstruate and reproduce.

Thus, large game, such as deer and peccaries, which are the most important protein sources in the Ka'apor diet, and which are strongly attracted to domesticated and non-domesticated plants found exclusively in settlements and swiddens, are not so quickly depleted in the local habitat. The only game species that is depleted (or becomes micro-locally extinct) near settlements is the yellow-footed tortoise. But tortoises still occur in the Ka'apor habitat and are present in the vicinity of new settlements, whereas tortoises are absent in zones of Western colonization and occupation on the present borders of Ka'apor lands. In other words, local extinctions of tortoises in the Ka'apor habitat are not coterminous with regional extinction of this species, yet Ka'apor ritual hunting has taken place in this region for more than one hundred years.

Effects on Plants

Although the Ka'apor ritual system and practices of other Indians may have led to a kind of semidomestication of certain game species in Amazonia, animal husbandry per se never developed. Game management, moreover, was ancillary to or a by-product of plant management and domestication (see Harris and Hillman 1989:2). Many of the domesticated species found in Ka'apor swiddens are from the Neotropics. Several of these are putatively of Amazonian origin: cotton (Stephens 1973; cf. Pickersgill

1989:436); chile peppers, *Capsicum annum* and *C. frutescens*, (Pickersgill 1989:433); annatto (Schultes 1984:21); peanut, *Arachis hypogaea*, (David Williams, personal communication, 1988; cf. Brücher 1989:104); and tobacco, *Nicotiana tabacum* (Brücher 1989:181). (See also Clement 1989).

Many other indigenous domesticates, if not from Amazonia, are believed to be from tropical South America. These species include: pineapple (Pickersgill 1976:16; Schultes 1984:22); cocoyam (Coursey 1968; Hawkes 1989:4; Plucknett 1976; Sauer 1950); guava (Chan 1983); sweet potato (Hawkes 1989:488; Yen 1976); purge nuts, *Jatropha* (Brücher 1989:121); the American yam, *Dioscorea trifida* (Hawkes 1989:489; Rehm and Espig 1984:53); cashew (Mitchell and Mori 1987); and Indian shot, *Canna indica* (Bailey et al. 1976:219). Still other plant domesticates are thought to have originated elsewhere in the Neotropics, either from Middle or South America. Included here are: maize, calabash, and manioc (Allem 1987; Ugent et al. 1986; Hawkes 1989:485–487; Brücher 1989:10); crookneck squash, *Cucurbita moschata* (Heiser 1989:472; Nee 1990); and papaya (Leon 1984; Brücher 1989:222; Rehm and Espig 1984:180).

In most cases, wild ancestors are unknown for these domesticates. Ancestral extinctions may have occurred, but the evidence is inconclusive. In any case, more than one hundred plant species were domesticated in the Neotropics, more than in any other world region. This domestication itself represents a major contribution to current plant biodiversity (Pickersgill and Heiser 1977; Brücher 1989:1–2). The antiquity of several domesticates in tropical South America is great. For example, the cocoyam (*Xanthosoma* sp.) may be between 4,000 and 7,000 years old (Plucknett 1976) and the American yam (*Dioscorea trifida*) is deemed older than 5,000 years (Coursey 1976:71; Hawkes 1989:489). In terms of Amazonia, the most significant recent agricultural finding is the occurrence of maize at least 5,300 years ago, which implies that indigenous people dependent on agriculture have been altering Amazonian environments since antiquity (Bush et al. 1989; Piperno 1990). These alterations have contributed to the biological and ecological diversity of Amazonian forests today.

Precisely how and where this diversification first came about remains unclear. Probably the Amazon basin and surrounding lowlands were home to several independent crop domestications. Amazonia may have been a "noncenter" (Harlan 1971), whereby agriculture evolved concurrently over vast expanses. Clement (1989) made a good case for an Upper Amazonian center of domestication for several tree crops, yet it is unlikely that any of the major domesticated tubers and herbs of lowland South America would have originated there. It seems likely therefore that polyagrogenesis took place, that is, independent domestication of plants in more than one area (Bonavia and Grobman 1989:456).

Domestication occurs "when the reproductive system of the plant population has been so altered by sustained human intervention that the domesticated forms—genetically and/or phenotypically selected—have become dependent upon human assistance for their survival" (Harris 1989:19). It is not always certain, however, that weedy plants coexistent with domesticates have been so altered (cf. Pinkley 1973).

Piperno (1989:549) pointed out that the biodiversity of wild plant resources actually increased in the surroundings of agricultural settlements in the lowland forests of Panama. Pickersgill (1989:436) suggested that weedy species rarely evolve from crop plants; rather, weeds normally precede crops. In any case, it is clear that whether many semidomesticated species are escapes or ancestors of modern domesticates, these species are frequently found across the Amazon basin and into the adjoining lowlands. Pinkley (1973) called these plants anthropophytes; I have referred to them as semidomesticates (Balée 1988b, 1992b). Semidomesticates in Amazonia are not merely weedy, herbaceous "camp followers" (e.g., Anderson 1969; Harlan 1975), but are most often large trees. In this context, one may wish to speak of "weedy trees" (Jack Harlan, personal communication, 1993). I refer here especially to species in old fallow (*taper*) that are not to be found in the primary forest (*ka'a-te*) or other undisturbed habitats. The occurrence of these species in the *terra firme* appears to be due to human intervention, even if this was directed primarily at the cultivation of herbaceous domesticates (see below), not these trees. It is possible that the genetics of these semidomesticates, like the domesticates themselves, have been altered, although there is no direct supporting evidence for such an assertion (Pinkley 1973).

In any case, unless and until proven otherwise, the domesticates and semidomesticates of Amazonia may be seen as having contributed toward an increase, not a reduction, in biodiversity. Evidence exists, moreover, for an indigenous effect on the ecological diversity of Amazonia.

Fallow Versus High Forest and Description of Study Plots

Terra firme forests are the well-drained forests of Amazonia. At least with respect to ethnobotany, one must distinguish two kinds of *terra firme* forest that occur in the region inhabited by the Ka'apor: old fallow (or simply fallow) and high forest. Fallow refers to sites that were once used for agriculture but which are now forested; the disturbance must have occurred from forty to a hundred or more years ago. High forest is not necessarily "higher" than fallow. It is, rather, a forest of the *terra firme* which displays a primary character; it must not have been disturbed for agriculture within the last two hundred or three hundred years, if ever.

Between 1985 and 1990, I carried out inventories of eight separate hectares of *terra firme* forest in the habitats of the Ka'apor, Guajá, and Tembé peoples. Four of the study plots were of high forest, four were of fallow. The methods I used are comparable for each of the eight hectares: (1) all trees greater than or equal to 10 centimeters in diameter at breast height (dbh) [using a standard of 1.3 meters] on each plot were measured, tagged, collected (except in a few isolated instances), and identified; (2) all plots were subdivided into 40 sampling units (or subplots) of 10m X 25m, in order to sample relative diversity (see below); (3) all plots are narrowly rectangular in dimension, being either 10m X 1000m or 20m X 500m; (4) all plots are situated near indigenous villages, but none were the objects, at the time of study, of agricultural activity. These inventory methods are identical to those of many other studies in Amazonia (e.g., Campbell et al. 1986; Boom 1986; Salomão et al. 1988; Gentry 1988).

The eight hectares span a linear distance of 150 km, from the left bank of the Rio Pindaré to the left bank (and Pará side) of the Rio Gurupi. This study area can be considered part of the pre-Amazonian forest (e.g., Projeto Gurupi 1975). Analysis of these

TABLE 6.1 *Eight Study Plots of Terra Firme Forest in Pre-Amazonia (Each Study Plot is One Hectare)*

Plot and Location	Type	Number of Individuals	Number of Families	Number of Species	Basal Area(m²)
(1) P.I. Awá/R. Pindaré	Fallow	506	38	157	22.1
(2) P.I. Guajá/R. Turiaçu	Fallow	563	41	125	21.1
(3) P.I. Guajá/R. Turiaçu	High forest	521	45	145	27.2
(4) Urutawi/(Turiaçu basin)	High forest	519	41	126	25.3
(5) Urutawi/(Turiaçu basin)	Fallow	451	36	95	30.3
(6) Gurupiuna (Gurupi basin)	High forest	467	43	123	30.3
(7) Gurupiuna (Gurupi basin)	Fallow	497	43	141	23.3
(8) P.I. Canindé/R. Gurupi	High forest	475	41	144	34.5
Averages for high forest		496	43	135	29.3
Averages for fallow		504	40	130	24.2
Averages for all plots		500	41	132	26.8

hectares permits one to calculate several important aspects of the pre-Amazonian forest, such as floristic composition, species richness, and physiognomy.

For all eight hectares, and any combinations thereof, one can calculate total basal area (that is, the number of square meters of aggregate plant tissue or wood attributable to individuals of at least 10 cm dbh). One can also calculate basal area of individual species and families, relative frequency (the number of individuals of a species ÷ all individuals of the plot, X 100), relative diversity (the number of sampling units in which a species occurs ÷ all occurrences of all species, X 100), and relative dominance (the basal area of a species ÷ total basal area of the plot). In addition, on the basis of these measures, one can calculate the ecological importance value for each species.

The ecological importance value has been defined as the sum of relative frequency, relative density, and relative dominance (see Greig-Smith 1983:151; Campbell et al. 1986; Salomão et al. 1988). By extension, it is possible to apply a similar logic to determine the importance value of plant families. This would be the sum of family relative diversity (the number of species in a family ÷ total number of species on the plot), relative density (the number of individuals in a family ÷ total number of trees on the plot), and family relative dominance (basal area of a family ÷ total basal area of all individuals on the plot). (See Mori et al. 1989; Campbell et al. 1986; Salomão et al. 1988.)

Following is a brief description of each of the eight hectares used as study plots for determining diversities, densities, frequencies, and economic importance values of tree and liana species and their families. Table 6.1 presents a summary of results.

HECTARE 1

Location: near P. I. (Posto Indígena) Awá, occupied by Guajá, within Reserva Indígena Caru, left bank of Rio Pindaré, approximately 46°2′W, 3°48′S. Forest type: Fallow. The dimensions of this site were 500m X 20m. The site had 506 individuals, 157 species in 38 families, and a total basal area of 22.1 m². The ten ecologically most important species of trees in descending order of importance were babaçu palm (*Orbignya phaler-*

ata), munbaca palm (*Astrocaryum munbaca*), *Myrciaria obscura*, wild papaya (*Jacaratia spinosa*), breu manga (*Tetragastris panamensis*), jeniparana (*Gustavia augusta*), *Pouteria bilocularis*, sapucaia (*Lecythis pisonis*), wild cacao (*Theobroma speciosum*), and cecropia (*Cecropia ilicifolia*). Several of these species are important sources of food for the Guajá, such as babaçu, wild papaya, breu manga, sapucaia (of the Brazil nut family), and wild cacao. The five most important families were, in descending order, Brazil nut family (Lecythidaceae), palm family (Arecaceae), sapote family (Sapotaceae), caesalpinia family (Caesalpiniaceae), and myrtle family (Myrtaceae). The Guajá of the site described it as *ka'a-ate* ('high forest'), but it is clear that the site is a fallow. (The Guajá do not appear to have an equivalent expression to the Ka'apor *taper* 'fallow'.) This fallow would have resulted from agricultural activities of the Guajajara—an indigenous society not to be confused with the Guajá—probably more than 100 years ago. These activities altered what was 'high forest' on the site; a preliminary archaeological survey of the plot yielded (in addition to many potsherds) a large charcoal sample identified as wood of *Dinizia excelsa*, an enormous mimosaceous tree of the high forest which I have not encountered previously in Maranhão (Kipnis 1990). Most charcoal in the soils of Amazonian *terra firme* forests appears to have come from swidden fires, not lightning strikes (Piperno 1990).

HECTARE 2

Location: near P. I. Guajá, region occupied by Guajá, right bank of upper Rio Turiaçu, within Reserva Indígena Alto Turiaçu, 45°58'W, 3°6'S. Forest Type: Fallow. This site was 500m X 20m. It had 563 individuals, 125 species in 41 families, and a total basal area of 21.1 m². The ten ecologically most important species in descending order were babaçu palm (*Orbignya phalerata*), jeniparana (*Gustavia augusta*), joão mole (*Neea* sp. 1), joão mole (*Pisonia* sp. 2), pau d'arco roxo (*Tabebuia impetiginosa*), tapereбá or hog plum (*Spondias mombin*), farinha seca (*Lindackeria latifolia*), jutaípororoca (*Dialium guianense*), inajá palm (*Maximiliana maripa*), and amarelão (*Apuleia leiocarpa*). Babaçu, hog plum, jutaípororoca, and inajá palm are important food species to the Guajá. The five most important families, in descending order, were palm family (Arecaceae), caesalpinia family (Caesalpiniaceae), four o'clock family (Nyctaginaceae), mimosa family (Mimosaceae), and Brazil nut family (Lecythidaceae).

This site had been a Ka'apor village in the 1940s, according to oral testimony from Ka'apor inhabitants of the village of Urutawi (which is 19 km due east) as well as from a non-native visitor, Major, a long time employee of FUNAI. Potsherds and remains of Ka'apor ceramic manioc griddles have been found in this fallow. This was near the site where the Guajá were first peaceably (and officially) contacted by the federal government around 1973. The Guajá describe this forest as *wa'i̵-tu* 'babaçu grove'; the Ka'apor call it either *yetahu-'i̵-ti̵* ('babaçu grove') or *taper* 'fallow'.

HECTARE 3

Location: about 1.5 km southeast of Hectare 2. Forest Type: High forest. This site measured 10m X 1000m. The site had 521 individuals, 145 species in 45 families, and a basal area of 27.2 m². The ten most important species in descending order were matá-matá (*Eschweilera coriacea*), caçador (*Lecythis idatimon*), *Sagotia racemosa*, toari (*Couratari guianensis*), breu manga (*Tetragastris altissima*), pau de cachimbo (*Mabea caudata*), cupiúba (*Goupia glabra*), breu branco (*Protium pallidum*), açaí palm (*Euterpe oleracea*), and *Pithecellobium gongrijpii*. These species are mostly important to the Ka'apor for lashing material (*Eschweilera*, *Lecythis*, and *Couratari*), resin for illumination (*Protium*), and pipe stem (*Mabea caudata*). Only one of these is an important food species, açaí. The five most important families were Brazil nut family (Lecythidaceae), bursera family (Burseraceae), spurge family (Euphorbiaceae), mimosa family (Mimosaceae), and tropical rose family (Chrysobalanaceae). There was no evidence for disturbance on this site, and it was classified as 'high forest' (*ka'a-te*) by Ka'apor informants as well as by Guajá informants.

HECTARE 4

Location: near village of Urutawi, region inhabited by Ka'apor, within the Reserva Indígena Alto Turiaçu, on a minor tributary of right bank of Rio Turiaçu, 19 km due east of P. I. Guajá. Forest type: High forest. This site measured 500m X 20m. It had 519 individuals, 126 species in 41 families, and a basal area of 25.3 m². The ten

FIGURE 6.1 Inajá palm (*Maximiliana maripa*) in a field of arrow cane (*Gynerium sagittatum*) on the edge of a Ka'apor village. Inajá palms are frequently indicative of past human disturbance.

most important species, in descending order, were matámatá *(Eschweilera coriacea)*, caçador *(Lecythis idatimon)*, *Sagotia racemosa*, *Tetragastris altissima*, pau de remo *(Chimarrhis turbinata)*, *Dodecastigma integrifolium*, breu *(Protium trifoliolatum)*, pau de cachimbo *(Mabea caudata)*, breu branco *(Protium pallidum)*, and abiu *(Pouteria jariensis)*. These species are mostly useful as lashing material, resin for illumination, and firewood *(Sagotia* and *Tetragastris*, in particular, are considered to be excellent firewoods by the Ka'apor). Two are but minor food species *(Pouteria* and *Tetragastris)*. The five most important families in descending order were Brazil nut family (Lecythidaceae), sapote family (Sapotaceae), bursera family (Burseraceae), spurge family (Euphorbiaceae), and mimosa family (Mimosaceae). The site was classified as *ka'a-te* ('high forest') by Ka'apor informants and there was no evidence for disturbance.

FIGURE 6.2 Babaçu palms near FUNAI Post P.I.
Awá, habitat of the Guajá Indians in the basin
of the Pindaré River. This photograph was
taken near the hectare-1 test site; most of the
immediate environs appears to be fallow forest.

HECTARE 5

Location: about 2 km east of Hectare 4, region inhabited by Ka'apor. Forest Type: Fallow. This site measured 500m X 20m. It had 451 individuals, 95 species in 36 families, and a total basal area of 30.3 m² (this basal area is artificially high for reasons explained below). The ten most important species, in descending order, were wild papaya (*Jacaratia spinosa*), tucumã (*Astrocaryum vulgare*), hog plum (*Spondias mombin*), abiu (*Pouteria macrophylla*), *Platypodium elegans*, jeniparana (*Gustavia augusta*), inajá palm (*Maximiliana maripa*), *Guarea guidonia*, paraparaúba (*Jacaranda duckei*), and serve p'rá tudo (*Simaba cedron*). These species are useful as food (*Jacaratia, Astrocaryum, Spondias, Pouteria, Maximiliana*), medicine (*Gustavia* and *Simaba,* respectively, serve as snakebite and fever remedies), and in handicrafts (*Astrocaryum* and *Maximiliana* are used to manufacture, respectively, rings and combs), among other uses. The five most important families, in descending order, were palm family (Arecaceae), mimosa family (Mimosaceae), papaya family (Caricaceae), Brazil nut family (Lecythidaceae), sapote family (Sapotaceae). This site was a Ka'apor swidden in the late 1940s to early 1950s, according to oral testimony. It is the most recent of the four fallows in the sample. The Ka'apor classify it, nevertheless, as *taper* 'fallow', not *taperer* 'old swidden' or 'young secondary forest'. The basal area of 30.3 m² is artificially high because of a large number of caespitose individuals, especially the armed tucumã palm (*Astrocaryum vulgare*), which alone occupied a basal area of 5.9 m², or about 20 percent of the total basal area. As this fallow ages, it is likely that caespitose palms and multistemmed young trees (such as *Gustavia augusta*) will become less dominant and the basal area will actually diminish. This hectare, at 95 species, was the least diverse of the eight hectares. Its species/area curve (figure 6.3), unlike those of all others, is already becoming asymptotic after only one-half hectare. This confirms the widely held assumption that the younger a fallow (or secondary forest), the lower its diversity.

HECTARE 6

Location: near village of Gurupiuna, region occupied by Ka'apor, drained by tributary of the Igarapé Gurupiuna which empties into right bank of Rio Gurupi, 46°20′W, 2°40′S. It is about 60 km north of P. I. Guajá and exactly 280 km south-southeast of Belém (based on measured air distance). Forest Type: high forest. This site measured 10m X 1000m. It had 467 individuals, 123 species in 43 families, and a basal area of 30.3 m². The ten most important species, in descending order, were breu manga (*Tetragastris altissima*), matámatá (*Eschweilera coriacea*), *Pourouma minor*, breu (*Protium trifoliolatum*), andiroba (*Carapa guianensis*), *Pourouma guianensis*, breu vermelho (*Protium decandrum*), breu branco (*Protium pallidum*), cecropia (*Cecropia obtusa*), tatajuba (*Bagassa guianensis*). These species are principally useful as resin for illumination (*Protium* spp.), firewood (*Tetragastris, Eschweilera, Protium* spp.), medicinal oil (*Carapa*), and lashing material (*Eschweilera*). The five most important families were bursera family (Burseraceae), cecropia family (Cecropiaceae), mimosa family (Mimosaceae), Brazil nut family (Lecythidaceae), and mahogany family (Meliaceae). The site was classified as *ka'a-te* ('high forest') by the Ka'apor of Gurupiuna and no evidence for human disturbance was encountered.

HECTARE 7

Location: about 2.5 km WNW of Hectare 6. Forest Type: Fallow. The site measures 10m X 1000m. It had 497 individuals, 141 species in 43 families, and a basal area of 23.3 m². The ten most important species, in descending order, were bacuri (*Platonia insignis*), jatobá (*Hymenaea parvifolia*), joão mole (*Neea* sp. 2), *Mouriri guianensis*, *Maytenus*, *Pouteria* sp. 2, angelim (*Pithecellobium racemosum*), joão mole (*Neea* sp. 1), an indeterminate Meliaceae, and jitó (*Trichilia verrocosa*). Of these species, *Platonia*, *Hymenaea*, and *Pouteria* are food species, with *Platonia* being a very important food item. The five most important families were sapote (Sapotaceae), legume (Caesalpiniaceae), four o'clock (Nyctaginaceae), myrtle (Myrtaceae), and clusia (Clusiaceae). The site was described both as *taper* 'old fallow' (based on age of disturbance) and *pakuri-ti* ('bacuri-grove') (based on the perceived, relative abundance of the bacuri tree). The site was disturbed most likely in the 1870s, which was when the Ka'apor expelled Afro-Brazilian refugee slaves from the area (see chapter 3). Considerable quantities of potsherds and charcoal are found on this site. The oldest Ka'apor of the village of Gurupiuna remember the site from their childhoods (c. 1920s) as having been a bacuri grove; it has the same appearance today, according to these informants, as it did then. In other words, this fallow has not been significantly disturbed probably since the Ka'apor arrived in the region during the 1870s. The importance of the site to the Ka'apor today lies in its source of highly prized bacuri fruits.

HECTARE 8

Location: about 5 km W of P. I. Canindé, region inhabited by Tembé, left bank of Rio Gurupi (state of Pará), 26 km WNW of hectare 7, 254 km ESE of Belém (based on measured air distance). The dimensions of this site were 10m X 1000m. Forest type: high forest. The site had 475 individuals, 144 species in 41 families, and a basal area of 34.5 m². The ten most important species, in descending order, were matámatá (*Eschweilera coriacea*), coumarurana (*Taralea oppositifolia*), *Sagotia racemosa*, breu vermelho (*Protium decandrum*), macucu (*Couepia guianensis*), abiu (*Pouteria virescens*), breu manga (*Tetragastris altissima*), breu branco (*Protium giganteum*), abiu (*Pouteria oppositifolia*), and visgueiro (*Parkia pendula*). These species are mostly useful as sources of dye (*Couepia*), lashing material (*Eschweilera*), and firewood (*Tetragastris, Eschweilera*, and *Protium*). The five most important families were sapote (Sapotaceae), Brazil nut (Lecythidaceae), bursera (Burseraceae), and two legume families (Mimosaceae and Fabaceae). The Tembé classified the site as *ka'a-ete* ('high forest'), as did Ka'apor informants who accompanied me to the site. No evidence of prior human disturbance was encountered here.

Results of Forest Inventories

Table 6.1 shows the survey results of the eight study plots. The high-forest study plots (hectares 3, 4, 6, and 8) are comparable in tree and liana species diversity to other eastern Amazonian "primary" forests. It is clear, moreover, that study plots only one hectare in size are insufficient for calculating the true diversity of *terra firme* forests. Figures 6.3 and 6.4 show that the species-per-area curves for all four one-hectare study plots of high forest and for three of the four one-hectare study plots of fallow forest are still steeply rising after one hectare. Nevertheless, studies that used similar methodologies for evaluating one-hectare samples of *terra firme* forest elsewhere in the Amazonian region can provide useful comparisons with my aggregated results (unlike my work, those studies did not distinguish fallow and high forest).

Surveys on ten one-hectare plots scattered throughout Breves (Marajó Island), Belém, the middle Xingu basin, and the Serra dos Karajás, suggest that the average number of tree and liana species per hectare in what the researchers consider "primary" forests of southeastern Amazonia is 126 (Campbell et al. 1986; Salomão et al. 1988). By comparison, the three plots of high forest I sampled in Maranhão (study plots 3, 4, and 6) have an average of 131 species per hectare. Adding hectare 8, from the left bank of the Rio Gurupi in Pará state, this average for the total of high forest plots climbs to 135 species per hectare. The combined average for all 8 plots, including fallow, is 132 species per hectare (table 6.1).

It may be concluded that although pre-Amazonian forests are less species rich than upper Amazonian high forests in general (see Gentry 1988), pre-Amazonian forests are clearly comparable in terms of diversity to the native forests of Rondônia in southwestern Amazonia (Salomão et al. 1988; Lisboa et al. 1991) and eastern Amazonia, even if, in terms of plant species present, little if any endemism appears to exist.

Remote sensing does not yet supply a basis for distinguishing between fallow and high forest in pre-Amazonia. Satellite imagery, to date, portrays both as being high forest, that is, primary vegetation. The pre-Amazonian forests of the Gurupi in the north

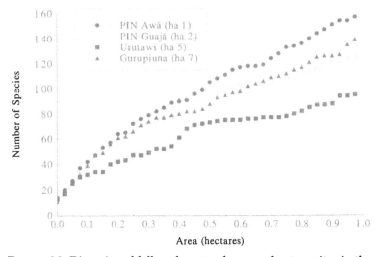

FIGURE 6.3 Diversity of fallow forest at four one-hectare sites in the Ka'apor habitat.

to the Pindaré in the south, regardless of their heterogeneity, are customarily mapped as *floresta densa* ('dense forest'). What, then, might be ground truth criteria for proposing such a distinction?

One possible criterion for maintaining the distinction is indigenous classification. The Ka'apor call fallow forests *taper* and high forest *ka'a-te*. A difficulty arises, however, if one is depending exclusively on indigenous knowledge. A shortness of historical memory is widespread in these small-scale, egalitarian, nonliterate societies, as discussed in the preceding chapter. Consider, for example, the terminology used by other speakers of the Tupí-Guaraní language family in the region. The Araweté and Asurini, for example, called old fallow forests of their region, respectively, by the terms *ka'ã-hete* and *ka'a-te*, which mean 'high forest' (Balée and Campbell 1990).

The Guajá of the P. I. Awá on the Rio Pindaré described the forest sample in the hectare 1 study plot as *ka'a-ate* ('high forest')—whereas my own observations led me to inventory it as fallow. This disagreement implies that if a site is left fallow long enough, some forest dwellers will consider it to be high forest, not recognizing that it was once occupied by agricultural peoples. More to the point, some forest dwellers do not appear to recognize a temporal succession from fallow to high forest—indeed, the Ka'apor evidently do not (Balée and Gély 1989), even though they distinguish the two forest types linguistically.

An important physiognomic difference between fallow and high forest concerns aggregated basal area. The data from the fallow forests, with the exception of hectare 5 (which is abnormally high in basal area because of a very high frequency and abundance of large but nonwoody caespitose species), show a consistently lower basal area than the high forests. The fallows, including hectare 5, average 24.2 m^2 in aggregate basal area per hectare; excluding hectare 5, this average is 22.1 m^2. The high forest plots, in contrast, average 29.3 m^2 in aggregate basal area per hectare. It seems that fal-

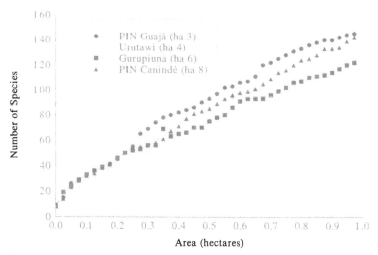

FIGURE 6.4 Diversity of high forest at four one-hectare sites in the Ka'apor habitat.

lows are typically within a range of about 18–24 m², whereas high forests range from 25–40m² (Pires and Prance 1985; Balée and Campbell 1990; Boom 1986; Saldarriaga and West 1986:364). These differences in basal area between high forests and fallow appear to be at least partly diagnostic.

In identifying a plot of forest as fallow, from a strictly botanical point of view, a great deal also typically hinges on the presence of "disturbance indicator" species. It would be inaccurate to consider disturbance indicators as synonymous with "pioneer" species (cf. Brown and Lugo 1990). Pioneer species tend to be short-lived. Disturbance indicators, in contrast, may be long-lived, as with the babaçu palm and Brazil nut tree (see Balée 1989a, 1989c). For Van Steenis (1958), disturbance indicators were biological nomads. They occurred as isolated or even rare individuals in a primary forest until a disturbance, such as fire in the service of agriculture, opened space for them. Van Steenis proposed no quantitative measure to determine when a biological nomad becomes an indicator of human activity—i.e., how diverse, frequent, abundant or ecologically important nomad plant species must become to qualify as an indicator of agricultural perturbation on a given plot of forest. In fact, to date, no good measures for disturbance indicators have been proposed (see Brown and Lugo 1990).

Some species, by their presence alone, appear to indicate agricultural activity in the past. These include ecologically important species such as the babaçu palm (*Orbignya phalerata*), the tucumã palm (*Astrocaryum vulgare*), wild papaya (*Jacaratia spinosa*), hog plum (*Spondias mombin*), inajá palm (*Maximiliana maripa*), and the enormous legumes *Platypodium elegans* and *Apuleia leiocarpa*, none of which occur in high forest of the *terra firme*. Yet mere presence/absence data may be misleading in determining disturbance. Upon combining all four fallow plots as one forest type (appendixes 1 and 2) and all high forest plots as another forest type (appendixes 3 and 4), the two different forest types share 18 percent of their species. (See below for data on species held in common by different study plots within the same forest type.) Were one to use the criterion of mere presence or absence, a very large number of species (253) would be taken to be disturbance indicators (360 total species in fallow minus 107 species shared with high forest); conversely, numerous species (234) would be indicators of undisturbed forest (341 total species of high forest minus 107 species shared with fallow). But this criterion alone is clearly inadequate.

Table 6.2 shows species often considered to be disturbance species, and which occur on one or more of my four fallow plots. Yet nearly half of these also occur in one or more of the four high forest plots. Except for a few species, which do appear to be unique to fallow within the size range greater than or equal to 10 cm dbh, it is clear that species which indicate disturbance cannot be understood simply on the basis of their presence. Indication of disturbance, from a botanical point of view, should be a measure of degree or a probability, not an absolute factor.

There are good phytosociological reasons for the separation of high forest hectares from the fallow hectares and for recognizing two different composite types of *terra firme* forest in pre-Amazonia. The eight study plots of fallow and high forest in this sample yield 28 pairs of study plots (see table 6.3). The similarity of these pairs can be systematically compared using the Jaccard coefficient, which is simply the number of

TABLE 6.2 *Known Disturbance Indicators Present on One or More Fallow Sites*

Family and Species	Sources
ARECACE AE	
Astrocaryum vulgare	Wessels Boer 1965:132
Orbignya phalerata	May et al. 1985
Maximiliana maripa	Schulz 1960:222
ANACARDIACEAE	
Spondias mombin	Ducke 1946:20; Lisboa et al. 1987:55
Tapirira guianensis	Huber 1909:162
ANNONACEAE	
Annona montana	Ducke 1946:4
Rollinia exsucca	Lisboa et al. 1987:55
ARALIACEAE	
*Schefflera morototoni**	Huber 1909:161; Schulz 1960:239
BIGNONIACEAE	
*Jacaranda copaia**	Fanshawe 1954:75; Huber 1909:161
CARICACEAE	
Jacaratia spinosa	Lisboa et al. 1987:55
CAESALPINIACEAE	
Apuleia leiocarpa	Ducke 1949:112
Cassia fastuosa	Ducke 1949:115
Cassia lucens	Ducke 1949:118; Lisboa et al. 1987:55
*Dialium guianense**	Ducke 1949:112; Huber 1909:162
*Hymenaea courbaril**	Ducke 1949:97
CECROPIACEAE	
*Cecropia sciadophylla**	Lisboa et al. 1987:55
COCHLOSPERMACEAE	
Cochlospermum orinocense	Lisboa et al. 1987:55
FABACEAE	
Andira retusa	Ducke 1949:201
Derris spruceana	Ducke 1949:199
Bowdichia nitida	Ducke 1949:59
FLACOURTIACEAE	
Caesaria javitensis	Fanshawe 1954:78
MELIACEAE	
Trichilia schomburgkii	Fanshawe 1954:78
MIMOSACEAE	
*Inga alba**	Ducke 1949:26
*Inga auristellae**	Ducke 1949:26
Inga falcistipula	Ducke 1949:27
Stryphnodendron spp.*	Ducke 1949:59
MORACEAE	
*Bagassa guianensis**	Huber 1909:161
NYCTAGINACEAE	
Neea spp.*	Fanshawe 1954:78
OLACACEAE	
*Minquartia guianensis**	Fanshawe 1954:75
SAPOTACEAE	
Pouteria guianensis	Fanshawe 1954:74
*Pouteria venosa**	Fanshawe 1954:75
SIMARUBACEAE	
*Simaruba amara**	Huber 1909:161; Schulz 1960:218
STERCULIACEAE	
*Theobroma speciosum**	Ducke 1953:14
VIOLACEAE	
Rinorea flavescens	Fanshawe 1954:78

*Species also occurs on one or more high forest plots.

TABLE 6.3 *Jaccard Coefficients of Similarity Among Study Plots*

Pairs of Study Plots	# Species in Common	Total Species	Coefficient of Similarity (%)
High Forest Pairs			
3, 6	49	219	22.4
3, 4	53	218	24.3
3, 8	48	241	19.9
6, 4	50	199	25.1
6, 8	50	217	23.0
4, 8	49	221	22.2
Fallow Pairs:			
1, 2	44	238	18.5
1, 5	35	217	16.1
1, 7	40	258	15.5
2, 5	37	183	20.2
2, 7	43	223	19.3
5, 7	28	208	13.5
Fallow/High Forest Pairs			
1, 3	30	272	11.0
1, 6	25	255	9.8
1, 4	27	256	10.5
1, 8	24	277	8.7
2, 3	28	242	11.6
2, 6	25	223	11.2
2, 4	28	223	12.6
2, 8	29	240	12.1
5, 3	29	211	13.7
5, 6	25	193	13.0
5, 4	21	200	10.5
5, 8	23	216	10.6
7, 3	23	263	8.7
7, 6	25	239	10.5
7, 4	28	239	11.7
7, 8	23	262	8.8

species divided by the total number of species in the sample (i.e., the sum of the total number of species of each plot minus the shared species) expressed as a percentage (Greig-Smith 1983:151). On average, the coefficient of similarity for pairs of hectares of high forest and fallow is only 10.9 percent. In contrast, the average coefficient of similarity for pairs of hectares that are both high forest is 22.8 percent, which is very significantly higher. Also, the average coefficient of similarity for pairs of hectares that are both fallow is 17.2 percent, which is also very significantly higher than the mixed forest average but not significantly lower than the high forest average. In other words, wherever they occur, fallow forests resemble other, distant tracts of fallow forest, not nearby high forests. Likewise, high forests resemble other, distant tracts of high forest, not nearby fallow forests. Fallow forest and high forest are two fundamentally different types of vegetation.

If one compares the high forest near Gurupiuna (hectare 6) to the fallow (hectare 7) near Gurupiuna (the two plots are separated by about 2.5 km), the coefficient of similarity is only 10.5 percent. Likewise, comparing the high forest (hectare 3) with the fallow (hectare 2) near P. I. Guajá in the Turiaçu basin (the two plots are separated by only about 1.5 km), the coefficient of similarity is only 11.6 percent. The

Gurupiuna sites on one hand and the P. I. Guajá sites on the other are separated by a linear distance of about 60 km. It is interesting, therefore, that the fallows of the P. I. Guajá and Gurupiuna (hectares 2 and 7) have a similarity coefficient of 19.3 percent, which is very significantly higher than those of the two pairs of nearby hectares. The high forest hectare of Gurupiuna and the P. I. Guajá (hectares 6 and 3), with a similarity coefficient of 22.4 percent, are also much more similar to each other than either is to its nearby fallow forest. As would be expected, the pairings Gurupiuna high forest with P. I. Guajá fallow (hectares 6 and 2) and P. I. Guajá high forest with Gurupiuna fallow (hectares 3 and 7) have low similarity coefficients, respectively, of 11.2 percent and 8.7 percent. It can thus be concluded that the overriding factor that accounts for divergence in floristic composition between forest stands in this eight hectare sample of pre-Amazonia is not linear distance, but rather past agricultural perturbation.

Traditional Indigenous Activity as an Enhancer of Regional Biodiversity

When plotted on a species-area curve as separate forest parcels, the fallow forest and the high forest accumulate diversity at similar rates (see figure 6.5). These plots are organized along a continuum; the fallow curve represents increasing diversity from hectares 1, 2, 5, and 7 in that order (i.e., from the fallow plot of the P. I. Awá, through the Turiaçu fallows, to the basin of the Gurupi). The forty 10m X 25 m sampling units of each hectare are, moreover, seen as a continuum, as with the high forest plots. The high forest plots are represented in this order: hectares 4, 3, 6, 8 (i.e., from the Turiaçu to

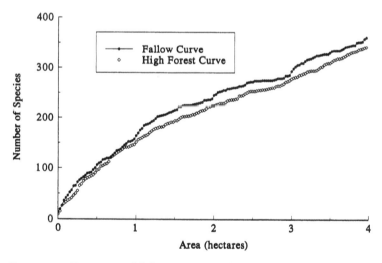

FIGURE 6.5 Diversity of fallow and high forest compared in the Gurupi, Turiaçu, and Pindaré basins.

the Gurupi). Interestingly, in my survey of the forests of the Pindaré, I saw no evidence for any plot of *terra firme* to be high forest. This is not surprising, considering a probably much higher density of Guajajara settlements and overall population in the Pindaré than there ever were for the Ka'apor of the Turiaçu basin.

These curves and the evidence of floristic composition support the familiar notion that for secondary forests, "within a span of 80 years or less, the number of species approaches that of mature forests" (Brown and Lugo 1990:6). In my investigation, moreover, not only does fallow "approach" high forest in species diversity, the diversity between the two forest types is insignificantly different. This alone is powerful evidence that these fallow forests represent reforestation of a once-disturbed habitat. The species richness of high forests in particular sites is being replaced by an equivalently rich secondary forest *because of human activity*. This is an indigenous contribution to regional biodiversity.

The most astounding difference, in terms of species, concerns ecologically important species. By ecologically important species I mean those with the highest sums of values indicating relative dominance, frequency, and diversity. In comparing the 30 ecologically most important species of the four fallows with those of the four high forest plots, the two forest types are seen to share but a single species, the *matámatá* tree, *Eschweilera coriacea* (see table 6.4). This yields a coefficient of similarity of only 1.7 percent for the 30 most important species (number of shared species [1] + total number of species on the two plots [(30 + 30) - 1] X 100). This difference is extremely significant. The average coefficient of similarity for the 30 most important species in pairs of high forest is 16.4 percent; the average index of similarity for the 30 most important species in pairs of fallow is 11 percent; the average index of similarity for the 30 most important species of mixed pairs, however, is consistently less than 2 percent. This permits one to conclude that the important species between fallow and high forest are significantly and consistently different.

In this quantitative sense, the 30 most important species of fallow less *Eschweilera coriacea* may be considered to be indicators of disturbance; similarly, the 30 most important species of high forest less *E. coriacea* may be considered to be indicators of non-disturbance. Although from this vantage point, *Eschweilera coriacea* is a facultative species, in three of the four high forest hectares (hectares 3, 4, and 8) it is *the* ecologically most important species, whereas in any fallow forest hectare it does not attain a rank higher than 14. In fallow forests, the babaçu palm (*Orbignya phalerata*) is the ecologically most important species on two hectares (1 and 2), yet it does not approach the extremely high values that *E. coriacea* does on high forest plots.

For reasons that militate against the grain of received wisdom, the fallow forests are actually less dominated by a few species than are the high, presumably primary forests. For example, the average total of the two ecologically most important species (with a total possible value of 300) of the four fallows is only 40.3, whereas the comparable figure for the four high forest hectares is 60.3, which is significantly higher. For analytical purposes, I consider a species to be rare if it occurs only once, regardless of whether it is present in one of the fallow hectares or one of the high forest hectares. By this criterion, the high forest has 199 species (or 59 percent of the total) that are rare; fallow forest has 139 species (or 38.6 percent of the total) that are rare. Both forest types,

TABLE 6.4 *Comparison of the Thirty Ecologically Most Important Species From High Forest And Fallow Inventories in the Pre-Amazonian Forest*

High Forest Species	I.V.[*]	Fallow Species	I.V.[*]
Eschweilera coriacea	37.83	Jacaratia spinosa	11.40
Lecythis idatimon	14.53	Gustavia augusta	10.41
Sagotia racemosa	12.67	Orbignya phalerata	9.37
Tetragastris altissima	11.60	Astrocaryum vulgare	7.76
Protium trifoliolatum	7.76	Spondias mombin	6.53
Protium decandrum	7.07	Neea sp. 1	6.26
Protium pallidum	6.78	Pisonia sp. 2	6.25
Carapa guianensis	5.69	Pouteria macrophylla	5.71
Couepia guianensis	5.07	Maximiliana maripa	5.40
Pourouma minor	4.54	Platypodium elegans	5.02
Taralea oppositifolia	4.51	Platonia insignis	4.32
Mabea caudata	4.06	Simaba cedron	4.26
Pourouma guianensis	3.28	Hymenaea parvifolia	4.17
Dodecastigma integrifolium	3.10	Trichilia quadrijuga	4.06
Couratari guianensis	2.77	Lecythis pisonis	3.56
Oenocarpus distichus	2.72	Dialium guianense	3.32
Sterculia pruriens	2.65	Astrocaryum munbaca	3.31
Bagassa guianensis	2.65	Eschweilera coriacea	3.19
Cecropia obtusa	2.60	Theobroma speciosum	3.11
Newtonia psilostachya	2.47	Lindackeria latifolia	3.05
Chimarrhis turbinata	2.40	Tabebuia impetiginosa	2.85
Simaruba amara	2.39	Myrciaria obscura	2.75
Euterpe oleracea	2.37	Neea sp. 2	2.64
Lecythis chartacea	2.25	Hymenaea courbaril	2.61
Parkia pendula	2.23	Protium heptaphyllum	2.59
Protium polybotryum	2.22	Tetragastris panamensis	2.56
Apeiba echinata	2.19	Apuleia leiocarpa	2.53
Fusaea longifolia	2.18	Mouriri guianensis	2.49
Protium giganteum	2.12	Cupania scrobiculata	2.40
Tachigali myrmecophila	2.11	Pouteria bilocularis	2.34
Total High Forest I.V.:	166.81	Total Fallow I.V.:	136.22

[*]I.V. refers to Importance Value (itself the sum of relative density, relative frequency, and relative dominance).

in other words, harbor significant quantities of rare species. (Many other rare species, of course, were not sampled in the study plots themselves, but in the general species collections I made of the whole region.) The data I have collected thus offer only partial support for the statement "A large number of species in mature forests is due to the presence of rare species. In contrast, secondary forests are usually composed of common species" (Brown and Lugo 1990:7).

Aside from the fallow and high forest differences in age, basal area, overall floristic composition, and ecologically important species, another difference to which I have alluded, being strictly related to utilitarian concerns, deserves mention here: *Fallows are indigenous orchards, whether consciously planted or not.* Of the 30 ecologically most important species of fallow, 14 are significant food species, either for the Ka'apor or other indigenous peoples, whereas for the 30 ecologically most important species of high forest, there are only 6 important food species. Some significant food species of the fallow include babaçu palm (*Orbignya phalerata*), hog plum (*Spondias mombin*), tucumã palm (*Astrocaryum vulgare*), inajá palm (*Maximiliana maripa*), bacuri (*Platonia insignis*), and jatobá (*Hymenaea* spp.). These orchards would not exist would it not have been for help from the Indians. Neither forest type, of course, will survive without help from the state.

What Do They Know, and When Did They Know It?

It would be erroneous to conclude that human activity as an enhancer, rather than a degrader, of regional biodiversity is only feasible in societies with small population densities and high mobility (e.g., Johnson 1989; Redford 1991). This enhancement of biodiversity occurred in pre-Columbian times, when indigenous population densities and settlement occupations in Amazonia and eastern Brazil were, by several orders of magnitude respectively, higher and longer than they are today (Denevan 1976; Balée 1984a, 1988a; Dean 1984; Piperno 1990; Roosevelt 1987, 1989a, 1989b, 1991). Although there is no evidence for indigenous states in lowland South America, many societies were organized into paramount chiefdoms (Balée 1984a; Roosevelt 1987, 1989b, 1991). These societies would have exerted greater effects on biotic and abiotic components of the environment than modern indigenous swiddeners, yet these exertions did not apparently result in extinctions and environmental degradation.

In other words, pre-Columbian indigenous resource management, with its potential for extinctions, would have been, if anything, more intensive than modern indigenous management—yet there is no evidence to attribute to it a worse environmental record than more mobile, less populated, acephalous indigenous societies of today. With specific reference to the Ka'apor, this heritage of plant management dates from at least 1,800 to 2,000 years ago, the probable age of the mother tongue, proto–Tupí-Guaraní (Aryon D. Rodrigues, personal communication, 1988; Migliazza 1982). Plant management among the ancestral Ka'apor is probably even older, as words for 'swidden' and 'digging stick' reconstruct in proto-Tupí, the mother language of the stock to which Tupí-Guaraní pertains and which is some four to five thousand years old (Rodrigues 1988). The proto–Tupí-Guaraní plant lexicon included names at least for maize, manioc, yam, sweet potato, pineapple, peanut, gourd, calabash, cashew, annatto, and the bromeliaceous fiber plant caroá (cf. Lemle 1971; Rodrigues 1988). Probably many words for semidomesticated trees have also descended from proto–Tupí-Guaraní to its contemporary daughter languages (Balée and Moore 1991).

The historically known, sixteenth-century coastal Tupinambá, who were organized into paramount chiefdoms, cultivated maize, manioc (of which there were 28 named varieties—Métraux 1928:65–67), banana, peanut, chile pepper, crookneck squash, yam, sweet potato, pineapple, papaya, calabash, cashew, gourd, lima bean, passion fruit, cotton, annatto, caroá, and tobacco (see Balée 1992a). Several semidomesticated trees, which were probably artifacts of Tupinambá swiddening, were also present, such as the mountain soursop *Annona montana* (Vasconcellos 1865:136), hog plum (Lisboa 1967:113; Sousa 1974:101; Vasconcellos 1865:138), monkey pot tree (Léry 1960:157; Lisboa 1967:106; Vasconcellos 1865:133), inajá palm (Lisboa 1967:125), copal trees (Vasconcellos 1865:133), and *Inga* spp. (Sousa 1974:101). Thus, even at a time of much higher population density and far more intensive utilization of biological resources (Dean

1984), forest fallow appears to have existed near Tupinambá settlements. But fallow is absent in the virtually monocultural rice fields and cattle pastures of modern state society in the Amazon today.

The modern Ka'apor are, to a measurable degree, heirs to the plant management expertise of their early Tupí-Guaraní–speaking forebears, who were no doubt similar in sociopolitical organization and technology to the Tupinambá. Yet the extent to which this ancient inherited expertise is cognized by Ka'apor adults today remains questionable.

Rindos (1984:99) pointed out that early agricultural peoples may have been aware, in a retrospective sense, that their dependence on plant management entailed a different lifestyle from that of other hunting and gathering cultures. This is certainly the case with the Ka'apor, who readily distinguish the neighboring, foraging Guajá Indians as a people who do not 'swidden' (*kupiša moú 'im*). They indicate that the Guajá are *purara* 'poor' because they depend mainly on babaçu nuts and other nondomesticates, instead of on the domesticated tubers, corms, and fruits found, for example, in Ka'apor swiddens. The Guajá themselves readily admit 'we do not plant' (*Ijə tum awa*). To paraphrase Rindos (1984:99), the Ka'apor and the Guajá know who they are, as distinguished from each other. But whether they know who they will become, as their lands are sucked into the vortex of an expanding state's frontier (which was most certainly not by *their* design), is quite a different issue.

It seems unlikely, moreover, that today's forest dwellers know exactly who they were. The Guajá, for example, have no historical memory of ever having planted swidden fields (Ricardo Cordeiro Nassif, personal communication, 1991), yet linguistic and other evidence suggests that they lost control of these species during a phase of agricultural regression (see chapter 8).

More to the point of the subject of this chapter, do the Ka'apor know what they are doing? That is, do they intend the effects of their environmental interventions to be somehow beneficial to more distant, ensuing generations of their own kind? I believe the answer is no, and for the same reasons that Ka'apor informants cannot often remember grandparents' names, do not distinguish more than a few of the many poisonous principles in their habitat, avoid consuming some fruits that are otherwise edible, and believe that stone axeheads in their forests are 'thunder-seeds' (*tupã-ra'i*) that were never used for felling trees. This conclusion does not necessarily imply that the beneficial effects of Ka'apor interactions with their surroundings are purely happenstance. Historical-ecological mechanisms can produce over the generations a kind of intelligence that is not consciously exercised.

In contrast, a major case has been made for incorporating indigenous knowledge into rational development schemes for Amazonia (Posey 1983, 1984, 1985; Posey et al. 1986). The argument is partly based on observational evidence that traditional forest dwellers, despite thousands of years of habitation, have not, by and large, damaged Amazonian forests, polluted streams, or otherwise degraded natural ecosystems. Proponents of this alternative viewpoint argue that the seeming custodial care of nature is deliberate. Specifically, in the case of the Kayapó, Posey has argued that the 'forest islands' (*apêtê*) of fruit trees and other utilitarian, nondomesticated plants were actually planted with the *intention* of benefiting future Kayapó generations.

One may find Posey's argument basically inapplicable, however, to the Ka'apor. Consider: the major long-term beneficiaries of past forest management by the Ka'apor Indians have not been Ka'apor descendants. Rather, the beneficiaries have been their traditional enemies, the foraging Guajá, who traditionally rely heavily on the palms and other disturbance indicator plants found in Ka'apor fallows, not the Ka'apor of today (Balée 1988b, 1992a).

In spite of a universal and immediate sense that they are swidden horticulturalists, whose subsistence activities affect the distributions of plants, the Ka'apor exhibit no rationalistic knowledge concerning the remote acquisition or domestication of neotropical crops found in their swiddens today. If the ancestral Ka'apor (and earlier Tupí-Guaraní forebears) had transmitted an entirely rationalistic knowledge of Amazonian plants to their descendants, one would expect this to be reflected in modern Ka'apor speech and behavior. Rational phylogenetic knowledge would seem especially important concerning the origins of plants that, to persist and thrive, *must* remain under human supervision and interference. Yet even the most significant traditional domesticates are not perceived as having been derived from any related plant— although closely related undomesticated congenerics of these domesticates are to be found throughout the Ka'apor habitat and one need not employ a microscope to note the salient, overall resemblances.

For example, at least two species of the genus *Anacardium* (*A. giganteum* and *A. parvifolium*) occur in mature forests of the Ka'apor habitat. Both are closely related to their domesticated congener cashew, *A. occidentale*, (see Mitchell and Mori 1987). Ka'apor informants say domesticated cashew trees came into being when the culture hero, Ma'ir, planted branches of *yaši-amir* (*Lecythis idatimon*), a profoundly unrelated tree in the Brazil nut family seen in the high forest. Sweet potatoes (*Ipomoea batatas*) are said to have first sprouted and bloomed when Ma'ir planted *iwi-pu'a* ('round soil', a clayey loam); yet at least four species of nondomesticated morning glories (*Ipomoea*), closely related to sweet potatoes, are found in the Ka'apor habitat, usually on the edges of swiddens. Finally, it is said that manioc (*Manihot esculenta*) originated when Ma'ir planted the (unspecified) branches of high forest trees; yet three closely related, nondomesticated species of *Manihot* are common in Ka'apor swiddens (see below). In fact, these species can be intercrossed with *Manihot esculenta* (Jennings 1976:81; Rogers and Appan 1973). Interestingly, the Waimiri Atroari claim that the origins of manioc can be traced to a palm (Milliken et al. 1992:22).

These explanations for the origins of traditional domesticates are associated with a lexical and classificatory dichotomy between traditional domesticates and nondomesticates to be taken up in the next chapter. In any case, these explanations evince a lack of knowledge of the principle of hybridization in plants. They also militate against a concept of semidomestication, which is also not encoded linguistically in Ka'apor. Yet many nondomesticated, opportunistic species of Ka'apor swiddens and fallows occur nowhere else, or are ecologically important only there, and are perhaps best understood as being semidomesticated. Thus, over the centuries the Ka'apor have effectively been breeders of plants (testimony to this is to be seen in landraces of certain domesticates and semidomesticates in Ka'apor swiddens and fallows not to be found elsewhere), but they were not engaged in plant breeding by long term design.

The Ka'apor do exhibit a sophisticated knowledge of the anatomical parts, life processes, and techniques for manipulating and harvesting many individual plants; the plant lexicon, moreover, incorporates names for many hundreds of plants (see chapter 7 and appendix 9). Yet many of these types of plants have generational times far shorter than the life span of the average Ka'apor who survives into adulthood. A knowledge of these plants, their habits, and requirements does not perforce imply a rationalistic knowledge of *long-term* ecological and successional processes that involve several human generations. Although one can live to see that, as the elders say, red brocket deer eat wild manioc leaves, no one lives long enough to observe a young swidden become a very old fallow. Likewise, no one lives long enough to see yellow clay become transformed into anthropogenic black earth (cf. Smith 1980).

To expect such knowledge would be tantamount to overlooking significant limitations on the oral transmission of information about events, people, and places in lowland South America. This does not deny to the Ka'apor, however, a material role in the aspects of forest management that unfold over the course of several scores of years. Rather, Ka'apor forest management appears to be incidental to developmental processes associated with semisedentary, indigenous Amazonian society—it is, simply, not the product of long-term design, indigenous or otherwise.

Settlement History and Ecological Diversity

Virtually nothing is known of the prehistory of the Ka'apor habitat. But it has been occupied by several forest-dwelling peoples briefly since the early 1800s, including maroon colonies of Afro-Brazilian refugees (i.e., the *quilombos*) of the 1830s to 1870s, the Tenetehara (who migrated through it when relocating from the Pindaré to the Gurupi and Guamá basins in the 1850s), the Kren-Yê Timbira, and the Guajá (see chapter 3). Stone axeheads (*tupã-ra'ĩ* 'thunder-seeds') not of Ka'apor manufacture occur in the habitat, indicating agriculture in the remote past. But probably the habitat was never as intensively settled as some of the other peopled regions of prehistoric Amazonia, such as the Amazon River proper, the Xingu, the Tapajós, and Marajó Island. This is because the fertile soil called *Terra Preta do Índio* does not occur in the region, nor is there evidence for major prehistoric settlements, mounds, and highly elaborated pottery as elsewhere (cf. Roosevelt 1987, 1989a, 1989b, 1991). Evidently, in terms of major cultural developments in Amazonia, the region currently occupied by the Ka'apor Indians was probably always somewhat marginal. Hence, much of the forest described as *taper* 'fallow' is probably recent; i.e., more than 40 but less than about 200 or 250 years have passed since its last burning. Most fallows in this part of pre-Amazonia are artifacts of historically known horticultural societies, such as the maroon colonies, the Tenetehara, the Timbira, and, most important of all, the Ka'apor ancestors themselves, all of whom fled here at different times yet for similar reasons.

In modern times, Ka'apor plant management is associated with the birth, expansion, and decay of agricultural settlements (Balée and Gély 1989). A settlement comes into being while another is slowly depopulated, usually for ecological reasons that are immediately apparent to the actors themselves and easily stated. In the historical past, the Ka'apor sometimes fled their settlements because of invasions by hostile state militia or the spread of epidemic diseases (Balée 1988a; Ribeiro 1951). More recently, several villages in the Paruá River basin relocated 25 km or more to the west because of encroaching pastoral and agricultural settlement that ultimately deforested that basin (Balée 1990).

Although at present the peculiarities of each Ka'apor village are different, the four village relocations unrelated to external social pressures that I have witnessed suggest certain constant features. Without external threats to security, the move to a new settlement is prompted by declining yields of tortoises (yaši), a perceived 'excess of old swidden' (taperer heta), infestation of 'roaches' (tarawe) and 'scorpions' (yawayïr), and the depletion of certain nondomesticated, yet useful plants.

An absence of tortoises is probably the most significant indicator of an impending relocation, since without these, the Ka'apor ritual and ecological system that, in effect, manages major mammalian game species cannot operate. Although Ka'apor men do not say they hunt tortoises to manage other game supplies, but rather to feed their menstruating and postpartum women, compliance with this ritual injunction has a management effect. A village whose tortoise resources are to be found only five or more kilometers away is an old village, which is probably decaying in other respects, such as by pest infestation and an overrepresentation of old swiddens.

Two plants whose absence indicates advancing settlement age are the fish poison Derris utilis (a large liana whose active principle is rotenone) and the preferred palm used for roofing thatch, ubim (Geonoma baculifera). Derris utilis 'never flowers and fruits' (putïr 'ïm, i'a 'ïm), a fact recognized by Ka'apor informants. In the upper Amazon, it is seen only as a domesticate in swiddens, where it is propagated by cuttings (Ducke 1949:197–198). Ka'apor informants further state that where Derris utilis has been cut and harvested, it never reappears. For unclear reasons, Geonoma baculifera, although it is a flowering plant capable of setting seed by itself, is also quickly depleted near older settlements as the Indians harvest it for thatching and rethatching their houses. In fact, an increasing problem near older Ka'apor settlements is an utter absence of Derris utilis and Geonoma baculifera within a range of some five to seven kilometers, due to nonsustainable harvesting over many years.

In the village of Gurupiuna, which is about twenty-five years old, for example, informants claim upa ya-monok ('we cut and harvested them all') in reference to both Derris and Geonoma. Instead of relocating the village, however, they now deploy mainly the domesticate Clibadium sylvestre for fishing expeditions and have substituted the leaves of the inajá palm (Maximiliana maripa) for Geonoma baculifera. Although G. baculifera is a native Amazonian species, Derris utilis is also in Africa (where it was classified as Lonchocarpus nicou) and it may even be of African origin (cf. Lathrap 1977). If so, maroon settlements of the 1830s to 1870s may have introduced it into the Ka'apor habitat. This is not an unfeasible proposition, given that elsewhere in the Amazon Derris is known only as a cultivated species, and hence may be a domesticate incapable of self-propagation.

Derris utilis appears to have an uneven distribution in the Amazon basin, further suggesting the possibility of human intervention and diffusion. For example, it does not occur, to my knowledge, in the Xingu River basin. After the Ka'apor expelled the

Afro-Brazilian refugees in the 1870s, they did not manage the plant. It and *Geonoma baculifera* are the only plant species I know of that are directly threatened with micro-local extinction by traditional Ka'apor forest utilization; of course, both species are still present in the Ka'apor habitat seen from a regional perspective. But from a micro-local, village point of view, the depletion of these species constitutes one of the material indicators that it is time to relocate.

A new settlement begins as a swidden (*kupiša*), not a bona fide village. It begins when several related nuclear families, aware of decaying conditions in the old settlement, clear a field five kilometers or further from the village. This distance is increasing among populations where beasts of burden, such as donkeys and mules, are present. Such is the case, for example, at Urutawi. For families or entire settlements without these investments (i.e., who live traditionally), going to and from the swiddens on foot is the only option available. By related nuclear families, I refer specifically to the residential unit. A residential unit is normally composed of a core of mothers, daughters, and both real and classificatory sisters. With the exception of the headman (*kapitā*), who tends to remain with his own kinsmen upon marriage, the men who have married the women usually hail from elsewhere. In other words, a residential unit is a unilocal (specifically, uxorilocal) unit of families that acts like, and is in fact coterminous with, small Ka'apor villages. Large villages, in contrast, contain more than one residential unit, and these unilocal groups tend to arrange marriages among themselves. Small villages, however, are the norm—average Ka'apor village size, in the recent past, was only about 33 and small villages had anywhere from 15 to about 27 residents with a single headman each. This is the same size as the typical group of people that first establishes a new settlement site.

The site is selected based on ample availability of *terra firme* forest, either primary or old fallow. The soil type most prevalent should be *ɨwɨ-te* ('true soil') or *ɨwɨ-šu'i* ('sandy soil'), exclusive, therefore, of hard scrabble (*ita-kururu-tɨ* 'rock-toad–place of'), seasonally inundated soils (*ɨapo-ran*), swamp soils (*tu'um*), and riverine clays (*tuyuk*). The most important food species in Ka'apor culture, manioc, requires well-drained sandy soils—where 'there is no soft [poorly drained] land' (*memek nišoi*).

In general, vegetation is a good indicator of soil types appropriate for agriculture. Tree species of well-drained sections of high forest, such as *Eschweilera coriacea*, *Lecythis idatimon*, *Sagotia racemosa*, *Tetragastris altissima*, and several species of *Protium*, all indicate appropriate agricultural soils. Indicator tree species of arable fallows include wild papaya (*Jacaratia spinosa*), *Gustavia augusta*, babaçu palm (*Orbignya phalerata*), bacuri tree (*Platonia insignis*), hog plum (*Spondias mombin*), and copal trees (*Hymenaea* spp.), among many others. Fallows recognized as fallows (*taper*) are, however, infrequently converted to cultivation, for reasons explained below.

In addition, those establishing a new swidden seek a site no further than one kilometer from a small, yet 'perennial' (*tɨpa 'ɨm*) stream. Usually, the site is well known to hunters, who would have visited it in all seasons and who are intimately familiar with general conditions in the surrounding forest.

A new communal swidden, depending on the size of the residential unit, is normally more than five hectares. In contrast, the swiddens developed later, as the village ages, are dispersed, familial swiddens measuring only between 0.5 and 2.5 hectares.

(Larger swiddens, even up to about five hectares, have recently been cleared for commercial rice and manioc production, respectively, near Urutawi and Gurupiuna.) During the season of tree-felling at the new swidden, and after it has been decided that a residential unit 'will move' (*tirik-ta*), the new arrivals build temporary shelters (*šipa*) in the swidden. After the manioc has been planted, the men build more permanent houses (*ok*). But they and their families permanently take up residence in the new swidden only after manioc tubers have matured, usually by about December of the following year. Meanwhile, they continue to harvest manioc and other domesticates and semidomesticates from the old village. The old village (*taperer*) and its managed environs, including old familial swiddens (also *taperer*), serve as repositories of crop germplasm. The full repertoire of intensively managed species and cultivars from the prior site is not apparent at the new site until about three years after the initial planting, since the villagers keep returning to the old site for many years, often collecting seeds or cuttings of some of these species and cultivars for planting and transplanting at the new settlement (Balée and Gély 1989:131).

The act of 'moving' (*tirik*) a settlement, nevertheless, sometimes results in the 'loss' (*kãyĩ*) of a few domesticates. Probably such losses of crop germplasm are far more severe and frequent with trekking peoples than with semisedentary peoples such as the Ka'apor. There is a delicate interplay between the advantages and disadvantages of sedentism. Settlement moves are major changes. The costs and benefits must be carefully weighed beforehand. Possible costs involve loss on food investments (still productive old swiddens will be less harvested by humans and increasingly predated by animals), whereas possible benefits include better tortoise hunting and, within a few years, greater biological and ecological diversity at the new site than was present at the old site at the time of its abandonment.

Anthropic changes in the newly settled landscape increase in tandem with the size and age of the village. After a while, distinct vegetational zones can be distinguished, in principle, on the basis of human (i.e., Ka'apor) management. Others, however, owe to the presence of naturally occurring groves of salient species. The names for 'groves' of salient species incorporate the bound morpheme *ti* in final position. These groves may be either managed or unmanaged. For example, açaí groves (*wasaí-ti*) and those of the spiny *Bactris maraja* palm (*maraya-ti*) are generally seen in swamp forest, which is not managed by the Ka'apor. Clusters of these two species are important in the Ka'apor economy, not only because both bear edible fruits (the fruits of açaí are quite important in the Ka'apor diet, while *Bactris maraja* fruits are a secondary food source, i.e., snack food), but because both attract important game animals. Red brocket deer, for example, visit açaí groves to eat fallen fruit, and white-lipped peccary herds are said to revel in the muddy and spiny milieu of *Bactris maraja* groves, where human hunters surround and kill them.

On the other hand, bacuri groves (*pakuri-ti*) and babaçu groves (*yetahu-'i-ti*) are virtually always associated with old fallow (*taper*). Within swiddens, areas with high densities of a single domesticate may likewise be denoted by incorporating the *ti* morpheme, hence: *mani'i-ti* 'manioc-grove' and *waya-ti* 'guava grove'. Still other zones, based on one or more factors not necessarily implying human management, include 'vine-grove' (*sipo-ti*) [said to be a good area for hunting yellow-footed tortoises], the 'treefall gap' (*mira-re*), and 'burned forest' (*ka'a-kai*). The full range of vegetational communities in the environs of a village will thus be found in a combination of managed and nonmanaged zones, either in old swiddens and fallows or in high forests and swamp forests (see Balée and Gély 1989:131–132).

Named vegetational zones which specifically imply either the presence or absence of the human factor are more commonly heard in everyday speech of the Ka'apor. These include high forest (*ka'a-te*), swamp forest (*iapo*), young swiddens (*kupiša*), dooryard gardens (*kar*), and fallow (*taper*). The high forest (*ka'a-te*) is located, by Ka'apor definition, on 'well-drained soils' (*iwi-te*) and includes the species listed in appendix 3. Many species appear to be unique to high forest; of the 30 ecologically most important species, only one is shared between high forest and the other corresponding forest type, old fallow.

Another feature that distinguishes *ka'a-te* in cognitive terms is that, according to Ka'apor informants, no swiddens have ever existed there in historical memory. Even though one occasionally finds in the high forest a stone axehead, which would have been appropriate at least for girdling if not also for felling trees, the Ka'apor do not believe such axeheads were ever used in swiddening. They claim that their ancestors, during the time of warfare (*parahi-wa-rahã*) felled forests with steel axeheads pillaged on raids against settlers. In fact, the only use of these stone axeheads in the past, according to informants, was to impart courage to the warriors on such raids. On the eve of a hostile expedition, the men would put such a stone axehead in the coals of the fire, allow it to become piping hot; then each man, in turn, would clutch the axehead against his breast for as long as he could stand it, to be 'unafraid' (*kiye-'im*) during the coming attack. In modern times, hot stone axeheads are pressed on hunting dogs' chests so that, upon encountering a jaguar, they will bravely hold their ground.

Although I found no 'charcoal' (*tatapũi*) in randomly dug pits on high forest sites, nor were any potsherds apparent, the axeheads I did find remain puzzling and suggest the possibility, at least, that some of these forests were at one time in the remote past felled for swiddens. But if so, it is likely that such felling occurred long before the arrival of the Ka'apor in this habitat, perhaps upwards of two hundred years ago. In any event, the lack of charcoal and soot in the soils as well as a lack of species generally recognized as disturbance indicators suggest that *ka'a-te* may be considered, for analytical purposes, to be primary.

Another kind of primary forest is also recognized. This is swamp forest (*iapo*), which denotes any forest on seasonally or permanently inundated soils. The soils of permanently inundated forests are called 'mud' (*tu'um*), whereas those of seasonally inundated forests are called *iapo-ran*. These forests abound in spiny vines (such as *Uncaria guianensis*) and palms (such as *Astrocaryum* spp., *Bactris maraja*, *Desmoncus* spp., and açaí), as well as indicator trees (such as *Inga* spp., *Anacardium giganteum*, *Myrciaria dubia*, and *Tovomita* spp.—see table 2.1).

As with the managed vegetational zones of dooryard garden, young and old swidden, and fallow, the edges of the unmanaged zones are encoded linguistically with the postposed bound morpheme *mi'i* 'edge'. *Ka'a-te-mi'i*, for example, denotes 'high forest edge' and, depending on location, is bordered either by swamp forest or managed zones such as swiddens and fallows. In these ecotones, nearly 30 percent of the game meat entering the diet is taken (Balée 1985). Many of these ecotones are anthropogenic, specifically, those that border the managed units (such as 'young swidden edge' *kupiša-mi'i* and 'old swidden edge' *taperer-mi'i*—Balée and Gély 1989).

Young familial swiddens (*kupiša*) and other managed zones comprehended in terms of known age, and persons associated with them, are different from unmanaged units. Young swiddens are very personal plots of land during the period of their use, but are not property in a Western legal sense. In addition, of course, young swiddens harbor plant species and varieties not found elsewhere. In particular, the fast-growing variety of manioc known as *tikuwi* is seen only here. Familial swiddens are not 'owned' in the sense of a commodity. Land itself is not a commodity. Rather, people possess usufruct rights to the land they clear and to the crops they plant. Although the forest has no human 'owner' (only the divinity *kurupir*, for example, is believed to 'own' and control game supplies in the forest), young and old swiddens are designated by the names of people who cleared and planted them. For example, *Aramoro'i-kupiša* is a swidden that was cleared and planted by Aramoro'i and his household. He and other adult household members are said to be the *yar* ('usufruct-owner') of the plot. This usufruct right fades after young swidden becomes old swidden (*taperer*). Although the plot may be reused by other families after it has become old swidden, the domesticates therein are of Aramoro'i and his household and should not be harvested without their permission (see also Balée and Gély 1989:131).

A young swidden is identified on the basis of its known age, being no more than about two years old since initial planting. Old swiddens (*taperer*) are of greater age (up to thirty years), and their former owners are also known. What was *Aramoro'i-kupiša* (Aramoro'i's young swidden) becomes *Aramoro'i-taperer* (Aramoro'i's old swidden). Old swiddens contain many more trees greater than 10 cm dbh than do dooryard gardens and young swiddens. The *taperer* which was, in fact, an abandoned village site includes a section that is not found in old familial swiddens. This is the *kaŋwer-rena* 'bone-village', i.e., cemetery. Here the Ka'apor bury the dead from the current village, some of whom may have lived in the *taperer* when it was occupied as a settlement. The trails leading to the *taperer* are marked by deliberately felled spiny trees (such as *Bactris* spp., *Zanthoxylum* spp., *Solanum* spp.), bramble, and vines (such as *Amphilophium, Gurania, Senna,* and *Cayaponia*) to deter earth-bound souls of the *kaŋwer-rena* from leaving their resting place. My informants were virtually unanimous in saying that if a Ka'apor person were to witness an *ãyaŋ*, he or she would die. Except for burying the dead, therefore, the Ka'apor tend to avoid the *kaŋwer-rena* within the *taperer*. This may ease hunting pressure and permit natural replenishment of game in this zone (Balée and Gély 1989).

Old familial swiddens near villages are sometimes felled again, usually after at least ten years have passed, when many secondary growth trees have become fairly tall and thick, and all bitter manioc has long ago been harvested. Here, people normally do not plant bitter manioc, but rather sweet manioc—which is evidently less demanding of the soil. People may reclear the old familial swidden, incidentally, without asking permission from the family that originally cleared the plot, another indication that land is not a commodity and permanent ownership does not ensue from having originally cleared and planted a plot of land. Once a familial old swidden has been recleared, reburned, and replanted, moreover, it becomes the *kupiša* of the new usufruct-owner.

The *taper* 'fallow' is recognized according to several criteria. First, fallows, unlike old swiddens, abound in many species of trees (appendix 1). In a sense, to the Ka'apor, the essential distinction to be made explicit is between fallow and high forest, not fallow and old swiddens. Second, a fallow is not named for a person, unlike the new and old familial swiddens and dooryard gardens. Rather, the fallow is referred to simply as *taper* and distinguished from other fallows in terms of locational markers as well as the species present or absent therein. The *taper* is not attached to a person usually because its last usufruct-owner is likely dead and may not even be remembered by any of the living.

Even so, in the case of fallows where I carried out inventories, some history of each was remembered. For example, hectare 2, a sample of fallow dominated by babaçu palms near P. I. Guajá, was known to have been a Ka'apor village in the 1940s; one of the two headmen of Urutawi, now living 19 km away, claimed to have lived there as a boy. Hectare 5, a fallow dominated by wild papaya and tucumã palms some 4 km from Urutawi, was known to have been a Ka'apor village whose headman had long ago died. Hectare 7, a fallow dominated by bacuri trees near Gurupiuna was believed by the oldest man and woman in the village to have been a colony of refugee Afro-Brazilian slaves. If so, this fallow is older than a hundred years, since maroon colonies were expelled by the Ka'apor from this region in the 1870s. It is certainly old, because both the oldest man and woman in Gurupiuna, who were about seventy years old in 1990, and who have known the bacuri grove since they were children, told me that it had always been a bacuri grove in their memory.

In all fallows I have examined, Ka'apor informants also note the presence of potsherds, charcoal, and darker soils. Although they assume that a cemetery is present, they cannot usually indicate where it is and they seem less concerned about inadvertently walking into a spiritually dangerous situation in *taper* than in *taperer* (where often relatives known to them are interred). Informants also say that certain fallows are easier to fell than high forest, having fewer large trees (an indigenous observation that well accords with lower aggregated basal area for fallow forests). On the other hand, people tend to avoid felling fallows where babaçu is present, such as the fallow near P. I. Guajá. This is because, they say, trying to fell babaçu palms wears down quickly and may even 'break' (*mu-pen*) one's axe. Other fallows are not felled because of the presence of numerous trees highly productive of important fruits. Although the bacuri grove, for example, would be ideal for a swidden—being only about 4 km from Gurupiuna and devoid of babaçu but rich in dark soils—the Ka'apor of Gurupiuna claim that no one fells it because of the highly desirable bacuri fruit trees. The *taper* near Gurupiuna is a kind of community property. Bacuri fruit trees are, incidentally, virtually absent in high forest, and groves of them, associated with fallow, are sparse in the Ka'apor habitat.

In addition to material indicators of fallow, such as potsherds and charcoal, the Ka'apor also distinguish fallow by the presence of indicator plant species. One of these common species, hog plum (*Spondias mombin*), is named *taper-iwa-'i* 'fallow-fruit-stem', i.e., fruit tree of the fallow. This and many other species (see appendix 1) are seen as indicative of fallow. Also, although it is sometimes present, informants distinguish fallow from high forest by its relative (or absolute) lack of *parawa'i* (*Eschweilera* spp.), which is otherwise extraordinarily common in the high forest (see appendix 3).

It may be that fallows in which *Eschweilera* individuals are increasingly common are gradually becoming mature forest, yet the Ka'apor do not recognize any processual connection between fallow and high forest. The Ka'apor distinction between fallow and high forest, however, is sophisticated. Ka'apor knowledge of this distinction is more discriminating, for example, than that of many phytogeographers and cartographers of eastern Amazonia, who have tended to lump fallow and high forests together as "primary" (Balée and Campbell 1990).

In spite of such discrimination between vegetational zones, based on their knowledge of the "ground truth," the Ka'apor in general do not closely monitor fallow formation, which begins with the communal swidden and the first dooryard gardens, the *kar*, of a new Ka'apor settlement.

Management of Dooryard Gardens

The dooryard garden, called *kar*, began as *terra firme* forest (either *ka'a-te* or *taper*) and became part of a new communal swidden (*kupiša*). This is the space immediately surrounding the house; in a mature settlement, the *kar* implies a zone that has been intensively 'weeded' (*ka'a-ra'i-muma*). Often, the only nondomesticates herein were deliberately planted (cf. Harris 1989). During the life of the settlement, the *kar* remains relatively clear of unplanted species. Thus, the *kar* represents an interruption of the usual successional process intervening between young and old swidden. In a sense, the Ka'apor enact a "scorched earth policy" in the dooryard garden. Here they have not only burned the site once for the initial communal swidden and village itself, but they also maintain their ever-burning hearths and manioc furnaces here. They dispose of their garbage, including the charred remains of game and fish, along its edges, which contributes to a darkening (and enrichment) of the surrounding soils.

The *kar* begins after the first manioc plants have all been harvested. Manioc is not planted again here, even though in another young swidden not destined to become *kar* it is often 'replanted' (*yïtïm-we*) as each plant is harvested. As the manioc is harvested and people have begun to settle into their new homes, the *kar* takes form as the space immediately around each house is planted in numerous domesticates and nondomesticates. The *kar* is planted mainly in food crops and plants used in tool and utensil making, spices, and medicinals (see table 6.5). Food crops include cashew, mango, papaya, mountain soursop, sweet potato, pineapple, cupuaçu, watermelon, crookneck squash, peanut, West Indian gherkin, coconut, avocado, banana and plantain, guava, citrus fruits, passion fruit, and cacao. The principal plants used as utensils or adornment include *kirawa* (*Neoglaziovia variegata*), calabash tree, annatto, cotton, and genipapo. Spices include lemon grass, turmeric, basil, chile peppers, and *Eryngium*. Medicinals include tobacco, ginger root, wormseed, kalanchoë, purge nuts, and turmeric.

The *kar* is also one of the few vegetational zones where the Ka'apor cultivate species that are not domesticated—that is, where wild plants and semidomesticated forest plants are occasionally planted and protected, without any evident effects on their genotypes. A number of authors have pointed to the dooryard garden, with its concatenation of domesticates and nondomesticates, as a locus of agricultural evolution, hybridization, and domestication (Anderson 1969; Rindos 1984:133–135; Harris 1989:20). In fact, Ka'apor selection in the *kar* is very intense. The Indians intensively 'weed' (*ka'a-ra'i-muma*) here, claiming that this will reduce snakebite incidents. Yet certain nondomesticates are also deliberately planted here.

Some people plant *Caladium picturatum*, an herb otherwise of the *terra firme* dense forest floor. This plant is believed to ward off evil divinities. As for food plants, the semidomesticated cacao (*Theobroma speciosum*), which is mainly seen in old fallows, is sometimes deliberately planted in dooryard gardens. The Indians claim they plant the seeds only when the fruit exocarp turns 'yellow' (*i'a-tawa-rahã*), otherwise the seeds would be 'unviable' (*pihĩk-'im*). The cupuaçu tree, a relative of cacao and which is most commonly encountered in the high forest, may be planted by men by means of swallowing the seed: "Through my feces it germinates" (*hẽ puši pe har ke hiwõi*). *Inga alba* and *Inga brevialata*, which are used for painting calabash bowls and which contain an edible white pulp, are also planted in the *kar* (Balée and Gély 1989).

In one village (Wasaisĩ, near Gurupiuna), a man planted the seeds of hog plum (*Spondias mombin*) many years ago; today there is a grove of these trees standing on the edge of the present village. With the exception of *Caladium* and cupuaçu, the above-listed species are very common in old fallows and one is tempted to infer that their frequency and density there are the result of human propagation. But the Ka'apor claim that these species are common because *Cebus* monkeys eat their fruits during the rainy season and 'toss' (*omor*) the seeds away, of which many 'germinate' (*hiwõi*). *Cebus* monkeys invade swiddens for maize (Balée 1985), but they also forage in recently abandoned settlements for fruits. Thus although human agents begin the propagation of these species at a specific village site, monkeys will later effect their dispersal into the surrounding old swiddens, which later develop into fallow. Other animals are, according to the Ka'apor, responsible for the high frequency and density of other species in fallows (see below).

In a fallow forest, one can presume that what was formerly the *kar* and the entire settlement site itself (*hena*) would have been the most intensively managed zone. It is the only zone in which intensive weeding took place. And unlike short-lived familial swiddens, here people continuously burned home fires, enriching the soil. They tossed their garbage, rich in organic waste, in pits about the settlement. Long after abandonment, when the vegetation cloaking the old *kar* may be indistinguishable from the vegetation on the rest of the fallow, the former site of the *kar* can usually be assumed to comprise the central portion of old fallow (*taper*).

The dooryard garden of the Ka'apor is an intermingling of "weedy camp followers" and their domesticated congenerics. In this milieu, unconscious experimentation is certainly taking place with traditional crop and noncrop germplasm (Alcorn 1981). What is striking, however, is the high number of introduced domesticates—with no close, weedy congenerics—that the Ka'apor cultivate in dooryard gardens. Some 25 of these species are of Old World origin (table 6.5). Others are certainly or at least probably of tropical American origin, such as Indian shot (*Canna indica*—Bailey et al.

TABLE 6.5 *Plants of Dooryard Gardens (Kar)*

Scientific name	English name	Ka'apor name	Main use
	Traditional Domesticates		
ANACARDIACEAE			
Anacardium occidentale	cashew	**akayu**	food
BIGNONIACEAE			
Crescentia cujete	calabash	**kwi**	container
BIXACEAE			
Bixa orellana	annatto	**uruku**	body paint
BROMELIACEAE			
Ananas comosus	pineapple	**nana**	food
Neoglaziovia variegata	rope plant	**kɨrawa**	rope
CARICACEAE			
Carica papaya	papaya	**mãmã**	food
CONVOLVULACEAE			
Ipomoea batatas	sweet potato	**yɨtɨk**	food
CUCURBITACEAE			
Cucurbita moschata	crookneck squash	**yurumũ**	food
FABACEAE			
Arachis hypogaea	peanut	**manuwi**	food
MALVACEAE			
Gossypium barbadense	cotton	**maneyu**	thread
MUSACEAE			
Musa sp.1	plantain	**pako**	food
Musa sp.2	banana	**pako**	food
MYRTACEAE			
Psidium guajava	guava	**waya**	food
SOLANACEAE			
Capsicum annum	chile pepper	**kɨ'ʔ-hu**	spice
Capsicum fructescens	chile pepper	**kɨ'ʔ-awi**	spice
Nicotiana tabacum	tobacco	**pɨtɨm**	smoking material
	Introduced Domesticates		
AMARANTHACEAE			
Amaranthus oleraceus	amaranth	**ka'a-memek**	spice
ANACARDIACEAE			
Mangifera indica	mango	**mã**	food
Rollinia mucosa	soursop	**mirima**	food
APIACEAE			
Eryngium foetidum		**ka'a-piher**	spice
ARACEAE			
Colocasia esculenta	taro	**taya-ran**	food
ARECACEAE			
Cocos nucifera	coconut	**kuk**	food
CAESALPINIACEAE			
Adenanthera pavonina	bead tree	**pu'ipiraʔ**	beads
CANNACEAE			
Canna indica	canna	**awaí**	beads
CHENOPODIACEAE			
Chenopodium ambrosioides	wormseed	**motoroi**	remedy
CRASSULACEAE			
Bryophyllum sp.	kalanchoë	**kure-nami**	remedy
CUCURBITACEAE			
Citrullus lanatus	watermelon	**waraši**	food
Cucumis anguria	West Indian gherkin	**waraši-ran**	food
Luffa cylindrica	bath sponge	**u'i-hu-ruwɨ**	shotgun shell filter
CYPERACEAE			
Cyperus corymbosus	—	**pipir-iwa**	fragrance
EUPHORBIACEAE			
Jatropha curcas	purge nut	**piyã**	remedy
Jatropha gossypifolia	purge nut	**piyãpihun**	remedy
Ricinus comunis	castor	**karapatu**	hair cream
Desmodium adescendens	tick clover	**ka'a-pe**	fodder
LABIATAE			
Ocimum micranthum	basil	**arapawak**	spice
Ocimum sp. 1	basil	**arapawak**	spice

TABLE 6.5 *Continued*

Scientific name	English name	Ka'apor name	Main use
LAURACEAE			
Persea americana	avocado tree	**apakasi**	food
MALPIGHIACEAE			
Malpighia sp. 1	West Indian cherry	**ma'e-ɨwa-pɨraŋ-ran**	food
Malpighia sp. 2	West Indian cherry	**ma'e-ɨwa-pɨraŋ-ran**	food
MALVACEAE			
Hibiscus rosa-sinensis	hibiscus	**tupã-ka'a**	ornamental
MYRTACEAE			
Syzygium malaccencis	rose apple	**ma'e-ɨwa-pɨraŋ-ran**	food
OXALIDACEAE			
Averrhoa carambola	carambola	**ɨwa-mẽ-'ɨ**	food
POACEAE			
Brachiaria umidicula	braquiaria	**kãpɨ**	fodder
Coix lacryma-jobi	Job's tears	**pu'ɨ-risa**	beads
Cymbopogon citratus	lemon grass	**kãpɨ-piher**	tea
RUBIACEAE			
Coffea arabica	coffee	**kase**	coffee
RUTACEAE			
Citrus aurantiifolia	lime tree	**irimã-te**	food
Citrus sinensis	orange tree	**narãi-'ɨ**	food
SAPINDACEAE			
Cardiospermum halicacabum	balloon vine	**awa-i-ran**	beads
STERCULIACEAE			
Theobroma cacao	cacao	**kaka**	food
ZINGIBERACEAE			
Curcuma sp. 1	turmeric	**tawa**	remedy
Zingiber officinalis	ginger root	**marakatai**	remedy

Cultivated Nondomesticates

Scientific name	English name	Ka'apor name	Main use
ACANTHACEAE			
Justicia pectoralis		**pɨra-pisɨ-ka'a**	magic
Justicia spectabilis		**pɨra-pisɨ-ka'a**	magic
ANACARDIACEAE			
Spondias mombin	hog plum	**taper-ɨwa-'ɨ**	food
ANNONACEAE			
Annona montana	mountain soursop	**arašiku-ran**	food
Rollinia exsucca		**ape-'ɨ-pihun**	food
ARACEAE			
Caladium picturatum	caladium	**ka'a-ro-pinɨ**	magic
CAESALPINIACEAE			
Dialium guianense	jutaɨpororoca	**yurupepe-'ɨ**	food
CUCURBITACEAE			
Posadea sphaerocarpa		**yere'a**	bowl
EUPHORBIACEAE			
Euphorbia sp. 1	spurge	**kupɨ'ɨ-pu'a**	remedy
PASSIFLORACEAE			
Passiflora edulis	passion fruit	**murukuya**	food
PHYTOLACCACEAE			
Petiveria alliacea	guinea hen weed	**mɨkur-ka'a**	remedy
RUBIACEAE			
Genipa americana	genipapo	**yenipa-'ɨ**	body paint
STERCULIACEAE			
Theobroma grandiflorum	cupuaçu	**kɨpɨ-hu-'ɨ**	food
Theobroma speciosum	wild cacao	**kaka-ran-'ɨ**	food

Noncultivated Nondomesticates

Scientific name	English name	Ka'apor name	Main use
AMARANTHACEAE			
Amaranthus spinosus		**kururu-ka'a**	none
EUPHORBIACEAE			
Phyllanthus niruri		**teyu-pɨtɨm-ran**	none
RUBIACEAE			
Borreria verticulata		**kururu-ka'a**	none
SCROPHULARIACEAE			
Scoparia dulcis		**tapiša**	remedy

1976:217), tick clover (*Desmodium adescendens*), *Eryngium foetidum* (Bailey et al. 1976:443), purge nut (*Jatropha* spp.—Bailey et al. 1976:612; Brücher 1989), balloon vine (*Cardiospermum halicacabum*—Bailey et al. 1976:222), West Indian cherry (*Malpighia* spp.—Milliken et al. 1992:86; Purseglove 1968 2:637), avocado (Bergh 1976:149), cacao, soursop (*Rollinia mucosa*—Clement 1989:626; Cavalcante 1988:58), wormseed (Purseglove 1968 2:632), and *Cyperus corymbosus*. Yet these are nevertheless recent introductions to the Ka'apor. Some of these American species, including purge nut, balloon vine, West Indian cherry, tick clover, and avocado, are so new to the Ka'apor that they are called *karaí-ma'e* ('that which is of non-Indians'). The large number of introduced species is, in fact, one of the distinguishing features of the dooryard garden; the space of young swiddens, in contrast, is mostly given to manioc and other traditional domesticates.

Introduced species have arrived to the Ka'apor from different times and places. The origins of old world species introduced soon after contact (around 1500), are explained in mythical terms. The coming of watermelons and three banana varieties, for example, are related in a myth fragment I collected among Ka'apor men in the village of Oriru in the Turiaçu basin:

> Once upon a time a boy was very sad because his father had recently died. The shaman told the boy to close his eyes, which he did, while the shaman chanted. Then the shaman commanded the boy to climb up a monkey ladder vine (*Bauhinia* sp.) to see his father. The boy did so and when he got to the top of the vine, he saw a trail leading through a forest. A large dead soul pulled him off the vine and put him onto the trail, which the boy followed until he arrived at the place of the dead. He slept there for ten days. He saw his father. The dead gave him *waraši-te* ('watermelon-true') and *waraši-pinĩ* ('watermelon-striped'), *pako-hu* ('banana-big'), *pako-tawa* ('banana-yellow'), and *pako-nerĩ* ('banana-?'). The boy returned home with these. Ten days later he died.

Although some banana varieties were probably aboriginal (*pako-te*, for example), at least three varieties were introduced in the 1950s: *katumẽ*, called banana maçã in Portuguese, *pako-parawa*, and *pako pĩraŋ*. Watermelons are certainly African in origin, but their introduction to the Ka'apor probably dates back several hundred years. Most of the other introduced species and varieties, however, are recent introductions, and they are known to the Ka'apor to be *karaí-ma'e* ('that which is of non-Indians'). According to Darcy Ribeiro (1955), Ka'apor raiders of the nineteenth century, in addition to plundering the steel goods of settlers, also made off with lime tree seedlings. Older Ka'apor men themselves say that their ancestors (prior to 1928 and during the time of war) cut shoots of sugarcane in settlers' fields and planted the nodes upon returning home.

After 1928, SPI, its successor FUNAI, missionaries of the Summer Institute of Linguistics (SIL), nearby settlers, and even the Tembé and Guajajara Indians passed on to the Ka'apor other foreign domesticates. SPI and FUNAI, for example, introduced mango, coconut, cacao, purge nuts, avocado, hibiscus, lemon grass, orange trees, and rice between 1930 and 1980. The SIL missionaries introduced soursop, taro, fodder species (braquiaria, *Desmodium*, and *Bryophyllum*) for the recently introduced mules and donkeys, West Indian cherries, rose apple, and carambola between 1962 and 1985.

TABLE 6.6 *Probable Origins of Old World Domesticates in Ka'apor Dooryard Gardens*

Species Name	English Name	Place of origin	Sources
	Food Crops		
Averrhoa carambola	carambola	Indonesia	Harlan 1975:75; Purseglove 1968(2):368
Citrullus lanatus	watermelon	Kalahari Desert	Whitaker and Bemis 1976:67
Citrus spp.	citrus	SE Asia	Cameron and Soost 1976:261-262
Cocos nucifera	coconut	SE Asia	Whitehead 1976:222
Colocasia esculenta	taro	Indo-Malaysia	Plucknett 1976:11
Cucumis anguria	West Indian gherkin	Africa	Purseglove 1968 (1):108
Mangifera indica	mango	Indo-Malaysia	L. B. Singh 1976:8
Syzygium malaccensis	rose apple	SE Asia	Harlan 1975:75
	Spices, Stimulants, Teas		
Amaranthus oleraceus	amaranth	N. India	Harlan 1975:75
Coffea arabica	coffee	SW Ethiopia	Ferwerda 1976:258
Cymbopogon citratus	lemon grass	Sri Lanka, S. India	Bailey et al. 1976:354
Curcuma sp.1	turmeric	SE Asia	Smith 1976:323
Ocimum spp.	basil	Old World tropics	Purseglove 1968 (2):636
Zingiber officinalis	ginger root	China	Smith 1976:324
	Tools, Utensils, Personal Adornment, Remedy		
Adenanthera pavonina	bead tree	SE Asia	Bailey et al. 1976:24
Bryophyllum sp. 1	kalanchoë	Africa	Bailey et al. 1976:24
Coix lacryma-jobi	Job's tears	SE Asia	Smith 1976:308
Hibiscus rosa-sinensis	hibiscus	Pacific Rim	Purseglove 1968 (2):365
Luffa cylindrica	bath sponge	India	Smith 1976:306
Ricinus comunis	castor	Old World tropics	D. Singh 1976:85
Tephrosia sinapou	fish poison	Old World tropics (?)	Bailey et al. 1976: 1101; cf. Purseglove 1968 (1):222
	Fodder for Mules and Donkeys		
Brachiaria umidicula	brachiaria	Old World	Bailey et al. 1976:175

In recent years, from surrounding settlers, the Ka'apor obtained West Indian gherkin, castor bean, bead trees for the red seeds used in bead ornaments, and bath sponge (all of Old World origin). Finally, the neighboring Tembé and Guajajara Indians passed on several other species from across the ocean: spice plants (*Amaranthus oleraceous, Eryngium foetidum*, and basil), coffee, balloon vine (a vine whose tiny black seeds are used in bead ornaments), and wormseed (see table 6.6).

Even today, the Ka'apor eagerly request seeds and seedlings of edible foreign domesticates, often demanding that visitors, upon their return, bring plantable parts of the visitors' foods they enjoyed sharing in. For example, when I introduced some

Ka'apor men, who were helping me with plant collections, to canned peaches, which they unanimously described as being 'delicious' (*katu-te*), they asked if I could bring them viable seeds (knowing that the seeds from the peaches in the can were inviable). I informed them that peaches most likely would not be viable here, requiring, as they do, a temperate climate. This was taken with good humor by the men, since they know also that this is the case with other temperate plants that they relish only on occasion, usually when visitors drop by, such as onion and garlic, which they call *ma'e-wa-nem* 'some-fruit-fetid.'

The dooryard garden is a laboratory of the exotic, yet it is home to the Ka'apor. When the Ka'apor move, introduced domesticates are frequently 'lost' (*kãyĩ*), i.e., do not survive in the emerging fallow and are not replanted or transplanted at the new site. This occurred, for example, with avocado trees and cacao trees (the latter introduced by FUNAI) in the village of Šoanĩ in the early 1980s. The now deceased headman, Ori-ru, asked me where he could obtain seeds of avocado, a fruit he and his kinsmen had once relished, because after their move in 1985 to the new site some thirty kilometers distant, the avocado trees in the old village had failed and died. The native trees of dooryard gardens, such as hog plum and wild cacao, and a select few introduced domesticates, especially lime trees and mangos rarely become *kãyĩ* to any particular group of Ka'apor. But avocados and other introduced tree crops, including custard apple, soursop, and West Indian cherry, are short-lived and compete poorly with the emergent vegetation of old swiddens.

Given that traditional domesticates are, in some sense, most familiar to the Ka'apor, it is remarkable that these species are outnumbered by introduced domesticates. Table 6.7 compares the plants that tend to be found in dooryard gardens, young swiddens and old swiddens. In the dooryard garden, traditional domesticates (that is, neotropical crops not introduced to the Ka'apor within the last hundred years) number 16 species. Introduced domesticates (that is, Old World crops or neotropical crops introduced to the Ka'apor within the last hundred years) account for 36 species. There are 14 cultivated nondomesticates (that is, wild plants or plants spontaneously associated with swidden fallows that are occasionally planted). Noncultivated nondomesticates (which are wild plants or plants spontaneously associated with areas of human disturbance that are never planted) account for only 4 species.

Management of Young Swiddens

'Young swiddens' (*kupiša*), unlike dooryard gardens, are not weeded. The sole exception is rice-producing swiddens; rice is quite recently introduced (Balée and Gély 1989). Although the number of traditional domesticates considerably outnumbers the introduced domesticates in Ka'apor swiddens, it is intriguing that a few *varieties* of traditionally domesticated species are, in fact, introduced. In addition to the three banana varieties mentioned in the myth above, this is the case with a variety of bitter manioc. The introduced variety of bitter manioc, *tikuwi*, likewise does not incorporate the folk generic name *mani'i*, common to the other bitter manioc varieties, yet it, too, is classified as a kind of *mani'i*. *Tikuwi* is a borrowing from *tiquía*, local Portuguese for this vari-

TABLE 6.7 *Summary of Plant Types in Dooryard Gardens and Swiddens*

	Dooryard Garden	Young Swidden (Numbers of Species)	Old Swidden
Traditional Domesticates	16	17	12
Introduced Domesticates	36	9	5
Cultivated Nondomesticates	14	0	2
Noncultivated Nondomesticates	4	42	72

ety which was evidently first bred by neo-Brazilians in the lower Amazon region south of the Amazon River proper. It was introduced to the Ka'apor by SPI agents in the 1950s. This variety today tends to occupy more swidden space on average than any other bitter manioc variety. It is the fastest maturing variety of bitter manioc to be found in Ka'apor swiddens, becoming harvestable after only eight months (Balée and Gély 1989:133). *Tikuwi* tends to be the first variety planted, in the swidden center (*kupiša-piter*). It is a poor competitor, vulnerable to leaf-cutter ants of the genus *Atta* (*sa'i*). A slow maturing traditional variety, *mani'itawa* ('yellow manioc'), is customarily planted along the swidden's edge (*kupiša-mi'i*). It is immune to attack by leaf-cutter ants (perhaps because of greater toxicity). The Ka'apor say that leaf-cutter ants tend to avoid the swidden center where the most vulnerable manioc variety is present, because of the unattractive (to the ants) manioc variety found along the swidden's edges. The non-Indian breeders of *tikuwi* invested in rapid growth but not protection for this plant (cf. Rindos 1984:109). On the other hand, the Ka'apor have devised a means for protecting it by using one of their own ancient cultivars, yellow manioc. This kind of pest control by judicious placement is reminiscent of pest control practices of the sixteenth century Tupinambá. They scattered certain useless leaves attractive to the ants along blind trails to detour leaf cutter ants from manioc swiddens (Sousa 1974:89).

The edge of the swidden also contains numerous nondomesticates. Weeding (*ka'a-ra'i-muma*) is rare here. Although the Ka'apor do not intentionally cultivate any nondomesticates in the swidden, many of the opportunistic species that invade the swidden edge are useful to the Ka'apor. These species tend to cluster along swidden edges. Among such volunteers are several species that attract game animals, in particular red brocket deer, one of the most important animals in the Ka'apor diet. Incidentally, some of the volunteer species, such as *Triumfetta*, used for making rope, appear to be of Old World origin (Purseglove 1968 (2),613), where they were domesticated. But in the Ka'apor habitat, these plants grow spontaneously.

Most of the noncultivated nondomesticates in Ka'apor swiddens are somehow useful (table 6.7). Several are important specifically because they attract game animals. For example, morning glories, the nondomesticated congenerics of sweet potato, are said to attract collared peccaries, which are frequent invaders of swiddens. Several nondomesticated manioc species, which are called 'red brocket deer–manioc' (*arapuha-mani'i*), do tend to be true to their name. These are woody species not eaten by the Ka'apor. The plants' energy is apparently invested more in stem lignin and fruit than in tubers, which are virtually absent—the Ka'apor claim that these species 'produce no

TABLE 6.8 *Plants of the Young Swidden (Kupiša)*

Scientific Name	English Name	Ka'apor Name	Main Use
	Traditional Domesticates		
ARACEAE			
Xanthosoma sagittifolium	cocoyam	**taya**	food
ANACARDIACEAE			
Anacardium occidentale	cashew	**akayu**	food
ASTERACEAE			
Clibadium sylvestre	cunumi	**kanamɨ**	fish poison
BROMELIACEAE			
Ananas comosus	pineapple	**nana**	food
CARICACEAE			
Carica papaya	papaya	**mãmã**	food
CUCURBITACEAE			
Cucurbita moschata	squash	**yurumũ**	food
Lagenaria siceraria	bottle gourd	**kawasu**	container
DIOSCOREACEAE			
Dioscorea trifida	yam	**kara**	food
EUPHORBIACEAE			
Manihot esculenta	manioc	**mani'ɨ, makaser maniaka**	food
FABACEAE			
Arachis hypogaea	peanut	**manuwi**	food
Phaseolus lunatus	lima bean	**kamana**	food
MALVACEAE			
Gossypium barbadense	cotton	**maneyu**	thread
MUSACEAE			
Musa X paradisiaca	plantain	**pako**	food
Musa X sapientum	banana	**pako**	food
POACEAE			
Gynerium sagittatum	arrow cane	**u'ɨwa**	arrows
Zea mays	maize	**awaši**	food
SOLANACEAE			
Nicotiana tabacum	tobacco	**pɨtɨm**	smoking material
	Introduced Domesticates		
CANNACEAE			
Canna indica	Indian shot	**awa-i**	beads
CUCURBITACEAE			
Citrullus lanatus	watermelon	**waraši**	food
Tephrosia sinapou	fish poison	**šimo-'i**	fish poison
Vigna adnantha	cowpea	**kamana-'i-tuwɨr**	food
POACEAE			
Coix lacryma-jobi	Job's tears	**pu'ɨ-risa**	beads
Oryza sativa	rice	**awaši-'i**	food
Saccharum officinarum	sugarcane	**kã**	food
ZINGIBERACEAE			
Curcuma sp. 1	turmeric	**tawa**	remedy
Zingiber officinalis	ginger root	**marakatai**	remedy
	Cultivated Nondomesticates		
NONE			
	Noncultivated Nondomesticates		
APOCYNACEAE			
Himatanthus sucuuba	—	**kanaú-'ɨ**	remedy
Tabernae-montana angulata	—	**kaŋwaruhu-mɨra**	remedy
ARACEAE			
Dracontium sp. 1	—	**anɨ-ran**	poison
ARECACEAE			
Astrocaryum munbaca	mumbaca	**yu**	food
Bactris setosa	—	**piri'a**	food
Maximiliana maripa	inajá	**inaya-'ɨ**	food
Oenocarpus distichus	bacaba	**pinuwa-'ɨ**	food

TABLE 6.8 *Continued*

Scientific Name	English Name	Ka'apor Name	Main Use
ASCLEPIADACEAE			
Asclepias curassavica	milkweed	ira-kiwa-ran	remedy
ASTERACEAE			
Conyza bonariensis	—	teyu-pitim	remedy
Eupatorium macrophyllum	—	ka'a-yuwa-ran	game food
Porophyllum ellipticum	—	teyu-pitim	remedy
Pterocaulon vergatum	—	teyu-pitim	remedy
BIGNONIACEAE			
Mansoa angustidea	—	sipo-nem	none
BORAGINACEAE			
Cordia multispicata	—	kurupi-'i-sipo	lashing material
CAESALPINIACEA E			
Bauhinia viridiflorens	—	ainumir-mira	bow, arrow point
CONVOLVULACEAE			
Ipomoea aff. squamosa	morning glory	yitik-ran	game food
Ipomoea phyllomega	morning glory	yitik-ran	game food
CUCURBITACEAE			
Gurania eriantha	—	kawasu-ran	remedy
CYPERACEAE			
Diplasia karataefolia	—	karaya-hi	blade for haircutting
Scleria secans	—	tiriri	game food
EUPHORBIACEAE			
Manihot leptophylla	wild manioc	arapuha-mani'i	game food
Manihot quinquepartita	wild manioc	arapuha-mani'i	game food
MALVACEAE			
Sida cordifolia	—	mira-kirawa-ran	rope
Sida santaremensis	—	mira-kirawa-ran	rope
Urena lobata	aramina fiber	mira-kirawa	rope
MENDONCIACEAE			
Mendoncia hoffmann-seggiana	—	pikahu-ãmpũi-rimo	magic
PASSIFLORACEAE			
Passiflora aranjoi	passion fruit	murukuya-ran	game food
Passiflora coccinea	passion fruit	murukuya-ran	game food
Passiflora sp. 2	passion fruit	murukuya-ran	game food
PHYTOLACACCEAE			
Phytolacca rivinoides	—	ka'a-riru	arrow shaft dye
PIPERACEAE			
Piper annonifolium	jamira	yamir-hu	none
POACEAE			
Digitaria insularis	—	u'i-wa-ran	none
Paspalum conjugatum	—	kupi'i-pe	game food
SOLANACEAE			
Physalis angulata	gooseberry	kamamu	food
Solanum crinitum	—	yuruwe-te	food
Solanum rugosum	—	ka'a-yuwar	game food
Solanum stramonifolium	—	yuruwe-piraŋ	food
Solanum subinerme	—	yuruwe-tawa	game food
TILIACEAE			
Triumfetta rhomboidea	—	mira-kirawa	rope
Triumfetta semitriloba	—	mira-kirawa	rope
URTICACEAE			
Laportea aestuans	stinging nettles	purake-ka'a	remedy
ZINGIBERACEAE			
Renealmia alpinia	canarana	kurupi-kã	body paint

tubers' (*hapo 'im*). According to informants, red brocket deer eat the large fruits of these wild manioc plants, digesting the pericarp and regurgitating the seeds. The regurgitated seeds may be viable and then grow into mature plants. To indicate their knowledge of how wild manioc plants are dispersed, the Ka'apor say jokingly that the deer 'plant' (*yitim*) wild manioc. The deer tend to feed on the swidden's edge, avoiding the open area of the swidden center. Nevertheless, red brocket deer are frequently taken in swiddens, often as they are fleeing from human hunters who surprised them at the swidden's edge (see Balée and Gély 1989:137).

The most significant food species are to be found in the sunny center of the swidden. In a representative young swidden at Šoanĩ in 1985, shown in figure 6.6, paired associations of bitter manioc with yams, sweet potatoes, and squash are seen in the swidden center. Yams are said to do best along 'fallen logs' (*mira-pe-wi*), where there is shade during some of the day. Squashes and watermelons are planted in the swidden center also. Their seeds must be 'planted' (*yitim*), not 'scattered' (*omor*), for otherwise, it is said, the sun would 'pop open' (*opok*) the seeds. But the germinated plants do well in the intense solar radiation at the swidden center.

Traditional domesticates of young swiddens are being rapidly supplanted in some Ka'apor villages near the eastern frontier of the Reserva Indígena Alto Turiaçu (Balée and Gély 1989). Agents of FUNAI encouraged these villages to plant rice for sale, so that they would have money to purchase tools, clothing, radios, and other goods. Rather than intensifying agriculture to support an increment of cash crop, these villagers have reduced the crop space available for traditional crops. They have sacrificed certain foods, spices, and other utilitarian plants for access to industrial commodities. Hence, many traditional domesticates are no longer to be found in swiddens in these villages.

Management of Old Swiddens and the Formation of Fallow

Old swiddens (*taperer*) are swiddens two years or older. Exclusive of the space of abandoned villages (also called *taperer*), old swiddens harbor most of the traditional domesticates found in young swiddens, with the exception of certain short-lived species. One of the major differences between *taperer* and *kupiša*, in terms of traditional domesticates, concerns the varieties of manioc found in each. The fast-growing cultivars of bitter manioc, especially *tikuwi*, are gone from old swiddens; moreover, *makaser* and *maniaka*, traditional types of sweet manioc whose tubers mature seven to eight months after planting, were all harvested when the swidden was still young. Rather, the only manioc cultivars remaining are traditional bitter varieties, including *mani'itawa, mani'ise,ararũmani'i, mani'ihowi,* and *mani'ipo,* the tubers of which mature some two years after initial planting.

Some replanting without burning occurs in the *taperer*, usually in areas where all manioc plants have been harvested—this is one of the places where people defecate and also one of the places where occasional tree planting occurs upon defecation. The *taperer* is fundamentally unlike the young swidden, which is more than 80 percent covered with manioc plants. A rule stated to visitors who ask where to defecate is 'not

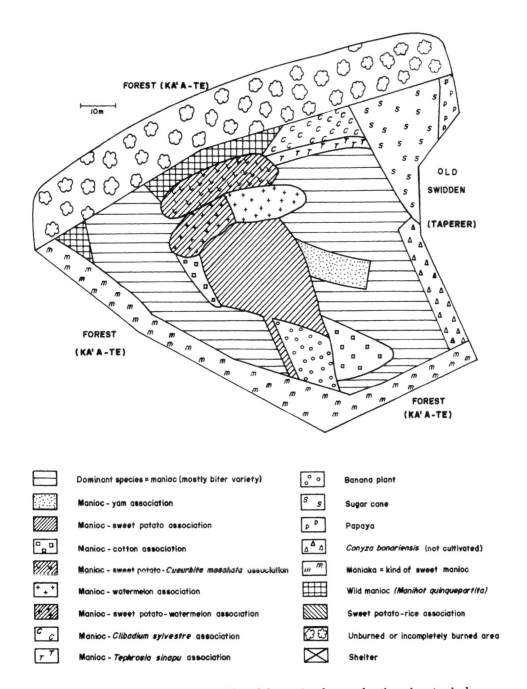

FOREST (KA' A -TE)

10m

FOREST
(KA' A -TE)

OLD

SWIDDEN

(TAPERER)

FOREST
(KA' A-TE)

	Dominant species = manioc (mostly biter variety)			Banana plant
	Manioc - yam association			Sugar cane
	Manioc - sweet potato association			Papaya
	Manioc - cotton association			*Conyza bonariensis* (not cultivated)
	Manioc - sweet potato-*Cucurbita maschata* association			Mäniaka = kind of sweet manioc
	Manioc - watermelon association			Wild manioc *(Manihot quinquepartita)*
	Manioc - sweet potato-watermelon association			Sweet potato-rice association
	Manioc -*Clibadium sylvestre* association			Unburned or incompletely burned area
	Manioc -*Tephrosia sinapu* association			Shelter

FIGURE 6.6 Layout of a young swidden. Schematic of crop plantings in a typical young swidden near the village of Šoanī (After Balée and Gély 1989:140).

around the manioc plants' or, in other words, not in the young swidden nor in the parts of the old swidden where manioc tubers are still to be found.

Old swiddens tend to be well stocked with noncultivated nondomesticates, including several palms, *Cecropia*, several wild cucurbits, wild cacao trees, *Heliconia* plants, trees of the melastome family, wild passion fruit, numerous spiny shrubs and trees of the tomato family, and species of the ginger family (table 6.9). Many investigators view this emerging vegetation as the first step toward forest regeneration.

Unlike young swiddens in which many species are not yet capable of producing seed, old swiddens harbor the crop germplasm that is transplanted to new settlements. Old swiddens and old villages, of course, also contain the big-seeded fruits of the species that will later dominate the fallow. Although the Ka'apor do on occasion consciously plant nondomesticated tree species, such as ingá, wild cacao, jutaípororoca (*Dialium guianense*), and wild soursop, most of the important tree species of the fallow are not the result of planting. When queried about fallow formation, informants tend to say that many of these species are introduced or dispersed by game animals attracted to food plants near human settlements. Agoutis disperse babaçu palm, inajá palm, and copal trees; deer disperse bacuri and hog plum; and *Cebus* monkeys disperse wild cacao and ingá. With a few individuals of certain species, some human planting is involved (wild cacao, ingá, hog plum). But informants nevertheless say that groves of these species were not entirely planted, but rather propagated from the originally planted tree. Many palms, such as bacaba and inajá, are said to be present because humans first scattered (*omor*) the seeds (while throwing them away on the edge of the dooryard garden or village), but animals subsequently dispersed them.

As for babaçu palm, one of the most important fallow species, I have found no evidence for human planting—the Ka'apor eat the fruit only as snack food, do not tend to return to the village with it, and do not plant it. Rather, its presence near long-abandoned human settlements probably owes to dispersal by agoutis attracted to the settlements.

In other words, human involvement in fallows dominated by babaçu palms takes place indirectly. This kind of fallow management is fundamentally unlike that described for the Gorotire Kayapó by Posey (1983, 1985) and Anderson and Posey (1985, 1989). In fact, many of the species intentionally planted on 'forest islands' (*apêtê*) by the Kayapó and which are also present in Ka'apor fallows are never planted by the Ka'apor. These include *Tapirira guianensis*, *Himatanthus sucuuba*, *Schefflera*, *Tabebuia serratifolia*, *Tetragastris altissima*, *Maytenus*, *Casearia*, *Sacoglottis* spp. 1, *Mascagnia* sp. 1, *Cecropia palmata*, *Neea* spp., *Coccoloba paniculata*, *Simaruba amara*, and *Vitex flavens* (cf. Anderson and Posey 1989:162–168; see appendix 1). The Kayapó forest island, which contains many species that in the Ka'apor habitat indicate 'fallow' (*taper*), exhibits 75 percent 'plantable' species according to Kayapó informants (Anderson and Posey 1989:169). Yet in Ka'apor fallows, the percentage of tree and vine species occasionally planted by the Ka'apor is only about 1 percent (or 4 out of a total of 360 tree and vine species collected in fallow inventories). The four species are *Dialium guianense*, *Rollinia exsucca*, hog plum (*Spondias mombin*), and wild cacao (*Theobroma speciosum*). Whereas the Kayapó may be consciously planting and propagating nondomesticated tree species (cf. Parker 1992), Ka'apor fallows represent one of the unintended (yet unsurprising) results of human/animal interactions. Many of the plant species therein are present, in other words, because of animals attracted to the

TABLE 6.9 *Plants of the Old Swidden (Taperer)*

Scientific Name	English Name	Ka'apor Name	Main Use
	Traditional Domesticates		
ANACARDIACEAE			
Anacardium occidentale	cashew	**akayu**	food
BIXACEAE			
Bixa orellana	annatto	**uruku**	body paint
BROMELIACEAE			
Ananas comosus	pineapple	**nana**	food
Neoglaziovia variegata	rope plant	**kirawa**	rope
CONVOLVULACEAE			
Ipomoea batatas	sweet potato	**yitik**	food
DIOSCOREACEAE			
Dioscorea trifida	yam	**kara**	food
EUPHORBIACEAE			
Manihot esculenta	manioc	**mani'i**	food
(bitter variety only)			
MALVACEAE			
Gossypium barbadense	cotton	**maneyu**	thread
MUSACEAE			
Musa X sp. 1	plantain	**pako**	food
Musa X sp. 2	banana	**pako**	food
MYRTACEAE			
Psidium guajava	guava	**waya**	food
POACEAE			
Gynerium sagittatum	arrow cane	**u'iwa**	arrow shaft
	Introduced Domesticates		
ANACARDIACEAE			
Mangifera indica	mango	**mã**	food
CANNACEAE			
Canna indica	Indian shot	**awa-í**	beads
POACEAE			
Saccharum officinarum	sugarcane	**kã**	food
RUTACEAE			
Citrus aurantiifolia	lime	**irimã-te**	food
Citrus medica-acida	citron	**irimã-te**	food
	Cultivated Nondomesticates		
ANACARDIACEAE			
Spondias mombin	hog plum	**taper-iwa-'i**	food
STERCULIACEAE			
Theobroma speciosum	wild cacao	**kaka-ran-'i**	food
	Noncultivated Nondomesticates		
ANNONACEAE			
Annonaceae sp. 1	—	**yaši-mukuku-'i**	remedy
APOCYNACEAE			
Himatanthus sucuuba	—	**kanaú-'i**	remedy
Tabernaemontana angulata	—	**kaŋwaruhu-mira**	remedy
ARECACEAE			
Bactris setosa	—	**piri'a**	food
Markleya dahlgreniana	—	**inaya-kami**	game food
Maximiliana maripa	inajá	**inaya-'i**	food
Oenocarpus distichus	bacaba	**pinuwa-'i**	food
Orbignya phalerata	babaçu	**yetahu-'i**	food
ASCLEPIADACEAE			
Schubertia grandiflora	—	**iraí-sipo**	remedy
ASTERACEAE			
Conyza bonariensis	—	**teyu-pitim**	remedy
BIGNONIACEAE			
Amphilophium paniculata	—	**sipo-tuwir**	none
Jacaranda heterophylla	—	**para-'i-ran**	none
Jacaranda paraensis	—	**para-'i-ran**	none

TABLE 6.9 Continued

Scientific Name	English Name	Ka'apor Name	Main Use
BORAGINACEAE			
Cordia polycephala	—	kurupi-kǐʔ-ran	lashing material
Cordia sp. 2	—	ape-'ɨ-howɨ	food
BROMELIACEAE			
Tillandsia usneoides	Spanish moss	ɨra-hu-ra-wi	none
Aechmea brevicellis	—	karawata	game food
CAESALPINIACEAE			
Senna sylvestris	senna	ǯimo-'i-ran	game food
CECROPIACEAE			
Cecropia palmata	cecropia	ama-'ɨ-puku	food
CHRYSOBALANACEAE			
Hirtella racemosa var. *hexandra*	—	mukuku-'ɨ-wi	firewood
CLUSIACEAE			
Vismia guianensis	lacre	wayaŋ-ɨ	body paint
COCHLOSPERMACEAE			
Cochlospermum orinocense	periquiteira	samo'ã-'ɨ	rope
CONVOLVULACEAE			
Ipomoea phyllomega	morning glory	yɨtɨk-ran	game food
Ipomoea setofera	morning glory	yɨtɨk-ran	game food
Ipomoea aff. *squamosa*	morning glory	yɨtɨk-ran	game food
CUCURBITACEAE			
Cayaponia sp. 1	—	kawasu-ran	remedy
Gurania eriantha	—	kawasu-ran	remedy
Posadea sphaerocarpa	—	yere'a	bowl
CYPERACEAE			
Diplasia karataefolia	—	karaya-hi	blade for cutting hair
Scleria secans	—	tɨrɨrɨ	game food
ERYTHROXYLACEAE			
Erythroxylum cf. *leptoneurum*	—	tapiǯa-'ɨ	game food
Erythroxylum citrifolium	—	mɨra-ãtã	game food
EUPHORBIACEAE			
Mabea angustifolia	—	kaǯima-'ɨ	pipe stem
Manihot brachyloba	wild manioc	arapuha-mani'ɨ	game food
Manihot leptophylla	wild manioc	arapuha-mani'ɨ	game food
Manihot quinquepartita	wild manioc	arapuha-mani'ɨ	game food
FLACOURTIACEAE			
Banara guianensis	—	mitũ-eha-'ɨ	firewood
Neoptychocarpus apodanthus	—	yaǯi-mukuku-'ɨ	firewood
HAEMODORACEAE			
Xiphidium caeruleum	—	ɨrakahu-ka'a	poison
HELICONIACEAE			
Heliconia bihai	heliconia	tayahu-pako-ro	game food
LAURACEAE			
Ocotea glomerata	—	maha-yuwa-'ɨ	firewood
LORANTHACEAE			
Struthanthus marginatus	mistletoe	ma'e-wɨra-puǯi	none
MALVACEAE			
Sida santaremensis	—	mɨra-kɨrawa-ran	rope
Urena lobata	aramina fiber	mɨra-kɨrawa-ran	rope
MELASTOMATACEAE			
Aciotis purpurescens	—	yu'ɨ-ka'a	game food
Bellucia grossularioides	goiabinha de anta	mu-'ɨ	game food
Myriaspora egensis	—	mu'ɨ-ran	game food
MYRTACEAE			
Eugenia patrissi	pitanga	ma'e-ɨwa-pɨtaŋ	food
Myrcia paivae	pitanga	u'ɨ-tɨma-ran	food
PASSIFLORACEAE			
Passiflora aranjoi	passion fruit	murukuya-ran	game food
Passiflora sp. 2	passion fruit	murukuya-ran	game food
Passiflora occinea	passion fruit	murukuya-ran	game food

TABLE 6.9 *Continued*

Scientific Name	English Name	Ka'apor Name	Main Use
POACEAE			
Imperata brasiliensis	sape	u'ɨ-wa-ran	none
POLYGONACEAE			
Polygonaceae sp. 1	—	tatu-ruwai	remedy
RHAMNACEAE			
Gouania pyrifolia	—	tapuru-ka'a	soap
RUBIACEAE			
Coussarea paniculata	—	yakamĩ-mira	game food
Psychotria sp. 1	—	akuši-nami	game food
RUTACEAE			
Galipea trifoliolata	—	teyu-ka'a	remedy
Zanthoxylum rhoifolium	—	wa-šĩ-šĩ-'ɨ	firewood
SAPOTACEAE			
Chrysophyllum sparsifolium	—	ara-kã-šĩ-'aran-'ɨ	food
SOLANACEAE			
Solanum crinitum	—	yuruwe-te	food
Solanum rugosum	—	ka'a-yuwar	remedy
Solanum stramonifolium	—	yuruwe-piraɨ	food
Solanum subinerme	—	yuruwe-tawa	game food
STERCULIACEAE			
Guazuma ulmifolia	—	api'a-'ɨ	food
Sterculia alata	—	šimo-hu-'ɨ	game food
TILIACEAE			
Triumfetta semitriloba	—	mira-kirawa	rope
ULMACEAE			
Trema micrantha	trema	kuru-mi'u-'ɨ	food
URTICACEAE			
Laportea aestuans	stinging nettles	purake-ka'a	remedy
VERBENACEAE			
Lantana camara	lantana	kanami-ran	none
ZINGIBERACEAE			
Renealmia alpinia	canarana	kurupi-kã	body paint
Renealmia floribunda	canarana	kurupi-pitim	remedy

intensively human-managed domain of dooryard gardens and producing swiddens—not because of having been planted by human beings.

One need not plant that which will be predictably 'planted' by animals or which will germinate and grow upon scarification by swidden burning (as with copal trees—Balée and Gély 1989). The vast majority of time the Ka'apor spend gardening is not directed to planting nondomesticated trees, therefore, but to cultivating domesticated starchy tuber plants and domesticated herbs, vines, and shrubs. Despite this minimal management of nondomesticated tree species, at least compared to the Kayapó, it is still logical to conclude that fallows are human creations. Fallows—and fallows of the precise tree species characteristic of the Ka'apor habitat—would not occur and persist were it not for the Ka'apor management of other plants and their creation of diverse anthropogenic zones, including old and new villages as well as old and new swiddens, which by themselves attract the principal dispersal agents of many fallow forest tree species. Fallows are indigenous orchards, but to a large extent they are also the unintended artifacts of the manipulation of herbaceous, i.e., nontree vegetation.

States, Nonstates, and Bioecological Diversity

Indigenous plant management unquestionably changed the aboriginal forest profile of Amazonia and pre-Amazonia. High forests, nevertheless, still endure in many long-occupied indigenous areas, and these coexist with fallow forests. (In this context, William Denevan, in an otherwise insightful article, may have prematurely stated that "there are no virgin tropical forests today nor were there in 1492" [1992:375].) In many zones of recent penetration by civilization, however, neither high nor fallow forest has survived.

Indigenous fallows contrast markedly with cattle pastures and monocultural swiddens. This contrast is evident not only to ground-based observers, but also to satellites. Fallows are, in a way, the logical opposite of deforestation, because they are reforested patches of vegetation. They are the result of indigenous, nonstate forest management, whether intentional or not. Slash-and-burn agriculture and so-called sustainable agriculture are, by this evidence, not necessarily opposed categories. Yet the U.S. Environmental Protection Agency concluded that to reduce sources of greenhouse gases, managers of tropical forests should substitute "sustainable agriculture" for slash-and-burn (see Tirpak 1990:42). Modern state societies indeed show no indication of being capable of truly sustainable forms of slash-and-burn regimes, but indigenous, nonstate societies of Amazonia and pre-Amazonia evidently did and continue to do so.

Pre-Columbian horticultural peoples altered the topography as well as the vegetation. Evidence reported by Roosevelt (1987, 1989b, 1991) includes enormous mounds, earthworks, and differential burials on Marajó Island at the mouth of the Amazon and near Santarém, at the mouth of the Tapajós River. Anthrosols (*Terra Preta do Índio*— 'Indian Black Earth'), which occur widely throughout the Amazon Basin, also evince pre-Columbian human manipulation of the natural landscape (Smith 1980). Many landscapes, soils, and forests today in the Amazon, then, intimate a very old human factor, specifically, one that was not associated with a state society.

In some ways, modern Amazonian Indians use and manage the forest differently from their pre-Columbian forebears. Present-day hunter-gatherer societies of the forests of eastern Amazonia, such as the Guajá, who do not traditionally fell and burn forest for swidden fields, exert less influence on forest composition than did the pre-Columbian chiefdoms of Amazonia—some of which may have been close to becoming state societies at the time of the European Conquest. Modern hunter-gatherers also, of course, have less effect on forest composition than do modern horticultural village societies, such as the Ka'apor.

It is a remarkable yet well-documented fact that many of the hunter-gatherer societies lost the art of fire-making. The hunting and gathering Guajá Indians of pre-Amazonia, for example, do not know how to make fire from a drill and hearth, an art that is otherwise known widely to horticultural peoples of the Amazon basin. They used torches from the combustible sap of the *Manilkara huberi* tree (maçaranduba, sapote family) to travel from camp to camp and maintained camp fires with slow-burning woods, such as *Sagotia racemosa* of the spurge family, known to them as the *mirikə-'i* 'woman-stem'. Unlike their horticultural neighbors, the Ka'apor, the Guajá did not recently fell and burn forests for swiddens. Yet in the remote past, the ancestral culture of the Guajá Indians certainly had a suite of domesticated crops, given their affiliation with the Tupí-Guaraní family of languages. The Guajá lost horticulture and their plant domesticates probably because of the introduction of Old World diseases and attendant marked depopulation that followed the Conquest. Severe depopulation destabilized the society and made it increasingly nomadic (Balée 1992a). In the tropics, full-time nomadism appears to be incompatible with horticulture.

Even if today's hunter-gatherers exert a negligible effect on forest composition, their ancestors did manipulate many of the forests often erroneously assumed today to be primary (Balée 1987, 1989a, 1989c). One would be mistaken, however, to consider these largely unintentional manipulations of past environments as simply deforestation, since the artifactual fallows the Indians left behind, wittingly or not, are a kind of forest, too. In any case, the pre-Columbian chiefdoms of the past, together with the traditional hunting-and-gathering and horticulturalist Indian peoples of today in the Amazon have more in common with each other than any of them do with modern nation-states, in terms of effects on biological and ecological diversity. Among these various forms of societies, only the state type is a cause of the massive biotic depletions of today.

Most of the crops and the species that forest dwellers exploit in primary forest and fallow, despite recent introductions of alien species to swiddens and dooryard gardens, are neotropical. Most Amazonian peoples continue to be like their forebears— they are still Indians—certainly in terms of their plant resources and forest management practices. The resource management practices of the indigenous farmers and foragers of Amazonia of today are less destructive of the environment, by any measure, than are the rapacious nation-states with economies based on the burning of fossil fuels. Pre-Columbian chiefdoms and modern horticulturalist village societies, such as the Ka'apor, did and continue to alter the natural environment. But to archaeologists of the future, indigenous burial mounds and other earthworks (such as stockades) will represent a qualitatively different kind of environmental manipulation than will mega-projects such as the hydroelectric dams of Tucuruí, Balbina, and, if ever constructed, Kararaô (on the Rio Xingu).

The trails and highways that linked indigenous Amazonian villages—some of which, such as the Tapajós Indian village at the mouth of the Rio Tapajós, were incipient urban centers (Roosevelt 1989b)—will never rival the Belém-Brasília, Belém-São Luís, and Trans-Amazon highways in terms of habitat conversion. The carbonized plant remains from prehistoric Indian villages, swiddens, and fallows will suggest far more species of trees than will the strata indicative of extensive cattle pastures and monospecific stands of rice. Finally, there is really no indigenous equivalent for rivers and lakes poisoned by mercury in the modern Amazon gold rush.

Indigenous forest management, such as that of the Ka'apor, rather than reducing natural bioecological diversity, appears to increase it. Fallows contain a great number of species not present in high forest, and they contain an even greater number that are only minimally present in high forest. This diversification has taken place over many centuries. Such a long heritage of plant manipulation, even though many of its positive results may have been unintended, has evidently affected the ways in which the Ka'apor cognize, classify, and talk about plants.

SEVEN

Plant Nomenclature and Classification

The preceding chapter argued for the primacy of long-term historical, unconscious factors underlying modern Ka'apor plant management. This chapter adduces linguistic evidence for these factors.

That the practices which lead to fallow formation and increased plant biodiversity, however they may be lacking in conscious design, are truly ancient, can be inferred from patterns revealed in the Ka'apor language itself. Specifically, Ka'apor plant nomenclature evinces highly determinate mechanisms that, in effect, dichotomize the plant domain into traditional domesticates versus nondomesticates (whether truly wild or semidomesticated). Folk generic names for traditional domesticates appear to always be literal, simple, primary lexemes with unitary meanings (words such as 'oak', 'pine', and 'carrot' may serve as folk English analogues). Although folk generic terms for nondomesticates may also be literal and simple primary lexemes, many are not. In addition, the folk classification of plants appears partly to bear out these nomenclatural patterns: domesticates (including introduced as well as traditional ones) do not seem to be subsumed under major life form taxa ('trees', 'herbs', and 'vines'), yet the great majority of nondomesticates pertains to these.

These and related findings suggest an intimate relationship between language and culture indicative of historical interdependence rather than borrowing. At least in the Ka'apor language, but probably in many other languages associated with traditional horticultural societies as well, one can observe the effects of many generations of plant management activities. This is not simply based on the fact that the plant vocabulary contains specific words for domesticates as well as words for managed vegetational zones, horticultural tools, and forest management activities themselves. This relationship between language and culture is to be apprehended most conclusively, rather, from the evidence for an *unconscious* nomenclatural dichotomy between traditional domesticates and other plants, which itself is partly reflected in the elicited structure of Ka'apor plant classification. The most significant observation to be made here is that thousands of years of horticultural activity have evidently affected Ka'apor principles of plant classification as well as patterns of plant nomenclature.

Ethnolinguistic Methods

Many linguists, in attempting to describe the phonology or develop a dictionary of an unwritten language, may use only a few informants, or sometimes only one. Such a procedure is unavoidable where a society has been so reduced that only one or a few informants are still surviving, as with Ishi of anthropological fame. It is also arguably legitimate for the discovery of minimal pairs and phonemes or for ascertaining basic vocabulary words for such extremely salient phenomena as 'water', 'sun', and 'cloud'. But with regard to the plant domain, especially in an area as botanically rich as Amazonia, a people's knowledge of specific plant names as well as the uses and proper classification thereof is perforce of a very technical kind. Botanical knowledge is also of a local or cultural kind in Amazonia, since vocabulary often denotes many species of limited distribution. For these reasons, at least, historical reconstruction of dead languages and protolanguages tends to exclude names of local flora and fauna, which are often more restricted in their distribution than the languages being compared (see, for example, Kaufman 1990:18).

In the case of the Ka'apor, a linguistic study that places a special emphasis on plants simply must make use of more than a few informants. The extraordinary complexity of the Ka'apor habitat, while it may be perceived in the same general ways by many informants, is seldom dissected into identifiable parts that are named and classified with unanimity. Therefore, my study is based on data obtained mainly through twenty-three adult informants (seventeen men and six women).

In spite of some variation among Ka'apor informants who occupy different villages and even among informants who live in the same village, it is axiomatic that the language, culture, and, more specifically, nomenclatural and classificatory scheme of plants are unitary, not multiple phenomena. Simply stated, there are "correct" and "incorrect" ways to name and classify plants, according to Ka'apor custom. It is my aim in this chapter (and in appendixes 9 and 10) to characterize the proper Ka'apor classification and nomenclature of plants, which are components of those shared phenomena, language and culture, as I understand them. This understanding is based upon the specific methods I employed and the data obtained.

The methodology used in determining Ka'apor plant nomenclature and classification consisted of two basic steps: elicitation of plant names and collection and identification of the referents of those names. This two-part methodology did not demand, however, a sequential ordering, since most often plants were collected prior to the elicitation of their names. The original objectives of the research, moreover, were not primarily linguistic. Nevertheless, I took care to obtain names and activity contexts from several informants for all plants I collected.

Ka'apor informants were initially selected for their reputed knowledge of plants. In fact, I soon learned that all adults, regardless of sex, are believed to be knowledgeable about plants. It should be noted that, unlike the Mundurucu (Murphy and Murphy 1985) and some other lowland South American Indians, Ka'apor society is not fundamentally divided along gender lines. Any division of labor is primarily in the marriage, not in sex per se. Although it is true that only women weave hammocks and virtually only men weave baskets, few other activities are as profoundly exclusive to one sex as are these (see chapter 4). Just as each sex knows the other's activities, because each sex participates, however insignificantly, in most of the activities of the other, it may also be said that neither sex holds a monopoly on any sort of plant knowledge. In the only case of a plant whose "secret" principle was supposedly known only to one sex, males, I learned that women were equally conversant about it. Specifically, women as well as men regarded *pira-piši-ka'a* (*Justicia* spp. and *Ardisia guianensis*) as a plant fetish used by men to attract sexual partners. Both sexes declared without the slightest hesitation that these plants were useful for *inamohar-puhan* 'girlfriend medicine'. In brief, few if any secrets, in contrast to the closely guarded ritual paraphernalia of men's houses in some other societies, can be properly attached to any one sex in Ka'apor society. It is a society in which knowledge is freely transmitted within and between the sexes, and in more than one form, including gossip about contemporaries, ritual incantations of ancestral spirits, and communication referring to the names, activity contexts, and classification of botanical referents in the immediate habitat.

Ka'apor society is generally egalitarian in the sense that there are no significant political discriminations other than by age and sex criteria. There are headmen, but their special privileges are negligible and their responsibilities are mainly to guarantee the uxorilocal nexus of mothers, sisters, and daughters. A few shamans in Ka'apor society exert spiritual influences on their followers, but they demonstrate no knowledge of plants unique to their office. As for gender-based distinctions, women tend to be subordinate to men in the public domain (only one 'headwoman' [*kapitã*] is known to have existed in recent Ka'apor history) and they spend more of their time in certain activities, such as child care, than do men). But women are not deemed to possess any more or any less knowledge about plants than do men. I found that women's knowledge of the specific names of plants varied insignificantly in comparison with men's knowledge of the same. Hence, I do not consider it to be an analytical problem that, in the determination of plant names and classification, I used more male than female informants.

With respect to age differences, adults (understood here to be married or widowed people) are believed to possess the most knowledge about plants. For example, when a teenager is asked "Who knows about trees here?" he or she invariably replies with something like "The elders do" (*Tamũi-ta ukwa*), wherein 'elders' includes both adult sexes and does not denote any name or specific status, such as that of a headman or shaman.

In eliciting the Ka'apor names and classificatory system of plants, I spoke mainly with twenty-three adult Ka'apor informants (seventeen men and six women), each on an individual basis. My method was to show the informant all specimens from my daily collections and then to elicit the Ka'apor name and the activity context of each. In the village, informants were shown both sterile and fertile specimens of all herbs, vines, and trees. Specimens of the same species were often collected more than once and hence were reidentified over the course of this research. This procedure increased the proportion of species collected in a fertile (fruiting and flowering) state, which therefore augmented the overall accuracy of scientific determinations made by myself and the botanists who later studied the material.

Because many species were collected more than once and in different regions of the Ka'apor habitat, and because (after multiple interviews) informants tended toward agreement on species names, overall informant accuracy in this study is probably very high. At least two (and usually more than two) informants were present during my collection of virtually every plant specimen. Informants, therefore, saw each plant in its living, integral state and in its local vegetational context. This considerably strengthened the accuracy of the indigenous plant identifications I obtained. All plant specimens were later shown to more informants for reidentification in the villages where I worked. Reidentification usually confirmed the name most cited by the field informants who had seen the plant before it was pressed.

Even though my collection contains a significant proportion (about 35 percent) of sterile plants, Ka'apor informants demonstrated considerable precision in identifying even these. Informants proved themselves experts at identifying (in their own language) tree seedlings, for example, which several of the systematic botanists I consulted had never seen before and could not begin to identify. Whereas many systematists usually must content themselves with studying specimens inside the herbarium, Ka'apor informants have direct and daily access to the living plants of their habitat; they know these plants from seed to reproductive adult. The Ka'apor may not boil flowers, measure cotyledons, or count chromosomes, but throughout their lives they do see, touch, smell, and taste the living botanical diversity of their homeland.

In identifying many plants, the visible characteristics of floral and fruiting parts seemed far less important to most informants than dendrological and architectural traits. For many plants, most informants used native characters associated with the size, shape, color, and texture of leaves; the color, texture, smell, and sometimes taste of the bark; gestalt appearance of the living stem and any bole, crown, branch, or buttress; size, shape, and texture of any spines; the copiousness, viscosity, and sometimes taste of any latex, resin, or sap; the size, shape, color, and texture of roots and any tubers, rhizomes, or corms; and the type of habitat in which the plant occurred, taking into deliberate account factors such as neighboring plants, stratum of the forest, archaeological remains, and soil quality.

The exact ethnotaxonomic criteria and procedures any one informant employs to identify any given plant are often difficult to specify either by the informant or by his or her interlocutor—a point that has been well and often made (e.g., Berlin et al. 1974). At the same time, the entire cognitive apparatus of Ka'apor taxonomic criteria appears to be holistic and nonarbitrary. This cognitive apparatus encompasses, in principle, plant species never before seen and which are even alien to the Ka'apor habitat. For example, although the Western Amazonian bacaba palm (*Oenocarpus bacaba*) does not occur in the Ka'apor habitat, one of my male informants, Parasarïru, identified its herbarium exemplar at the Goeldi Museum in 1987 as *pinuwa'i* (*Oenocarpus distichus*), a closely related palm which does grow in the Ka'apor habitat.

The method I used to designate a name for every plant collected was by majority. If a majority of the informants gave name X to plant Y, then name X is the one used in appendix 9 and elsewhere in this book where indigenous names of botanical species are given (also see Berlin et al. 1974).

Although not all twenty-three informants saw my entire collection, the overlap in observations was great enough for me to feel confident about designating names. About 97 percent of the species given Ka'apor names in appendix 9 were identified as such by at least two informants. I am therefore confident of the accuracy of this list. Moreover, I believe that the folk generic, specific, and varietal names in appendix 9, in addition to the life form labels, are essentially exhaustive of the present botanical lexicon of Ka'apor.

I did not consider multiple names to be valid for a given plant except where a majority of the informants who saw the plant volunteered the information that the plant "has two names" (*makōi-herr*). In such cases, I recorded both names as synonyms. There were no cases in which more than two plant names were synonymous for a given botanical referent. In addition, exemplars of certain species were seen by the same informants more than once; subsequent identifications of the same species by the same informants tended to confirm the first identification and to validate further the name derived from majority response.

Of the 768 botanical species (including taxonomically valid subspecies) listed in appendix 5, 272 species (or 36 percent) were collected more than once. Many of these, in fact, were collected numerous times in the study-plot inventories or while making general collections. Only 25 (or 3 percent) of the 768 species in appendix 5 are not associated with voucher specimens and collection numbers. These include 12 domesticated species that were, however, reliably determined in the field: *Cocos nucifera, Jatropha gossypifolia, Bactris gasipaes, Phaseolus lunatus, Hibiscus rose-sinensis, Musa* sp.1, *Musa* sp.2, *Passifora edulis, Saccharum officinarum, Zea mays, Coffea arabica,* and *Theobroma cacao.*

One domesticate, the avocado (*Persea americana*), existed in one or two villages in the recent past, but the trees died and no new ones have been introduced. As the Ka'apor say, the avocado was 'lost' (*kāyī*) to their inventory of domesticates. I know that the species in question was truly avocado because two bilingual informants referred both to its Portuguese name (*abacate*) and its native name. There is only one species of *abacate* in Brazilian Portuguese, *Persea americana*.

Because of the huge number of specimens I collected, I encountered some difficulty in maintaining every collected specimen with its appropriate voucher number. Three specimens (*Allamanda cathartica, Astrocaryum javari,* and *Tabebuia impetiginosa*) were identified before I lost them and their collection numbers. I lost three other specimens identified by the Ka'apor informants, which I am calling Indeterminate genus 2, Indeterminate genus 5, and Indeterminate genus 89, prior to obtaining scientific determinations. The Ka'apor names for these lost collections, however, have been retained in my notes.

Three Amazonian species that do not occur in the Ka'apor habitat, and which were therefore not collected, nevertheless do have distinct Ka'apor names. These are Brazil nut tree (*Bertholletia excelsa*), *Acrocomia* palm, and moriche palm (*Mauritia flexuosa*). The Brazil nut tree does not grow in the Ka'apor habitat, but informants have seen, handled, and tasted its seeds, brought to them by myself and other visitors. They called these seeds *kātāi*, a borrowing from *castanha*, itself the Portuguese name for Brazil nut. Informants averred that these seeds belonged to the *kātāi'i*, i.e., Brazil nut tree. I am convinced that *kātāi'i* was borrowed from Portuguese *castanheira* long before the Ka'apor migrated into Maranhão, perhaps as early as the final phase of the mission period of the late 1700s. The Brazil nut tree figures in their folklore, which is not surprising given that the Ka'apor ancestors lived in eastern Pará, a region rich in Brazil nut groves, from the late 1700s to the 1870s. In folklore *kātāi'i* is described as a huge emergent with a 'sweeping crown' (*iākā-tiha-te*)—an accurate description of this magnificent tree.

The spiny *Acrocomia* palm also does not occur in the present habitat of the Ka'apor. But it, like the Brazil nut tree, grows in the fallow forests of eastern Pará, persisting in pastures. The Portuguese term for the edible *Acrocomia* is *mucajá*; Ka'apor informants claim that their ancestors once ate the fruits of a spiny, orthotropic, palmlike plant which the informants call *mukaya'i*. The Portuguese *mucajá*, as with names for other palms, was no doubt derived from an indigenous, Tupí-Guaraní language (Balée and Moore 1991).

The moriche palm (*Mauritia flexuosa*), too, is absent from the floodplains of the rivers draining the Ka'apor habitat. But informants have seen and tasted its oily fruits, brought by nearby settlers who are now cultivating this massive, solitary palm. The Ka'apor term for moriche palm is *miriši'i*, whereas the Portuguese name is *buriti*. Although this species does not appear in Ka'apor folklore, it is doubtful that the name *miriši'i* was borrowed from the Portuguese *buriti*; that *buriti* derived from an indigenous language is much more likely, given the cognates for Ka'apor *miriši'i* in several other Tupí-Guaraní languages (Baleé and Moore 1991).

I argue that the native term for *Mauritia flexuosa* was never lost in the Ka'apor language, even though, as the Ka'apor ancestors fled eastern Pará (where it did occur) they entered pre-Amazonian Maranhão, where the native term no longer had an immediate botanical referent. The Ka'apor ancestors also left behind the Brazil nut tree and the *Acrocomia* palm, yet their language has retained the earlier names for these species. In part, this name retention is due to the persistence of ethnobotanical fragments in Ka'apor folklore. But it is also likely that the Ka'apor have not been separated for long from these plants.

Until as recently as 1928, their daunting raids on settlers often took Ka'apor warriors into the lower reaches of the Guamá river basin in eastern Pará, where they would have seen and probably eaten the fruits of Brazil nut trees, *Acrocomia* palms, and moriche palms. Even after peace was achieved in 1928, Ka'apor men sometimes undertook the long journey from Maranhão through the forests of eastern Pará to Belém in order to secure tools and other goods at the local headquarters of SPI, Indian Protection Service (Expedito Arnaud, personal communication, 1990). In so doing they would have traversed Brazil nut groves (such as that which once stood in the city of Castanhal, enroute near Belém), *Acrocomia* palms (which are extremely common in forest clearings of eastern Pará), and moriche palm groves (which abound in the Amazonian estuary near Belém). Thus, although these three Amazonian species never pertained to the modern homeland of the Ka'apor, it is not surprising that they have a continued presence in Ka'apor culture and language. I, therefore, consider the names *kātāi'i*, *mukaya'i*, and *miriši'i* to be valid components of the Ka'apor botanical lexicon, despite the absence of living referents of these names in the current Ka'apor habitat.

Three non-Amazonian species also have distinctive Ka'apor names. Two of these, onion (*Allium cepa*) and garlic (*Allium sativa*), are collectively called *ma'e-wa-nem* 'some-fruit-fetid'. Although no one cultivates them, bulbs of these plants are familiar to the Ka'apor through the visits of non-Indians. The third non-Amazonian species that has a distinctive Ka'apor name is marijuana (*Cannabis sativa*). It is called *mara-ran* 'malaria-false'. Elderly informants claimed to have seen marijuana plants growing in fields cultivated by settlers downstream along the Gurupi River (the past cultivation of marijuana there is confirmed in Huxley 1957:22) as well as in Guajajara villages to the south. One informant, the late headman Oriru, told me in 1985 that he once smoked some marijuana at the behest of a military policeman while visiting São Luís. Oriru claimed that after inhaling the smoke, he felt *ka'u* 'insane'. He never tried it again. The Ka'apor do not use marijuana nor do they cultivate it. But I take its name *mara-ran*, as well as the name *ma'e-wa-nem* for onions and garlic, to constitute a legitimate part of the Ka'apor botanical lexicon.

As for the *classification* of plants, almost all interviewing was conducted in the Ka'apor language itself, in which, during several years of exposure and practice, I acquired good speaking ability. With the exceptional bilingual informant, interviews took place in Portuguese, but Ka'apor words for plants, uses, and their classification were always obtained and recorded. For each plant I showed from the collection and on location, informants were first asked "What is its name?" I also asked a series of standard questions on plant utility, such as: "Is it edible? Is it a remedy? Is it good for firewood?"

I determined life form labels on the basis of general discussions, when it became obvious that the words *mira* 'tree', *ka'a* 'herb', and *sipo* 'vine' were frequently used to refer to macro-level discontinuities in the plant world. The authenticity of these three life form labels is supported by evidence from other, genetically related languages. In a little-known but pioneering study of Tupí-Guaraní botanical nomenclature, the great Brazilian naturalist João Barbosa Rodrigues wrote, "The Indians divide plants, as we have seen, into *ybyrá* or *mbyrá*, trees, *kaá*, herbs or shrubs, and *ycipo* or *cipo*, climbers" (Barbosa Rodrigues 1905:47). In a more recent ethnobiological work, Pierre Grenand (1980:34–35) indicated the same life form labels for Wayãpi (*wila* 'trees,' *ka'a* 'herbs', and *ipo* 'lianas').

In addition to soliciting from many informants the Ka'apor names for plants I collected, I asked four principal informants (Parasarĩru, Tĩmaparu, Kaŋwa, and the late Tanuru) to provide the ethnotaxonomic status of plant names in the absence of actually collected specimens. This exercise not only helped uncover the structure of Ka'apor plant classification, it also led me to discover and later collect plants that were missing from the collection.

To elicit folk generic names from these four principal informants I began by asking them to supply names belonging to life forms and other categories. With regard to trees I asked, "Tell me all tree names" (*Eme'u upa mira herr-ta*). For each generic name then elicited, I inquired whether other members of the taxon existed—a procedure which revealed folk specific and varietal names. For example, with respect to the folk genus *ama'i* (a kind of *mira* 'tree'), I asked each informant whether other types of *ama'i* existed (*A'e-ta amũ ama'i?*). If any folk genus were monotypic, as with the enormous tree *iwahu'i* (*Micropholis melinoniana*), the informant's answer was typically "no" (*ani*) or "there are no other types of *iwahu'i*" (*nišoi amũ iwahu'i; iwahu'i āyõ*). In the actual case of *ama'i*, however, which denotes five botanical species of *Cecropia*, my inquiries determined that the folk genus encompassed three folk specific taxa: *ama'ite* (referring to *Cecropia pupurescens* and *C. sciadophylla*), *ama'ipuku* (referring to *C. palmata*), and *ama'ituwir* (referring to *C. concolor* and *C. obtusa*). Informants indicated that *ama'ite*, *ama'ipuku*, and *ama'ituwir* "are kinds of *ama'i*" (*A'e-ta ama'i ti*).

In addition to these oral interviews, one youthful informant from Gurupiuna, Hosi (who was literate in his language, having learned to write it from the SIL missionary), supplied me with a complete (i.e., as complete as he was capable) written list of generic and specific names for botanical life form taxa. Much of Hosi's list did conform with the information presented by the four adult informants. Finally, I conducted a further check on the classification of generic names by asking informants the life form status of each generic term, the generic taxon to which each specific name pertained, and the specific taxon covering each varietal name (varietal names are rare). In calculating the number of folk generic and folk specific plant names in Ka'apor, synonyms were included.

General and Special Purpose Taxonomies

Many specialists in folk biological classification distinguish between two basic types of taxonomy: general and special purpose. General purpose taxonomies, with respect to plants, dissect the domain into groupings based on gross morphological discontinuities (e.g., Berlin 1973; Berlin et al. 1973, 1974; C. H. Brown 1977, 1984; Conklin 1954; Hays 1979, 1983). These gross morphological groupings are called *life forms*, and the distinctive features that define each one are considered to be "natural" and perceptible to any normal adult human being. General purpose schemes of classification stress similarities and differences between organisms, in terms of their singularity, rather than any relationship between the organisms and extraneous factors, such as their habitat or human beings (Atran 1983:60, 1985:308). Special purpose classifications, on the other hand, tend to emphasize these contextual factors. In particular, special purpose classifications isolate one "powerful factor" in a given domain in order to comprehend that factor alone, not the entire domain (Gilmour 1961:34; cf. Atran 1985:304–305). In the domain of plants, for example, a general purpose taxonomy might distinguish among (1) 'trees', plants that are orthotropic (erect-growing), woody, and tall (≥ 2 m in height); (2) 'herbs', plants that are orthotropic or prostrate, short (< ? m in height), and nonwoody; and (3) 'vines', plants that are plagiotropic (non-erect–growing) or prostrate, woody or nonwoody, and whose stems are of variable length.

A special purpose taxonomy might isolate plants according to some perceived, utilitarian feature, such as "food plants," "domesticates," "medicinals," "lashing material," "firewood," and so on. For example, the Ka'apor classify under the term *kanei* all resins, latices, and saps that are combustible (Balée and Daly 1990). Although the generic name *kanei'i* (which, with incorporation of the postposed '*i* ['stem'] denotes an organism) encompasses nine species of burseraceous trees in the genus *Protium* (see appendix 9), *kanei* refers not only to the combustible resins of these species, but also to those of other, botanically unrelated species as well. Specifically, the saps, resins, and latices of the clusiaceous *irati'i* (*Symphonia globulifera*), the sapotaceous *irikiwa'i*

(*Manilkara huberi*), and the caesalpiniaceous *tarapa'i* (*Hymenaea* spp.) are also considered to be types of *kanei*. The resins of these species are burned at night to supply illumination. As the late headman Oriru once jokingly remarked to me, "*kanei a'e yane rampari tikwer*" (*kanei* is our kerosene), in reference to the similarity of function between *kanei* and the kerosene lamp used by non-Indians (including myself) where there is no electricity.

The essential identity of *combustible* resins, saps, and latices, according to this classification, is further revealed in the origin myth of *kanei* (quoted in Balée and Daly 1990:29):

> Long ago, no one possessed any *kanei*. The settlement, therefore, was dark at night and people had to eat game meat in the dark. There was a man named Kanei-yar [lit., *kanei* owner]. Another man, Sarakur [lit., Wood rail], wanted light, so he would need not fear snakes while dining at night. So Sarakur killed Kanei-yar by scoring his neck with a machete. Sarakur then reached into the opened nape of Kanei-yar and pulled out pieces of bone, which were really pieces of *irikiwa-hik* [latex from *Manilkara huberi*]. Then, Sarakur ignited these pieces and the settlement, henceforth, was well-illuminated at night.

Besides the obvious parallels—bone with oxidized sap and neck with tree—this myth fragment shows that the latex of *Manilkara huberi* is cognized as being a kind of *kanei*, even though *irikiwa'i* (the organism, *Manilkara huberi*) and *kanei'i* (nine *Protium* species) are fundamentally different folk genera of 'trees'. They are, of course, quite different *botanical* species as well.

Lévi-Strauss (1966:12) suggested a similar possible "special" classification, which would perhaps be appropriate in a European folk botany: "On intuitive grounds alone we might group onions, garlic, cabbage, turnips, radishes and mustard together even though botany separates liliaceae [*sic*] and crucifers. In confirmation of the evidence of the senses, chemistry shows that these different families are united on another plane: they contain sulphur." Accordingly, even though many types of *kanei* contain combustible terpenes, this chemical principle is evidently not common to all of them (Douglas Daly, personal communication, 1989). The common criterion for membership in the special taxon *kanei* may be more generally stated, however, as 'combustible resin, latex, or sap' (Balée and Daly 1990:28). In the same way, *iwir*, although it denotes the fibrous barks of 40 tree species in 10 botanical families, may be glossed as 'stratified phloem'. Thus, combustible plant fluids in the first instance and fibrous bark in the second are each the "powerful factor" isolated in two examples of special purpose classification among the Ka'apor.

Atran (1990:24–25) has implied that the dichotomy between special and general purpose classifications is artificial, since in reality, all schemes of classification are artifacts of mind and, as such, not necessarily intrinsic to nature. Regardless of the questionable validity of an analytic distinction between general purpose and special purpose classifications, it is clear that the Ka'apor classify plants qua organisms, on the one hand, and also classify them according to highly specific, utilitarian criteria, on the other. Because the Ka'apor classification of plants as separate organisms encompasses far more species of the Ka'apor habitat than does any classification based on a single utilitarian property, the focus of this chapter is properly the domain of plants qua organisms, as this is cognized by the Ka'apor. The most comprehensive classification in Ka'apor ethnobotany, then, is based on a very few features that essentially define plants in terms of their anatomy and life processes, as distinct from other comparable domains, such as animals. This does not mean, however, that the same criteria relating to anatomy and life processes are used for distinguishing among individual plants themselves, i.e., at less comprehensive levels of the ethnobotanical hierarchy.

The Semantic Domain of Plants

Before examining the specifics of Ka'apor plant classification and nomenclature, it is useful to determine whether 'plant' constitutes a distinctive semantic domain and, if so, how precisely this domain is delineated vis-à-vis comparable domains. It should be noted that, first of all, no single expression marks 'plants' as understood in the most inclusive Linnaean sense of this word. (For that matter, no single word in Ka'apor denotes 'animals' or 'minerals' either.) In cognitive terms, however, the existence of 'plants' as a broad, semantic domain in Ka'apor is real. This reality is reflected immediately in the lexicon of plant parts (see table 7.1), wherein many specific words denote only aspects of the domain ordinarily understood as 'plants', especially vascular plants, in Linnaean taxonomy. In the felicitous phrasing of Berlin (1976:383–384) and Berlin et al. (1973:214; 1974:30), the domain of plants is "covert," i.e., unnamed. Its reality is attested to by terms for parts and certain life processes inherent to plants.

First, plants may be either 'living' (*šuwe*) or 'dead' (*manõ*), which separates them from inanimate objects 'not having life' (*šuwe 'ĩm*). Seeds of plants may 'germinate' (*hiwõi*), and they are the only living things that do so. The Ka'apor associate visits of bees and birds (especially hummingbirds and manakins) to flowering plants with subsequent fruiting, saying that these animals 'eat the flower nectar' (*putïr tikwer u'u*)—there appears to be no separate word for 'pollination', yet the concept is clearly understood.

Plants 'grow straight up' (*uhem* or *tiha*), as with many trees, or in 'creeping' fashion (*sururuk*), as with many vines. As with animals and human beings, sex in plants is recognized in principle. For example, the Ka'apor distinguish male papaya plants (*mãmã-sawa'e* ['papaya-male']) from female papaya plants (*mãmã-kuyã* ['papaya-female']). It should be mentioned that these words do not represent folk specifics of the generic name, *mãmã* 'papaya.' The four folk specifics of *mãmã* are *mãmã-te* 'papaya-authentic', *mãmã-howi* ['papaya-blue/green'], *mãmã-hu* ['papaya-big'], and *mãmã-pinĩ* ['papaya-striped'], each of which includes male and female exemplars. In other words, the status of sex (as well as specific life conditions, such as relative age and phenology) is not indicated in folk specific names for plants.

Many words for plant parts are not extended (in terms of range of meaning) beyond the domain of plants. Examples include *ha'ĩ* 'seed', *harï* 'rachis', *hupe* 'pod' (as of a bean pod), *ho* 'leaf', *hapo* 'root', *yu* 'spine', *hãšĩ* 'many small spines clustered together', *yihïk* 'sap', and *kanei* 'resin'. These and other terms unique to describing plants and plant parts indicate the cognitive reality of 'plants' as a unique beginner (i.e., 'kingdom'),

TABLE 7.1 *Ka'apor Names of Plant Parts*

	Ka'apor Name	Gloss
	Reproductive Parts	
Flower	putɨr	—
sepal	himo-pɨta	?-staying place
calyx	himo-pɨta	?-staying place
petal	putɨr	—
corolla	putɨr-pukwek-ha	flower-wrap-agentive
stamen	himo-pɨta	?-staying place
anther	i'a-ra'i-ătă	fruit-small-hard
style	putɨr-yapɨr	flower-?
stigma	hĕ'ĕ-ra'i	sweet-small
Fruit	iwa, i'a	—
exocarp	pirer-i'a	bark-fruit
mesocarp	haʔ-rukwer	seed-pulp
endocarp	pɨter-pe-har	center-in-place
seed	haʔ	—
endosperm	haʔ-pɨter	seed-center
rachis (fruits in bunches)	harɨ	—
bean pod	hupe	—
	Nonreproductive Parts	
stem, trunk	i'ɨ	—
bark	pirer	skin
leaf	ho	—
narrow leaf	ho-pu'i	leaf-thin
broad leaf	ho-ipe	leaf-flat
primary nerve	ho-šaŋwer	leaf-bone
secondary nerve	ho-yɨwa-yɨwa	leaf-arm-arm
leaflet	ho-owok	leaf-?
petiole	yɨwa-kaŋwer	arm-bone
pulvinus	uywaraha	—
base of leaf (palm)	šupe	—
old leaf	ho-maner	leaf-?
new leaf	ho-kɨr	leaf-unripe
wood	mɨra	tree
heartwood	i'ɨ-pɨter; hukwer	stem-center
palm heart	hĕ'ĕ	sweet
leaf sheath	ho-pirer	leaf-skin
spathe	ho-'ă-pirer	leaf-image-bark
root	hapo	—
buttress	pɨpĕ	claw
sulcus	pɨpĕ-ra'ir	claw-little
hairs on leaf	ha	—
spine	yu	—
many small spines	hăšĭ	—
branch	hăkă	head
insect gall	ho-šuru,	leaf-?
	ka'a-ro-šĭ-tă	herb-leaf-white-hard
resin	kanei	—
stem water, nectar	yihɨk, tɨkwer	—

despite the absence of an actual word for the category that botanists call plant. One curious omission in the above list of Ka'apor words that apply strictly to plant parts is what botanists regard as 'flower'. For the Ka'apor, *putɨr* 'flower' is polysemous, denoting not only flowers in the conventional sense, but also 'featherwork'.

The Ka'apor domain of 'plant' is not limited to vascular plants, but includes the nonvascular bryophytes as well. The domain appears also to subsume some visible fungi (which, according to recent classificatory schemes in Western botany, pertain to a different kingdom from plants). For example, the word for a luminescent fungus species of the forest floor, *ka'a-tuwir* 'herb-white' (aff. *Rhizomorpha* sp.1), incorporates the life form label (*ka'a*) that mostly refers to herbaceous vascular plants, 'herbs'. That fungus, moreover, is classified by the Ka'apor as a kind of 'herb'. On the other hand, *urupe*, which refers to all mushrooms (the foregoing example is not a mushroom in form), is not classified as to life form, and its parts are not ascribed plant part terms; its status, therefore, remains unclear.

In addition, it cannot be firmly determined whether *mira-meyu*, which refers to mosses in the Pterobryaceae and Sematophyllaceae families, is considered by the Ka'apor to be a 'plant'. Like mushrooms, these mosses are not designated by any life form term and their organs are not accorded plant part terms. Yet because at least some nonvascular species are assigned unequivocal plant status, and given the lack of any single term synonymous with 'plant', I have here included in the corpus of Ka'apor generic plant names the dubious terms (as to classificatory status) *urupe* and *mira-meyu*.

Not surprisingly, all vascular plants of the habitat, including flowering and non-flowering types, are subsumed under the concept 'plant'. Even the primitive vascular plant *Selaginella*, a kind of club moss, is named *yaŋwate-ka'a* 'spotted jaguar–herb' and is classified, moreover, as a kind of 'herb' (*ka'a*). Although not all ferns incorporate the life form label *ka'a* 'herb' (as with *wari-ruwai* 'howler monkey-tail' [*Lomariopsis japurensis*]), most ferns do; they are called *yakare-ka'a* 'alligator-herb', and they are classified as kinds of *ka'a* 'herb'. Although cycads, conifers, and the Gnetinae vines (represented elsewhere in Amazonia by the genus *Gnetum*) would probably be classified as 'plants', none were confirmed to exist in the Ka'apor habitat.

Figure 7.1 shows that beneath the unique beginner 'plant', the Ka'apor classificatory system includes several taxonomic categories for further discriminating among organisms. These categories, which were briefly addressed above, are, in terms of decreasing inclusiveness: (1) the three life forms of tree and herb and vine, (2) folk generic, (3) folk specific, and (4) folk varietal. In addition, a taxonomic category equivalent in inclusiveness to life form also appears to be real: this involves the folk generics and folk specifics which, according to Ka'apor criteria, are neither trees, herbs, nor vines, or which are ambiguously associated with more than one of these specified life forms.

The three life form taxa are *mira*, *ka'a*, and *sipo*. Up to now, I have used these terms rather loosely to refer, respectively, to 'tree', 'herb', and 'vine'. I will continue to do so, but first must qualify that these terms cannot be glossed precisely in English. *Mira* means not only 'tree' in a conventional sense, but also 'wood' and numerous finished wood products; *ka'a* refers not only to 'herbs', in the folk English sense of 'nonwoody herbaceous plants' but also to 'shrubs', 'forest', and even 'defecate'; *sipo* encompasses not only woody and nonwoody vines, but also certain forms of lashing material used by the Ka'apor in post-and-beam construction (Balée 1989b:9–10). Polysemous terms for plant life form taxa, incidentally, have been widely noted in other languages (e.g., Alcorn 1984a:265; Hunn 1982:837–839).

Figure 7.1 Model of Ka'apor Plant Classification

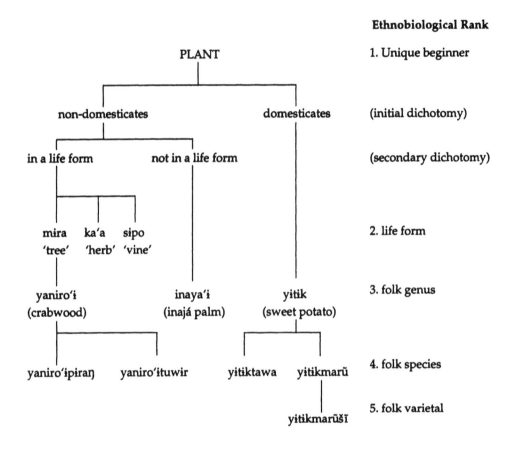

Ethnobiological Rank

PLANT — 1. Unique beginner

non-domesticates / domesticates — (initial dichotomy)

in a life form / not in a life form — (secondary dichotomy)

míra 'tree' / ka'a 'herb' / sipo 'vine' — 2. life form

yaniro'i (crabwood) / inaya'i (inajá palm) / yitik (sweet potato) — 3. folk genus

yaniro'ipiraŋ / yaniro'ituwir / yitiktawa / yitikmarũ — 4. folk species

yitikmarũšĩ — 5. folk varietal

Another Ka'apor word, *kāpĩ*, which covers numerous grasses, sedges, and other small, nonwoody plants, would seem, upon initial inspection, to qualify as a life form label also. This is because *kāpĩ* encompasses a large range of botanical species, and Ka'apor informants consider *kāpĩ* not to be a constituent of any of the other three life form taxa. The taxon *kāpĩ*, however, is monotypic in Ka'apor, evidently containing no contrast sets (Kay 1971). In other words, *kāpĩ* is an "empty taxon" (Hunn 1982:834; Turner 1974:34–35, 40). Folk botanical life form labels, on the other hand, are polytypic, entailing, by definition, two or more folk generic names (Atran 1985:307; Berlin et al. 1973:215; Randall and Hunn 1984:330; cf. C. H. Brown 1977:319–320). The term *kāpĩ*, therefore, must be classified here as a folk generic name that does not pertain to any life form taxon, in other words, a generic name for certain unclassified nondomesticates.

The Dichotomy of Domesticates and Nondomesticates

Ka'apor life form labels can be only imprecisely glossed by English plant words for another important reason: they refer only to many (but not all) nondomesticates. Domesticated plants, whether traditional or introduced, are not classified as *mira*, *ka'a*, or *sipo*. There is thus an emergent conceptual dichotomy between domesticates and nondomesticates. It may appear dubious, especially from the point of view of modern folk English, to propose such a dichotomy, yet the distinction is transparent in the data themselves.

First, domesticates do not occur on any of the many lists of the generic constituents of *mira*, *ka'a*, and *sipo*, which were collected among Ka'apor informants. Perhaps the most convincing proof of a classificatory dichotomy between domesticates and non-domesticates is rendered by a comparison of domesticates with their nondomesticated, local congenerics. Informants consistently segregate domesticates from nondomesticated congenerics in sorting tasks. (Unlike the Ka'apor, the Peruvian Aguaruna appear to link such plants in their classification—Berlin 1992:153–163.) When I asked an informant to list all specific types of a domesticated folk genus, he or she responded by naming the domesticated specific types but consistently omitted mention of very closely related nondomesticates.

The botanical genera conducive to a comparison of domesticates and congeneric nondomesticates are *Anacardium, Theobroma, Ipomoea, Bixa, Dioscorea, Manihot,* and *Passiflora*. Life form labels are applied to none of the domesticates of these genera. Perhaps most telling, the domesticated cashew tree, *akayu* (*Anacardium occidentale*), is not classified as a kind of *mira* 'tree', whereas the nondomesticated local members of the same genus, *Anacardium giganteum* and *Anacardium parvifolium*, are so classified by informants. These latter two species are referred to by the term *akayu'i*, which is a separate folk genus from *akayu* (see appendix 9). Linguistically, the only difference between the terms is that the nondomesticated congenerics of *Anacardium occidentale* incorporate the postposed morpheme *'i*, meaning 'stem'. The Ka'apor name two folk specifics of the domesticated cashew tree *akayu-piraŋ* 'L-red' and *akayu-tawa* 'L-yellow', where 'L' means 'Literal plant term'. They also distinguish two folk species of nondomesticated *akayu'i,* namely, *akayu-'i* (*-te*) 'L-stem-(authentic)' and *akayu-pinar-'i* 'L-fishhook-stem' (also called by the synonym *akayu-mena-'i* 'L-?-stem').

It should be noted that the origin myth of domesticated cashew further implies a dichotomy in the folk classification of the genus *Anacardium* (see chapter 6). One elderly male informant from the Gurupiuna area, Kaŋwa, stated that the cashew of the swidden, *akayu,* is not a *mira* 'tree' because of its relatively small size: *A'e pišik— amũ ka'a rupi har, a'e mira saka, tiha* 'It's small—the other [*akayu'i*] of the forest is like a tree, being large.' But it is unlikely that the criterion of relative size represents a principle by which domesticated trees are classified separately from their nondomesticated congenerics. For example, another tree genus that is split, according to Ka'apor criteria, into domesticated and nondomesticated types is *Theobroma.* Yet height and diameter differences are insignificant among the species of this genus. The folk genera *kakaran'i* (*Theobroma speciosum*) and *kipi* (*Theobroma grandiflorum* and *T. subincanum*) are not considered to be wild types of domesticated cacao (*kaka*) [*Theobroma cacao*], nor are any of these species believed to be derived from any of the others. Whereas *kakaran'i* and *kipi* occur on elicited lists of 'trees' (*mira*), *kaka* does not. In other words, *kaka* is not a 'tree' in Ka'apor classificatory terms, yet this is not because it is too small to qualify as such. Rather, domesticates are not 'trees', regardless of habit. Note: An early report (Balée 1989b:19) mistakenly indicated that *kaka-ran-'i* is considered to be a type of *kaka.*

Yet another genus of tree, *Bixa,* is divided into domesticated and nondomesticated folk genera. In fact, the Ka'apor here differentiate between botanical varieties of a single species, *Bixa orellana* (annatto). These varieties are *uruku,* which refers to domesticated annatto and *urukuran'i,* which denotes nondomesticated annatto trees (usually encountered in flooded forest). In addition, certain 'herb' and 'vine' genera containing domesticated and nondomesticated species are also so divided by the Ka'apor. For the folk genus *yitik,* sweet potatoes (*Ipomoea batatas*), Ka'apor informants supplied seven folk specific names and one folk varietal name (see appendix 9). Yet names for nondomesticated species of *Ipomoea,* many of which are called *yitikran,* were not considered to be constituents of the folk genus *yitik.* The nondomesticated *yitikran* is classified as a *sipo;* yet *yitik* (i.e., sweet potatoes) were not elicited as types of *sipo,* nor as types of any other life form.

Likewise, although *kara,* referring to yams (*Dioscorea trifida*), contains four folk species, none of the nondomesticated species of *Dioscorea* were included as kinds of *kara.* Rather, the nondomesticates were referred to by the terms *kararan* and *yakamĩtimakaŋwer,* and they were classified as 'vines' (*sipo*). The genus *Passiflora* (passion fruit) has both domesticated and nondomesticated species in the Ka'apor habitat. The nondomesticated species are *murukuyaran, murukuyaransipo,* and *murukuyatawa* and they are classified as 'vines' (*sipo*); the domesticated *Passiflora edulis,* called *murukuya,* on the other hand, is not considered to be a kind of 'vine' (*sipo*).

Finally, evidence for this classificatory dichotomy between domesticates and nondomesticates is powerfully evident in the realm of herbs, genus *Manihot.* An informant from the Gurupiuna area replied emphatically to the question of whether domesticated manioc (the folk genus name is *mani'i*) was an 'herb' (*ka'a*): *Ani, mani'i ka'a 'im— mani'i a'e mani'i* 'No, manioc is not an herb—manioc is manioc.' Yet he as well as other informants classified *arapuha-mani'i* 'brocket deer manioc', which refers to *Manihot brachyloba, Manihot leptophylla,* and *Manihot quinquepartita,* as a kind of *ka'a* 'herb.' This is a typical response to similar queries about all other domesticated plants and, where extant, their local nondomesticated congenerics.

I do not mean to suggest the absence of an intermediate group of plants, namely, semidomesticates, nor that such a group holds no implications for Ka'apor plant vocabulary. Semidomesticates (the anthropophytes of Pinkley 1973), which are plants that colonize anthropogenic zones yet do not depend on cultivation in order to survive and reproduce, tend to be assigned stable, literal names, as is the case with domesticates, especially traditional (i.e., nonintroduced) ones (Balée and Moore 1991). The point is that the artifactual class of plants best considered to be 'semidomesticated', although somewhat transparent from the nomenclature, lacks any proper classificatory status itself.

For example, although Ka'apor informants do not readily explain why names for semidomesticated trees, like domesticates, tend to be literal (as opposed to metaphorical—see below), they automatically classify many of these as 'trees', 'herbs', or 'vines'. Semidomesticates are, with respect to life form identification, more closely associated with nondomesticates than with domesticates. Ka'apor classificatory procedures thus appear to be binary (a plant is, first, either domesticated or not), yet unconscious nomenclatural patterns reflect at least three categories along a gradient of domestication (viz., 'wild', 'semidomesticated', and 'domesticated') [see Balée and Moore 1991].

In any case, it can be argued that nomenclature, at least partly, is reflected in Ka'apor plant classification, even if it is not a "near-perfect guide" (Berlin et al. 1973:216) to that classification. In classificatory terms, the essential distinction to be drawn at the level immediately subordinate to the covert taxon 'plant' is between domesticated and nondomesticated plants. At the next level, for nondomesticates a distinction is recognized between plants classified as to life form and plants not so classified. Below that are the levels folk genera, folk species, and folk varietals (see figure 7.1).

Classification of Folk Genera, Species, and Varietals

The immediate constituents of life form taxa are folk genera. These are taxa which may be monotypic or polytypic, in terms of the number of subordinate folk taxa (folk species and varietals) that they contain. In rural Louisiana, for example, 'oak', 'pine', and 'magnolia' would be, for many speakers, folk genera of the life form *tree* (Holmes 1990). Likewise, 'live oak' (*chêne vert*), 'water oak' (*chêne gris*), and 'cow oak' (*chêne vache*) would probably represent, for the same speakers, folk species of the folk generic taxon 'oak' (or *chêne* to French speakers). In the Ka'apor plant lexicon, there are 257 genera of 'trees' (*mira*), 58 genera of 'herbs' (*ka'a*), 60 genera of 'vines' (*sipo*), 67 genera of domesticates, and 41 genera of other plants unclassified as to life form. The total number of Ka'apor folk botanical genera, then, is 483. (Ka'apor folk generics, species, and varietals, together with their botanical referents, are given in appendix 9). These 483 generic names refer, moreover, to 768 botanical species (see appendix 5). In addi-

TABLE 7.2 *Correspondences Between Folk Genera and Botanical Species*

	'Trees'	'Herbs'	'Vines'	Domesticates	Other	Total
One-To-One Correspondence	153	44	39	57	27	320
Underdifferentiation	104	14	21	9	14	162
Overdifferentiation	0	0	0	1	0	1
Totals	257	58	60	67	41	483

tion, monotypic and polytypic folk genera (i.e., folk genera which contain two or more folk species) contain a total of 636 folk species. In gross numbers of terms, therefore, the Ka'apor specific plant lexicon is 83 percent as specific with regard to the 768 species in appendix 5 as is Western taxonomic botany. In comparison with the Hanunóo of the Philippines, who name at least 1,625 specific taxa to the 1,300 named in the Linnaean system for the same flora (Conklin 1954:116, cited in Berlin 1992:69), the Ka'apor may be said to be "lumpers." Nevertheless, the actual number (483) of folk genera named by the Ka'apor lies near the average of 390 (Berlin 1992:100) and beneath the upper limit of "nature's fortune 500+" for folk botanies in general (Berlin 1992:96).

Ka'apor nomenclature thus slightly underdifferentiates the flora in the Ka'apor habitat compared with the Linnaean system. Berlin et al. (1974:101–103) suggested a systematic comparison of folk botanies and Western taxonomic botany with regard to an identical flora in terms of the mapping of folk generic names onto scientific species' names. For example, a one-to-one correspondence is recognized for a folk generic name when it denotes only one botanical species; underdifferentiation by a folk generic name occurs when the name in question refers to two or more botanical species; and overdifferentiation is noted where more than one folk generic name refers to a single botanical species.

In Ka'apor ethnobotany, there is only one overdifferentiated botanical species, namely, *Manihot esculenta* 'manioc' which is named by the following three genera: *mani'i* (roughly, 'bitter manioc'), *makaser* (roughly, 'sweet manioc'), and *maniaka* (a kind of watery, genuinely sweet-in-taste manioc). Table 7.2 shows that for the named flora as a whole, 66 percent of Ka'apor generic names map onto a single species. This supports the observation by Berlin et al. (1973, 1974) that folk generic names tend to map onto single species' names. It can be concluded, then, that Ka'apor folk generic names tend to map onto botanical species' names, as was earlier formulated by Berlin et al. (1973, 1974) with respect to folk nomenclature in general.

Another pertinent observation made by Berlin and his colleagues concerned the relationship between polytypic folk genera that refer to a *single* botanical species. They claimed that this "can be expected to occur predominantly with plants of supreme cultural significance" (Berlin et al. 1974:103; also see Berlin 1992:118–119). My findings also bear this out, although not all plants of "supreme cultural significance," however one chooses to define this, are so named. Cultural significance, moreover, as a general utilitarian category including both useful nondomesticates and domesticates is apparently unrelated, in principle, to plant nomenclature and retention of plant names through time (Balée and Moore 1991).

Only under the life form 'trees' (*mira*) and the category 'domesticates' does one find cases of polytypic genera referring to a single species. Specifically, five 'tree' species and seventeen domesticates are each named by more than one polytypic folk genus. The five polytypic genera naming a single tree species are *irati* (*Symphonia globulifera*), *pakuri'i* (*Platonia insignis*), *tayi* (*Tabebuia impetiginosa*), *yaniro'i* (*Carapa guianensis*), and *yawi'i* (*Xylopia nitida*). According to informants, *irati* is subdivided into *irati-ātā* ('hard *irati*'), which is found only in the high forest (*ka'a-te*), and *irati-te* ('true *irati*'), which is encountered only in flooded forest (*iapo*); the latter folk species is, moreover, the most suitable for obtaining glue in the manufacture of arrows. *Pakuri'i* is subdivided into four folk species: *pakuri'i-te* (=*pakuri'i-tawa* ['L-yellow']), *pakuri'i-pu'a* ('round bacuri'), *pakuri'i-hāšī* ('spiny bacuri'), and *pakuri'i-pihun* ('black bacuri'). The names directly refer to qualities of the highly esteemed fruits of these different folk species. *Tayi* is subdivided into *tayi-te* ('true pau d'arco'), *tayi-tawa* ('yellow pau d'arco'), and *tayi-pihun* ('black pau d'arco'). The criteria for distinguishing among them depend on dendrological characteristics of the plants, for to both the Ka'apor and the systematic botanist, there appears to be no difference in the sexual parts, i.e., flowers and fruits. *Tayi-te* has a 'white' stem, *tayi-tawa* has a 'yellow' stem, and *tayi-pihun* has a 'black' stem. All these folk species, on the other hand, have 'purplish' (*pihuŋwer*) flowers. Only *tayi-te* and *tayi-pihun* are deemed to be appropriate for making bows, with *tayi-te* the most preferred because of its high 'elasticity' (*hayik*).

Likewise, *yawi'i* (*Xylopia nitida*), which is one of the most desired species for posts and beams in house construction, is divided into *yawi'i-tuwir* ('white' *yawi'i*) and *yawi'i-piran* ('reddish' *yawi'i*) based on color of the stem. The crabwood tree, *Carapa guianensis* in the mahogany family, is named by two folk species of the taxon *yaniro'i*: *yaniro'i-piraŋ* ('reddish crabwood') and *yaniro'i-tuwir* ('white crabwood'). The oil of both folk species is important in indigenous medicine. According to Ka'apor informants, *yaniro'ipiraŋ*, which grows in the high forest, has a reddish 'heartwood' (*hukwer*); *yaniro'ituwir*, which grows in flooded forest (*iapo*), has a white heartwood. The Ka'apor are not alone in explicitly recognizing and naming this subdivision of a single botanical species according to the color of its heartwood. According to a taxonomic specialist on the neotropical Meliaceae: "Two types of *Carapa* timber are recognized by foresters, red and white. Red or Hill crabwood is said to be superior to white and is obtained from trees growing on higher land. White crabwood is derived from those growing on swampy flat ground" (Pennington 1981:414). This observation suggests, of course, that Ka'apor 'splitting' of a botanical taxon is likely based, in principle, on empirically verifiable differences within such a taxon.

Of the 67 folk generic names for domesticated plants, 19 (28 percent) are polytypic—having more than one folk species while referring to one botanical species. As for folk genera with four or more folk specifics, there are *mani'i* (with 18 folk species), *pako* (with 6 folk species in reference to *Musa* sp.1 alone), *yitik* (with 7 folk species), *kwi* (with 6 folk species), *kā* (with 6 folk species), *kara* (with 4 folk species), *makaser* (with 4 folk species), and *māmā* (with 4 folk species).

Finally, the category of folk varietal, which is immediately subordinated to the category of folk species, has representative taxa only in the group of domesticated plants. In fact, in the entire corpus of Ka'apor plant names, there are only *two* folk varietals.

These are (1) *katumẽ-howi* ('blue/green *katumẽ*'), a kind of *katumẽ* (called *banana maçã* by local Brazilians), which is, in turn a folk species of *pako* (which covers bananas and plantains) and (2) *yitïkmarũ-šĩ* ('white *marũ* sweet potato'), a kind of *marũ* sweet potato, itself a kind of *yitïk* (which covers sweet potatoes).

In general, regardless of the number of botanical species denoted by given folk generic names, it is clear that polytypic genera are proportionally much more represented in the 'domesticates' group than in any other. As far as polytypy in this broad sense is concerned, only 34 'tree' genera (13 percent), 3 'herb' genera (5 percent), 4 'vine' genera (7 percent), and 5 unclassified nondomesticated genera (12 percent) are represented. But 23 domesticated plant genera (34 percent) are polytypic. It is apparent, therefore, that domesticates as a group, in contrast to all the other recognized categories of plants, are the most psychologically salient.

Ka'apor Plant Lexemes

The psychological salience of domesticates, in particular traditional (nonintroduced) domesticates, in the classification of plants is further revealed in the plant nomenclatural system of the Ka'apor language. Ka'apor plant names themselves may be classified in terms of their semantic and syntactic structure. In the analysis of Ka'apor plant nomenclature, it is necessary first to specify the units.

The basic unit in any discussion of folk nomenclature is the *lexeme* (Conklin 1969; Berlin et al. 1973, 1974). Berlin (1992:26–35) appears to have dropped his use of "lexeme" in favor of simply "name." I prefer to retain the use of "lexeme" for analytical purposes.

A lexeme is an expression whose meaning cannot necessarily be determined simply by knowing the meaning of its constituent morphemes or words, as in the expression *hot dog* (Frake 1969; Goodenough 1956). In addition, lexemes may be either monomorphemic or polymorphemic, since single free morphemes are by definition lexemes also (cf. Conklin 1969:43). Lexemes occur in cultural contexts and can only be rendered in translation by reference to these contexts (Conklin 1969; Lounsbury 1956). For example, the domain 'plant' in Ka'apor ethnobiological taxonomy involves an array of lexemes, the referents of which conform in specific ways to the defining features of the domain. Lexemes can be divided into several types: simple primary, productive primary, unproductive primary, and secondary lexemes (see Berlin et al. 1974). These correspond to Berlin's more recent use of primary and secondary "names" (1992:26–35).

For Ka'apor ethnobotany, simple primary lexemes tend to be "one-word" expressions that denote plants; these expressions do not incorporate life form labels, although many of the constituents of the class "simple primary lexemes" pertain to one or more of the life forms. An example of a simple primary lexeme in English is 'oak'. Productive primary lexemes, in contrast, do incorporate life form labels. The referents of these lexemes, moreover, are classified as pertaining to the superordinate taxa (life form labels) incorporated in them. For example, a 'pine tree' in folk English is a productive primary lexeme, insofar as the referent is a kind of 'tree'.

Unproductive primary lexemes generally either incorporate a life form label to which the referent does not belong in the classification system *or* they contain no nominal components that, taken alone, denote a 'plant'. In English folk botany, for example, 'poison oak' is not a kind of 'oak' and 'cattails' denote neither 'cats' nor 'tails'.

All Ka'apor folk generic and life form labels may be semantically analyzed in terms of one or more of these types of primary lexemes. Secondary lexemes, which are by definition productive (as with 'live oak' and 'water oak' in some folk English taxonomies), describe the semantic structure of many folk specific names. In addition to discriminating types of lexemes, any outline for a classification of Ka'apor plant names, moreover, should account for a distinction between literal and metaphorical names, which appear to crosscut typologies based on lexemes (see appendix 10).

Literal Generic Names

All literal plant names are, in terms of the linguistic structure of Ka'apor plant classification, simple primary lexemes, although not all simple primary lexemes referring to plants are literal plant names. In addition, literal plant names incorporate principal nominal components with unitary meanings (see Berlin et al. 1974:27). In other words, the principal morpheme(s) in a literal plant name refers to nothing other than a specific plant or specific range of plants (Balée and Moore 1991:231). The 'L' in many plant glosses given in this book indicates 'literal plant name'. These are the "opaque" or "linguistically unanalyzable" terms in Berlin's usage (1992:256). Literal plant names may be subsumed under any of the life form categories or under unnamed (artificial) groupings, such as unclassified domesticates and unclassified nondomesticates (i.e., plants that are unclassified as to life form, hence, "unique"—see Berlin et al. 1974:27).

For example, under the life form label *mira* 'tree', *kumaru'i* (*Dipteryx odorata* or tonka bean tree) and *kumaru'iran'i* (*Taralea oppositifolia*) are both literal plant names. Although the latter incorporates the morpheme for 'false' (*ran*), its principal nominal component, *kumaru*, is a literal term. Both names, moreover, pertain to the life form category *mira*. The morpheme *ran* would change the classification of a name only if its referent pertained to a life form different from that of the model.

In other words, *kumaru'iran'i* would not be considered a literal name if it did not refer to 'tree', as does *kumaru'i*. In fact, both names denote closely related members of the Fabaceae, although neither is a subtype of the other—both are separate primary lexemes. *Kumaru'iran'i* and other such names may be considered examples of superficial modeling of a name. In most cases, with regard to literal names incorporating *ran*, moreover, the models are closely related biologically to their analogues. For example, *ayu'i* and *ayu'iran'i* denote only lauraceous trees; *iŋa* and *iŋaran* refer to leguminous trees mostly of the mimosaceous type; *kirihu'i* and *kirihu'iran'i* refer to members of three closely related families, namely, Burseraceae, Anacardiaceae, and Sapindaceae; *kupa'i* and *kupa'iran'i* refer to members of a single genus (*Copaifera*), as do *para'i* and *para'iran'i* (*Jacaranda*). Similarly, with regard to 'herbs' (*ka'a*), *anī* and *anīran* refer to closely related aroids.

It is extremely difficult to determine the criteria that are employed to distinguish the focal model from its closely related analogue. It is also likely that literal plant names, as defined here, which incorporate the morpheme 'false' are recently emerged folk generics (also see Berlin et al. 1974).

The status of primary lexemes terminating in *ran* 'false' needs further explication, since this suffix does not well accord with findings from other languages, in particular, the Tzeltal Maya. For example in Tzeltal Maya, *bac'il lima* 'authentic lime' and *kaslan lima* 'foreign lime' are classified within the same folk generic, *lima* (Berlin et al. 1974:43, 173). Yet in Ka'apor, informants consistently segregate words terminating in *ran* from their focal models, which may or may not terminate in *te* 'authentic'. For example, *iŋa-ran* ('ingá-false') (also called *tayahu-iŋa* ['white-lipped peccary–ingá']) and *akuši-iŋa* ('agouti-ingá') are not kinds of *iŋa* in Ka'apor folk classification. Perhaps the factor of edibility accounts for the disjuncture, since all other names incorporating *iŋa* and which are classified as specific types of *iŋa* refer to edible plants while these do not.

The term *ran* is also postposed in names for several introduced domesticates (as *kaslan* 'foreign' is preposed in Tzeltal names). Examples are custard apple (*arašiku-ran* '*Annona* spp.–false'), taro (*taya-ran* 'cocoyam-false'), and West Indian gherkin (*waraširan* 'watermelon-false'), but these are not folk specifics of *arašiku* ('*Annona* spp.'), cocoyam, and watermelon, respectively. Although the polymorphemic construction of these terms imitates that of many secondary lexemes (which in classificatory terms are folk specifics and varietals), these terms are independently folk generic names. (Berlin's point [1992:27] that primary lexemes may be "complex"—that is, polymorphemic—is well taken here.) A few other primary lexemes also imitate the structure of secondary lexemes (such as *iratawahu'i* and *awaši'i*—see appendix 9), but because they are not considered to be subtypes of their models, they are here classified as folk generic names (i.e., primary lexemes).

The 155 literal generic names represent 32 percent of the entire corpus of Ka'apor plant generics (see table 7.3). These 155 literal names are distributed throughout all plant categories, but unevenly. Literal generic names are proportionally most prominent in the categories of 'domesticates' and 'nondomesticates unclassified as to life form' and least so in the categories of 'vines' (*sipo*), 'herbs' (*ka'a*), and 'trees' (*mira*). Specifically, with respect to the 155 literal names, 68 denote 'trees', which represents 26 percent of all 'tree' names and 44 percent of all literal names; 8 refer to 'herbs', which constitutes 14 percent of all 'herb' names and 5 percent of all literal names; 10 refer to 'vines', which accounts for 17 percent of all 'vine' names and 6 percent of all literal names; 46 denote 'domesticates', which represents 69 percent of all 'domesticate' names and 30 percent of all literal names; and 23 refer to 'nondomesticates unclassified as to life form', which constitutes 56 percent of all names for 'nondomesticates unclassified as to life form' and 15 percent of all literal names.

Metaphorical Generic Names

The class of metaphorical names is much more differentiated structurally and semantically than that of literal names. Metaphorical names are subdivided linguisti-

TABLE 7.3 *Numbers of Ka'apor Plant Names by Type of Name and Type of Plant*

Plant Type	Literal Names	Metaphorical Names			Row Totals
		Simple Primary Lexemes	Productive Primary Lexemes	Unproductive Primary Lexemes	
'Tree'	68	130	49	10	257
'Herb'	8	0	24	26	58
'Vine'	10	5	34	11	60
Domesticate	46	0	0	21	67
Nondomesticate unclassified as to life form	23	2	0	16	41
Column Totals	155	137	107	84	483

cally into simple primary lexemes, productive primary lexemes, and unproductive primary lexemes, which, in turn, are further subdivided into several subtypes.

SIMPLE PRIMARY LEXEMES

The first category, metaphorical simple primary lexemes, may at first seem to be self-contradictory. How can a simple primary lexeme, as defined by Berlin et al. (1974:27) [i.e., a "single word expression," such as 'oak' or 'pine'], be metaphorical? In fact, such terms are one-word expressions, even if they are mostly polymorphemic in Ka'apor.

In the domain of plants, with the exception of many names for domesticates and a few names for herbs, vines, and others, the majority (402 of 483 names, or 83 percent) of names are polymorphemic, even if many, as with literal tree names terminating in *'i*, are only superficially so (Balée 1989b). A majority (42, or 66 percent) of 64 genuinely monomorphemic generic names designating plants refers to domesticates. Examples include *akayu* 'cashew', *kase* 'coffee', and *pako* 'banana'. A smaller segment (12, or 19 percent) refers to nondomesticates unclassified as to life form, such as *kāpī* ('grass'), *owi* (the palm used for roofing thatch, *Geonoma baculifera*), and *urupe* (many mushroom species). Only 10 denote members of life form taxa; in other words, of 375 folk generic names for plants pertaining to a life form taxon, which account for 78 percent of all Ka'apor generic plant names, a mere 2 percent are truly monomorphemic. The data indicate, therefore, that the concept of "simple primary lexeme" should be broadened to account for superficially binomial, literal names and even for other polymorphemic constructions that cannot be classified logically as either unproductive or productive primary lexemes (also see Berlin 1992:27).

In most cases, with regard to literal simple primary lexemes, incorporation of a postposed bound morpheme such as *'i* or *-rimo* (with regard to vines) is obligatory—many plant names, then, are constituted by two or more bound morphemes. The initial morpheme which serves as the principal nominal component (e.g., *ama* in *ama'i*, a generic name denoting five botanical species of *Cecropia*) is bound as well. Such terms have been described as "superficially binomial" (Hunn and French 1984:77; Balée 1989b:11).

Metaphorical simple primary lexemes, on the other hand, incorporate free morphemes in addition to one or more bound morphemes such as -'*i*. These free morphemes are not productive primary lexemes insofar as appropriate life-form head
terms, which by definition are free morphemes, are not incorporated. Nor can they be
classified as unproductive primary lexemes, because they do not incorporate constituents that would indicate a life form to which the referents do not belong and
because they *do* incorporate either a bound morpheme, referring generally to plants,
such as -'*i*. Some name modeling occurs in these types of names. For example, several
names are modeled on an undomesticated plant generic, plant part, or utilitarian
aspect of plants (such as taste, edibility) and may be possessed or unpossessed.

Examples of possession by animals include *akuši-yu-'i* 'agouti–*Astrocaryum
gynacanthum*–stem', *ira-muru-'i* 'bird–*Astrocaryum murumuru*–stem', *parawa-mi'u-'i*
'Mealy parrot–food–stem', and *yaši-wa-'i* 'yellow-footed tortoise–fruit–stem', all of
which refer to 'trees' (*mira*). The models are in different life forms from their analogues. For example, *yu'i* and *muru'i* in the above examples refer to two species of
palms, which, as with all palms, are not considered to pertain to any of the life form
categories, including 'tree'. In one case, possession is by a habitat (*taper-iwa-'i* 'fallow-
fruit-stem'). Perhaps a more familiar (although slightly less accurate) gloss would be
'fruit tree of the fallow'.

Unpossessed terms for trees include *iwa-hu-'i* 'fruit-big-stem', *wa-pini-'i* 'fruit-
stripe-stem', *šimo-'i* ('*Derris utilis*–stem'), and *tata-'iran-'i* 'fire-false-stem'. *Derris utilis*
(*šimo*), which is one of the principal species used for poisoning fish, and *Guatteria scandens* (*tata'i*), which is the most important species used in making fire drills and hearths,
are both classified as 'vines' (*sipo*), unlike their analogues here. In all cases, nevertheless, because of the head term '*i*, none of these names can be considered to be unproductive or productive primary lexemes.

The same holds in reference to several metaphorical simple primary lexemes modeled by analogy on generic names for domesticated plants. Examples include *akayu-'i*
('cashew-stem') and *māmā-ran-'i* ('papaya-false-stem'). All these generic names, which
belong to the class *mira* 'tree', include generic models that do not pertain to that class
(since domesticates are not considered to be members of any of the life form taxa).
Rather, these generic names incorporate the head term '*i*, which does not violate the
concept 'tree', even if it does not mean exactly 'tree'.

Other metaphorical simple primary lexemes simply merge a word for an animal,
animal part, or animal characteristic (such as color, hide texture, feather quality) with
a bound head term (such as '*i* for many 'tree' names or *rimo*, as in the case of many
'vine' names). Examples of such possessed names for 'trees' (*mira*) include *kururu-'i*
'toad-stem', *merai-'i* 'anteater-stem', and *mitū-eha-'i* 'curassow-eye-stem'. Examples of
such names for 'vines' (*sipo*) include *araruhu-wai-rimo* 'Blue-green macaw–tail–stem'
and *pikahu-ampūi-rimo* 'Ruddy ground dove–beak–stem'. An example of such a term
for an unclassified nondomesticate is *yawar-'i* 'jaguar-stem', which refers to the well-
armed javari palm (*Astrocaryum javari*) of swamps and rivers in the Gurupi drainage.

Two metaphorical simple primary lexemes that are also possessed are modeled on
divinities, being *kiki'i* and *kurupi'i*. *Kiki* is a mythical, hairy, arboreal quadruped
believed to have blue bones; *Kurupir* is a bipedal forest midget, thought to influence

game supplies. Another type of simple metaphorical primary lexeme denotes the location or preferred place of an animal. Two names are of this sort, and they both refer to tortoises: (*karume-pita-'i* 'red-footed tortoise–staying-place-stem' and *yaši-pita-'i* 'yellow-footed tortoise–staying-place-stem'). Finally, two specific types of metaphorical primary lexemes for plants are modeled on an inanimate object or event—natural or cultural. These are unpossessed terms. As for cultural terms, all of which denote 'tree', cultural objects featured in these names include: comb (e.g., *ayaŋ-kiwa-'i* 'divinity-comb-stem' and *uru-kiwa-'i* 'pheasant-comb-stem'); hammock (e.g., *inamu-kiha-wi-'i* 'tinamou-hammock-small-stem' and *yanu-kiha-wi-'i* 'spider-ham-mock-small-stem'); beer (*wari-kawï-'i* 'howler monkey–beer–stem'); trail (*karaí-pe-'i* 'non-Indian–trail–stem'); feather box (*patuwa-'i*); wax (*irati-'i*); lashing material (e.g., *iwir-'i*); resin (*kanei-'i*); pipe (*kašima-'i*); spoon (*kuyer-'i*); remedy (e.g., *kuru-pusan-'i* 'skin eruption–remedy–stem'); fish hook (*pina-'i*); arrow point (e.g., *piwa-'i*); war club (*tamaran-'i*); arrow shaft (e.g., *u'i-tima-'i* 'arrow-shaft-stem'); oil (*yaniro-'i*); broom (*tapiša-'i*); manioc press (*tapeši-'i*); beads (e.g., *pu'i-piraŋ-'i* 'bead-red-stem'); and mirror (e.g., *waruwa-'i*).

With regard to names modeled on natural (but nonliving) objects or events, there are three 'tree' names and one 'vine' name. The 'vine' name is *tata-'i* 'fire-stem'. The 'tree' name models refer to rain (*aman-'i*), the sound of wind in the forest (*ka'aro-tiapu-'i* 'leaf-noise-stem'), and a scream or cry (*ya-pukai-'i* 'we-scream-stem').

Overall, metaphorical simple primary lexemes, of which there are 137, describe the structure of 130 'tree' names (51 percent of all tree names, 95 percent of all such lexemes), no 'herb' name, 5 'vine' names (8 percent of all vine names, 4 percent of such lexemes), no name for a domesticate, and 2 names for nondomesticates unclassified as to life form (5 percent of all such names, 1 percent of such lexemes). In terms of all 483 Ka'apor plant generics, metaphorical simple primary lexemes account for 137 or 28 percent of the total. The great majority of these (130, or 95 percent), in turn, designate 'trees'(*mira*).

PRODUCTIVE PRIMARY LEXEMES

Productive primary lexemes incorporate, by definition, life form labels that do in fact refer to the life form taxa to which the denotata belong in Ka'apor classification. In other words, productive primary lexemes must incorporate one of the following terms, *mira* 'tree', *ka'a* 'herb', or *sipo* 'vine'. These terms usually fall in final (head) position. Many different types of names modeled on analogy occur with productive primary lexemes. Two tree names are modeled on a domesticated plant name. In one case, the name is modeled on a folk specific (*kwi-ra'i-mira* ['calabash-small-tree']); in the other, the name is modeled on a folk generic (*waya-mira* ['guava-tree']). Both these plants are nondomesticated; the significance of this type of modeling will be discussed below.

Several other names, including names for 'trees', 'herbs', and 'vines', are modeled on the name for a nondomesticated plant generic or plant part. Examples include *tak-wari-mira* ('bamboo-tree') (the tree species denoted, *Coccoloba paniculata*, actually has a hollow stem, like bamboo), *ka'a-ro* 'herb-leaf', and *tiriri-sipo* ('kind of grass–vine'). One productive primary lexeme is modeled on a different plant life form from that to which the denotate pertains, *sipo-mira* 'vine-tree'. It is classified as a tree, although the habits of the two species (*Coccoloba latifolia* and *Coccoloba racemulosa*) referred to by this name are tortuous and reminiscent of a plagiotropic vine.

Many productive primary lexemes referring to 'trees', 'herbs', and 'vines' incorporate animal names. These are linguistically possessed names. Examples of 'tree' names in this category include *akuši-mira* 'agouti-tree', *arapuha-mira* 'red brocket deer–tree', *tatu-mira* 'armadillo-tree', and *yanu-mira* 'spider tree'. These 'trees' are possessed by archetypal animals. It is said, for example, that "*the* spider"—not just any spider—"owns" the *yanu-mira* tree. Examples of 'herb' names in this category include *irakahu-ka'a* 'weasel herb', *kururu-ka'a* 'toad-herb', *teyu-ka'a* 'skink-herb', and *yaŋwate-ka'a* 'spotted jaguar–herb'. Examples of 'vine' names in this category include *arapuha-sipo* 'red brocket deer–vine', *ira-i-sipo* 'bird-little-vine', *maha-sipo* 'gray brocket deer–vine', and *wiriri-sipo* 'swift-vine'.

One productive primary lexeme is modeled on the name for a divinity, *kurupi-'i-sipo* 'divinity-little-vine'. Others are modeled on inanimate cultural objects or events, such as *tarara-mira* 'manioc griddle rake–tree', *ka'a-riru* 'herb-vessel', and *mišik-sipo* 'roast in coals–vine'. A few names are modeled on a personal office or a kin relation, including *pa'i-mira* 'priest-tree', *anã-mira* 'relative [or 'sister', female speaking]–tree', and *anã-sipo* 'idem-vine'.

Many other names are modeled on a quality such as relative hardness, color, taste, motion, weight, smell, shape, effect upon touching, relative size, smoothness, authenticity, and falseness. In most cases with this type of productive primary lexeme, the life form label occurs in initial position, unlike the great majority of other productive primary lexemes. This is probably because the postposed adjuncts referring to qualities are usually not nouns. Examples include *mira-piraŋ* 'tree-reddish', *mira-pirer-hẽ'ẽ-'i* 'tree-bark-sweet-stem', *mira-tawa* 'tree-yellow', *ka'a-tuwir* 'herb-white', *ka'a-yuwar* 'herb-itch', *ka'a-piši'u* 'herb–fishy smell', *sipo-memek* 'vine-weak', *sipo-ãtã* 'vine-hard', *sipo-nem* 'vine-fetid', *sipo-tai* 'vine-spicy', and *sipo-tawa* 'vine-yellow'. The 107 productive primary lexemes are distributed as follows: 49 'tree' names (19 percent of all tree names, 46 percent of such lexemes), 24 'herb' names (41 percent of herb names, 22 percent of such lexemes), and 34 'vine' names (57 percent of vine names, 32 percent of such lexemes). Because domesticates and obviously nondomesticates unclassified as to life form do not belong to any life form, there are no productive primary lexemes to be found in these categories.

UNPRODUCTIVE PRIMARY LEXEMES

Unproductive primary lexemes incorporate head terms that refer to either a plant life form or plant generic name to which the denotate does not belong—or which is "obscure" in the sense that none of the morphemes incorporated in the name refer to any plant or plant part but rather to nonbotanical phenomena (see Balée 1989b; Berlin et al. 1974:39). Several names from all life forms and nondomesticated generics unclassified as to life form are modeled on generic names for domesticates and may be either possessed or unpossessed. As for possessed names, 'tree' names include *ara-ki'ĩ* 'macaw–chile pepper' and *ãyaŋ-ruku* 'divinity-annatto'; 'herb' names include *arapuha-mani'i* 'red brocket deer–manioc' and *tayahu-manuwi* 'white-lipped peccary–peanut'; 'vine' names include *a'ĩhu-pako* 'sloth-banana' and *yuruši-ki'ĩ* 'ruddy ground dove–chile pepper'. None of the species denoted by these terms is a domesticate, nor are any classified as affiliates of a generic name for a domesticate. In other words, just as in folk English 'poison oak' is not a kind of 'oak' and 'prairie dog' is not a kind of 'dog', in Ka'apor 'macaw–chile pepper' is not a kind of chile pepper, 'red brocket deer–manioc' is not a kind of manioc, 'sloth-banana' is not a kind of banana, and so on.

One of these possessed names is a nondomesticate unclassified as to life form, *kurupi-nana*, which refers to a bromeliad that is not considered to be a kind of 'pineapple' (*nana*). As for unpossessed names, an example of a 'tree' name is *mira-ki'ĩ* ('tree–chile pepper'). Unpossessed 'herb' names include *pako-sororo* ('banana-hunger'), *kanami-ran* ('*Clibadium sylvestre*-false'), and *kara-ran* ('yam-false'). Because the *ran* 'false' morpheme is incorporated in the names of these last two plants, they also are examples of plants that do not belong to the same life form as the model; they are unproductive primary lexemes. In other words, the referents of 'false yam' and 'false *Clibadium*' are not domesticates, but 'vines'; yet neither 'yams' nor the fish poison *Clibadium* pertain to any life form. In contrast, as noted above, where the *ran* morpheme is affixed to a generic name which pertains to the same life form as 'generic name + *ran*', such names are not unproductive.

There is one exception to this rule. Although generic names for domesticated introduced plants may incorporate the morpheme *ran*, and although they pertain to the unnamed class 'domesticates', these are not kinds of the generics on which they are modeled. For example, *waraši-ran* ('watermelon-false')—the introduced West Indian gherkin—does refer to a domesticate, but it is not considered to be a type of *waraši* ('watermelon'). The same is true with regard to *awaši'i* ('maize-little') which refers to rice and not a folk species of *awaši* ('maize'). As for an unproductive primary lexeme modeled on a generic name for a domesticate that designates an unclassified nondomesticate, there is but one example: *mira-kirawa* ('tree-caroá'). It is not a folk species of *kirawa* (a domesticated bromeliad used for rope-making), nor is it a 'tree' (*mira*), hence its structure as an unproductive primary lexeme.

Several unproductive primary lexemes are modeled on generic names for nondomesticates or plant parts. 'Tree' names include *tayahu-iŋa* ('white-lipped peccary–inga') and *akuši-iŋa* ('agouti-inga'), which are not folk species of the generic *iŋa*. Herb names of this type include *muyu-ka'aro* ('mythical snake–arrow root plants') and *tatu-ka'aro* ('armadillo–arrow root plants'). Several names for domesticates are modeled on names for nondomesticates as well. These include *pu'i-piraŋ-'i-karai-ma'e* 'bead–reddish–stem–non-Indian his/her' (another gloss could be the 'non-Indians' reddish bead tree'), which refers to an introduced cultivated tree [*Adenanthera pavonina*]. It is not considered to be a kind of *pu'ipiraŋ'i*, its nondomesticated forest model. The same is true for *arašiku-ran* ('araticum-false') which refers to soursop, for *šimo-'i* ('*Derris utilis*-little') which refers to the cultivated fish poison *Tephrosia sinapou*, and for *tukumã-ran* ('tucumã palm–false') which refers to peach palm. All these plants are recent introductions to the Ka'apor and are not classified as folk species of the generic names upon which they are modeled (for example, 'peach palm' is not classified as a kind of the uncultivated *tukumã* ['tucumã' palm]).

A few unproductive primary lexemes incorporate life form labels (as opposed to generic plant names) to which the referents do not belong in the Ka'apor plant classification system (see also Berlin et al. 1974:39). For example, *akuši-ka'a* 'agouti-herb' is classified as a 'tree' (*mïra*), not an 'herb' (*ka'a*); *ita-mïra* 'stone-tree', which refers to certain tiny, succulent species of *Phyllanthus* in the spurge family, is classified as an 'herb' (*ka'a*), not a 'tree' (*mïra*); *tapuru-ka'a* 'larva-herb' is classified as a 'vine' not an 'herb'; and, as for domesticates, *ka'a-memek* 'herb-weak', *ka'a-pe* 'herb-flat', *ka'a-piher* 'herb-aromatic', *tupã-ka'a* 'thunder-herb', and *mïkur-ka'a* 'opossum-herb' are not classified as kinds of herbs, but are, rather, unaffiliated with any life form taxon. Also, the domesticated lemon grass called *kãpï-piher* ('grass–sweet smelling') is not classified as a kind of *kãpï* ('grass'). It is important to note that all these names refer to recently introduced species of dooryard gardens. This appears to be related to a fundamental dichotomy in the nomenclature which segregates traditional domesticates from other plants.

Another plant whose life form constituent is misleading concerns the species *Solanum leucocarpon*, which may be designated by either of two synonyms (*ãyaŋara-mïra* and *ãyaŋaraka'a*), yet it is classified neither as a 'tree' nor as an 'herb'—it corresponds to the concept of a generic taxon that lies on the "boundary" of two life form taxa (see Berlin et al. 1974:30–31). This species is believed to be extremely dangerous to the touch.

Finally, the class of unproductive primary lexemes includes several "obscure" names (which contain no plant morphemes) in all life forms and unclassified generic categories. Tree names include *yaši-amïr* 'yellow-footed tortoise–deceased' and *ma'e-ira-pïraŋ* 'some-bird-reddish'; 'herb' names include *kupi'i-pe* 'termite-flat', *wari-ruwai* 'howler monkey–tail', *ka'uwa-pusan* 'insanity-remedy', and *tapiša* 'broom'; 'vine' names include *akuši-nami* 'agouti-ear', *ãyaŋ-ampũi-putïr* 'divinity-nose-flower', and *surukuku-yu-rãšï* 'bushmaster-yellow-spine'; names for domesticates in this category include *awa-i* 'person-little', *pu'ï-risa* 'bead-cold', *tawa* 'yellow', and *kure-nami* 'pig-ear'.

It is significant that all names for domesticates in this category refer to recently introduced plants, another indication of nomenclatural dichotomy. Other unclassified nondomesticates in this category are referred to, for example, by the names *ira-hu-ra-wi* 'bird-big-down-light', *ma'e-wïra-puši* 'some-bird-feces' and *mïra-meyu* 'tree–manioc bread'. This latter name, although it includes a plant morpheme, *mïra* 'tree', is linguistically classified as obscure because the plant morpheme is not the head term, the denotate is not classified as a 'tree' (*mïra*), and *meyu* 'manioc bread' is not a plant or plant part semantically (although it might be so classified in a different cultural context, for example, in U.S. Customs), but rather, a cultural product.

In summary, the total number of unproductive primary lexemes is 84. There are 10 tree names (4 percent of all tree names), 26 'herb' names (45 percent of all herb names), 11 'vine' names (18 percent of all vine names), 21 names for domesticates (31 percent of all names for domesticates), and 16 names for nondomesticates unclassified as to life form (39 percent of all names for unclassified nondomesticates). Perhaps it is most interesting that a large proportion of such names should occur with domesticates; what is of outstanding significance here is that all such 21 unproductive primary lexemes refer to introduced species. These words indicate something about the origins of names for plants more generally. In other words, it may be hypothesized that all literal names (which are the most common type of name for domesticates, less so for other plant categories) begin either as borrowings from other languages or as metaphorical names—the latter including linguistically simple primary lexemes (such as *kururu-'i* 'toad-stem'), productive primary lexemes (such as *arapuha-mira* 'red brocket deer–tree'), and unproductive primary lexemes (such as *tawa* 'yellow').

This development of literal names does not mean that no traditional species is accorded a metaphorical name. Rather, the first name a nondomesticated species of the habitat receives in a Tupí-Guaraní language is metaphorical; the same is true with introduced domesticates, unless a name from another language is borrowed with the plant referent itself. Metaphorical words are in flux, either progressing to become new metaphorical words or new literal words.

To some extent this is evident in some of the free variation for certain plant names (free variants have not been included here in plant word calculations, although synonyms were included). For example, in the Turiaçu river basin, several large tree species of the Sapotaceae family are referred to by the term *ira-tawa-'i* 'bird-yellow-stem'; remarkably, these same species are denoted by the term *iwa-tawa-'i* 'fruit-yellow-stem' in the village of Gurupiuna. These two plant words differ in but a single sound in the initial morpheme of each word; since *w* and *r* are different phonemes, the meanings of the initial morphemes *ira* and *iwa* are different ('bird' versus 'sky'). What is most interesting here is that the name of this species appears to be in flux, insofar as there is a kind of semantic tension between the name bestowed in Turiaçu versus that used in Gurupiuna. In other words, this may represent a metaphorical term in transition to another (albeit related) metaphorical term—it is not clear, however, which word was used first to designate these species.

Other metaphorical words appear, by similar reasoning, to be in transition to literal terms. For example, *yani-ro-'i* 'oil-bitter-stem' denotes the crabwood tree (*Carapa guianensis*) of the mahogany family in the Turiaçu basin, whereas the term *yini-ro-'i* 'L-bitter-stem' is used in reference to the same species in Gurupiuna. Literal words, on the other hand, do not appear to become metaphorical words. Literal words are more stable through time than are metaphorical words (Balée and Moore 1991).

Overall, unproductive primary lexemes, with a total of 84, constitute 17 percent of all Ka'apor folk generic plant names. Names for traditional domesticates are literal simple primary lexemes. Names for 'trees' tend to be metaphorical, especially of the type metaphorical simple primary lexeme. Linguistically, names for herbs strongly tend to be metaphorical, with productive and unproductive primary lexemes being most common. Names for vines also tend to be metaphorical, especially of the subtype productive primary lexeme. Names for nondomesticates not classified under life form labels, on the other hand, tend to be literal. Many of these refer to palms (Balée and Moore 1991).

Secondary Lexemes and Names for Domesticated Plants

Pierre Grenand (1980:43) described a cognitive barrier between domesticated and nondomesticated plants in Wayãpi ethnobotany as an "uncrossable frontier." The

Table 7.4 *Generic Names for Nondomesticates that are Modeled on Names for Domesticated Plants in Ka'apor*

Ka'apor Name	Gloss	Botanical Referent	Botanical Model
		'Trees'	
akayu'ĭ	cashew	*Anacardium giganteum; A. parvifolium*	*Anacardium occidentale*
ara-kiʔ	macaw-chile pepper	*Aparisthmium cordatum*	*Capsicum* spp.
irimã-'ĭ-ran	lime-stem-false	*Brunfelsia guianensis*	*Citrus* spp.
kaka-ran-'ĭ	cacao-false	*Theobroma speciosum*	*Theobroma cacao*
kara-miri-'ĭ	yam-small-stem	*Micropholis guyanensis* spp. *guyanensis; Pouteria engleri*	*Dioscorea trifida*
kara-miri-ran-'ĭ	yam-small-false-stem	*Pouteria sagotiana; Pouteria* sp. 5	*Dioscorea trifida*
kase-ran-'ĭ	coffee-false-stem	*Casearia javitensis*	*Coffea arabica*
kwi-ra'i-mira	calabash-small-tree	*Eugenia egensis*	*Crescentia cujete*
mãmã-ran-'ĭ	papaya-false-stem	*Jacaratia spinosa*	*Carica papaya*
mani'ĭ-ran-'ĭ	manioc-false-stem	*Stryphnodendron polystachyum*	*Manihot esculenta*
mira-kiʔ	tree-chile pepper	*Myrciaria tenella*	*Capsicum* spp.
murukuya-'ĭ	passion fruit-stem	*Sloanea* spp.	*Passiflora edulis*
pitim-'ĭ	tobacco-stem	*Couratari* spp.	*Nicotiana tabacum*
pitim-inem-ran-'ĭ	tobacco-fetid-false-stem	*Eschweilera ovata*	*Nicotiana tabacum*
uruku-ran-'ĭ	annatto-false-stem	*Bixa orellana*	*Bixa orellana* (domesticated variety)
waya-mira	guava-tree	Indeterminate genus 2	*Psidium guayava*
waya-taper-rupi-har	guava-fallow-through-place	cf. *Calyptranthes* sp. 1	*Psidium guayava*
		'Herbs'	
a'ĭhu-pako	two-toed-sloth-banana	*Trigonidium acuminatum*	*Musa* sp. 1; *Musa* sp. 2
arapuha-mani'ĭ	red brocket deer-manioc	*Manihot brachyloba; M. leptophylla; M. quinquepartita*	*Manihot esculenta*
kanamĭ-ran	kind of fish poison-false	*Lantana camara*	*Clibadium sylvestre*

Wayãpi, he pointed out, distinguish no genealogical relationship between domesticated and nondomesticated species of *Manihot*, which occupy the same habitat and outwardly appear similar (a chief difference being that the nondomesticated species are dispersed by nonhuman agents). Likewise, the Ka'apor distinguish domesticated manioc (*Manihot esculenta*), the "bitter" forms of which usually incorporate the generic head *mani'ĭ*, from nondomesticated manioc (*Manihot brachyloba, M. leptophylla,* and *M. quinquepartita*), which are generically termed *arapuha-mani'ĭ* 'red brocket deer–manioc'.

As noted earlier, *arapuha-mani'ĭ* is an unproductive primary lexeme, for it incorporates as the head term a generic name to which it does not belong as a constituent. That is, 'red brocket deer–manioc' is *not* a kind of manioc. Although generic names for nondomesticates may be modeled on names of domesticated species (see table 7.3), names for traditional domesticates are never modeled by analogy on nondomesticated plants. Similar linguistic patterns appear to hold for other Tupí-Guaraní languages.

Despite superficial similarities, there should be no confusion concerning complex

TABLE 7.4 *Continued*

Ka'apor Name	Gloss	Botanical Referent	Botanical Model
kara-ran	yam-false	*Dioscorea* sp. 2	*Dioscorea trifida*
kawasu-ran	bottle gourd-false	*Cayaponia; Gurania* spp.	*Lagenaria siceraria*
kurupi-kã	divinity-sugar cane	*Renealmia alpinia*	*Saccharum officinarum*
kurupi-nana	divinity-pineapple	*Ananas nanas*	*Ananas comosus*
kurupi-piłim	divinity-tobacco	*Renealmia floribunda*	*Nicotiana tabacum*
mira-kirawa	tree-caroá	*Triumfetta rhoimbodea; T. semitriloba; Urena lobata*	*Neoglaziovia variegata*
mira-kirawa-ran	tree-caroá-false	*Sida* spp.	*Neoglaziovia variegata*
murukuya-pinim	passion fruit-striped	*Passiflora* sp. 3	*Passiflora edulis*
murukuya-ran	passion fruit-false	*Passiflora aranjoi, P. coccinea, P.* sp. 2	*Passiflora edulis*
murukuya-tawa	passion fruit-yellow	*Passiflora* sp. 1	*Passiflora edulis*
nana-ran	pineapple-false	*Ananas nanas*	*Ananas comosus*
pako-sororo	banana-hunger	*Phenakospermum guianensis*	*Musa* sp. 1; *Musa* sp. 2
tapi'i-kanami	tapir-fish poison	*Psychotria poeppigiana*	*Clibadium sylvestre*
tayahu-manuwi	white-lipped peccary-peanut	*Ctenanthe* sp. 1; Marantaceae sp. 1	*Arachis hypogaea*
tayahu-manuwi-ran	white lipped peccary-peanut-false	Marantaceae sp. 2	*Arachis hypogaea*
tayahu-pako-ro	white lipped peccary-banana-leaf	*Costus arabicus; Heliconia* spp.	*Musa* sp. 1; *Musa* sp. 2
teyu-piłim	skink-tobacco	*Conyza banariensis; Porophyllum ellipticum; Pterocaulon vergatum*	*Nicotiana tabacum*
teyu-piłim-ran	skink-tobacco-false	*Phyllanthus niruri*	*Nicotiana tabacum*
yiłik-ran	sweet potato-false	*Ipomoea* aff. *squamosa; I. phyllomega*	*Ipomoea batatas*
yuruši-kiʼī	ruddy ground dove-chile pepper	*Geophila repens*	*Capsicum* spp.

primary lexemes and secondary lexemes in terms of their internal structure. For example, the unproductive primary lexemes that take as their model a domesticated plant, which, in turn, is possessed by an animal name, are fundamentally unlike secondary lexemes denoting domesticates that also involve animal possession. As for the folk generic names possessed by animals in table 7.4, the animals denoted in each such name are, in fact, ecologically associated with the nondomesticated referents. Specifically, *arapuha* 'red brocket deer' do disperse the seeds of *arapuha-muni'i* ('red brocket deer–manioc'); *ara* 'macaws' do eat the fruits of *araki'ī* ('macaw–chile pepper'); *tayahu* 'white-lipped peccaries' do forage on the rhizomes of *tayahu-manuwi* ('white-lipped peccary–peanut') and *tayahu-manuwi-ran* ('white-lipped peccary–peanut–false'); *tapi'i* 'tapirs' do browse on the leaves of *tapi'i-kanami* ('tapir–*Clibadium* fish poison'); *teyu* ('skinks') are, in fact, common to swiddens and other clearings, where one also typically finds *teyu-piłim* ('skink-tobacco'); *yuruši* 'ruddy ground doves' are indeed known to eat the tiny, globose, crimson fruits of *yuruši-ki'ī* ('ruddy ground dove–chile pepper'); and *a'ihu* ('two-toed sloths') do feed on the leaves of *a'ihu-pako* ('two-toed sloth–banana'). Of course, none of these plants is a domesticate and, despite the incorporation of domesticated generic head terms, none is classified as a domesticate.

But such ecological relationships appear to never be connoted in secondary lexemes (folk species) referring to domesticates. Folk specific names for domesticates may incorporate preposed animal attributives, but the animals are not ecologically associated with the plants themselves (Balée 1989b:14–15). In the Ka'apor language, only one folk generic name lends itself to such prepositioning (and possession). This is *mani'i* ('bitter manioc'). Five of the 18 folk species of bitter manioc named by the Ka'apor incorporate preposed animal attributives. These are *yararak-mani'i* ('fer de lance–manioc'), *yaši-mani'i* ('yellow-footed tortoise–manioc'), *sarakur-mani'i* ('Wood rail–manioc'), *ararū-mani'i* ('Hyacinthine macaw–manioc'), and *šimokape-mani'i* ('Black vulture–manioc'). Fer de lances, rails, tortoises, macaws, and vultures do not browse on manioc in swiddens (fer de lances and vultures do not, of course, consume any kind of vegetation). With the notable exception of fer de lances, these species are, moreover, rarely encountered in Ka'apor swiddens.

This disjunction between domesticates and nondomesticates with respect to animal associations is evident in a comparison of the domesticated 'fer de lance–manioc' with the nondomesticated 'skink tobacco'. Whereas skinks are only seen in swiddens and other areas of anthropogenic perturbation, fer de lances, which inhabit both fallow and high forest, cannot be said to be so strongly associated with any human factor. Hence, while skinks are genuinely associated in habitat with *teyu-pitim* ('skink-tobacco'), which refers to three herbaceous composites, fer de lances are more facultative in terms of habitat preference than is 'fer de lance-manioc'—a true prisoner of swidden fields.

Hence, folk specifics for cultivated plants do not evoke ecological relationships as do unproductive primary lexemes incorporating preposed animal names and postposed generic names for domesticates. In other words, unproductive primary lexemes incorporate animal attributives in semantically different ways from secondary lexemes denoting folk species of domesticated plants. This principle can be extended to other Tupí-Guaraní languages. For example, the only name for a folk species of yam (*Dioscorea trifida*) modified by a preposed animal name among the Tembé is *yowoi-kara* ('boa constrictor–yam'). The boa constrictor, like all snakes, is strictly carnivorous, and it is not typically encountered in swiddens.

Two powerful principles appear to distinguish the corpus of generic names for traditional domesticates from the corpus of generic names for other plants: (1) productive primary lexemes refer only to nondomesticates and (2) unproductive primary lexemes refer to some nondomesticates and some introduced domesticates, but never to traditional domesticates. What remains to be explained is why any domesticates at all receive unproductive primary lexemes as generic names. I refer to names such as *waraširan* ('watermelon-false' [West Indian gherkin]), *awaši-'i* ('maize-little' [rice]), *arašikuran* ('wild soursop–false' [soursop]), *šimo-'i* ('*Derris utilis*–little' [*Tephrosia sinapou*]), *tukumā-ran* ('tucumā palm–false' [peach palm]), *tupā-ka'a* ('thunder-herb' [hibiscus]), *ka'a-memek* ('herb-weak' [*Amaranthus oleraceous*]), *kāpī-piher* ('grass-aromatic' [lemon grass]), *tawa* ('yellow' [turmeric]), and so on for a total of some 21 domesticated generic names of this type.

I hinted earlier that the semantic structure of these names may indicate something more general concerning processes associated with the origins of plant names. For it must be admitted that, except possibly for certain names borrowed from Portuguese or another indigenous language, the names for introduced species must be younger than the names for traditional species. The possible cognitive sequence involved probably depends on an initial recognition than an introduced species is indeed a domesticate. Even if the Ka'apor and their ancestors never cultivated it before, someone did. And regardless of the ethnocentric strictures that attend most non-Ka'apor as persons, the same is not true with regard to their material and agricultural artifacts, some of which, such as steel tools and sugar cane, have been readily absorbed into Ka'apor culture. So it is likely that any introduced plant is automatically classified as a domesticate when it is initially given a name. That name is either a borrowing of a literal term (as with *irimã* ['lime'] from Portuguese) or one of several types of metaphorical constructions.

In several cases, as with *taya-ran* 'cocoyam-false' for the introduced taro, the new name is modeled on a folk generic name for an existing domesticate. Yet the new name represents a separate folk generic, for despite the phenotypic similarities (as between cocoyam and taro, which are closely related aroids of the same genus), the newly introduced plant is still not so well known as to be considered an exemplar of its model. Intro-

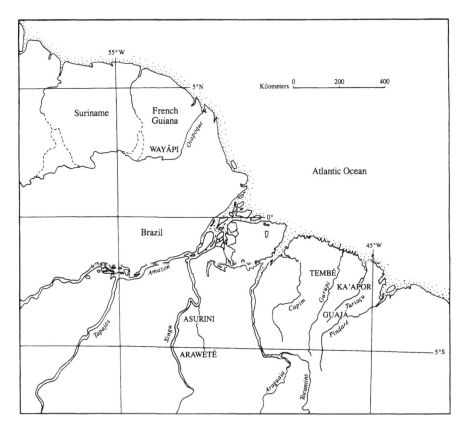

Some Tupí-Guaraní societies of eastern Amazonia.

TABLE 7.5 *Generic Names for Nondomesticates that are Modeled on Names for Domesticated Plants in Other Tupí-Guaraní Languages*

Indigenous Name	Gloss	Botanical Referent	Botanical Model
TEMBÉ:			
arapʰa-mani'iw	red brocket deer-stem	*Manihot leptophylla*	*Manihot esculenta*
kána-ran	sugar cane-false	*Renealmia alpinia*	*Saccharum officinarum*
kunami-ran	[fish poison]-false	*Bertieria guianensis*	*Clibadium sylvestre*
		Clibadium sylvestre	
mãmã-ran	papaya-false	*Paquira aquatica*	*Carica papaya*
maɲ-'iw-ran	mango-stem-false	*Tovomita brasiliensis*	*Mangifera indica*
murukuza-ran	passion fruit-false	*Cucurbitaceae* sp.;	*Passiflora edulis*
		Cissus erosa	
		Passiflora acuminata;	
paku-ran-'iw	banana-false-stem	*Heliconia bihai*	*Musa* spp.
tamari-kawaw	tamarind monkey-calabash	*Alibertia edulis*	*Crescentia cujete*
tamari-kawaw-ran-'iw	tamarind monkey-calabash-false-stem	*Ixora pubescens*	*Crescentia cujete*
taza-ran	cocoyam-false	*Costus arabicus*	*Xanthosoma sagitti-folium*
uruku-ran-'iw	annatto-false-stem	*Sloanea guianensis;*	*Bixa orellana*
		Christiana africana;	
		Aparisthmium cordatum	
wawarew-kawaw	squirrel-calabash	*Memora flavida*	*Crescentia cujete*
waya-ran	guava-false	*Eugenia muricata*	*Psidium guayava*
wira-ki'iw-ran	tree-chile pepper-false	*Conceveiba guianensis*	*Capsicum* spp.
wira-kurawa	tree-caroá	*Urena lobata*	*Neoglaziovia variegata*
wira-mani'akaw-ran-'iw	tree-mandiocaba [kind of sweet manioc]-false-stem	*Pouteria caimito*	*Manihot esculenta* (variety)
zitik-ran	sweet potato-false	*Ipomoea squamosa*	*Ipomoea batatas*
ARAWETÉ:			
aní-kirawã	divinity-caroá	*Vriesia* subgenus *Alcantarea* sp. nov.	*Neoglaziovia variegata*

duced plants are subjected to a kind of casual experimentation, in a sense, and for this reason are most likely to be encountered very near human habitation, where they can be closely observed and cared for (see the discussion of dooryard gardens in chapter 6).

Besides linguistic borrowing and modeling on names for existing domesticates, three other types of names given to new species involve (1) modeling on a folk generic name for an existing nondomesticate (as with *arašiku-ran* 'Annona spp.–false' for the introduced soursop, *Annona montana*); (2) incorporation of life form labels to which the referents do not pertain (as with *tupã-ka'a* 'thunder-herb', which refers not to an 'herb' in the classificatory sense, but to ornamental hibiscus, *Hibiscus rosa-sinensis*); and (3) obscure lexemes, i.e., those without any plant terms, such as *kure-nami* 'pig-ear' for the introduced plant of African origin *Bryophyllum* sp. and *tawa* 'yellow' for the Southeast Asian turmeric (*Curcuma*). In other words, unproductive primary lexemes may refer to domesticated plants in several specific ways, but only to *introduced* domesticated plants (table 7.4).

Precisely the same kind of modeling occurs in the botanical vocabularies of other Tupí-Guaraní languages with which I have had the opportunity to work and to

TABLE 7.5 *Continued*

Indigenous Name	Gloss	Botanical Referent	Botanical Model
kaní-əhə	sugar cane-big	*Costus spiralis; Renealmia alpinia*	*Saccharum officinarum*
komán-'i-'i	bean-small-stem	*Cenostigma tocantinum*	*Phaseolus* sp.
kuru-pitlm-'i	divinity-tobacco-stem	*Rheedia gardneriana*	*Nicotiana tabacum*
tatétu-karã	armadillo-yam	*Dioscorea* sp.	*Dioscorea trifida*
ASURINI:			
awǎi-tu'a-wi-rana	maize-?-thin-false	*Casearia* spp.	*Zea mays*
kumamu-'iwa	bean-stem	*Cenostigma tocantinum*	*Phaseolus* sp.
kwi-arána-'iwa	calabash-false-stem	*Swartzia flaemengii*	*Crescentia cujete*
kwi-pia-'iwa	calabash-egg-stem	*Coccoloba latifolia*	*Crescentia cujete*
uruku-irana	annatto-false	*Mollia lepidota*	*Bixa orellana*
waya-wa-'iwa	guava-fruit-stem	*Rinorea* spp.	*Psidium guayava*
yiti-rana	sweet potato-false	*Ipomoea* sp.	*Ipomoea batatas*
GUAJÁ:			
apiri-kowa-'i	parakeet-calabash-stem	*Pouteria eugennifolia;*	*Crescentia cujete*
araku-ran (=wuruku-ran)	annatto-false	*Bixa orellana* (forest variety); *Bixa arborea*	*Bixa orellana* (swidden variety)
*arapʰa-mani'i	red brocket deer-manioc	*Manihot leptophylla*	*Manihot esculenta*
haíra-kowa-'i	Blue headed parrot-calabash-stem	*Eriotheca globosa*	*Crescentia cujete*
*mani'o-'i	manioc tuber-stem	*Buchenavia parvifolia*	*Manihot esculenta*
ka'i-kowa-'i	*Cebus* monkey-calabash-stem	*Trichilia* spp.	*Crescentia cujete*
nana-'1	pineapple-not	*Bromelia goeldiana; Aechmea bromeliifolia*	*Ananas comosus Bromelia balansae*

*Mani'o is an archaic form; its meaning has evidently been lost in Guajá, since the current term for manioc (and, it should be noted, farinha made from manioc as well) is **tərəmã**. The example is included, nevertheless, since mani'o at one time could have meant nothing other than 'manioc.' Similar reasoning is applied in the inclusion of **arapʰa-mani'i** and its gloss.

assemble extensive linguistic collections (see table 7.5 for a list of words and the associated map for the locations of the various peoples). Several of the lexemes in other
Tupí-Guaraní languages have, by inspection, cognates, in which the referents are in
the same botanical species as in Ka'apor (compare tables 7.4 and 7.5). For example,
Ka'apor *yitik-ran* 'sweet potato–false' has cognates in Tembé (*zitik-ran*) and Asurini
(*yiti-rána*); Ka'apor *uruku-ran-'i* 'annatto-false-stem' has a cognate in Guajá (*araku-ran*
or *wuruku-ran*); Ka'apor *arapuha-mani'i* 'red brocket deer–manioc' has cognates in
Tembé (*arapʰa-mani'iw*) and Guajá (*arapʰa-mani'i*); Ka'apor *mira-kirawa* 'tree-caroá' has
a cognate in Tembé (*wira-kurawa*); and Ka'apor *kurupi-kã* 'divinity–sugar cane' has
cognates in Tembé (*kána-ran* 'sugar cane–false') and Araweté (*kaní-uhu* 'sugar
cane–big'). If referents are not restricted to botanical species, however, and are not limited to the level of 'nondomesticates', then other cognates may be noted. Examples
include Ka'apor *kanami-ran* (*Lantana camara*) and Tembé *kunami-ran* (*Bertieria guianensis*); Ka'apor *mãmã-ran* (*Jacaratia spinosa*) and Tembé *mãmã-rãn* (*Paquira aquatica*);
Ka'apor *murukuya-ran* (*Passiflora ananjoi, P. coccinea, P.* sp. 2, *P.* sp. 4, Cucurbitaceae sp.
1) and Tembé *murukuza-ran* (*Passiflora acuminata, Cissus erosa*, and Cucurbitaceae sp.);
Ka'apor *uruku-ran-'i* (*Bixa orellana*—forest variety), Tembé *uruku-ran-'iw* (*Sloanea guianensis, Christiana africana, Aparisthmium cordatum*), and Asurini *uruku-irána* (*Mollia lep-
idota*). Depending on how one wishes to count cognates and the constraints imposed

on referential specificity (see Balée and Moore 1991), other examples may be found as well.

The point of this exercise, however, is to reveal the essential nomenclatural pattern common to these languages. With the notable exception of the Guajá botanical lexicon (to be discussed later), names for some nondomesticates are modeled on generic names for domesticates in all these languages; conversely, names for traditional domesticates are never modeled on generic names for nondomesticates in any of these languages, based on the extensive corpus of plant names and referents I have collected in each (see also Balée and Moore 1991). This is independent evidence that, generally speaking, names for traditional domesticates are in fact older than names for nondomesticates.

I do not mean that there are no unproductive primary lexemes for domesticates in the botanical lexicons of these other Tupí-Guaraní languages. Such lexemes, however, appear to refer only to introduced, not traditional, species—as is also the case with Ka'apor. For example, in Tembé, *kumanu-'iran* ('bean-false'), which is not classified as a 'bean', refers to the introduced bath sponge (*Luffa cylindrica*). For the recently settled Guajá of FUNAI Post P. I. Awá on the Pindaré River, *tai-ran* ('chile pepper–false'), which is not classified as a 'chile pepper', denotes the even more recently introduced carambola tree (*Averrhoa carambola*). Even in the language of a foraging society, in other words, fragments of the lexicon reflect a horticultural past. In Tembé, the word for carambola tree is *akara-wira-'iw* 'kind of fish–tree–stem', but it is not a 'tree' (*iwira*), according to the list of 'tree' names supplied to me by Tembé informants—it is a domesticate, albeit introduced.

Some of the botanical models in tables 7.4 and 7.5 are introduced domesticates. Principles of Ka'apor nomenclature do not exclude such modeling, any more than they exclude modeling names of nondomesticated generics on the names of other nondomesticated generics. Names in Ka'apor, such as *irimã-'i-ran* ('lime-stem-false'), *kase-ran-'i* ('coffee-false-stem'), and *kurupi-kã* ('divinity–sugar cane'), take introduced domesticates as their models. Likewise, Tembé *maŋ-'iw-ran* ('mango-stem-false' [*Tovomita brasiliensis*]), *kuku-'iw-ran* ('coconut-false-stem' [*Alchornea brevistyla*]), *marakatai-ran* ('ginger root–false' [*Xiphidium caeruleum*]), and *kána-ran* ('sugar cane–false' [*Renealmia alpinia*]) also are modeled on generic names for introduced domesticates. And Arawueté *kaní-əhə* ('sugar cane–big') is also modeled on such an introduction.

But such name modeling does not constitute an exception to the nomenclatural pattern, which demonstrates that generic names for nondomesticates may be modeled on names for domesticates (either traditional or introduced), but that generic names for traditional domesticates may never be modeled on names for other types of plants. What does constitute an exception to this pattern, however, is some evidence from the Guajá language.

The Guajá Exception

Although, as is apparent from the above discussion and table 7.5, the Guajá language does contain words for nondomesticates that incorporate (and, hence, are modeled on) generic names for traditional domesticates of Tupí-Guaraní societies, the converse also occurs. Among the Tupí-Guaraní languages that have been the focus of this discussion, only Guajá evinces a name for a traditional domesticate that is clearly modeled on the name for a nondomesticate. This concerns the cashew. The Guajá name for domesticated cashew (*Anacardium occidentale*) is *akayu-rána*, which may be glossed as 'nondomesticated *Anacardium* spp.-false'.

Of course, the postposed and bound morpheme *rána* (Ka'apor *ran*), which always indicates modeling when it occurs in a generic plant name, is not incorporated in generic names for traditional domesticates in the other Tupí-Guaraní languages. In Guajá, the name for cashew, which is recently under cultivation at the instigation of FUNAI personnel, derives its model from nonanthropogenic high forests, swamps, and riverbanks. In addition to this apparent anomaly, the Guajá language encodes no lexical distinction between 'pineapple' and several nondomesticated bromeliads (*Aechmea bromeliifolia*, *Bromelia goeldiana*, and *B. balansae*). All, including the traditional domesticate of Tupí-Guaraní society called 'pineapple' in English, are denoted by the cover term *nana-'ï* (literally, 'not pineapple') in the Guajá language. This appears to violate the proposed dichotomy between the nomenclature of traditional domesticates and other plants. In the Guajá language, the two domains, which are otherwise rigorously segregated in other Tupí-Guaraní languages, seem to be conflated.

It should be recalled that the Guajá are a foraging people, only some of whom are now acquiring agriculture under the aegis of FUNAI and then only quite recently. If the historical trajectory of Guajá society is taken into account, it will be clear why they do not conform to the plant nomenclatural pattern I have adumbrated for other horticultural Tupí-Guaraní languages of the present. Guajá society has been, in its recent history, both a society of foragers and a society of horticulturalists, at different times of course.

Gellner (1988:16) pointed out that passage of an agrarian society to a hunting and gathering society, "though conceivable, is improbable, and rare." I argue that the historical trajectory of Guajá society, in terms of its relationship with plants, represents precisely one of these "improbable" and "rare" events, which, in lowland South America, may be much more common than received wisdom about the evolution of sociopolitical systems would grant. Ricardo Nassif (personal communication, 1992) has uncovered a detailed myth fragment among the Guajá of the Pindaré River to the effect that Maíra (the cognate culture-hero in many Tupí-Guaraní languages) never gave the Guajá manioc or any other cultigen, but rather scattered the fruits of the nondomesticated babaçu palms, the mainstay of their diet, across the land. The Guajá have no historical memory of ever having been horticulturalists.

But a lack of extant memory about any horticultural practices in a culture dependent on oral transmission is not in itself proof of an unbroken foraging history. The Guajá language itself, rather, suggests just the opposite. The Guajá language contains words for traditional domesticates, even if the people have not in their recent history cultivated these domesticates. The Guajá referents are similar to words for the same species in many other Tupí-Guaraní languages and that do not appear to have been borrowed from any other language. These words include *paku* (bananas and plantains), *kamana'i* (beans), *awači* or ᵏ*wači* (maize), *urumũ* (crookneck squash), *waya* (guava tree), and *kara* (yams). These words appear to be descended from proto–Tupí-Guaraní, not borrowed. A few words for traditional domesticates in Guajá are not cognate with words for the corresponding referents in other Tupí-Guaraní languages, such as *tərəmɜ̃* (manioc), *mačitu* (sweet potato), and *tai* (chile peppers, among other species). This is to be expected in a language associated with a society that may not have practiced horticulture for more than two hundred years. The missionary José Noronha, for example, mentioned a set-tled village society of the lower Tocantins basin in 1767 called the *Uayá* (Noronha 1856:8–9). In later accounts, the term *Uayá* denotes only the foraging Guajá society. Settled villages, of course, are incompatible with a full-time foraging lifestyle in lowland South America. It seems likely, given the available evidence, that the Guajá became foragers after 1767 and before about 1850. The Guajá appear to have given up a horticultural, settled way of life because of forces that will be discussed in the next chapter.

What requires explanation now concerns why the Guajá have any cognate words at all for traditional domesticates, not the general lack thereof. How could a people main-tain words for objects (in this instance, cultigens) which have been absent in their cul-ture for many generations? As with Ka'apor names for certain plants absent from the present Ka'apor homeland, such as Brazil nut trees, the answer is that these objects have not been entirely absent, even if the Guajá themselves did not cultivate them within historical memory. The Guajá and their antecedents from the mid–nineteenth century at the latest through contemporary times have raided the swiddens of their horticultural neighbors, both aboriginal and colonizing. It thus appears that a suffi-cient number of Guajá persons maintained just enough contact with domesticated plants over time so as not to lose all the words for these. Some names were lost, to be sure, but several were retained, even if their referents remained in but a peripheral relation to these now foraging nomads.

Because the Guajá probably experienced a history in which they passed from an agrarian to a hunting and gathering society, their botanical lexicon evinces this mixed heritage. In part, Guajá plant words are lexical "survivals" of an earlier age. For this reason, the Guajá plant lexicon deviates from the aboriginal Tupí-Guaraní model; specifically, whereas the other Tupí-Guaraní languages never model names for tradi-tional domesticates on generic names for wild plants, the Guajá language occasional-ly does so. Yet the Guajá language also contains words for nondomesticates generated by analogy with generic (and cognate) names for traditional domesticates, as do the other Tupí-Guaraní languages. The Guajá, then, represent an exception to the nomen-clatural principles (1) that unproductive primary lexemes never denote traditional domesticates and (2) that productive primary lexemes never refer to traditional domesticates.

The differences are derived from the fact that Ka'apor and their horticultural con-geners (for example, Wayãpi, Tembé, and Kagwahiv) have maintained a lifestyle based on forest management since the earliest of times, i.e., since the origins of proto–Tupí-Guaraní, whereas the Guajá, because of historical accidents, have not. The noted exceptions in the Guajá botanical nomenclature do not therefore undermine the proposition that in proto–Tupí-Guaraní, a nomenclatural and classificatory dichotomy existed between domesticates and nondomesticates and that names for domesticates, in general, must be of an older lexical corpus than names for nondomesticates. That this dichotomy and the words that reflect it have managed to survive the Conquest is itself a point worth pondering.

EIGHT

Toward a Comparative Ethnobotany of Lowland South America

Few anthropologists would argue that the Neolithic Revolution was not one of the great watersheds in human history. The technological, economic, and sociopolitical differences between agrarian and nonagrarian societies are so noteworthy that Ernest Gellner (1988:16) described the distinction as analogous to that between "species." Schemes that dichotomize human societies into "simple" and "complex," or that divide societies into more than two such groups (notably, those proposed by Gellner and Lewis Henry Morgan), have been intrinsically appealing for their help in situating comparative problems in human history. At the same time, however, any schema intended for universal application inevitably oversimplifies the differences within and between each category.

There is little doubt that the ethnobotanical systems of foragers are, in general, extremely unlike those of agrarian societies. As noted in the preceding chapter, these differences seem to be reflected in the nomenclatural and classificatory subsystems. (See C. H. Brown [1977], for example, on the typically few plant life form labels found in languages associated with simple societies versus the typically many in languages associated with complex societies.) Yet if "Agraria" constitutes one of the three great phases of the human historical enterprise (for Gellner, the other two are Foragers and "Industria"), certainly considerable diversity also exists within it.

Applied to South America, for example, "Agraria" (agricultural societies) would lump the Inca Empire, the Tupinambá chiefdoms, the trekking Araweté and Waorani, and the settled but autonomous villages of the Ka'apor. Even so, with the possible exception of state-level societies in South America, historical and linguistic connections do exist among societies as diverse as the foraging Guajá, the trekking Araweté, and the relatively settled Ka'apor. Such connections have implications for the explanation of similarities and differences among their respective ethnobotanies. *I propose that Ka'apor ethnobotany, as a total systemic relationship between a people and its plants, cannot be fully understood until it is compared, in a historical-ecological perspective, to the ethnobotanical systems of societies that exploit like environments.*

Comparing Horticulturalists with Foragers

If a transition to agriculture implied profound transformations of ecological, technological, and even cognitive aspects of human culture, the far less probable but no less documentable transition from agriculture to foraging, as it occurred among the

TABLE 8.1 *Comparison of Ka'apor and Guajá Folk Specific Names For Nondomesticated Species of Five Genera*

Species	Ka'apor Name	Guajá Name
Eschweilera apiculata	parawa'iwi	wiri'i
E. amazonica	parawa'ipihun	wiri'i
E. coriacea	parawa'ite	wiri'i
E. micrantha	parawa'iwi	wiri'i
E. ovata	pitiminemran'i	wiri'i
E. pedicellata	parawa'ite	waiha'i
Inga alba	iɲašiši'i	čičipe'i
I. auristellae	iɲaperě'i	čičipe'i
I. capitata	iɲahu'i	čičipe'i
I. fagifolia	kaŋwaruhuiɲa	čičipe'i
I. falcistipula	kaŋwaruhuiɲa	čičipe'i
I. heterophylla	iɲaperě'i	čičipe'i
I. marginata	iɲaperě'i	čičipe'i
I. miriantha	iɲaperě'i	čičipe'i
I. rubiginosa	tapi'iriɲa'i	čičipe'i
Ocotea caudata	ayu'ipihun	wayuwa'i
O. fasciculata	ayu'ituwir	wayuwa'i
O. opifera	mahayuwa'i	wayuwa'i
O. sp. 1	ayu'iran	wayuwa'i
Pouteria reticulata	wai'ite	wayu'i
P. hispida	kupapa'i	wayu'i
P. sagotiana	karamiriran'i	wayu'i
P. gongrijpii	arakāšī'i	wayu'i
P. jariensis	kupapa'i	wayu'i
P. caimito	iratawa'i	wayu'i
P. macrophylla	akušitiriwa'i	akučiterewa'i
P. trigonosperma	iratawa'i	akučiterewa'i
Protium aracouchinii	yawamira	yawar'a'i
P. decandrum	arakanei'i	yawar'a'i
P. heptaphyllum	arakanei'i	yawar'a'i
P. pallidum	kanei'ituwir	yawar'a'čï'i
P. giganteum	kanei'ituwir	yawar'a'čï'i
P. polybotryum	kaneiape'i	yawar'a'i
P. sagotianum	kaneiaka'i	yawar'a'čï'i
P. spruceanum	kanei'ituwir	yawar'a'čï'i
P. tenuifolium	kaneiape'i	yawar'a'i
P. trifoliolatum	sekātāi'i	yawar'a'i

indigenous peoples of a colonial South America, must also have involved profound transformations of the same aspects. In chapter 5, I suggested that dependence on oral versus written traditions may help explain the differences between the potentials of Ka'apor knowledge concerning plant utility and Western scientific approaches to the same. But I also pointed out that the limits of an oral tradition cannot logically be used to explain similarities and differences among the ethnobotanical systems of lowland South American societies where no aboriginal script was ever developed.

I submit that historical ecology holds a key to unlocking the reasons for similarities and differences among the extant indigenous cultures. In order to understand why Ka'apor ethnobotany is similar to that of some societies, yet different from that of others, a comparative approach, obviously, is needed. Although I earlier argued (in chapter 7) that certain similarities in plant name-modeling between the Guajá and Ka'apor are indicative of a horticultural past for both societies, I also pointed out anomalies in Guajá name-modeling that came about because of their long (albeit still recent) history of foraging. In fact, an examination of Guajá folk generic names is (at 353, of which only 22, or 6 percent, are polytypic) even more suggestive of the linguistic effects characteristic of a remote transition from horticulture to foraging.

In comparing the botanical vocabularies of the Ka'apor and Guajá languages, it seems that Guajá possesses relatively few folk specific (binomial) names, a point also noted for the foraging Agta of the Philippines (Headland 1983) as well as for several other foraging groups (Berlin 1992:283–285). Stated another way, most Guajá folk generics are monotypic, "empty" taxa. This is immediately clear in reference to traditional domesticates, which in Guajá are named only by monotypic generic names. Even sweet and bitter manioc plants are not yet distinguished lexically, both being called by the monotypic generic name *taɨamã* (which, of course, functions simultaneously as a folk specific name). In contrast, Ka'apor has three folk generics and a total of 24 folk specific names for *Manihot esculenta*.

This underdifferentiation of generic domesticates is to be expected in a foraging society that only recently has begun to adopt (or readopt) a horticultural way of life. Yet it might be further anticipated that Guajá should evince a correspondingly higher number of folk specific names for wild plants. If it is true for Amazonia that "most hunter-gatherers have an exhaustive knowledge of the resources of their habitat" (Meggers et al. 1988:281), we might expect such knowledge to be reflected in the richness of the vocabulary for that most important resource, plants. But applied to the Guajá, such an assumption appears invalid.

I fully agree that the specific lexicon of any semantic domain in a language is, in principle, unrelated to perceptual abilities of its speakers (e.g., Berlin and Kay 1991:139). The size of a lexicon for a given semantic domain, on the other hand, is probably positively correlated with its economic or cultural importance (Berlin and Kay 1991:16; Brown 1977, 1986). While the Guajá indeed do know thoroughly their environment, not all of this knowledge is so important as to be compartmentalized in encyclopedic entries of that part of the lexicon concerning the plants.

The relative paucity of folk specific names for wild plants in Guajá ethnobotany can be illustrated by comparing the number of folk specific names in Guajá with that of the Ka'apor in reference to botanical species of given botanical genera. In making this comparison, I am also replying to the implicit challenge that one cannot conclude that foragers habitually have fewer plant names than horticulturalists until foragers and horticulturalists of the same habitat are compared (see Berlin 1992:284). For convenience, I compare names within each of five genera that are represented by four or more species in my collections. The genera are *Eschweilera* (Lecythidaceae), *Inga* (Mimosaceae), *Ocotea* (Lauraceae), *Pouteria* (Sapotaceae), and *Protium* (Burseraceae). There are 37 species common both to the Ka'apor and Guajá in these genera. It is obvious from table 8.1 that the Ka'apor language distinguishes with different folk specific names far more botanical species than does the Guajá language. For six shared species of *Eschweilera*, Ka'apor has four names, Guajá two; for nine species of *Inga*, Ka'apor has five names, Guajá one; for four species of *Ocotea*, Ka'apor has four names, Guajá one; for eight species of *Pouteria*, Ka'apor has six names, Guajá two; and for ten species of *Protium*, Ka'apor has six names, Guajá two. For the corpus of 37 nondomesticated botanical species, the Ka'apor language encodes 25 folk specifics, whereas Guajá encodes merely 8. There is no single botanical genus common to both the Ka'apor and the Guajá, moreover, which Guajá differentiates more than Ka'apor. It may be predicted for other genera, therefore, that the Ka'apor language differentiates folk species generally to a higher degree than that of the Guajá language, domesticates aside.

In a noteworthy paper, C. H. Brown (1985) observed that small-scale agrarian societies tend to have much more extensive inventories of plant (and animal) names than do foragers. Brown's explanation for this unexpected finding centered on the relative fragility of horticultural systems. Agriculture creates a diversity of "ecotypes," and hence biodiversity tends to be higher in the habitats of horticulturalists than of foragers. Horticulturalists cannot afford to lose knowledge about wild plants because crops may from time to time fail. Then too, the fact of plant failures brings horticulturalists to a new level of intimacy with and curiosity about plants in general that may lead to a deeper and more extensive knowledge of wild plants as well.

Although Brown's explanation is attractive, it does not fully apply to a comparison of the Ka'apor and the Guajá. First, both languages and cultures are descended from a mother language/culture that was associated with horticultural society; Brown's analysis appears to be based on the assumption that the foragers in his sample are pristine foragers. Second, the Guajá do not depend mostly on wild resources; it is more accurate to say that the basic species in their economic botany are *semidomesticates*. Third, the habitat of the Guajá, like that of the Ka'apor, is also composed of many ecotypes owing to horticultural perturbations of the past—that is, the Guajá do exploit anthropogenic habitats, even if not of their own creation. Nevertheless, Brown's initial set of observations on the relatively truncated nature of botanical inventories of foraging societies is, in fact, well supported with data from a Ka'apor and Guajá comparison.

Because borrowing has been a relatively minor influence in the Ka'apor language, it is probable that the Ka'apor corpus of specific plant names is not appreciably greater than the corpus of specific plant names that would have existed in proto–Tupí-Guaraní. It is, however, highly likely that the Guajá corpus of folk specific names is appreciably smaller than that of proto–Tupí-Guaraní because the parent language was associated with a horticultural society. While there has surely been a decline, therefore, in the total number of plant words in Guajá, a corresponding extension in range of meaning for the folk generic and specific plant names that have remained has also occurred.

What accounts for this loss of names and increasing tendency toward taxonomic and nomenclatural lumping? In terms of plant management, lexical encoding of specific plants, and plant "uses," it would seem that horticulturalists, paradoxically, are generalists, whereas foragers are specialists. This is a paradox not without reasonable explanation—an explanation that depends on knowledge of the historical ecology of the two kinds of groups.

Knowledge of historical ecology reveals that the transition from horticulture to foraging in lowland South America was processual, not abrupt. Lathrap (1968) supplied a glimpse of such a process (which he termed "devolution" or "degradation") as it may have occurred in pre-Columbian times for the Upper Amazon. In comparison with the settled, seriously horticultural Panoan-speaking Shipibo of the Ucayali River floodplains, the linguistically related Amahuaca of the Peruvian montaña were regarded as "degraded descendants of peoples who at one time maintained an advanced form of tropical forest culture" (Lathrap 1968:25). They lost their horticultural practices because they were forced out of the fertile floodplains by warfare with more powerful groups. The limits of agriculture in the interfluves, in other words, were lower than those of the floodplains, given aboriginal technology.

With the forced migration, hunting became more important for the Amahuaca, and fishing less so. Greater dependence on hunting (with consequent local extinctions of game) made permanent settlements a thing of the past. Thus ensued the "devolution" of this now "degraded" people (Lathrap 1968; also see Carneiro 1970). Amahuaca society can be considered degraded in the sense that it lost certain domesticated plant species and other artifacts (such as polychrome pottery), which were otherwise retained by other Panoan-speaking Indians of the floodplains. Another sign of devolution is that the average length of settlement occupation was reduced to less than two years (Carneiro 1970). Also, regional population density declined to a mere .04 persons/km^2 (Carneiro 1970), whereas the average for present Amazonian societies is .2/km^2, and the aboriginal density of the white-water floodplains, no doubt, was even much higher (Denevan 1976; Roosevelt 1991). Average settlement size declined from more than 50 to a mere 15 persons, and any institutional statuses, even those of headman and shaman, all but disappeared (Carneiro 1970).

The reasons postulated by Lathrap for such degradation relate essentially to conditions of the landscape—specifically, to a widely held dichotomy between the subsistence potential of the white-water floodplains versus the *terra firme* interfluves (for summaries, see Roosevelt 1987, 1991). Lathrap's argument is not, however, the full story. The population decline of the Amahuaca from some 6000–9000 in 1900 to a mere 500 in 1960 and the inferred decline in population density from a possible high of .19/km^2 to .04/km^2 (Carneiro 1970) were likely caused not by difficulties intrinsic to the landscape, but by introduced disease, warfare, or some other palpably historical-ecological factor.

Although the dichotomy between white-water floodplain and interfluvial *terra firme* societies has been heuristically useful in generating an understanding of Amazonian ecology (having sparked more than a generation of scholarly work), the dichotomy cannot fully explicate notable similarities and differences among the ethnobotanical systems of aboriginal societies limited to one or the other type of habitat. But a historical-ecological perspective can supply a more refined approach for such explanations. A historical-ecological perspective can be used to reconstruct aboriginal relationships between culture and environment as well as the historical events that have, by and by, transformed these relationships. In order to comprehend how the Ka'apor and Guajá (whose languages descend from common roots and who occupy *terra firme* forests of the same region) developed such different ethnobotanical systems, the historical and ecological paths of each must be examined.

It may be taken as a given that proto–Tupí-Guaraní society cultivated numerous domesticated plants. Around the time of contact in 1500, early records show that the coastal Tupinambá and Guarani cultivated manioc, bananas, yams, capsicum peppers, crookneck squash, sweet potatoes, and pineapples (see chapter 6); they had available to them, moreover, many other kinds of crops for the satisfaction of diverse needs beyond food consumption. The Tupinambá- and Guarani-speaking societies of late pre-Columbian times appear to have had large, nucleated villages, along with high population densities and multivillage sociopolitical organizations that are suggestive of paramount chiefdoms (Balée 1984a; Clastres 1973; Dean 1984). None of the early chroniclers reported the small horticultural villages of a type so common at present, as with the Ka'apor, let alone any Tupí-Guaraní–speaking societies dependent exclusively on foraging, as with the Guajá.

It is therefore intriguing to note that several Tupí-Guaraní societies are now nomadic foragers. In addition to the Guajá, these include the Héta, Aché (or Guayaki), Avá-Canoeiro, and Šeta. Were it not for the fact that proto–Tupí-Guaraní words for many domesticates, including those mentioned above, can be reconstructed, one might be led to the erroneous conclusion that horticulture arose independently among diverse Tupí-Guaraní societies and at different times, but did not emerge among all of them.

One might speculate, further, that these modern foragers indeed never had domesticates prior to their recently documented contact. Yet the botanical lexicons of these foraging groups suggest otherwise, namely, that they do have cognates for some domesticates and that they somehow regressed from an aboriginal lifestyle based on horticulture (Clastres 1973:271; Martin 1969). The nomadic Aché of Paraguay, for example, use the term waté for maize (Clastres 1968:51–53; Vellard 1939:90–91), which reconstructs as *aßati in proto–Tupí-Guaraní (A. D. Rodrigues, personal communication, 1990). There is some scattered historical evidence, moreover, that the Aché cultivated maize, if not also other crops, in the eighteenth century (Métraux and Baldus 1946:436; Clastres 1972:143). Cognate terms for maize have also been documented for the Avá-Canoeiro of the Tocantins basin (Rivet 1924:177; Toral 1986) and the Šeta (Loukotka 1929:374,392), an extinct society of southern Brazil. As for the Héta of Paraná state in extreme southern Brazil, it appears that they "practiced some plant cultivation" before they became authentic foragers (Kozák et al. 1979:366). Finally, and as previously stated, several Guajá words for domesticated plants indicate descent from a language associated with a horticultural society. These words include wači or ᵏwači ('maize'), kara ('yams'), and paku ('bananas'). The reconstructed proto–Tupí-Guaraní terms for these plants, respectively, would be: *aßati, *kara, and *pakößa.

Sociopolitical forces associated with the Conquest appear to be responsible for the regression of semisedentary horticultural societies into foraging nomads. Epidemic diseases of Old World origin, slave raids, and colonial warfare are the forces that not only devastated the Tupinambá, Omáguas, and Guarani chiefdoms but that also induced the splintering of society into atomistic levels for which horticulture became increasingly costly to enjoy. The Guajá lived in settled villages of the lower Tocantins in the 1760s (Noronha 1856:8–9). Yet from that time into the mid-nineteenth century, this region experienced successive waves of intensive warfare instigated by colonial militia and their acculturated Indian mercenaries (including some of the Mundurucu and Turiuára) against indigenous societies. The pandemonium that afflicted the lower Amazon in these years culminated in 1835–1836, with the final phase of the great and bloody Civil War known as the Cabanagem.

By 1872, the clandestine Guajá, who had relocated to the upper Gurupi drainage (somewhat upstream from the region the Ka'apor would shortly thereafter flee to) were described as being "persecuted" by all surrounding indigenous groups and as lacking the "arts" of settled, horticultural life (Dodt 1939:177). According to a Ka'apor man from the Turiaçu basin, the Ka'apor ancestors would not have permitted the Guajá to establish permanent settlements, Ka'apor warriors also being much more numerous than those of the Guajá (Balée 1988b). Although the Ka'apor also suffered military defeats and were forced to flee the encroaching effects of civilization during the nineteenth century, they managed to retain their horticultural practices. It may thus be surmised that the effects of contact were far from unitary in eastern Amazonia—some groups having been affected more profoundly than others (a point made more generally by Newson 1992b). Once an indigenous society lost horticulture, moreover, it would have had great difficulty regaining such practice in the face of hostile actions by neighboring groups that had managed to maintain settled villages despite the odds against doing so.

Comparisons with Trekkers

In the southern cone of South America, conquest had similar effects on Tupí-Guaraní–speaking societies. The Héta and Aché felt the influences of missionaries and the ravages of epidemic disease and slave raids (Métraux 1948; Clastres 1972:143–144). After the *bandeirantes* from São Paulo destroyed the Jesuit missions of the region in the eighteenth century, the aboriginal Guarani abandoned their villages and fled into central Paraguay (Chmyz and Sauner 1971:10; Maack 1968:42–44). Gê-speaking Kaingáng (Xokleng), who did maintain some horticulture, evidently prevented the Héta from doing the same (Kozák et al. 1979:366). Similarly, the Ka'apor, Tembé, and Guajajara impeded the development of any kind of permanence to Guajá camps.

Agricultural regression, the term I use to describe a transition from "Agraria" to foraging, is gradual insofar as not all domesticates are usually 'lost' (*kãyĩ*) at once. Rather, cultivation of domesticates diminishes over time, until only one or two remain, and finally, none. One of the longest retained domesticates appears to be maize. Although maize cultivation is closely associated with the pre-Columbian Tupí-Guaraní chiefdoms (and probably with chiefdoms elsewhere in lowland South America—see Roosevelt 1987), maize cultivation is also practiced by extant *trekking* peoples. Trekkers may be defined as groups that spend six months or more per year away from a village site to which they ultimately return. Trekking, thus, constitutes one more analytic distinction within "Agraria," which also applies to chiefdoms and to semisedentary autonomous villages (as with the Ka'apor, Kagwahiv, Wayãpi, and Mundurucu).

The Araweté, who until the mid-1980s trekked on average six months per year (Viveiros de Castro 1986:271), are highly dependent on maize rather than tuber crops.

TABLE 8.2 *A Model of Agricultural Regression in Lowland South America*

	Phase 1	Phase 2	Phase 3	Phase 4
crop inventory through time	maize bananas sweet manioc bitter manioc other crops	maize bananas sweet manioc other crops	maize other crops (?)	no crops

→

INCREASING DEPENDENCE ON NON-DOMESTICATED PLANTS

Adapted from Balée (1992a)

Viveiros de Castro (1986:26) noted that, with respect to the Araweté, "Maize aggregates people; it is practically the only force that does so. Innumerable other forces work toward dispersion." Maize aggregated the Araweté during planting and harvest seasons. In the other seasons, people tended to forage in the liana forests that abound in their habitat (see Balée and Campbell 1990).

The maize harvest also united the societies of several macro-Gê peoples, who were often described as "marginal tribes" or "foot nomads" in the early ethnographic literature. At other times, the tribesmen dispersed themselves into small groups that gathered, hunted, and fished (Galvão 1979:245; Maybury-Lewis 1967:47). Maize dependence combined with trekking also characterize the Panoan-speaking Amahuaca, one of Lathrap's "degraded" societies (Carneiro 1970) and the linguistically isolated Hotí of southern Venezuela (Coppens 1975:68; Coppens and Mitrani 1974:36). Many trekkers traditionally cultivated other food plants, especially bananas (including plantains) and sweet manioc, one or more of which may have been calorically more significant than maize in the diet. But they cultivated little or no bitter manioc (e.g., Coppens and Mitrani 1974:136; Smole 1976:119).

Manioc is, of course, the caloric staple of most nontrekking horticultural village societies of lowland South America. In a very useful summary, archaeologist J. P. Brochado (1977:57) found that of 553 indigenous peoples of the tropical forest who cultivate manioc, 478 (86 percent) employ it as a primary food source and 75 (14 percent) use it only as a supplementary resource. For 60 (80 percent) of these latter groups, the dietary staple is maize, bananas, peanuts, or sweet potatoes.

Trekking peoples seem to utilize maize rather than bitter manioc as a staple because of its relatively low start-up costs, simple processing requirements, low transport costs, and fast maturation time (Balée 1985; Galvão 1979:245). Where trekking peoples of the tropical forest do cultivate manioc, it tends to be only the sweet type. Trekkers that cultivate only sweet manioc include the Hotí (Coppens and Mitrani 1974:136), Yanoama of Venezuela (Smole 1976:119), Campa (Denevan 1971), the Amahuaca, and the Waorani (see chapter 5). Sweet manioc demands no elaborate procedures for detoxification nor special installations in order to render it fit for human consumption. Brochado (1977:64) observed that of the 75 groups that employ manioc, the named nontoxic varieties outnumber the toxic ones by almost two to one (cf. Lathrap 1970:53). For the Ka'apor, this ratio is 18 to 5, or almost four to one (see appendix 9: folk species of *mani'i* versus those of *makaser* and *maniaka*).

One can reasonably propose that, of the two staples (maize and bitter manioc), bitter manioc is 'lost' (*kãyĩ*) first in the transition from horticulture to foraging (see table 8.2). That the Guajá language (as well as languages of several other Tupí-Guaraní–speaking foragers, such as the Aché [Clastres 1968:53; Vellard 1939:90–91]) has retained a cognate for maize (*awači*), but not one for manioc (*tərəmã̃*), supports this hypothesis. In addition, the recent currency of the Guajá word for manioc is further indicated by the fact that its focal referent appears to be the edible mesocarp of the babaçu palm, one of the principal botanical food sources in the aboriginal Guajá diet (Balée 1988b; Balée 1992a); the term *tərəmã̃* is also extended in range of meaning to any kind of flour, including manioc flour. Although the Guajá have not in memory cultivated maize, their language nevertheless reflects that it was cultivated more recently than manioc.

Even though the Guajá have for more than one hundred years raided Ka'apor swiddens for food, it may be surmised that they soon started to avoid harvesting manioc tubers there of any type. One might object to this statement upon cursorily examining comparable ethnographic evidence from the Aché. According to Vellard (1939:90), the Aché raided other group's swiddens for sweet manioc; but he added in the same sentence that sweet manioc was "practically the only kind cultivated" in the region. It thus

FIGURE 8.1 Araweté woman opening a babaçu nut to remove edible grub inside.

appears that the Aché probably ran few or no risks of stealing bitter manioc tubers by mistake (for which, of course, they would have had no processing technology). Bitter manioc, in fact, appears to be restricted to the more tropical regions of South America, especially in the equatorial zone of eastern South America (McKey and Beckerman, ms.); the Aché habitat is subtropical.

Even if the Guajá had knowledge of manioc toxicity (all varieties contain some degree of prussic acid), they would not have retained a technology for detoxifying the bitter varieties when they became full-fledged foragers. Such technology demands knowledge about constructing and using manioc sheds, griddles, and presses. In fact, no known foraging society has retained this technological ability. It is most likely, therefore, that Guajá swidden raiders avoided all manioc plants. At present, the Guajá folk specific vocabulary does not distinguish between poisonous and relatively non-poisonous manioc types (such a distinction is, however, likely to arise soon with the Guajá's increasing dependence on horticulture). On the other hand, the Guajá of the past could have harvested from other peoples' swiddens domesticated yams (*kara*) and maize (*awačĩ*), the cognates for which have been retained.

In late September 1987, an isolated family of Araweté Indians came to the attention of FUNAI. They were living some 200 kilometers from the closest Araweté village (of 155 people), having become separated from their kinsmen in the 1960s as a conse-

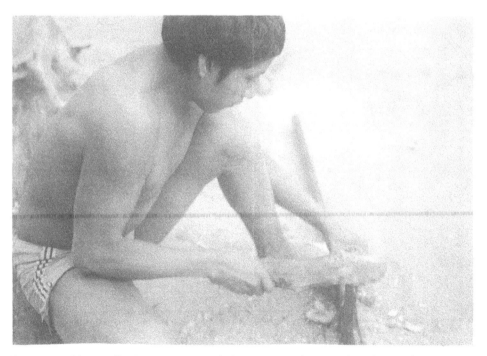

FIGURE 8.2 Young Guajá man opens a babaçu nut with a wooden club and a steel axe. Babaçu appears to have been the most important plant food for the Guajá when they were predominantly foragers.

quence of a major raid by Kayapó Indians (Viveiros de Castro 1992:26–29). Early in September 1987, Xikrin-Kayapó Indians attacked this isolated family of seven, killing one man and one boy in the melee. One elderly Araweté man escaped into the forest. The remaining hostages were taken to the Kayapó village of Cateté (near the Rio Ita-caiunas, midway between the Tocantins and Xingu Rivers). Once apprised of the hostage "crisis," FUNAI authorities, together with two Araweté Indian interpreters from the Araweté village, departed in search of the missing Araweté man. They also entered into negotiations with the Kayapó captors for the release of the four hostages, which eventually took place. The search party made contact after a few days with the missing Araweté man, convincing him to return with them. All the surviving Araweté of this isolated group were later returned to live among the Araweté villagers (the missing man had a sister in the Araweté village).

In October 1987, I interviewed these survivors, their interpreters, and members of the FUNAI search party, when they all arrived in the city of Marabá on the Tocantins River, in transit to the Araweté village. The reconstructed scenario of their camp is as follows. At the site of the raid stood a house in a small swidden. The only crop in the swidden was maize. It should be pointed out here that the Araweté traditionally grow a number of species, including maize, manioc, papaya, cotton, annatto, caroá (*Neoglaziovia variegata*), and yams (Viveiros de Castro 1986:150–153). The isolated family was subsisting, according to evidence found at the site, on maize, babaçu meso-

FIGURE 8.3 An Araweté nuclear familiy. The infant carrying strap is made of cotton.

carps, and Brazil nuts (Balée 1987). These last two botanical resources are often present because of past horticultural influences (Balée 1989; Ducke 1946:8).

Although the Araweté traditionally make their hammocks from cotton, the hammocks of the isolated family were woven from the soft bast fibers of a nondomesticated tree, *Couratari guianensis* of the Brazil nut family. Whereas the Araweté usually fashion their bowstrings from the domesticated bromeliad caroá (*Neoglaziovia variegata*), the bowstring of the missing man was from the sword leaf fibers of tucumã palm (*Astrocaryum vulgare*), a common denizen of old fallows. Evidently, this group had become so small they could not maintain an inventory of more than one domesticate, maize. These Araweté thus seem to have been living on the edge between trekking and foraging. For my purposes here, their story provides further confirmation that maize is the last domesticate a trekking society loses before becoming true foragers.

The Role of Semidomesticates in Agricultural Regression

For a society undergoing agricultural regression, it makes sense to develop a growing economic dependence on nondomesticated plants. Yet the species that tend to become primary resources (as food, technology, medicine, and so on) are not readily

FIGURE 8.4 Ka'apor hammock made of cotton. The Ka'apor have long enjoyed the benefits of a sedentary way of life.

TABLE 8.3 *Uses of Palms by the Guajá*

Species	Guajá Name	Portuguese Name	Uses
Astrocaryum munbaca	**yu**	*mumbaca*	edible seed
Astrocaryum vulgare	**takamã**	*tucumã*	sword leaf fibers for bowstring, hammock,women's skirt, infant-carrying strap, rope, string for bead necklace, foreskin string
Bactris humilis	**wãʔã**	*marajá*	petiole for arrow shaft
Bactris maraja	**mariawa**	*marajá*	edible mesocarp
Bactris major	**kiripirihu**	*marajá*	edible mesocarp
Bactris sp.	**kiripiri**	*marajá*	edible mesocarp
Euterpe oleracea	**pinuwapihun**	*açaí*	edible exocarp
Geonoma baculifera	**yowoʔ**	*ubim*	seeds for bracelet
Maximiliana maripa	**inaya'i**	*inajá*	edible mesocarp and seed; spathe for bowl
Oenocarpus distichus	**pinuwa'i**	*bacaba*	edible exocarp
Orbignya phalerata	**waʔï**	*babaçu*	edible kernels and mesocarp; leaves for roofing thatch
Markleya dahlgreniana	**parana'i**	?	edible kernels and mesocarp

FIGURE 8.5 Guajá hammock made of tucumã fibers. The tucumã palm, *Astrocaryum vulgare*, is native to fallows.

characterized as being "wild." Rather, these plants, in terms of their ecological importance (see chapter 6), rank among the principal markers of fallow. They are, in other words, semidomesticates.

The semidomesticates represent a subtle yet important economic link between foragers and horticulturalists in eastern Amazonia (as well as in southern Brazil and Paraguay, evidently). This link is subtle, insofar as it is unlike the much-documented economic exchanges that take place between the riverine Tukanoans and the interfluvial Makú of the Northwest Amazon (e.g., Jackson 1983). And it is important, since foragers of these regions heavily rely on the botanical artifacts of their horticultural neighbors.

Although the Guajá of recent times (as well as other foragers such as the Aché and Héta) occasionally raided young producing swiddens, they were—and are—far more dependent on the semidomesticates of fallows left behind by their horticultural neighbors (Balée 1988b, 1992a). The Guajá display a very high dependence on palms for food, shelter, clothing, tools, and utensils (see table 8.2; Balée 1988b, 1992a). The economically most important palms in Guajá subsistence, moreover, are also major species of old fallows (see appendix 1 and chapter 6). These palms include food species such as babaçu

FIGURE 8.6 Infant carrying strap (Guajá) made from sword-leaf fibers of the tucumã palm.

(*Orbignya phalerata*), inajá (*Maximiliana maripa*), and the hybrid *Markleya dahlgreniana*. These palms all have edible mesocarps and kernels for the Guajá. In addition to species used for food, the semidomesticated *tucumã* palm (*Astrocaryum vulgare*) is the most important source of fibers in traditional Guajá manufactures. The sword leaf fibers of this species supply raw materials for infant carrying straps, hammocks, women's skirts, bowstrings, rope, foreskin string, and string for armbands (Balée 1988b:50; 1992a). As for horticultural societies (including trekkers such as the Araweté) of eastern Amazonia, domesticated cotton is used for these purposes.

Comparative evidence suggests that dependence on semidomesticates, notably the palms, may indeed be widespread among foragers, at least for those of the Tupí-Guaraní family of languages. Although the Aché obtain most of their calories from game animals, not plants (Kaplan and Hill 1985:228; Clastres 1972:160), the most important botanical resource—both as food and for making many tools and utensils—is the jerivá palm (*Syagrus romanzoffiana*). For the Aché, the buds, flour, and weevil larvae of jerivá are edible; in addition, from this species the Aché made bow wood, bowstring, roofing thatch, matting, fans, and a case for keeping feather art (Clastres 1972:161; Métraux and Baldus 1946:436; Vellard 1934:232, 240). The jerivá palm was also a mainstay for the Héta, for whom the mesocarp, kernel, and palm heart were all edible. The Héta also used the spathe as a receptacle; all basketry, mats, and fishing line came from the fiber or sword leaves (Loureiro Fernandes 1959:25; 1964:41–43; Kozák et al. 1979:395).

The jerivá palm appears to have been as important to the Héta and Aché as is the babaçu and other palms to the Guajá. In addition, jerivá palms, as with babaçu palms, occur in a habitat that appears to be secondary—that is, an artifact of past horticultural occupations, possibly by the sedentary Guarani (Balée 1992a). Comparative evidence suggests, then, that Guajá historical ecology has parallels elsewhere in lowland South America.

These parallels are essentially threefold. First, contemporary Tupí-Guaraní foragers are descendants of societies that underwent a process of agricultural regression, probably losing cultivation of maize after they had lost other domesticates. Second, among botanical families, none is more important in subsistence, technology, and construction than palms. While it may be superficially true that palms are the "cereals of the tropics," and that they are important among horticulturalists, my point is that palms are qualitatively and quantitatively more significant in the economic life of foraging societies than in the economic life of any other society, including trekkers. Third, several of these most important palm species—such as jerivá and *Acrocomia* (mucajá) in southern Brazil and Paraguay, or babaçu, inajá, and tucumã in the habitat of the Guajá—are, in their relative ecological importance, dominance, frequency, or density, indicators of the past presence of horticultural societies.

In fact, these palms are associated with other suites of dicotyledonous trees which themselves are either exclusive to or in some other way resultant from the formation of fallow forests (see chapter 6 and appendixes 1 and 2). Foragers, thus, are dependent on fallows; they appropriate semidomesticated plants of fallows for some of the same objects (such as textiles) that horticulturalists produce only from domesticates. The final cultural product of agricultural regression, as I have indicated elsewhere (Balée 1992a), is a people of the fallow—however innocent they may be of the techniques of horticulture and the historical ecology by which they have been constrained. Foragers do not practice horticulture; but they are dependent upon those that did. This dependence, which may not be a common attribute of foraging societies elsewhere in the world, is perhaps one of the most intriguing discoveries of ethnobotanical research in lowland South America.

Implications of Agricultural Regression

An earlier section of this chapter concluded that, for the foraging Guajá, there has been a withering of what was formerly a rich ethnobotanical vocabulary. Does this loss also apply to the subdomain of plant activity signatures? Do the Guajá have fewer uses for plants or fewer useful plant species than do the Ka'apor?

The answer to both questions appears to be yes. First and most obvious is the fact that the Ka'apor inventory of tools and utensils is considerably larger than that of the Guajá. While the Guajá use no plant in a way that does not have a counterpart in Ka'apor ethnobotany, the Ka'apor use certain plants for objectives that have no counterpart in Guajá culture.

For example, the Guajá do not have a traditional technology for making fire (obtained from lightning strikes), whereas the Ka'apor utilize fire drills and hearths (from annatto and *Guatteria scandens*—see appendix 8, "HHI01"). In addition, the Guajá make no pottery, so they could not have had any use for sherd tempers (which the Ka'apor obtain from *Licania* spp.—see appendix 8, "HHI21") or for pottery slips (for which the Ka'apor use copal resins from *Hymenaea* spp.—see appendix 8, "HHI22"). The Guajá, who traditionally camp beneath lean-tos built around standing fallow specimens of *arakowã'i* (*Poecilanthe effusa*), have no use for house posts (which the Ka'apor obtain from ten species), nor for rafters, tiebeams, and cross beams (which the Ka'apor obtain from 48 species—see appendix 8, "PBC01" and "PBC02"). Unlike the Ka'apor, who utilize the bark or leaves from *Enterolobium* sp. nov., *Gouania pyrifolia*, papaya, and *Parkia guianensis* as soap for washing their cotton clothing and hammocks, the Guajá have no soap species for cleaning their clothing and hammocks, made from tucumã fibers.

Because they do not grow tobacco or smoke, the Guajá would have no use for a pipe. The Ka'apor, in contrast, make pipe stems from *Mabea* spp. (appendix 8, "HHI20") and cigar rolling papers from the bark of *Couratari guianensis* (appendix 8, "HHI19'). Because the Guajá do not cultivate the fish poisons, *Tephrosia* and *Clibadium* (they use only *Derris utilis* and *Serjania lethalis*, which are also used by the Ka'apor), the Guajá have fewer fish poisons than have the Ka'apor. Since they make no fences around their lean-tos, nor do they construct benches, the Guajá make less intensive use of 11 species that are consistently used by the Ka'apor for making these same objects (appendix 8, "PBC01" and "PBC08"). Because they lack bottle gourds and calabashes, the Guajá also did not have a technology for dying these, for which the Ka'apor employ six species (appendix 8, "PAC08").

TABLE 8.4 *Ka'apor Activity Signatures for Useless Plants of the Guajá*

Useless Guajá Plant	Ka'apor Activity Signature*
ANACARDIACEAE	
Astronium lecointei	TLW03; HHI06; FUE01
ANNONACEAE	
Ephedranthus pisocarpus	MAG02
Xylopia nitida	PBC01; PBC07; ANF02; FUE01
ARACEAE	
Monstera supinata	ANF02; POI09; POI12
ARECACEAE	
Desmoncus polyacanthos	TLW13
ASTERACEAE	
Conyza bonariensis	MED02
Eupatorium maximilliani	ANF02
BIGNONIACEAE	
Tabebuia impetiginosa	TLW12
BOMBACACEAE	
Ceiba pentandra	ANF02; MYT03; MED31
BROMELIACEAE	
Aechmea bromeliifolia	ANF01; MED10
Bromelia goeldiana	ANF01; MED10
CAESALPINIACEAE	
Bauhinia viridiflorens	TLW12; TLW15
Copaifera duckei	MED14; MED24
Macrolobium acaciaefolium	ANF01; FUE01
Tachigali myrmecophila	MED02; MED08; MED32; POI11
CECROPIACEAE	
Cecropia palmata	SFS01
CELASTRACEAE	
Cheiloclinium cognatum	ANF02; FUE01; SFS01; MAG02
CLUSIACEAE	
Vismia guianensis	PAC07
CYPERACEAE	
Diplasia karataefolia	PAC12
EUPHORBIACEAE	
Conceveiba guianensis	ANF02
Croton matourense	PBC04; FUE02; MED06
Mabea caudata	ANF02; HHI19; FUE01
FABACEAE	
Andira retusa	PAC02; COM02
Platypodium elegans	ANF02; FUE01
Taralea oppositifolia	ANF01; FUE01
HAEMODORACEAE	
Xiphidium caeruleum	POI19

Perhaps most surprising is that the foraging Guajá do not have a larger number of edible nondomesticated species than the Ka'apor have. Whereas the Ka'apor consume 179 nondomesticated plants, the Guajá recognize only 49 as being edible, based on my research at P. I. Awá and P. I. Guajá.

These fundamental differences in the usage of plants between the Ka'apor and Guajá may be rendered more explicit by examining the species in each society that lack activity signatures altogether—i.e., plants that one may choose to call "useless." Of 768 botanical species about which they were queried, the Ka'apor encode no activity signature for 75 (or 10 percent); of 547 botanical species known to the Guajá, the Guajá

TABLE 8.4 *Continued*

Useless Guajá Plant	Ka'apor Activity Signature*
LAURACEAE	
Ocotea opifera	ANF01; MAG02
MALPIGHIACEAE	
Schwannia sp. 1	MED11
MARANTACEAE	
Calathea sp. 1	ANF02; HHI26
MELASTOMATACEAE	
Bellucia grossularioides	SFS01; ANF02; PAC08
MELIACEAE	
Guarea guidonia	ANF02; FUE01
MIMOSACEAE	
Acacia multipinnata	MAG08; FUE01
Parkia ulei	ANF01; FUE01
Newtonia psilostachya	ANF02; FUE01
MONIMIACEAE	
Siparuna amazonica	MAG09; MED02
MYRTACEAE	
Eugenia schomburgkii	ANF02; SFS01; FUE02
Myrciaria tenella	MED29
Eugenia muricata	SFS01; ANF02; FUE01
PHYTOLACCACEAE	
Phytolacca rivinoides	PAC09; MAG09
POACEAE	
Olyra caudata	ANF02
RUBIACEAE	
Geophila repens	ANF02
Psychotria racemosa	ANF02
Uncaria guianensis	MED25
SAPINDACEAE	
Pseudyma frutescens	ANF02; POI13
SAPOTACEAE	
Pouteria engleri	SFS01; ANF01;FUE01
TILIACEAE	
Apeiba tibourbou	TLW02; FUE01
VIOLACEAE	
Rinorea pubiflora	ANF02; TLW15; HHI07; FUE01
VOCHYSIACEAE	
Erisma uncinatum	ANF02; FUE01
ZINGIBERACEAE	
Costus arabicus	ANF02

*See Appendix 6 for codes to activity signatures listed here.

encode no activity signature for 129 (24 percent). Useless species in Guajá ethnobotany, then, are proportionally more than twice as numerous than in Ka'apor ethnobotany. Of these 129 species with no activity signature to the Guajá, the Ka'apor know 54. For only 5 of these 54 species, however, neither the Ka'apor nor the Guajá have an encoded activity signature (*Phenakospermum guianensis*, *Coccoloba latifolia*, *Connarus favosus*, *Memoria allamandiflora*, and *Lantana camara*). The Guajá do not encode an activity signature for any of the useless plants of the Ka'apor (appendix 8, "NONE"). Yet of the 129 useless plants of the Guajá, the Ka'apor have an activity signature for 49 (see table 8.4). In other words, while the Guajá do not recognize as useful any plant also not so recognized by the Ka'apor, the Ka'apor utilize 38 percent of the useless plants of the Guajá.

The lack of horticulture among foragers in Amazonia thus implies much more than a lack of domesticates per se. In ethnobotanical terms, it means fewer *uses* for plants also. Without certain domesticates, other plants used in their processing or finishing are unnecessary. Without horticulture, permanent houses are an absurd extravagance. Without horticulture, there is no need for axe handles, digging sticks, fire drills, and manioc presses, all of which are made from plants. Without horticulture, there is generally less need to know the specific properties, potential uses, or biological principles of a great range of plants, both domesticated and not. Hence, there is an apparent specializing tendency of the Guajá and probably of other foragers of lowland South America, with regard to plants. These peoples focus on and heavily exploit the plants that they do know well and which serve the purposes of their hunting and gathering technology. Coincidentally, most of their focal plants are denizens of fallows. Though they are foragers in their daily round, they thus depend upon the past horticultural activities of their neighbors.

Ka'apor ethnobotany, by now, should be fairly clear in its context, having been compared to systems of similar linguistic and cultural origins of peoples that inhabit like environments. All these modern ethnobotanies of the Tupí-Guaraní family share certain resemblances, especially in terms of the plant lexicon but also with regard to certain material objects made from plants (such as hammocks and infant carrying straps). All of these ethnobotanies suggest a historical beginning that was inextricably tied to an agroforestry complex indigenous to Amazonia. The modern foragers evince this agroforestry complex in but muted form, where it surfaces in a few items of the lexicon, and, indirectly, in terms of their continuing material dependence on old fallows. The horticultural, sedentary societies, such as the Ka'apor, Tembé, and Wayãpi, on the other hand, have managed to maintain much of this agroforestry complex despite contact, colonial warfare, forced migrations, and conquest. Because these horticultural societies did not undergo agricultural regression (even if their forebears did abandon the institutions of chief, lineage, and conquest warfare, and even if they did suffer depopulation and dispersal), they also kept a very extensive ethnobotanical system (on the Wayãpi, for example, see Grenand 1980).

The Ka'apor also are associated with a higher overall population, greater population density, longer settlement occupation spans, more permanent housing, a greater proliferation of material artifacts, and finally, a more extensive plant system than are the Guajá. These differences, which lend themselves to a historical ecological construct, would be otherwise astounding were one to consider only that the Guajá and the Ka'apor speak closely related languages descended from a common mother tongue and that they exploit the same relatively small, well-bounded province known as pre-Amazonia. Rather, the foundation for all these differences is the agricultural regression experienced by the Guajá and somehow avoided by the Ka'apor.

Despite common origins and a common habitat, the historical and ecological paths of the Guajá and the Ka'apor diverged considerably after the conquest of lower Amazonia. Their footprints, in a sense, followed markedly different paths—the one group maintaining a horticultural life and the other suffering agricultural regression. The fact of that divergence is already seen to have disfavored the forager in the struggle to persist as a people, but the horticulturalist too faces an uncertain future in the changing world of Amazonian forests. And both urgently need protection by an enlightened state, which is distinctly possible in Brazil.

Ecological Importance Values of Fallow Species

Family and Species	Ind.[1] (#)	S.U.[2] (#)	Basal[3] Area (m²)	Density[4] (%)	Freq.[5] (%)	Domi.[6] (%)	I.V.[7]
ANACARDIACEAE							
Anacardium giganteum	2	2	0.0573	0.10	0.13	0.06	0.29
Astronium lecointei	2	2	0.1281	0.10	0.13	0.13	0.36
Astronium obliquum	1	1	0.0177	0.05	0.06	0.02	0.13
Spondias mombin	46	17	3.0543	2.28	1.09	3.16	6.53
Tapirira guianensis	7	6	0.5312	0.35	0.39	0.55	1.29
Thyrsodium paraense	1	1	0.0133	0.05	0.06	0.01	0.12
ANNONACEAE							
Annona montana	6	5	0.0672	0.30	0.32	0.07	0.69
Annona sericea	1	1	0.0079	0.05	0.06	0.01	0.12
Duguetia marcgraviana	3	2	0.0388	0.15	0.13	0.04	0.32
Duguetia surinamensis	13	12	0.2318	0.64	0.77	0.24	1.65
Ephedranthus pisocarpus	10	10	0.2107	0.50	0.64	0.22	1.36
Fusaea longifolia	1	1	0.0227	0.05	0.06	0.02	0.13
Oxandra sessiflora	1	1	0.0539	0.05	0.06	0.06	0.17
Rollinia exsucca	8	8	0.2368	0.40	0.51	0.24	1.15
Xylopia barbata	1	1	0.1419	0.05	0.06	0.15	0.26
Annonaceae sp. 2	1	1	0.1244	0.05	0.06	0.13	0.24

NOTE: Data are drawn from the aggregate of four one-hectare study plots of fallow. Author inventoried all trees and lianas on those study plots that were ≥ 10 cm dbh. Some species listed here were collected only in the immediate environs of the Tembé or the Guajá peoples and hence are not listed in Appendix 5, which lists species collected only among the Ka'apor. See Chapter 6 for a description of the four study plots of fallow.

[1] Number of individuals inventoried (number of occurrences).

[2] Number of sampling units (s.u.) on which the species occurs, out of a total number of 160 sampling units.

[3] Aggregate basal area (at breast height) of the species.

[4] Relative density (number of sampling units of occurrence divided by total number of occurrences [1556] inventoried).

[5] Relative frequency (number of individuals of species divided by total number of individuals [2018] inventoried).

[6] Relative dominance (aggregate basal area of species divided by total basal area of all species [97.7666 m²] inventoried).

[7] Ecological importance value (I.V.), which is the sum of relative density, relative frequency, and relative dominance. A monoculture would have an I.V. of 300.

Family and Species	Ind. (#)	S.U. (#)	Basal Area (m^2)	Density (%)	Freq. (%)	Domi. (%)	I.V.
APOCYNACEAE							
Aspidosperma album	1	1	0.0209	0.05	0.06	0.02	0.13
Aspidosperma cylindrocarpon	1	1	0.0080	0.05	0.06	0.01	0.12
Aspidosperma desmanthum	3	2	0.1561	0.15	0.13	0.16	0.44
Aspidosperma verruculosum	1	1	0.0143	0.05	0.06	0.01	0.12
Himatanthus sucuuba	5	5	0.1224	0.25	0.32	0.13	0.70
Apocynaceae sp. 1	1	1	0.0296	0.05	0.06	0.03	0.14
AQUIFOLIACEAE							
Ilex parviflora	1	1	0.0079	0.05	0.06	0.01	0.12
ARALIACEAE							
Schefflera morototoni	4	4	0.3004	0.20	0.26	0.31	0.77
ARECACEAE							
Astrocaryum sp. 1	2	1	0.0409	0.10	0.06	0.04	0.20
Astrocaryum gynacanthum	32	16	0.6682	1.59	1.03	0.69	3.31
Astrocaryum vulgare	16	13	5.9330	0.79	0.84	6.13	7.76
Bactris major	1	1	0.2827	0.05	0.06	0.29	0.40
Euterpe oleracea	2	2	0.0226	0.10	0.13	0.02	0.25
Maximiliana maripa	37	26	1.8382	1.83	1.67	1.90	5.40
Oenocarpus distichus	13	10	0.4882	0.64	0.64	0.50	1.78
Orbignya phalerata	49	35	4.5393	2.43	2.25	4.69	9.37
BIGNONIACEAE							
Anemopaegma setilobum	4	4	0.4003	0.20	0.26	0.04	0.50
Cuspidaria sp. 1	1	1	0.0152	0.05	0.06	0.02	0.13
Jacaranda copaia	5	5	0.0811	0.25	0.32	0.08	0.65
Jacaranda duckei	9	6	0.9400	0.45	0.39	0.97	1.81
Mussatia sp. 1	2	2	0.0221	0.10	0.13	0.02	0.25
Stizophyllum riparium	1	1	0.0108	0.05	0.06	0.01	0.12
Tabebuia impetiginosa	10	8	1.7787	0.50	0.51	1.84	2.85
Tabebuia obscura	7	7	0.1318	0.35	0.45	0.14	0.94
Tabebuia serratifolia	9	8	0.3094	0.45	0.51	0.32	1.28
Tabebuia sp. 1	1	1	0.0547	0.05	0.06	0.06	0.17
BOMBACACEAE							
Ceiba pentandra	3	3	0.6111	0.15	0.19	0.63	0.97
BORAGINACEAE							
Cordia scabrida	21	9	0.5446	1.04	0.58	0.56	2.18
Cordia scabrifolia	7	6	0.1039	0.35	0.39	0.11	0.85
Cordia sellowiana	1	1	0.0113	0.05	0.06	0.01	0.12
Cordia sp. 1	1	1	0.0495	0.05	0.06	0.05	0.16
Cordia trachyphylla	1	1	0.0278	0.05	0.06	0.03	0.14

Family and Species	Ind. (#)	S.U. (#)	Basal Area (m²)	Density (%)	Freq. (%)	Domi. (%)	I.V.
BURSERACEAE							
Protium decandrum	1	1	0.0252	0.05	0.06	0.03	0.14
Protium giganteum	1	1	0.0370	0.05	0.06	0.04	0.15
Protium heptaphyllum	20	18	0.4229	0.99	1.16	0.44	2.59
Protium sagotiana	2	2	0.0225	0.10	0.13	0.02	0.25
Protium spruceanum	2	2	0.0314	0.10	0.13	0.03	0.26
Protium tenuifolium	2	2	0.0697	0.10	0.13	0.07	0.30
Tetragastris altissima	1	1	0.0104	0.05	0.06	0.01	0.12
Tetragastris panamensis	15	12	1.0167	0.74	0.77	1.05	2.56
Trattinickia rhoifolia	5	5	0.0711	0.25	0.32	0.07	0.64
CAESALPINIACEAE							
Apuleia leiocarpa	15	11	1.0466	0.74	0.71	1.08	2.53
Bauhinia acreana	1	1	0.0308	0.05	0.06	0.03	0.14
Bauhinia corniculata	2	2	0.0223	0.10	0.13	0.02	0.25
Bauhinia coronata	3	3	0.0358	0.15	0.19	0.04	0.38
Bauhinia guianensis	4	3	0.0927	0.20	0.19	0.10	0.49
Cassia apoucouita	1	1	0.0252	0.05	0.06	0.03	0.14
Cassia fastuosa	5	4	0.7884	0.25	0.26	0.81	1.32
Cassia lucens	5	5	0.0948	0.25	0.32	0.10	0.67
Cassia sylvestris	2	2	0.4985	0.10	0.13	0.52	0.75
Copaifera duckei	5	4	0.3834	0.25	0.26	0.40	0.91
Copaifera reticulata	1	1	0.0257	0.05	0.06	0.03	0.14
Dialium guianense	23	18	0.9870	1.14	1.16	1.02	3.32
Hymenaea courbaril	13	11	1.2164	0.64	0.71	1.26	2.61
Hymenaea parvifolia	21	19	1.8436	1.04	1.22	1.91	4.17
Hymenaea reticulata	1	1	0.1146	0.05	0.06	0.12	0.23
Senna chrysocarpa	1	1	0.0087	0.05	0.06	0.01	0.12
Swartzia brachyrachis	1	1	0.0097	0.05	0.06	0.01	0.12
Swartzia flaemengii	1	1	0.0419	0.05	0.06	0.04	0.15
Swartzia sp. 1	2	2	0.0256	0.10	0.13	0.03	0.26
Tachigali myrmecophila	13	12	0.2175	0.64	0.77	0.22	1.63
Tachigali paniculata	1	1	0.0102	0.05	0.06	0.01	0.12
Zollernia paraensis	7	7	0.6809	0.35	0.45	0.70	1.50
CARICACEAE							
Jacaratia spinosa	102	43	3.4720	5.05	2.76	3.59	11.40
CECROPIACEAE							
Cecropia concolor	12	9	0.2072	0.59	0.58	0.21	1.38
Cecropia ilicifolia	15	7	0.3167	0.74	0.45	0.33	1.52
Cecropia palmata	3	3	0.0633	0.15	0.19	0.07	0.41
Cecropia purpurescens	1	1	0.0093	0.05	0.06	0.01	0.12
Cecropia sciadophylla	1	1	0.0866	0.05	0.06	0.09	0.20
Coussapoa sp. 1	1	1	0.0125	0.05	0.06	0.01	0.12

Family and Species	Ind. (#)	S.U. (#)	Basal Area (m^2)	Density (%)	Freq. (%)	Domi. (%)	I.V.
CELASTRACEAE							
Maytenus sp. 1	2	2	0.0228	0.10	0.13	0.02	0.25
Maytenus sp. 2	1	1	0.0079	0.05	0.06	0.01	0.12
Maytenus sp. nov.	20	10	0.5952	0.99	0.64	0.62	2.25
Tontelea laxiflora	1	1	0.0145	0.05	0.06	0.01	0.12
CHRYSOBALANACEAE							
Couepia guianensis	7	7	0.5533	0.35	0.45	0.57	1.37
Hirtella bicornis	5	3	0.3597	0.25	0.19	0.37	0.81
Hirtella fasciculata	1	1	0.0079	0.05	0.06	0.01	0.12
Hirtella macrophylla	1	1	0.0373	0.05	0.06	0.04	0.15
Hirtella triandra	1	1	0.0079	0.05	0.06	0.01	0.12
Licania heteromorpha	2	2	0.0394	0.10	0.13	0.04	0.27
Licania hypoleuca	1	1	0.0260	0.05	0.06	0.03	0.14
Licania kunthiana	17	13	0.6105	0.84	0.84	0.63	2.31
Licania octandra	1	1	0.0079	0.05	0.06	0.01	0.12
Licania sprucei	2	2	0.1883	0.10	0.13	0.19	0.42
CLUSIACEAE							
Platonia insignis	6	5	3.5825	0.30	0.32	3.70	4.32
Rheedia acuminata	1	1	0.0141	0.05	0.06	0.01	0.12
COCHLOSPERMACEAE							
Cochlospermum orinocense	1	1	0.0191	0.05	0.06	0.02	0.13
COMBRETACEAE							
Buchenavia parvifolia	1	1	0.6138	0.05	0.06	0.63	0.74
Terminalia amazonia	4	4	0.5207	0.20	0.26	0.54	1.00
Terminalia dichotoma	2	2	0.1433	0.10	0.13	0.15	0.38
EBENACEAE							
Diospyros melinoni	5	5	0.0695	0.25	0.32	0.07	0.64
ELAEOCARPACEAE							
Sloanea grandis	4	4	0.1229	0.20	0.26	0.13	0.59
Sloanea guianensis	3	3	0.3480	0.15	0.19	0.36	0.70
Sloanea parviflora	1	1	0.0547	0.05	0.06	0.06	0.17
Sloanea porphyrocarpa	3	3	0.0612	0.15	0.19	0.06	0.40
EUPHORBIACEAE							
Chaetocarpus sp. 1	1	1	0.1971	0.05	0.06	0.20	0.31
Conceveiba guianensis	1	1	0.0299	0.05	0.06	0.03	0.14
Croton cajucara	15	12	0.2158	0.74	0.77	0.22	1.73
Drypetes amazonica	1	1	0.0199	0.05	0.06	0.02	0.13
Drypetes variabilis	4	4	0.2068	0.20	0.26	0.21	0.67
Mabea caudata	3	3	0.0313	0.15	0.19	0.03	0.37
Margaritaria nobilis	1	1	0.2290	0.05	0.06	0.24	0.35

Family and Species	Ind. (#)	S.U. (#)	Basal Area (m²)	Density (%)	Freq. (%)	Domi. (%)	I.V.
Pera furruginea	1	1	0.0284	0.05	0.06	0.03	0.14
Sapium caspceolatum	1	1	0.0495	0.05	0.06	0.05	0.16
Sapium lanceolatum	11	10	0.4283	0.55	0.64	0.44	1.63
FABACEAE							
Andira retusa	1	1	0.0531	0.05	0.06	0.05	0.16
Andira sp. 1	1	1	0.3675	0.05	0.06	0.38	0.49
Bowdichia nitida	1	1	0.0113	0.05	0.06	0.01	0.12
Cenostigma macrophyllum	9	7	0.5280	0.45	0.45	0.55	1.45
Derris spruceana	2	2	0.1077	0.10	0.13	0.11	0.34
Dioclea reflexa	1	1	0.0000	0.05	0.06	0.00	0.11
Machaerium angustifolium	1	1	0.0082	0.05	0.06	0.01	0.12
Machaerium quinata	2	2	0.0255	0.10	0.13	0.03	0.26
Ormosia flava	2	2	0.1637	0.10	0.13	0.17	0.40
Ormosia paraensis	1	1	0.0189	0.05	0.06	0.02	0.13
Platypodium elegans	8	7	4.0351	0.40	0.45	4.17	5.02
Pterocarpus rohrii	2	2	0.0748	0.10	0.13	0.08	0.31
Fabaceae sp. 1	2	2	0.0986	0.10	0.13	0.10	0.33
FLACOURTIACEAE							
Banara guianensis	1	1	0.0123	0.05	0.06	0.01	0.12
Casearia commersoniana	1	1	0.0170	0.05	0.06	0.02	0.13
Casearia arborea	14	12	0.2514	0.69	0.77	0.26	1.72
Casearia decandra	1	1	0.0314	0.05	0.06	0.03	0.14
Casearia javitensis	1	1	0.0314	0.05	0.06	0.03	0.14
Casearia sylvestris	1	1	0.0113	0.05	0.06	0.01	0.12
Casearia ulmifolia	1	1	0.0299	0.05	0.06	0.03	0.14
Homalium sp. 1	1	1	0.0113	0.05	0.06	0.01	0.12
Laetia corymbulosa	2	2	0.0245	0.10	0.13	0.03	0.26
Lindackeria latifolia	23	21	0.5460	1.14	1.35	0.56	3.05
HUMIRIACEAE							
Sacoglottis amazonica	6	6	0.3674	0.30	0.39	0.38	1.07
Sacoglottis guianensis	3	3	0.1968	0.15	0.19	0.20	0.54
Schistostemon macrophyllum	1	1	0.0679	0.05	0.06	0.07	0.18
[INDETERMINATE 11]							
Indeterminate genus 15	1	1	0.1104	0.05	0.06	0.11	0.22
LAURACEAE							
Aiouea multiflora	1	1	0.0079	0.05	0.06	0.01	0.12
Aniba burchellii	1	1	0.0452	0.05	0.06	0.05	0.16
Licaria debilis	1	1	0.0092	0.05	0.06	0.01	0.12
Mezilaurus lindaviana	1	1	0.1720	0.05	0.06	0.18	0.29
Ocotea caudata	2	2	0.1141	0.10	0.13	0.12	0.35
Ocotea fasciculata	4	4	0.0672	0.20	0.26	0.07	0.53
Ocotea laxiflora	1	1	0.0137	0.05	0.06	0.01	0.12

Family and Species	Ind. (#)	S.U. (#)	Basal Area (m²)	Density (%)	Freq. (%)	Domi. (%)	I.V.
Ocotea marmellensis	1	1	0.0117	0.05	0.06	0.01	0.12
Ocotea silvae	8	6	0.6553	0.40	0.39	0.68	1.47
Systemonodaphne mezii	2	2	0.0947	0.10	0.13	0.10	0.33
Lauraceae sp. 4	1	1	0.0172	0.05	0.06	0.02	0.13
LECYTHIDACEAE							
Couratari guianensis	3	2	0.0641	0.15	0.13	0.07	0.35
Couratari oblongifolia	2	2	0.0449	0.10	0.13	0.05	0.28
Couratari sp. 1	3	3	0.0948	0.15	0.19	0.10	0.44
Eschweilera amazonica	4	3	0.1125	0.20	0.19	0.12	0.51
Eschweilera apiculata	5	4	0.2631	0.25	0.26	0.27	0.78
Eschweilera bracteosa	1	1	0.0099	0.05	0.06	0.01	0.12
Eschweilera coriacea	17	16	1.2761	0.84	1.03	1.32	3.19
Eschweilera micrantha	5	5	0.1380	0.25	0.32	0.14	0.71
Eschweilera ovata	11	10	0.4696	0.55	0.64	0.49	1.68
Eschweilera parviflora	1	1	0.0560	0.05	0.06	0.06	0.17
Eschweilera pedicellata	4	4	0.2905	0.20	0.26	0.30	0.76
Gustavia augusta	103	54	1.7849	5.10	3.47	1.84	10.41
Lecythis idatimon	8	7	0.2550	0.40	0.45	0.26	1.11
Lecythis jarana	5	5	0.3463	0.25	0.32	0.36	0.93
Lecythis lurida	1	1	0.0158	0.05	0.06	0.02	0.13
Lecythis pisonis	4	4	2.9954	0.20	0.26	3.10	3.56
MALPIGHIACEAE							
Banisteriopsis pubipetala	1	1	0.0079	0.05	0.06	0.01	0.12
Byrsonima stipulacea	1	1	0.0113	0.05	0.06	0.01	0.12
Mascagnia macrodisca	2	2	0.0165	0.10	0.13	0.02	0.25
Mascagnia sp. 1	1	1	0.0095	0.05	0.06	0.01	0.12
Tetrapterys styloptera	1	1	0.0106	0.05	0.06	0.01	0.12
MELASTOMATACEAE							
Bellucia dichotoma	1	1	0.113	0.05	0.06	0.01	0.12
Miconia serialis	10	10	0.1185	0.50	0.64	0.12	0.26
Miconia surinamensis	5	2	0.2701	0.25	0.13	0.28	0.66
Mouriri guianensis	18	12	0.8049	0.89	0.77	0.83	2.49
Mouriri huberi	2	1	0.1210	0.10	0.06	0.13	0.29
Mouriri sagotiana	13	9	0.2616	0.64	0.58	0.27	1.49
Mouriri trunciflora	1	1	0.0201	0.05	0.06	0.02	0.13
Mouriri vernicosa	1	1	0.0299	0.05	0.06	0.03	0.14
MELIACEAE							
Cedrela fissilis	2	2	0.2435	0.10	0.13	0.25	0.48
Guarea guidonia	13	8	1.0327	0.64	0.51	1.07	2.22
Guarea macrophylla	1	1	0.0123	0.05	0.06	0.01	0.12
Guarea sp. 2	8	7	0.3431	0.40	0.45	0.35	1.20
Trichilia areolata	4	3	0.0890	0.20	0.19	0.09	0.48
Trichilia cipo	5	3	0.0845	0.25	0.19	0.09	0.53

Family and Species	Ind. (#)	S.U. (#)	Basal Area (m²)	Density (%)	Freq. (%)	Domi. (%)	I.V.
Trichilia elegans	8	6	0.1127	0.40	0.39	0.12	0.91
Trichilia micrantha	2	2	0.0604	0.10	0.13	0.06	0.29
Trichilia quadrijuga	34	28	0.5588	1.68	1.80	0.58	4.06
Trichilia schomburgkii	1	1	0.0080	0.05	0.06	0.01	0.12
Trichilia smithii	1	1	0.0201	0.05	0.06	0.02	0.13
Trichilia tenuiramea	1	1	0.0573	0.05	0.06	0.06	0.17
Trichilia verrocosa	13	11	0.2260	0.64	0.71	0.23	1.58
Meliaceae sp. 1	16	11	0.2892	0.79	0.71	0.30	1.80
Meliaceae sp. 2	6	5	0.2923	0.30	0.32	0.30	0.92
Meliaceae sp. 3	1	1	0.0165	0.05	0.06	0.02	0.13
MIMOSACEAE							
Abarema cochleata	1	1	0.0243	0.05	0.06	0.03	0.14
Calliandra sp. 1	1	1	0.0079	0.05	0.06	0.01	0.12
Hymenolobium excelsum	1	1	0.0224	0.05	0.06	0.02	0.13
Inga alba	6	6	0.4023	0.30	0.39	0.42	1.11
Inga auristellae	5	5	0.0612	0.25	0.32	0.06	0.63
Inga capitata	3	3	0.0515	0.15	0.19	0.05	0.39
Inga cecropietorum	1	1	0.0104	0.05	0.06	0.01	0.12
Inga falcistipula	8	8	0.0767	0.40	0.51	0.08	0.99
Inga gracilifolia	7	6	0.1399	0.35	0.39	0.14	0.88
Inga heterophylla	1	1	0.0129	0.05	0.06	0.01	0.12
Inga longiflora	1	1	0.0106	0.05	0.06	0.01	0.12
Inga marginata	3	3	0.0236	0.15	0.19	0.02	0.36
Inga miriantha	5	5	0.0514	0.25	0.32	0.05	0.62
Inga nitida	2	2	0.0302	0.10	0.13	0.03	0.26
Inga nobilis	1	1	0.0177	0.05	0.06	0.02	0.13
Inga paraensis	1	1	0.0093	0.05	0.06	0.01	0.12
Inga rubiginosa	1	1	0.0269	0.05	0.06	0.03	0.14
Inga thibaudiana	1	1	0.0363	0.05	0.06	0.04	0.15
Inga sp. 1	3	2	0.2835	0.15	0.13	0.29	0.57
Newtonia psilostachya	10	10	0.5981	0.50	0.64	0.62	1.76
Newtonia suaveolens	1	1	0.0314	0.05	0.06	0.03	0.14
Parkia multijuga	7	6	0.2937	0.35	0.39	0.30	1.04
Parkia nitida	1	1	0.0346	0.05	0.06	0.04	0.15
Parkia pendula	5	5	0.2934	0.25	0.32	0.30	0.87
Parkia ulei	2	2	0.4884	0.10	0.13	0.50	0.73
Pithecellobium cochleatum	1	1	0.0284	0.05	0.06	0.03	0.14
Pithecellobium corymbosum	1	1	0.0254	0.05	0.06	0.03	0.14
Pithecellobium jupumba	3	2	0.0449	0.15	0.13	0.05	0.33
Pithecellobium niopoides	5	5	0.5174	0.25	0.32	0.53	1.10
Pithecellobium racemosum	18	12	0.4366	0.89	0.77	0.45	2.11
Stryphnodendron barbadetiman	2	2	0.4196	0.10	0.13	0.43	0.66
Stryphnodendron guianensis	5	4	0.1772	0.25	0.26	0.18	0.69
Stryphnodendron polystachyum	7	7	0.5294	0.35	0.45	0.55	1.35

Family and Species	Ind. (#)	S.U. (#)	Basal Area (m²)	Density (%)	Freq. (%)	Domi. (%)	I.V.
MORACEAE							
Bagassa guianensis	5	4	0.3679	0.25	0.26	0.38	0.89
Batocarpus amazonicus	2	1	0.0314	0.10	0.06	0.03	0.19
Brosimum acutifolium	1	1	0.1669	0.05	0.06	0.17	0.28
Brosimum guianense	6	6	0.1231	0.30	0.39	0.13	0.82
Brosimum lactescens	6	6	0.2842	0.30	0.39	0.29	0.98
Castilla ulei	3	2	0.1342	0.15	0.13	0.14	0.42
Chlorophora tinctoria	1	1	0.2743	0.05	0.06	0.28	0.39
Clarisia racemosa	1	1	0.0145	0.05	0.06	0.01	0.12
Ficus sp. 4	1	1	0.0327	0.05	0.06	0.03	0.14
Maquira guianensis	2	2	0.0434	0.10	0.13	0.04	0.27
Naucleopsis melo-barretoi	1	1	0.0449	0.05	0.06	0.05	0.16
Pseudolmedia murure	1	1	0.0079	0.05	0.06	0.01	0.12
MYRTACEAE							
Campomanesia aromatica	1	1	0.0353	0.05	0.06	0.04	0.15
Campomanesia grandiflora	9	8	0.6100	0.45	0.51	0.63	1.59
Eugenia cf. *stipitata*	1	1	0.0097	0.05	0.06	0.01	0.12
Eugenia egensis	17	16	0.2722	0.84	1.03	0.28	2.15
Eugenia eurycheila	2	2	0.0257	0.10	0.13	0.03	0.26
Eugenia exaltata	2	2	0.0233	0.10	0.13	0.02	0.25
Eugenia flavescens	1	1	0.0087	0.05	0.06	0.01	0.12
Eugenia lamberbiana	1	1	0.0082	0.05	0.06	0.01	0.12
Eugenia modesta	1	0	0.0079	0.05	0.00	0.01	0.06
Eugenia muricata	5	4	0.0481	0.25	0.26	0.26	0.56
Eugenia omissa	8	8	0.1150	0.40	0.51	0.12	1.03
Eugenia patrisii	2	2	0.0638	0.10	0.13	0.07	0.30
Eugenia protracta	2	2	0.0873	0.10	0.13	0.09	0.32
Eugenia pseudopsidium	1	1	0.0163	0.05	0.06	0.02	0.13
Eugenia schomburgkii	1	1	0.0363	0.05	0.06	0.04	0.15
Eugenia sinemariensis	2	2	0.0374	0.10	0.13	0.04	0.27
Eugenia sp. 1	5	4	0.0609	0.25	0.26	0.06	0.57
Eugenia sp. 2	1	1	0.0104	0.05	0.06	0.01	0.12
Eugenia sp. 3	1	1	0.0133	0.05	0.06	0.01	0.12
Eugenia sp. 4	1	1	0.0167	0.05	0.06	0.02	0.13
Myrcia paivae	2	2	0.0475	0.10	0.13	0.05	0.28
Myrcia splendens	2	2	0.2395	0.10	0.13	0.25	0.48
Myrciaria disticha	2	2	0.0286	0.10	0.13	0.03	0.26
Myrciaria flexuosa	3	3	0.0764	0.15	0.19	0.08	0.42
Myrciaria floribunda	5	5	0.1338	0.25	0.32	0.14	0.71
Myrciaria obscura	25	16	0.4681	1.24	1.03	0.48	2.75
Myrciaria spruceana	2	2	0.0394	0.10	0.13	0.04	0.27
Myrciaria tenella	5	4	0.1085	0.25	0.26	0.11	0.62
NYCTAGINACEAE							
Neea glameruliflora	3	3	0.0453	0.15	0.19	0.05	0.39
Neea oppositiflora	5	5	0.1703	0.25	0.32	0.18	0.75

Family and Species	Ind. (#)	S.U. (#)	Basal Area (m²)	Density (%)	Freq. (%)	Domi. (%)	I.V.
Neea ovalifolia	1	1	0.0314	0.05	0.06	0.03	0.14
Neea sp. 1	53	34	1.3978	2.63	2.19	1.44	6.26
Neea sp. 2	25	14	0.4848	1.24	0.90	0.50	2.64
Pisonia sp. 1	1	1	0.0079	0.05	0.06	0.01	0.12
Pisonia sp. 2	51	32	1.6104	2.53	2.06	1.66	6.25
OCHNACEAE							
Ouratea castaneaefolia	1	1	0.0165	0.05	0.06	0.02	0.13
Ouratea salicifolia	1	1	0.0165	0.05	0.06	0.02	0.13
OLACACEAE							
Dulacia candida	2	2	0.0599	0.10	0.13	0.06	0.29
Dulacia guianensis	1	1	0.0401	0.05	0.06	0.04	0.15
Minquartia guianensis	1	1	0.0104	0.05	0.06	0.01	0.12
OPILIACEAE							
Agonandra brasiliensis	3	3	0.0739	0.15	0.19	0.08	0.42
PHYTOLACCACEAE							
Seguieria amazonica	3	3	0.0523	0.15	0.19	0.05	0.39
POLYGONACEAE							
Coccoloba latifolia	3	3	0.0305	0.15	0.19	0.03	0.37
Coccoloba paniculata	3	2	0.1308	0.15	0.13	0.14	0.42
Coccoloba racemulosa	1	1	0.0095	0.05	0.06	0.01	0.12
QUIINACEAE							
Quiina guaporensis	1	1	0.0133	0.05	0.06	0.01	0.12
RHAMNACEAE							
Colubrina sp. 1	7	2	0.8830	0.35	0.13	0.91	1.39
Ziziphus itacaiunensis	9	8	0.4814	0.45	0.51	0.50	1.46
RUBIACEAE							
Coussarea ovalis	1	1	0.0095	0.05	0.06	0.01	0.12
Duroia sp. 1	2	2	0.3114	0.10	0.13	0.32	0.55
Faramea sessilifolia	15	12	0.2340	0.74	0.77	0.24	1.75
Guettarda divaricata	1	1	0.0087	0.05	0.06	0.01	0.12
Randia armata	1	1	0.0095	0.05	0.06	0.01	0.12
RUTACEAE							
Zanthoxylum chiloperone	1	1	0.1555	0.05	0.06	0.16	0.27
Zanthoxylum rhoifolium	1	1	0.0177	0.05	0.06	0.02	0.13
Zanthoxylum sp. 2	2	2	0.2062	0.10	0.13	0.21	0.44
SAPINDACEAE							
Cupania acutifolia	2	2	0.0616	0.10	0.13	0.06	0.29
Cupania scrobiculata	20	18	0.2421	0.99	1.16	0.25	2.40

Family and Species	Ind. (#)	S.U. (#)	Basal Area (m²)	Density (%)	Freq. (%)	Domi. (%)	I.V.
Cupania sp. 1	1	1	0.0104	0.05	0.06	0.01	0.12
Matayba arborescens	1	1	0.0121	0.05	0.06	0,01	0.12
Talisia acutifolia	1	1	0.0104	0.05	0.06	0.01	0.12
Talisia lacerata	4	3	0.0533	0.20	0.19	0.06	0.45
Talisia longifolia	1	1	0.0087	0.05	0.06	0.01	0.12
Talisia microphylla	1	1	0.0170	0.05	0.06	0.02	0.13
Talisia retusa	8	8	0.1177	0.40	0.51	0.12	1.03
Talisia sp. 1	3	3	0.0344	0.15	0.19	0.04	0.38
SAPOTACEAE							
Chrysophyllum argenteum	1	1	0.0360	0.05	0.06	0.04	0.15
Chrysophyllum pomiferum	1	1	0.0163	0.05	0.06	0.02	0.13
Chrysophyllum sparsiflorum	8	7	0.2245	0.40	0.45	0.23	1.08
Manilkara huberi	5	2	0.2093	0.25	0.13	0.22	0.60
Micropholis acutangula	1	1	0.0150	0.05	0.06	0.02	0.13
Micropholis guyanensis	1	1	0.0211	0.05	0.06	0.02	0.13
Micropholis venulosa	1	1	0.0456	0.05	0.06	0.05	0.16
Pouteria bangii	12	10	0.1547	0.59	0.64	0.16	1.39
Pouteria bilocularis	17	14	0.5799	0.84	0.90	0.60	2.34
Pouteria caimito	4	4	0.1339	0.20	0.26	0.14	0.60
Pouteria engleri	7	7	0.7478	0.35	0.45	0.77	1.57
Pouteria glomerata	6	6	0.1649	0.30	0.39	0.17	0.86
Pouteria gongrijpii	2	2	0.0195	0.10	0.13	0.02	0.25
Pouteria guianensis	1	1	0.0156	0.05	0.06	0.02	0.13
Pouteria hispida	8	4	0.3839	0.40	0.26	0.40	1.06
Pouteria jariensis	1	1	0.0079	0.05	0.06	0.01	0.12
Pouteria macrocarpa	1	1	0.0117	0.05	0.06	0.01	0.12
Pouteria macrophylla	24	15	3.4440	1.19	0.96	3.56	5.71
Pouteria penicillata	1	1	0.0085	0.05	0.06	0.01	0.12
Pouteria reticulata	2	2	0.0256	0.10	0.13	0.03	0.26
Pouteria sagotiana	4	4	0.1149	0.20	0.26	0.12	0.58
Pouteria sect. *Franchetella*	2	2	0.0835	0.10	0.13	0.09	0.32
Pouteria trigonosperma	5	3	0.929	0.25	0.19	0.10	0.54
Pouteria venosa	3	2	0.2246	0.15	0.13	0.23	0.51
Pouteria sp. 1	2	2	0.1270	0.10	0.13	0.13	0.36
Pouteria sp. 2	17	12	0.5542	0.84	0.77	0.57	2.18
Pouteria sp. 3	1	1	0.0765	0.05	0.06	0.08	0.19
SIMARUBACEAE							
Simaba cedron	36	30	0.5305	1.78	1.93	0.55	4.26
Simaba cuspidata	2	2	0.0264	0.10	0.13	0.03	0.26
Simaba guianensis	1	1	0.0125	0.05	0.06	0.01	0.12
Simaba paraensis	1	1	0.3982	0.05	0.06	0.41	0.52
Simaruba amara	4	3	0.1578	0.20	0.19	0.16	0.55
SOLANACEAE							
Brunfelsia guianensis	1	1	0.1024	0.05	0.06	0.11	0.22
Solanum leucocarpum	4	4	0.0746	0.20	0.26	0.08	0.54
Solanum vanhuerckii	1	1	0.0079	0.05	0.06	0.01	0.12

Family and Species	Ind. (#)	S.U. (#)	Basal Area (m²)	Density (%)	Freq. (%)	Domi. (%)	I.V.
STERCULIACEAE							
Sterculia pruriens	13	11	0.3127	0.64	0.71	0.32	1.67
Theobroma speciosum	25	18	0.6896	1.24	1.16	0.71	3.11
TILIACEAE							
Apeiba burchellii	2	1	0.0284	0.10	0.06	0.03	0.19
Apeiba echinata	1	1	0.0104	0.05	0.06	0.01	0.12
Hueheopsis duckeana	5	5	0.7265	0.25	0.32	0.75	1.32
ULMACEAE							
Ampelocera edentula	1	0	0.0350	0.05	0.00	0.04	0.09
VERBENACEAE							
Vitex flavens	1	1	0.0123	0.05	0.06	0.01	0.12
VIOLACEAE							
Amphirrhox surinamensis	13	8	0.1273	0.64	0.51	0.13	1.28
Rinorea flavescens	3	2	0.0309	0.15	0.13	0.03	0.31
Total	2018	1556	96.7666	100.28	99.43	100.00	299.71

Ecological Importance Values of Fallow Families

Family	Indiv.[1] Species (#)	Individuals[2] (#)	Basal Area[3] (m²)	Density[4] (%)	Freq.[5] (%)	Domi.[6] (%)	I.V.[7]
Arecaceae	8	152	13.8131	2.22	7.53	14.26	24.01
Sapotaceae	27	138	7.5393	7.50	6.68	7.82	22.18
Lecythidaceae	16	177	8.2169	4.44	8.79	8.51	21.74
Caesalpiniaceae	22	128	8.2003	6.11	6.35	8.49	20.95
Mimosaceae	33	120	5.2175	9.17	5.99	5.37	20.53
Myrtaceae	28	110	2.6483	7.78	5.48	2.75	16.01
Meliaceae	16	116	3.4464	4.44	5.75	3.56	13.75
Nyctaginaceae	7	139	3.7479	1.94	6.90	3.87	12.71
Fabaceae	13	33	5.4924	3.61	1.65	5.68	10.94
Caricaceae	1	102	3.4720	0.28	5.05	3.59	8.92
Bignoniaceae	10	49	3.3841	2.78	2.45	3.50	8.73
Anacardiaceae	6	59	3.8019	1.67	2.93	3.93	8.53
Burseraceae	9	49	1.7069	2.50	2.43	1.76	6.69
Chrysobalanaceae	10	38	1.8382	2.78	1.89	1.90	6.57
Melastomataceae	8	51	1.6374	2.22	2.53	1.69	6.44
Moraceae	12	30	1.5254	3.33	1.50	1.56	6.39
Annonaceae	10	45	1.1361	2.78	2.24	1.18	6.20
Euphorbiaceae	10	39	1.4360	2.78	1.94	1.47	6.19
Flacourtiaceae	10	46	0.9665	2.78	2.28	0.99	6.05
Lauraceae	11	23	1.2082	3.06	1.15	1.26	5.47
Sapindaceae	10	42	0.5677	2.78	2.09	0.59	5.46
Simarubaceae	5	44	1.1254	1.39	2.18	1.16	4.73
Clusiaceae	2	7	3.5966	0.56	0.35	3.71	4.62

NOTE: Data are drawn from the aggregate of four one-hectare study plots of fallow. Author inventoried all trees and lianas on those study plots that were ≥ 10 cm dbh. See Chapter 6 for a description of the four study plots.

[1] Number of individual species in the family inventoried.

[2] Number of individual trees and lianas ≥ 10 cm dbh in the family inventoried.

[3] Aggregate basal area (at breast height) of the family.

[4] Relative density (number of individual species divided by total number of species [360] inventoried).

[5] Relative frequency (number of individuals of family divided by total number of individuals [2018] inventoried).

[6] Relative dominance (aggregate basal area of family divided by total basal area of all families [96.7666 m²] inventoried).

[7] Ecological importance value (I.V.), which is the sum of relative density, relative frequency, and relative dominance. A monoculture would have an I.V. of 300.

Family	Indiv. Species (#)	Individuals (#)	Basal Area (m²)	Density (%)	Freq. (%)	Domi. (%)	I.V.
Cecropiaceae	6	33	0.6956	1.67	1.63	0.72	4.02
Boraginaceae	5	31	0.7371	1.39	1.54	0.76	3.69
Sterculiaceae	2	38	1.0023	0.56	1.88	1.03	3.47
Rubiaceae	5	20	0.5731	1.39	0.99	0.59	2.97
Celastraceae	4	24	0.6404	1.11	1.19	0.66	2.96
Rhamnaceae	2	16	1.3644	0.56	0.80	1.41	2.77
Apocynaceae	6	12	0.3513	1.67	0.60	0.36	2.63
Combretaceae	3	7	1.2778	0.83	0.35	1.32	2.50
Elaeocarpaceae	4	11	0.5868	1.11	0.55	0.61	2.27
Tiliaceae	3	8	0.7653	0.83	0.40	0.79	2.02
Humiriaceae	3	10	0.6321	0.83	0.50	0.65	1.98
Malpighiaceae	5	6	0.0558	1.39	0.30	0.06	1.75
Violaceae	2	16	0.1582	0.56	0.79	0.16	1.51
Rutaceae	3	4	0.3794	0.83	0.20	0.39	1.42
Polygonaceae	3	7	0.1708	0.83	0.35	0.18	1.36
Solanaceae	3	6	0.1849	0.83	0.30	0.20	1.33
Olacaceae	3	4	0.1104	0.83	0.20	0.11	1.14
Bombacaceae	1	3	0.6111	0.28	0.15	0.63	1.06
Araliaceae	1	4	0.3004	0.28	0.20	0.31	0.79
Ochnaceae	2	2	0.0330	0.56	0.10	0.04	0.70
Ebenaceae	1	5	0.0695	0.28	0.25	0.07	0.60
Opiliaceae	1	3	0.0739	0.28	0.15	0.08	0.51
Phytolaccaceae	1	3	0.0523	0.28	0.15	0.05	0.48
[Indeterminate 11]	1	1	0.1104	0.28	0.05	0.11	0.44
Polygalaceae	1	2	0.0182	0.28	0.10	0.02	0.40
Ulmaceae	1	1	0.0350	0.28	0.05	0.04	0.37
Cochlospermaceae	1	1	0.0191	0.28	0.05	0.02	0.35
Aquifoliaceae	1	1	0.0079	0.28	0.05	0.01	0.34
Verbenaceae	1	1	0.0123	0.28	0.05	0.01	0.34
Quiinaceae	1	1	0.0133	0.28	0.05	0.01	0.34
Total	360	2018	96.7666	100.04	100.28	100.00	300.32

APPENDIX 3

Ecological Importance Values of High Forest Species

Family and Species	Ind.[1] (#)	S.U.[2] (#)	Basal Area[3] (m^2)	Density[4] (%)	Freq.[5] (%)	Domi.[6] (%)	I.V.[7]
ANACARDIACEAE							
Anacardium giganteum	9	9	0.8903	0.45	0.59	0.76	1.80
Anacardium sp. 1	1	1	0.0503	0.05	0.07	0.04	0.16
Anacardium microsepalum	1	1	0.0360	0.05	0.07	0.03	0.15
Anacardium parvifolium	5	4	0.5380	0.25	0.26	0.46	0.97
Anacardium spruceanum	1	1	0.1772	0.05	0.07	0.15	0.27
Astronium gracile	1	1	0.0177	0.05	0.07	0.02	0.14
Tapirira peckoltiana	4	4	0.2344	0.20	0.26	0.20	0.66
Thrysodium spruceanum	1	1	0.0299	0.05	0.07	0.03	0.15
Thrysodium paraense	1	1	0.0165	0.05	0.07	0.01	0.13
ANNONACEAE							
Anaxagorea dolichocarpa	1	1	0.0154	0.05	0.07	0.01	0.13
Anaxagorea phaerocarpa	1	1	0.0079	0.05	0.07	0.01	0.13
Annona paludosa	1	1	0.0892	0.05	0.07	0.08	0.20
Duguetia riparia	1	1	0.0380	0.05	0.07	0.03	0.15
Fusaea longifolia	19	16	0.1892	0.96	1.06	0.16	2.18
Guatteria dielsiana	2	1	0.0580	0.10	0.07	0.05	0.22
Guatteria elongata	1	1	0.0133	0.05	0.07	0.01	0.13
Guatteria sp. 1	3	3	0.0747	0.15	0.20	0.06	0.41

NOTE: Data are drawn from the aggregate of four one-hectare study plots of high forest. Author inventoried all trees and lianas on those study plots that were ⩾ 10 cm dbh. Some species listed here were collected only in the immediate environs of the Tembé or the Guajá peoples and hence are not listed in Appendix 5, which lists species collected only among the Ka'apor. See Chapter 6 for a description of the four study plots of high forest.

[1] Number of individuals inventoried (number of occurrences).

[2] Number of sampling units (s.u.) on which the species occurs, out of a total number of 160 sampling units.

[3] Aggregate basal area (at breast height) of the species.

[4] Relative density (number of sampling units of occurrence divided by total number of occurrences [1515] inventoried).

[5] Relative frequency (number of individuals of species divided by total number of individuals [1981] inventoried).

[6] Relative dominance (aggregate basal area of species divided by total basal area of all species [117.4089m^2] inventoried).

[7] Ecological importance value (I.V.), which is the sum of relative density, relative frequency, and relative dominance. A monoculture would have an I.V. of 300.

Family and Species	Ind. (#)	S.U. (#)	Basal Area (m²)	Density (%)	Freq. (%)	Domi. (%)	I.V.
Pseudoxandra polyphleba	3	3	0.0471	0.15	0.20	0.04	0.39
Xylopia nitida	6	5	0.2464	0.30	0.33	0.21	0.84
APOCYNACEAE							
Ambelania acida	7	7	0.0902	0.35	0.46	0.08	0.89
Aspidosperma ateanum	1	1	0.0214	0.05	0.07	0.02	0.14
Aspidosperma excelsum	3	3	0.1078	0.15	0.20	0.09	0.44
Himatanthus articulata	1	1	0.0330	0.05	0.07	0.03	0.15
Lacmellea aculeata	8	8	0.1131	0.40	0.53	0.10	1.03
Landolphia boliviensis	1	1	0.0129	0.05	0.07	0.01	0.13
Parahancornia amapa	3	3	0.2775	0.15	0.20	0.24	0.59
ARALIACEAE							
Schefflera morototoni	2	2	0.0400	0.10	0.13	0.03	0.26
ARECACEAE							
Astrocaryum sp. 2	1	1	0.0177	0.05	0.07	0.02	0.14
Euterpe oleracea	12	8	1.4414	0.61	0.53	1.23	2.37
Oenocarpus distichus	21	19	0.4864	1.06	1.25	0.41	2.72
BIGNONIACEAE							
Clytostoma sp. 1	2	2	0.0670	0.10	0.13	0.06	0.29
Jacaranda copaia	3	3	0.0619	0.15	0.20	0.05	0.40
Lundia erionema	1	1	0.0079	0.05	0.07	0.01	0.13
aff. *Anemopaegma* sp. 1	1	1	0.0161	0.05	0.07	0.01	0.13
BOMBACACEAE							
Bombax globosum	1	1	0.0284	0.05	0.07	0.02	0.14
BORAGINACEAE							
Cordia bicolor	1	1	0.0499	0.05	0.07	0.04	0.16
Cordia lomatoloba	1	1	0.0113	0.05	0.07	0.01	0.13
Cordia scabrifolia	4	3	0.0674	0.20	0.20	0.06	0.46
Cordia sellowiana	3	3	0.1676	0.15	0.20	0.14	0.49
BURSERACEAE							
Protium altsoni	5	5	0.2145	0.25	0.33	0.18	0.76
Protium aracouchini	6	6	0.1678	0.30	0.40	0.14	0.84
Protium decandrum	52	41	2.0480	2.62	2.71	1.74	7.07
Protium giganteum	18	13	0.4167	0.91	0.86	0.35	2.12
Protium krukoffii	1	1	0.0293	0.05	0.07	0.02	0.14
Protium nodulosum	1	1	0.0327	0.05	0.07	0.03	0.15
Protium pallidum	55	41	1.5189	2.78	2.71	1.29	6.78
Protium paraense	9	8	0.2836	0.45	0.53	0.24	1.22
Protium polybotryum	17	15	0.4358	0.86	0.99	0.37	2.22
Protium robustum	1	1	0.1001	0.05	0.07	0.09	0.21
Protium sagotianum	7	7	0.4086	0.35	0.46	0.35	1.16

Family and Species	Ind. (#)	S.U. (#)	Basal Area (m²)	Density (%)	Freq. (%)	Domi. (%)	I.V.
Protium spruceanum	1	1	0.0079	0.05	0.07	0.01	0.13
Protium subserratum	1	1	0.0249	0.05	0.07	0.02	0.14
Protium tenuifolium	5	5	0.1500	0.25	0.33	0.13	0.71
Protium trifoliolatum	59	42	2.3609	2.98	2.77	2.01	7.76
Tetragastris altissima	76	50	5.2410	3.84	3.30	4.46	11.60
Tetragastris panamensis	2	2	0.0562	0.10	0.13	0.05	0.28
Trattinickia burserifolia	2	2	0.2300	0.10	0.13	0.20	0.43
Trattinickia rhoifolia	2	2	0.0292	0.10	0.13	0.02	0.25
Trattinickia sp. 1	2	2	0.0971	0.10	0.13	0.08	0.31
CAESALPINIACEAE							
Bauhinia guianensis	3	3	0.0338	0.15	0.20	0.03	0.38
Bauhinia rubiginosa	1	1	0.0079	0.05	0.07	0.01	0.13
Bauhinia sp. 1	11	10	0.1851	0.56	0.66	0.16	1.38
Bauhinia sp. 2	4	4	0.0893	0.20	0.20	0.08	0.48
Copaifera duckei	1	1	0.3217	0.05	0.07	0.27	0.39
Dialium guianense	2	2	0.0304	0.10	0.13	0.03	0.26
Dimorphandra sp. 1	1	1	0.0165	0.05	0.07	0.01	0.13
Hymenaea courbaril	1	1	0.1018	0.05	0.07	0.09	0.21
Macrolobium campestre	10	6	0.7818	0.50	0.40	0.67	1.57
Sclerolobium albiflorum	1	1	0.0616	0.05	0.07	0.05	0.17
Sclerolobium guianensis	2	2	0.4233	0.10	0.13	0.36	0.59
Sclerolobium paraense	1	1	0.0683	0.05	0.07	0.06	0.18
Sclerolobium sp. 1	1	1	0.0464	0.05	0.07	0.04	0.16
Sclerolobium sp. 2	1	1	0.4418	0.05	0.07	0.38	0.50
Tachigali macrostachya	8	6	0.2295	0.40	0.40	0.20	1.00
Tachigali myrmecophila	10	10	1.1099	0.50	0.66	0.95	2.11
Tachigali paniculata	1	1	0.0125	0.05	0.07	0.01	0.13
CARYOCARACEAE							
Caryocar glabrum	4	4	1.1054	0.20	0.26	0.94	1.40
CECROPIACEAE							
Cecropia obtusa	16	11	1.2470	0.81	0.73	1.06	2.60
Cecropia sciadophylla	2	2	0.1221	0.10	0.13	0.10	0.33
Cecropia sp. 1	1	1	0.0079	0.05	0.07	0.01	0.13
Cecropia sp. 2	3	2	0.5184	0.15	0.13	0.44	0.72
Pourouma bicolor	2	2	0.2198	0.10	0.13	0.19	0.42
Pourouma guianensis	23	17	1.1688	1.16	1.12	1.00	3.28
Pourouma minor	36	21	1.5610	1.82	1.39	1.33	4.54
CELASTRACEAE							
Cheiloclinium cognatum	3	3	0.0336	0.15	0.20	0.03	0.38
Cheiloclinium hippocrateoides	1	1	0.0095	0.05	0.07	0.01	0.13
Hippocratea sp. 1	2	1	0.0157	0.10	0.07	0.01	0.18
Hippocratea sp. 2	1	1	0.0079	0.05	0.07	0.01	0.13

Family and Species	Ind. (#)	S.U. (#)	Basal Area (m²)	Density (%)	Freq. (%)	Domi. (%)	I.V.
Pristimera tenuifolia	1	1	0.0087	0.05	0.07	0.01	0.13
Salacia insignis	2	2	0.0594	0.10	0.13	0.05	0.28
Salacia multiflora	1	1	0.0163	0.05	0.07	0.01	0.13
Tontelea fluminensis	1	1	0.0143	0.05	0.07	0.01	0.13
Tontelea sp. 1	1	1	0.0154	0.05	0.07	0.01	0.13
aff. *Cheiloclinium* or							
Salacia sp. 1	1	1	0.0104	0.05	0.07	0.01	0.13
Celastraceae sp. 1	3	3	0.0269	0.15	0.20	0.02	0.37
Celastraceae sp. 2	2	2	0.0177	0.10	0.13	0.02	0.25
CHRYSOBALANACEAE							
Couepia guianensis	40	30	1.2565	2.02	1.98	1.07	5.07
Exellodendron barbatum	5	5	0.0749	0.25	0.33	0.06	0.64
Hirtella glabrata	2	2	0.0579	0.10	0.13	0.05	0.28
Hirtella racemosa	1	1	0.0113	0.05	0.07	0.01	0.13
Licania apetala	1	1	0.0227	0.05	0.07	0.02	0.14
Licania canescens	14	14	0.4835	0.71	0.92	0.41	2.04
Licania glabriflora	1	1	0.0095	0.05	0.07	0.01	0.13
Licania heteromorpha	15	12	0.3006	0.76	0.79	0.26	1.81
Licania sp. 1	1	1	0.1662	0.05	0.07	0.14	0.26
Licania kunthiana	2	2	0.1808	0.10	0.13	0.15	0.38
Licania latifolia	1	1	0.0726	0.05	0.07	0.06	0.18
Licania macrophylla	2	2	0.5453	0.10	0.13	0.46	0.69
Licania membranacea	6	6	0.0955	0.30	0.40	0.08	0.78
Parinari rodolphii	3	3	0.1054	0.15	0.20	0.09	0.44
CLUSIACEAE							
Caraipa grandiflora	2	2	0.0649	0.10	0.13	0.06	0.29
Clusia sp. 1	1	1	0.0133	0.05	0.07	0.01	0.13
Platonia insignis	1	1	0.0133	0.05	0.07	0.01	0.13
Rheedia acuminata	1	1	0.0095	0.05	0.07	0.01	0.13
Rheedia brasiliensis	1	1	0.0194	0.05	0.07	0.02	0.14
Symphonia globulifera	10	10	0.6302	0.50	0,66	0.54	1.70
Tovomita brasiliensis	2	2	0.0272	0.10	0.13	0.02	0.25
Tovomita brevistaminea	1	1	0.0201	0.05	0.07	0.02	0.14
COMBRETACEAE							
Buchenavia capitata	1	1	0.0079	0.05	0.07	0.01	0.13
Buchenavia tetraphylla	2	2	0.5220	0.10	0.13	0.44	0.67
Combretum sp. 1	1	1	0.0113	0.05	0.07	0.01	0.13
Combretum laxum	2	2	0.0357	0.10	0.13	0.03	0.26
Terminalia amazonia	1	1	0.0222	0.05	0.07	0.02	0.14
Terminalia lucida	1	1	0.0177	0.05	0.07	0.02	0.14
CONVOLVULACEAE							
Calyobolus sp. 1	1	1	0.0113	0.05	0.07	0.01	0.13
Maripa glabra	1	1	0.0095	0.05	0.07	0.01	0.13

Family and Species	Ind. (#)	S.U. (#)	Basal Area (m^2)	Density (%)	Freq. (%)	Domi. (%)	I.V.
DILLENIACEAE							
Davilla aspera	1	1	0.0154	0.05	0.07	0.01	0.13
Davilla kunthiana	1	1	0.0363	0.05	0.07	0.03	0.15
Doliocarpus brevipedicellatus	1	1	0.0087	0.05	0.07	0.01	0.13
Doliocarpus dentatus	1	1	0.0109	0.05	0.07	0.01	0.13
Doliocarpus sp. 1	3	3	0.0379	0.15	0.20	0.03	0.38
Tetracera sp. 1	1	1	0.0083	0.05	0.07	0.01	0.13
Tetracera volubilis	3	3	0.0277	0.15	0.20	0.20	0.37
Tetracera willdenowiana	1	1	0.0087	0.05	0.07	0.01	0.13
EBENACEAE							
Diospyros duckei	5	5	0.1497	0.25	0.33	0.13	0.71
ELAEOCARPACEAE							
Sloanea grandiflora	1	1	0.5675	0.05	0.07	0.48	0.60
Sloanea guianensis	5	5	0.2528	0.25	0.33	0.22	0.80
EUPHORBIACEAE							
Alchornea brevistyla	1	1	0.4778	0.05	0.07	0.41	0.53
Amanoa guianensis	2	2	0.1324	0.10	0.13	0.11	0.34
Aparisthmium cordatum	18	12	0.2584	0.91	0.79	0.22	1.92
Conceveiba guianensis	3	3	0.0492	0.15	0.20	0.04	0.39
Croton matourensis	5	5	0.3786	0.25	0.33	0.32	0.90
Croton sp. 1	1	1	0.0129	0.05	0.07	0.01	0.13
Dodecastigma integrifolium	26	22	0.3991	1.31	1.45	0.34	3.10
Drypetes variabilis	4	4	0.0528	0.20	0.26	0.04	0.50
Hevea brasiliensis	1	1	0.0227	0.05	0.07	0.02	0.14
Hevea guianensis	1	1	0.1320	0.05	0.07	0.11	0.23
Hieronima sp. 1	1	1	0.1307	0.05	0.07	0.11	0.23
Mabea caudata	32	27	0.7745	1.62	1.78	0.66	4.06
Maprounea guianensis	1	1	0.0235	0.05	0.07	0.02	0.14
Margaritaria nobilis	1	1	0.0201	0.05	0.07	0.02	0.14
Sagotia racemosa	134	69	1.5984	6.76	4.55	1.36	12.67
FABACEAE							
Coumarouma micrantha	1	1	0.3848	0.05	0.07	0.33	0.45
Derris amazonica	2	2	0.0174	0.10	0.13	0.01	0.24
Diplotropis sp. 1	1	1	0.0935	0.05	0.07	0.08	0.20
Dipteryx odorata	3	3	1.2402	0.15	0.13	0.37	0.65
Poecilanthe effusa	6	5	0.1274	0.30	0.33	0.11	0.74
Pterocarpus rohrii	1	1	0.0079	0.05	0.07	0.01	0.13
Taralea opossitifolia	20	12	3.1835	1.01	0.79	2.71	4.51
Vaitarea guianensis	2	2	0.0406	0.10	0.13	0.03	0.26
FLACOURTIACEAE							
Laetia procera	3	3	0.3098	0.15	0.20	0.26	0.61

Family and Species	Ind. (#)	S.U. (#)	Basal Area (m²)	Density (%)	Freq. (%)	Domi. (%)	I.V.
GOUPIACEAE							
Goupia glabra	8	6	1.3485	0.40	0.40	1.15	1.95
HUMIRIACEAE							
Sacoglottis guianensis	4	4	0.2570	0.20	0.26	0.22	0.68
ICACINACEAE							
Emmottum acuminatum	1	1	0.1452	0.05	0.07	0.12	0.24
Humirianthera duckei	1	1	0.0079	0.05	0.07	0.01	0.13
[INDETERMINATE 4]							
Indeterminate genus 3	1	1	0.0143	0.05	0.07	0.01	0.13
[INDETERMINATE 5]							
Indeterminate genus 4	1	1	0.0133	0.05	0.07	0.01	0.13
[INDETERMINATE 6]							
Indeterminate genus 32	1	1	0.0260	0.05	0.07	0.02	0.14
[INDETERMINATE 7]							
Indeterminate genus 10	2	2	0.1155	0.10	0.13	0.10	0.33
[INDETERMINATE 8]							
Indeterminate genus 11	1	1	0.1419	0.05	0.07	0.12	0.24
[INDETERMINATE 10]							
Indeterminate genus 12	1	1	0.0594	0.05	0.07	0.05	0.17
LAURACEAE							
Acroclidium sp. 1	1	1	0.0755	0.05	0.07	0.06	0.18
Aniba burchellii	1	1	0.0121	0.05	0.07	0.01	0.13
Endlicheria sp. 1	1	1	0.0079	0.05	0.07	0.01	0.13
Endlicheria verticellata	1	1	0.0085	0.05	0.07	0.01	0.13
Licaria brasiliensis	1	1	0.0227	0.05	0.07	0.02	0.14
Licaria sp. 1	1	1	0.0163	0.05	0.07	0.01	0.13
Nectandra cuspidata	3	2	0.1071	0.15	0.13	0.09	0.37
Nectandra sp. 1	1	1	0.0483	0.05	0.07	0.04	0.16
Nectandra sp. 2	1	1	0.0377	0.05	0.07	0.03	0.15
Nectandra sp. 3	1	1	0.0330	0.05	0.07	0.03	0.15
Ocotea abbreviata	2	1	0.0531	0.10	0.07	0.05	0.22
Ocotea amazonica	1	1	0.0123	0.05	0.07	0.01	0.13
Ocotea canaliculata	1	1	0.0552	0.05	0.07	0.05	0.17
Ocotea caudata	4	4	0.2592	0.20	0.26	0.22	0.68
Ocotea laxiflora	1	1	0.0082	0.05	0.07	0.01	0.13
Ocotea rubra	4	4	0.8553	0.20	0.26	0.73	1.19
Ocotea sp. 1	2	2	0.0190	0.10	0.13	0.02	0.25

Family and Species	Ind. (#)	S.U. (#)	Basal Area (m²)	Density (%)	Freq. (%)	Domi. (%)	I.V.
Ocotea sp. 2	1	1	0.0087	0.05	0.07	0.01	0.13
Lauraceae sp. 5	1	1	0.0278	0.05	0.07	0.02	0.14
Lauraceae sp. 6	1	1	0.0079	0.05	0.07	0.01	0.13
LECYTHIDACEAE							
Couratari guianensis	8	8	2.1660	0.40	0.53	1.84	2.77
Eschweilera apiculata	1	1	0.0121	0.05	0.07	0.01	0.13
Eschweilera coriacea	259	124	19.4645	13.07	8.18	16.58	37.83
Eschweilera obversa	1	1	0.0235	0.05	0.07	0.02	0.14
Eschweilera piresii	1	1	0.5153	0.05	0.07	0.44	0.56
Gustavia augusta	4	4	0.0637	0.20	0.26	0.05	0.51
Lecythis chartacea	8	6	1.6997	0.40	0.40	1.45	2.25
Lecythis idatimon	128	67	4.2865	6.46	4.42	3.65	14.53
Lecythis pisonis	2	2	1.1948	0.10	013	1.02	1.25
MALPIGHIACEAE							
Byrsonima aerugo	1	1	0.0415	0.05	0.07	0.04	0.16
Byrsonima amazonica	3	2	0.2966	0.15	0.13	0.25	0.53
Byrsonima sp. 1	2	2	0.0453	0.10	0.13	0.04	0.27
Byrsonima laevigata	1	1	0.1257	0.05	0.07	0.11	0.23
Heteropterys multiflora	1	1	0.0095	0.05	0.07	0.01	0.13
MARCGRAVICEAE							
Souroubea guianensis	1	1	0.0235	0.05	0.07	0.02	0.14
MELASTOMATACEAE							
Miconia guianensis	1	1	0.0087	0.05	0.07	0.01	0.13
MELIACEAE							
Carapa guianensis	35	28	2.4359	1.77	1.85	2.07	5.69
Guarea kunthiana	10	9	0.3585	0.50	0.59	0.31	1.40
Trichilia sp. 1	1	1	0.0095	0.05	0.07	0.01	0.13
Trichilia micrantha	12	12	0.2086	0.61	0.79	0.18	1.58
Trichilia quadrijuga	6	6	0.1651	0.30	0.40	0.14	0.84
Trichilia schomburgkii	5	4	0.0718	0.25	0.26	0.06	0.57
Trichilia surinamensis	1	1	0.0133	0.05	0.07	0.01	0.13
MIMOSACEAE							
Acacia sp. 1	1	1	0.0222	0.05	0.07	0.02	0.14
Acacia multipinnata	2	2	0.0263	0.10	0.13	0.02	0.25
Inga alba	2	2	0.1176	0.10	0.13	0.10	0.33
Inga auristellae	2	2	0.0349	0.10	0.13	0.03	0.26
Inga breviata	1	1	0.0573	0.05	0.07	0.05	0.17
Inga capitata	12	10	0.2607	0.61	0.66	0.22	1.49
Inga fagifolia	4	4	0.1017	0.20	0.26	0.09	0.55
Inga gracilifolia	11	10	0.2233	0.56	0.66	0.19	1.41
Inga heterophylla	2	2	0.0425	0.10	0.13	0.04	0.27
Inga marginata	2	2	0.0396	0.10	0.13	0.03	0.26

Family and Species	Ind. (#)	S.U. (#)	Basal Area (m²)	Density (%)	Freq. (%)	Domi. (%)	I.V.
Inga rubiginosa	4	4	0.1904	0.20	0.26	0.16	0.62
Inga splendens	1	1	0.0353	0.05	0.07	0.03	0.15
Inga stipularis	3	2	0.0878	0.15	0.13	0.07	0.35
Inga thibaudiana	4	4	0.0614	0.20	0.26	0.05	0.51
Inga sp. 1	1	1	0.0232	0.05	0.07	0.02	0.14
Inga sp. 2	1	1	0.0123	0.05	0.07	0.01	0.13
Inga sp. 3	2	2	0.1103	0.10	0.13	0.09	0.32
Inga sp. 4	1	1	0.0113	0.05	0.07	0.01	0.13
Newtonia psilostachya	7	7	1.9444	0.35	0.46	1.66	2.47
Parkia gigantocarpa	1	1	0.0123	0.05	0.07	0.01	0.13
Parkia multijuga	3	3	0.0400	0.15	0.20	0.03	0.38
Parkia paraensis	1	1	0.2359	0.05	0.07	0.20	0.32
Parkia pendula	4	4	2.0815	0.20	0.26	1.77	2.23
Parkia sp. 1	1	1	0.1735	0.05	0.07	0.15	0.27
Pithecellobium cauliflorum	1	1	0.0290	0.05	0.07	0.02	0.14
Pithecellobium cochleatum	1	1	0.0707	0.05	0.07	0.06	0.18
Pithecellobium gongrijpii	5	4	1.0316	0.25	0.26	0.88	1.39
Pithecellobium jupumba	3	3	0.2382	0.15	0.20	0.20	0.55
Pithecellobium pedicellare	1	1	0.0243	0.05	0.07	0.02	0.14
Pithecellobium racemosum	6	6	0.4387	0.30	0.40	0.37	1.07
Stryphnodendron polystachyum	8	8	0.3533	0.40	0.53	0.30	1.23
MORACEAE							
Bagassa guianensis	2	2	2.8455	0.10	0.13	2.42	2.65
Brosimum acutifolium	2	2	0.0585	0.10	0.13	0.05	0.28
Brosimum guianense	3	3	0.1039	0.15	0.20	0.09	0.44
Brosimum paclescum	1	1	0.0246	0.05	0.07	0.02	0.14
Brosimum sp. 1	1	1	0.0272	0.05	0.07	0.02	0.14
Brosimum sp. 2	1	1	0.7854	0.05	0.07	0.67	0.79
Ficus sp. 4	1	1	0.0452	0.05	0.07	0.04	0.16
Helicostylis tomentosa	7	5	0.3405	0.35	0.33	0.29	0.97
Maquira guianensis	4	4	0.0789	0.20	0.26	0.07	0.53
Moraceae sp. 28	1	1	0.0079	0.05	0.07	0.01	0.13
MYRISTICACEAE							
Iryanthera juruensis	4	4	0.1444	0.20	0.26	0.12	0.58
Virola carinata	2	2	0.0824	0.10	0.13	0.07	0.30
Virola melinonii	3	3	0.1121	0.15	0.20	0.10	0.45
Virola michelii	5	5	0.2589	0.25	0.33	0.22	0.80
MYRTACEAE							
Calyptranthes or *Marlierea*	5	5	0.0611	0.25	0.33	0.05	0.63
Eugenia brachypoda	1	1	0.0087	0.05	0.07	0.01	0.13
Eugenia coffeifolia	1	1	0.0133	0.05	0.07	0.01	0.13
Eugenia florida	1	1	0.0113	0.05	0.07	0.01	0.13
Eugenia sp. 1	1	1	0.0143	0.05	0.07	0.01	0.13
Eugenia tapacumensis	1	1	0.0079	0.05	0.07	0.01	0.13

Family and Species	Ind. (#)	S.U. (#)	Basal Area (m²)	Density (%)	Freq. (%)	Domi. (%)	I.V.
NYCTAGINACEAE							
Neea glameruliflora	1	1	0.0123	0.05	0.07	0.01	0.13
Neea ovalifolia	1	1	0.0177	0.05	0.07	0.02	0.14
Neea sp. 1	4	4	0.0786	0.20	0.26	0.07	0.53
Neea sp. 2	1	1	0.0154	0.05	0.07	0.01	0.13
Pisonia sp. 1	1	1	0.0191	0.05	0.07	0.02	0.14
OCHNACEAE							
Ouratea cf. *coccinea*	1	1	0.0100	0.05	0.07	0.01	0.13
Ouratea cf. *discophora*	1	1	0.0079	0.05	0.07	0.01	0.13
OLACACEAE							
Dulacia candida	1	1	0.0095	0.05	0.07	0.01	0.13
Heisteria barbata	1	1	0.0152	0.05	0.07	0.01	0.13
Minquartia guianensis	4	4	0.1705	0.20	0.26	0.15	0.61
OPILIACEAE							
Agonandra brasiliensis	2	2	0.1173	0.10	0.13	0.10	0.33
POLYGALACEAE							
Moutabea guianensis	2	2	0.4713	0.10	0.13	0.40	0.63
Moutabea sp. 3	1	1	0.0201	0.05	0.07	0.02	0.14
POLYGONACEAE							
Coccoloba cf. *excelsa*	1	1	0.0083	0.05	0.07	0.01	0.13
Coccoloba cf. *scandens*	1	1	0.0092	0.05	0.07	0.01	0.13
QUIINACEAE							
Lacunaria jenmani	1	1	0.0150	0.05	0.07	0.01	0.13
Lacunaria sp. 1	2	2	0.0534	0.10	0.13	0.05	0.28
Lacunaria sp. 2	1	1	0.0093	0.05	0.07	0.01	0.13
RUBIACEAE							
Chimarrhis turbinata	2	2	2.5525	0.10	0.13	2.17	2.40
Faramea sp. 1	1	1	0.0150	0.05	0.07	0.01	0.13
Malanea sp. 1	1	1	0.1555	0.05	0.07	0.13	0.25
Remijia sp. 1	1	1	0.0143	0.05	0.07	0.01	0.13
RUTACEAE							
Ticorea sp. 1	1	1	0.0145	0.05	0.07	0.01	0.13
Zanthoxylum juniperina	2	2	0.1201	0.10	0.13	0.10	0.33
Zanthoxylum sp. 1	2	2	0.1318	0.10	0.13	0.11	0.34
Zanthoxylum tenuifolium	1	1	0.0794	0.05	0.07	0.07	0.19
SAPINDACEAE							
Cupania scrobiculata	3	3	0.0359	0.15	0.20	0.03	0.38
Talisia micrantha	1	1	0.0201	0.05	0.07	0.02	0.14
Talisia retusa	2	2	0.0658	0.10	0.13	0.06	0.29

Family and Species	Ind. (#)	S.U. (#)	Basal Area (m^2)	Density (%)	Freq. (%)	Domi. (%)	I.V.
SAPOTACEAE							
Chrysophyllum prieurii	3	3	0.8478	0.15	0.20	0.72	1.07
aff. *Chrysophyllum* sp. 1	3	3	0.5534	0.15	0.20	0.47	0.82
aff. *Chrysophyllum* sp. 2	1	1	0.1320	0.05	0.07	0.11	0.23
Manilkara huberi	2	2	0.4597	0.10	0.13	0.39	0.62
Manilkara paraensis	1	1	0.0712	0.05	0.07	0.06	0.18
Micropholis acutangula	6	4	0.3892	0.30	0.26	0.33	0.89
Micropholis guyanensis	8	7	0.4411	0.40	0.46	0.38	1.24
Micropholis melinoniana	1	1	2.0106	0.05	0.07	1.71	1.83
Micropholis venulosa	5	5	0.1867	0.25	0.33	0.16	0.74
Pouteria caimito	10	9	0.8769	0.50	0.59	0.75	1.84
Pouteria cuspidata ssp. *dura*	1	1	0.0100	0.05	0.07	0.01	0.13
Pouteria egregia	2	2	0.1781	0.10	0.13	0.15	0.38
Pouteria elegans	7	4	0.1813	0.35	0.26	0.15	0.76
Pouteria engleri	1	1	0.0330	0.05	0.07	0.03	0.15
Pouteria eugeniifolia	4	3	0.5781	0.20	0.20	0.49	0.89
Pouteria filipes	1	1	0.0191	0.05	0.07	0.02	0.14
Pouteria gongrijpii	5	5	0.0530	0.25	0.33	0.05	0.63
Pouteria guianensis	1	1	0.0665	0.05	0.07	0.06	0.18
Pouteria hispida	3	3	0.2095	0.15	0.20	0.18	0.53
Pouteria jariensis	15	11	0.3183	0.76	0.73	0.27	1.76
Pouteria macrocarpa	7	7	0.0825	0.35	0.46	0.07	0.88
Pouteria macrophylla	1	1	0.0165	0.05	0.07	0.01	0.13
Pouteria oppositifolia	7	7	1.2799	0.35	0.46	1.09	1.90
Pouteria reticulata	1	1	0.0243	0.05	0.07	0.02	0.14
Pouteria sagotiana	2	2	0.0343	0.10	0.13	0.03	0.26
Pouteria venosa	2	2	0.0197	0.10	0.13	0.02	0.25
Pouteria virescens	11	10	0.7913	0.56	0.66	0.67	1.89
Pouteria sp. 1	3	3	0.5339	0.15	0.20	0.46	0.81
Pouteria sp. 2	1	1	0.0119	0.05	0.07	0.01	0.13
Pouteria sp. 3	1	1	0.0177	0.05	0.07	0.02	0.14
Pouteria sp. 4	1	1	0.0117	0.05	0.07	0.01	0.13
aff. *Pouteria* sp. 1	1	1	0.1314	0.05	0.07	0.11	0.23
Sapotaceae sp. 4	1	1	0.0320	0.05	0.07	0.03	0.15
SIMARUBACEAE							
Simaba cedron	6	6	0.0671	0.30	0.40	0.06	0.76
Simaruba amara	11	10	1.3679	0.56	0.66	1.17	2.39
STERCULIACEAE							
Sterculia pruriens	16	14	1.0794	0.81	0.92	0.92	2.65
Theobroma grandiflorum	6	6	0.2225	0.30	0.40	0.19	0.89
Theobroma speciosum	3	3	0.0305	0.15	0.20	0.03	0.38
TILIACEAE							
Apeiba aspera	3	3	0.1265	0.15	0.20	0.11	0.46
Apeiba echinata	13	13	0.7828	0.66	0.86	0.67	2.19
Apeiba tibourbou	1	1	0.0079	0.05	0.07	0.01	0.13
Apeiba sp. 1	2	2	0.3845	0.10	0.13	0.33	0.56

Family and Species	Ind. (#)	S.U. (#)	Basal Area (m²)	Density (%)	Freq. (%)	Domi. (%)	I.V.
ULMACEAE							
Ampelocera edentula	3	3	0.5027	0.15	0.20	0.43	0.78
VIOLACEAE							
Paypayrola grandiflora	11	8	0.1259	0.56	0.53	0.11	1.20
Rinorea flavescens	3	3	0.0326	0.15	0.20	0.03	0.38
VOCHYSIACEAE							
Erisma uncinatum	7	7	1.4669	0.35	0.46	1.25	2.06
Total	1981	1515	117.4089	99.71	100.55	100.06	300.32

Ecological Importance Values of High Forest Families

Family	Indiv.[1] (#)	S.U.[2] (#)	Basal Area[3] (m^2)	Density[4] (%)	Freq.[5] (%)	Domi.[6] (%)	I.V.[7]
Lecythidaceae	9	414	29.4261	2.64	20.78	25.06	48.48
Burseraceae	20	322	13.8532	5.87	16.24	11.78	33.89
Sapotaceae	33	119	10.6046	9.68	5.97	9.04	24.69
Mimosaceae	31	98	8.1315	9.09	4.92	6.90	20.91
Euphorbiaceae	15	231	4.4631	4.40	11.65	3.79	19.84
Chrysobalanaceae	14	94	3.3827	4.11	4.74	2.87	11.72
Caesalpiniaceae	17	59	3.9616	4.99	2.96	3.40	11.35
Cecropiaceae	7	83	4.8450	2.05	4.19	4.13	10.37
Fabaceae	9	39	5.5298	2.64	1.96	4.71	9.31
Lauraceae	20	30	1.6758	5.87	1.50	1.44	8.81
Meliaceae	7	70	3.2627	2.05	3.53	2.78	8.36
Moraceae	10	23	4.3176	2.93	1.15	3.68	7.76
Anacardiaceae	9	24	1.9903	2.64	1.20	1.70	5.54
Annonaceae	10	38	0.7792	2.93	1.91	0.66	5.50
Celastraceae	12	19	0.2358	3.52	0.95	0.20	4.67
Arecaceae	3	34	1.9455	0.88	1.72	1.66	4.26
Clusiaceae	8	19	0.7979	2.35	0.95	0.69	3.99
Apocynaceae	7	24	0.6559	2.05	1.20	0.57	3.82
Rubiaceae	4	5	2.7373	1.17	0.25	2.32	3.74
Sterculiaceae	3	25	1.3324	0.88	1.26	1.14	3.28
Tiliaceae	4	19	1.3017	1.17	0.96	1.12	3.25
Dilleniaceae	8	12	0.1539	2.35	0.60	0.13	3.08

NOTE: Data are drawn from the aggregate of four one-hectare study plots of high forest. Author inventoried all trees and lianas on those study plots that were ≥ 10 cm dbh. See Chapter 6 for a description of the four study plots.

[1]Number of individual species in the family inventoried.

[2]Number of individual trees and lianas ≥ 10 cm dbh in the family inventoried.

[3]Aggregate basal area (at breast height) of the family.

[4]Relative density (number of individual species divided by the total number of species [341] inventoried).

[5]Relative frequency (number of individuals of family divided by total number of individuals [1981] inventoried).

[6]Relative dominance (aggregate basal area of family divided by total basal area of all families [117.4089 m^2] inventoried).

[7]Ecological importance value (I.V.), which is the sum of relative density, relative frequency, and relative dominance. A monoculture would have an I.V. of 300.

Family	Indiv. (#)	S.U. (#)	Basal Area (m²)	Density (%)	Freq. (%)	Domi. (%)	I.V.
Combretaceae	6	8	0.6168	1.76	0.40	0.53	2.69
Simarubaceae	2	17	1.4350	0.59	0.86	1.23	2.68
Myristicaceae	4	14	0.5978	1.17	0.70	0.51	2.38
Myrtaceae	6	10	0.1166	1.76	0.50	0.10	2.36
Malpighiaceae	5	8	0.5186	1.47	0.40	0.45	2.32
Nyctaginaceae	5	8	0.1431	1.47	0.40	0.13	2.00
Vochysiaceae	1	7	1.4669	0.29	0.35	1.25	1.89
Boraginaceae	4	9	0.2962	1.17	0.45	0.25	1.87
Goupiaceae	1	8	1.3485	0.29	0.40	1.15	1.84
Rutaceae	4	6	0.3458	1.17	0.30	0.29	1.76
Bignoniaceae	4	7	0.1529	1.17	0.35	0.13	1.65
Elaeocarpaceae	2	6	0.8203	0.59	0.30	0.70	1.59
Violaceae	2	14	0.1585	0.59	0.71	0.14	1.44
Caryocaraceae	1	4	1.1054	0.29	0.20	0.94	1.43
Olacaceae	3	6	0.1952	0.88	0.30	0.17	1.35
Sapindaceae	3	6	0.1218	0.88	0.30	0.11	1.29
Polygalaceae	2	3	0.4914	0.59	0.15	0.42	1.16
Quiinaceae	3	4	0.0777	0.88	0.20	0.07	1.15
Ulmaceae	1	3	0.5027	0.29	0.15	0.43	0.87
Ochnaceae	2	2	0.0179	0.59	0.10	0.02	0.71
Convolvulaceae	2	2	0.0208	0.59	0.10	0.02	0.71
Humiriaceae	1	4	0.2570	0.29	0.20	0.22	0.71
Polygonaceae	2	2	0.0175	0.59	0.10	0.02	0.71
Flacourtiaceae	1	3	0.3098	0.29	0.15	0.26	0.70
Ebenaceae	1	5	0.1497	0.29	0.25	0.13	0.67
Opiliaceae	1	2	0.1173	0.29	0.10	0.10	0.49
[Indeterminate 7]	1	2	0.1155	0.29	0.10	0.10	0.49
[Indeterminate 8]	1	1	0.1419	0.29	0.05	0.12	0.46
Araliaceae	1	2	0.0400	0.29	0.10	0.03	0.42
[Indeterminate 10]	1	1	0.0594	0.29	0.05	0.05	0.39
Marcgraviaceae	1	1	0.0235	0.29	0.05	0.02	0.36
Bombacaceae	1	1	0.0284	0.29	0.05	0.02	0.36
[Indeterminate 6]	1	1	0.0260	0.29	0.05	0.02	0.36
Melastomataceae	1	1	0.0087	0.29	0.05	0.01	0.35
[Indeterminate 5]	1	1	0.0133	0.29	0.05	0.01	0.35
[Indeterminate 4]	1	1	0.0143	0.29	0.05	0.01	0.35
Total	341	1981	117.4089	99.98	99.71	100.06	299.75

APPENDIX 5

Alphabetical List of Plant Species Known to the Ka'apor

Family and Species	Collection Numbers[1]	Family and Species	Collection Numbers[1]
ACANTHACEAE		*Duguetia echinophora* R.E.	
Justicia pectoralis Jacq.	0976	Fries	2664
Justicia spectabilis T.		*Duguetia marcgraviana*	
Anders. ex C.B. Clarke	4417	Mart.	3867
AMARANTHACEAE		*Duguetia riparia* Huber	2987
Amaranthus oleraceus L.	4422	*Duguetia surinamensis* R.E.	2677 4111 4205
Amaranthus spinosus L.	3088	Fries	4210 4211 4385
		Duguetia yeshidah Sandw.	2665
ANACARDIACEAE		*Duguetia* sp. 1	0545 0546
Anacardium giganteum		*Duguetia* sp. 2	0980
Hancock ex Engler	2282 2979 4426	*Ephedranthus pisocarpus*	
Anacardium occidentale L.	0856 0981	R.E. Fries	3895
Anacardium parvifolium	0225 0295	*Fusaea longifolia* (Aublet)	0155 1029
Ducke	0301 0407	Saff.	2885 3012
Astronium cf. *obliquum*		*Guatteria elongata* Benth.	2778
Griseb.	2209	*Guatteria scandens* Ducke	0793
Astronium lecointei Ducke	4041	*Guatteria* sp. 1	0231 0245 0316
Mangifera indica L.	1009 1059	*Rollinia exsucca* (DC. ex	0625 0704 0912
Spondias mombin L.	1006 2212	Dun.) A.DC.	2157 2169 2192
Tapirira guianensis Aublet	1069 2257 4259	*Rollinia mucosa* Baill.	3087
Tapirira peckoltiana Engler	0363 2910	*Unonopsis rufescens* (Baill.)	
Thyrsodium sp. 1	0620	R.E. Fries	3957
Thyrsodium spruceanum		*Xylopia barbata* Mart.	4136
Bentham	0437	*Xylopia nitida* Dunal	0344 2806
		Annonaceae sp. 1	1046
ANNONACEAE			
Anaxagorea dolichocarpa		APIACEAE	
Spr. ex Sandw.	0937 0985 4429	*Eryngium foetidum* L.	0941
Annona montana Macfad.	3090 0997 2205		
Annona paludosa Aublet	0066	APOCYNACEAE	
Annona sericea Dunal	4169	*Allamanda cathartica* L.	2090

[1]Collection numbers are on the series Balée. Herbarium vouchers are located at the New York Botanical Garden or the Museu Paraense Emílio Goeldi. "None" signifies no collection made of a species in the habitat.

Family and Species	Collection Numbers	Family and Species	Collection Numbers
aff. *Forsteronia*	0685	ARECACEAE	
Ambelania acida Aubl.	0989 2917 2964	*Acrocomia* sp. 1	NONE
Ambelania grandiflora Hubar	4423	*Astrocaryum gynacanthum* Mart.	2251
Aspidosperma cylindrocarpon Muell. Arg.	4115	*Astrocaryum javari* Mart.	NONE
Aspidosperma desmanthum Bth. Cx M. Arg.	4347	*Astrocaryum murumuru* Mart.	3972
Aspidosperma verruculosum M. Arg.	4304	*Astrocaryum vulgare* Mart.	2173
Himatanthus articulatus (Vahl.) Woodson	1063 2770 4097	*Astrocaryum* sp. 1	1032
		Bactris gasipaes Mart.	NONE
Himatanthus sucuuba (Spruce ex Muell. Arg.) Woods.	0832 2161 4097	*Bactris humilis* (Wallace) Burret	0679 3546
		Bactris major Mart.	0824
Lacmellea aculeata (Ducke) Monachino	0026 2011 2706	*Bactris maraja* Mart.	0825
		Bactris setosa Barb. Rodr.	1014
Masechites bicornulata (Rusby)Woods.	3935	*Bactris tomentosa* Mart.	3544
		Cocos nucifera L.	NONE
Parahancornia amapa Ducke	2895	*Desmoncus macroacanthos* Mart.	2301
Parahancornia fasciculata (Poir.) Benoist	0332	*Desmoncus polycanthus* Mart.	3930
Rauvolfia paraensis Ducke	3022	*Euterpe oleracea* Mart.	0378 0381 0382 2901
Tabernaemontana angulata Mart. ex Muell. Arg.	1015 3555	*Geonoma baculifera* Kunth.	0922 2675
		Geonoma leptospadix Trail	2965
ARACEAE		*Markleya dahlgreniana* Bondar	4404
Caladium picturatum C. Koch.	0909	*Mauritia flexuosa* Mart.	NONE
Colocasia esculenta (L.) Schott.	3083	*Maximiliana maripa* (Corr.Serr.) Drude	4071
Dracontium sp. 1	0914	*Oenocarpus distichus* Mart.	0530 2166 2824
Heteropsis longispathacea Engl.	0859	*Orbignya phalerata* Mart.	3543
Heteropsis sp. 1	0755	*Socratea exorrhiza* (Mart.) H. Wendl.	3541
Monstera cf. *pertusa* (L.) Vriesia	2299	*Syagrus* cf. *inajai* (Spruce) Becc.	2303
Monstera subpinata (Schott) Engl.	3912	*Syagrus* sp. 1	0921
Philodendron grandiflorum (Jacq.) Schott.	3961	ASCLEPIADACEAE	
Philodendron venustum Bunt.	0682	*Asclepias curassavica* L.	0897
		Schubertia grandiflora Mart.	0586 0588 1024
Philodendron sp. 1	0878	ASPIDIACEAE	
Xanthosoma sp. 1	3554	*Ctenitis pretensa* (Afz.) Ching	0783
ARALIACEAE		ASTERACEAE	
Schefflera morototoni (Aubl.) M.S.F.	0528 2784 4339	*Clibadium sylvestre* (Aubl.) Baill.	0795 0796 3045

Family and Species	Collection Numbers
Conyza banariensis (L.) Crang.	0952
Eupatorium macrophyllum L.	1039
Porophyllum ellipticum Cass.	3052
Pterocaulon vergatum (L.) DC.	3049
BIGNONIACEAE	
aff. *Memora bracteosa* (DC.) Bur. & K. Schum.	0741
Amphilophium paniculata (L.) H.B.K. var. *moller* (Cham. & Schl.) Gentry	1013
Anemopaegma setilobum A. Gentry	2184
Arrabidaea cf. *florida* DC.	0746
Arrabidaea sp. 1	0596
Crescentia cujete L.	0814
Cuspidaria sp. 1	4313
Jacaranda copaia (Aubl.) D. Don. ssp. *copaia*	0338 2284
Jacaranda copaia ssp. *spectabilis* (Mart. ex DC.) A. Gentry	1060 2874
Jacaranda duckei Vattimo	2178
Jacaranda heterophylla Bur. & K. Schum.	3032
Jacaranda paraensis (Huber) Vattim.	4127
Mansoa angustiden (DC.) Bur. & K. Schum.	3037
Memora allamandiflora Bur. ex Bur. et K. Schum.	3922
Memora bracteosa (DC.) Bur. & K. Schum.	0551
Memora flavida (DC.) Bur. & K. Schum.	3536
Rhabidalia sp. 1	0748 0750 0757
Styzophyllum riparium (H.B.K.) Sandw.	0727 0887 2225 4378
Tabebuia impegitinosa Standley.	NONE
Tabebuia serratifolia (Vahl.) Nichols	2189 4182
Tabebuia sp. 1	4349
Bignoniaceae sp. 1	0618

Family and Species	Collection Numbers
BIXACEAE	
Bixa orellana L.	0801 3101 3102
BOMBACACEAE	
Ceiba pentandra Gaertn.	0992 2262
Pachira aquatica Aubl.	3942
BORAGINACEAE	
Cordia exaltata Lam.	1034 2671
Cordia lomatoloba Johnst.	2886
Cordia multispicata Cham.	3048
Cordia polycephala (Lam.) Johnston	1018
Cordia scabrida Mart.	2249 4043 4047 4125
Cordia scabrifolia C. DC.	0190 2655 2290 2655 4332
Cordia sellowiana Chamb.	2294 2818 2875 2988
Cordia trachyphylla Mart.	4267
Cordia sp. 2	0802
BROMELIACEAE	
Aechmea brevicellis L.B. Smith	3068
Aechmea bromeliifolia Baker	3021
Ananas comosus L.	1020 1021
Ananas nanas (L.B. Smith) L.B.Smith	2680
Bromelia goeldiana L.B. Smith	3974
Guzmania lingulata (L.) Mez.	3948
Neoglaziovia variegata (Arr. Cam.) Mez.	0953
Tillandsia usneoides (L.) L.	3097
Bromeliaceae sp. 1	0959
BURSERACEAE	
Protium altsoni Sandw.	0184
Protium aracouchini (Aubl.)March.	0969 1002
Protium decandrum (Aublet) Marchand	0122 0313
Protium giganteum Engl. var. *giganteum*	1055 1061
Protium heptaphyllum ssp. *heptaphyllum* (Aublet) Marchand	2234 2266 2268 4066 4070 4095 4109 4282

Family and Species	Collection Numbers	Family and Species	Collection Numbers
Protium pallidum Cuatrec.	0005 0265	*Macrolobium acaciaefolium* Bth.	4406
Protium polybotryum (Turcz.) Engl.	0097 0098 0099	*Sclerolobium albiflorum* R. Ben.	2924
Protium sagotianum Engl.	0033 0855	*Sclerolobium guianense* Aubl.	2692 2984
Protium spruceanum (Benth.) Engl.	0843	*Sclerolobium paraense* Huber	0171
Protium tenuifolium (Engl.) Engl.	0157	*Sclerolobium* sp. 16	2976
Protium trifoliolatum Engl.	0020 0021 0413	*Sclerolobium* sp. 20	0096
Tetragastris altissima (Aublet) Swart.	0007 0008 0507 0598 1027 2689 2695 2709 2720 2722 2727 2728	*Senna chrysocarpa* (Desv.) Irwin & Barneby	2223 4410
Tetragastris panamensis (Engl.) O.K.	0451 4343 4382 4384 4398	*Senna pendula* (Lamk.) Irwin & Barneby	2656
Trattinickia burserifolia Mart.	0876 2896	*Senna sylvestris* (Vell.) Irwin & Barneby	3057
Trattinickia rhoifolia Willd.	2175 2204	*Swartzia brachyrachis* var. Harms	2674 3537
Trattinickia sp. 1	2855	*Swartzia brachyrachis* var. snethlageae (Ducke) Ducke	1054
CACTACEAE		*Tachigali* sp. 1	0548 0549 0569
Rhipsalis baccifera (J. Miller) Stearn.	0840	*Tachigali macrostachya* Huber	2700 2714 2734 2741 2811 2983 3017
Rhipsalis mijosurus K. Schum.	4077	*Tachigali myrmecophila* Ducke	0039 0201 0209 0348 2176 4251 4328
CAESALPINIACEAE		*Tachigali paniculata* Aublet	0188
Adenanthera pavonina L.	0797 1041	*Zollernia paraensis* Huber	2672 3921
Apuleia leiocarpa (Vogel) Spr. var. *molaris* (Spruce ex Benth.) Koeppen	2182 2275	CANNABIDACEAE	
Bauhinia rubiginosa Bong.	2750	*Cannabis sativa* L.	NONE
Bauhinia viridiflorens Ducke	3044	CANNACEAE	
Bauhinia sp. 1	0069	*Canna indica* L.	0798 0799
Bauhinia sp. 2	2887	CAPPARIDACEAE	
Cassia fastuosa Willd.	2150	*Capparis sola* Macbr.	0998
Copaifera duckei Dwyer	3883	CARICACEAE	
Copaifera reticulata Ducke	4172	*Carica papaya* L.	0918
Copaifera sp. 1	2186	*Jacaratia spinosa* (Aubl.) A.DC.	2158 2191 2685
Crudia parivoa DC.	3034	CARYOCARACEAE	
Dialium guianense (Aubl.) Sandw.	0635 0646 0733 0774 0881 1040 2179 2217 4122 4227	*Caryocar glabrum* (Aubl.) Pers.	0113 0114 2753
Hymenaea courbaril L.	1000 2273 2285 4295	*Caryocar villosum* Aubl.	1038
Hymenaea parvifolia Huber	0880 2793 4060 4146 4164 4177	CECROPIACEAE	
Hymenaea reticulata Ducke	4158	*Cecropia concolor* Willd.	2202

Family and Species	Collection Numbers	Family and Species	Collection Numbers
Cecropia obtusa Trecue	2783	*Hirtella* cf. *paraensis* Prance	0587
Cecropia palmata Willd.	0809 2670	*Hirtella eriandra* Benth.	2658
Cecropia purpurescens C.C. Berg	4184	*Hirtella racemosa* Lam. var. *racemosa*	1058
Cecropia sciadophylla Mart.	0408 2801	*Hirtella racemosa* var. hexandra (Willd. ex R.G.S.) Prance	3031
Coussapoa sp. 1	4363		
Pourouma guianensis Aubl. ssp. *guianensis*	0711 0968 2688 2690 2759 2903	*Licania apetala* (E. Mey.) Fritsch	0948
Pourouma minor Benoist	2693 2710 2747 2803 2904	*Licania canescens* R. Ben.	0044 0045 0046 0117 0125 0218 0426 0508 0975
Pourouma mollis Trec. ssp. *mollis*	0958	*Licania glabriflora* Prance	2943
CELASTRACEAE		*Licania heteromorpha* Benth. var. *heteromorpha*	0162
Cheiloclinium cognatum (Miers) A.C.	0269 0833 2729	*Licania* sp. 1	0652
aff. *Cheiloclinium* or *Salacia* sp. 1	0474	*Licania kunthiana* Hook. F.	0174 0581 4055 4113 4264 4316
Hippocratea ovata Lam.	3941	*Licania latifolia* Benth.	0421
Maytenus sp. 10	3869	*Licania macrophylla* Benth.	3943
Maytenus sp. nov.	4036 4057 4058 4114 4121 4129 4160 4165 4174 4186 4213	*Licania membranacea* Sagot. ex Laness.	0228
Peritassa huanuclara (Loes.) A.C.	0574	*Licania octandra* (Hoffmgg. ex R. & S.) Kuntze	0560 2288
Pristimera tenuifolia (Mart.) A.C.	0327	*Parinari excelsa* Sabine	0645
Salacia insignis A.C. Smith	0247	CLUSIACEAE	
Salacia multiflora (Lam.) DC.	0075	*Caraipa grandiflora* Mart.	3550
CHENOPODIACEAE		*Clusia* sp. 1	0964
Chenopodium ambrosioides L.	0821 0822 3084	*Platonia insignis* Mart.	3026 4056 4278
		Rheedia acuminata Pl. & Tr.	2957 2997 4337
CHRYSOBALANACEAE		*Rheedia brasiliensis* (Mart.) Pl. et Tr.	0233 1062
Couepia guianensis Aubl. ssp. *divaricata* (Hub.) Prance	2239 3056 3551 4225	*Symphonia globulifera* L.	0102 0194 0381 0577 1023 2756 2815 2860 2973
Couepia guianensis Aubl. ssp. *guianensis*	0017 0308 0664 2737 2829 2836 2882 2902 2925 2928 2929 2982	*Tovomita brasiliensis* (Mart.) Walp.	0938 4419
Couepia guianensis ssp. *glandulosa* (Mig.) Prance	3018	*Tovomita brevistaminea* Engl.	4418
Exellodendron barbatum Prance	0325 0326	*Vismia guianensis* (Aubl.) Choisy	0807 0854 3067
		COCHLOSPERMACEAE	
		Cochlospermum orinocense (H.B.K.) Steud.	4409
		COMBRETACEAE	
Hirtella bicornis Mart. ex Zucc.	4245 4247 4249 4252 4262	*Buchenavia* cf. *tetraphylla* (Aubl.) Howard	0185

Family and Species	Collection Numbers	Family and Species	Collection Numbers
Combretum laxum Jacq.	1064 2717 2773	CYPERACEAE	
Terminalia amazonia (J. Gmel.) Exell.	0173 3043 4298 4321 4359	*Cyperus corymbosus* L.F.	0961 3082
Terminalia dichotoma G. Meyer	4067	*Diplasia karataefolia* Rich.	0844
Terminalia lucida Hoffmgg.	2805	*Scleria secans* (L.) Urb.	2840 3940
COMMELIACEAE		*Scleria* sp. 1	0977
Dichorisandra affinis Mart. ex Roem. & Schultes	3915	*Scleria* sp. 2	0982
		Cyperaceae sp. 1	0640 0644
CONNARACEAE		DENNSTAEDTIACEAE	
Connarus favosus Planch.	3920	*Pteridium aquilinium* (L.) Kuhn.	3070
Connaraceae sp. 1	0547		
Connaraceae sp. 2	0612	DILLENIACEAE	
CONVOLVULACEAE		*Davilla kunthii* St. Hil.	0485
Calyobolus sp. 1	2754	*Davilla nitida* (Vahl.) Kubitzki	1050
Ipomea setofera Oir.	3065	*Doliocarpus* sp. 1	0321
Ipomoea aff. *squamosa* Choisy	2224	*Tetracera volubilis* L. ssp. *volubilis*	0030 0366 2785
Ipomoea batatas Lam.	0893 0804 0805	*Tetracera willdenowiana* Schlechtd. ssp. *willdenowiana*	2858
Ipomoea sp. 1	3540		
Ipomoea phyllomega (Velloso) House	0879	DIOSCOREACEAE	
Maripa sp. 1	1070	*Dioscorea trifida* L.	0927
CRASSULACEAE		*Dioscorea* sp. 1	0624
Bryophyllum sp. 1	3072	*Dioscorea* sp. 2	0899
CUCURBITACEAE		EBENACEAE	
Cayaponia sp. 1	0829	*Diospyros artanthifolia* Mart.	3938
Citrullus lanatus (Thunb.) Matsumi & Nakai	0803 3081	*Diospyros duckei* Sandw.	0038 0429 2302
Cucumis anguria L.	0895	*Diospyros melinoni* (Hiern.) A.C. Smith	4035 4236 4254 4354
Cucurbita moschata Duch.	0851 3046		
Gurania cissoides (Benth.) Cogn.	2669	ELAEOCARPACEAE	
Gurania eriantha (Poepp. & Endl.) Cogn.	0830 0831	*Sloanea grandiflora* Smith	0065 2857
Lagenaria siceraria (Molina) Standl.	0906 0817	*Sloanea grandis* Ducke	2276 4283 4362
Luffa cylindrica (L.) M.J. Roem.	0965	*Sloanea guianensis* (Aubl.) Benth.	0078 0131
Posadea sphaerocarpa Cogn.	0913	*Sloanea porphyrocarpa* Ducke	4045 4180
Cucurbitaceae sp. 1	3933	ERYTHROXYLACEAE	
CYCLANTHACEAE		*Erythroxylum* cf. *leptoneurum* O.E. Schulz	3060
Evodianthus funifer (Poit.) Lindm.	0960	*Erythroxylum citrifolium* A. St. H.L.	3023

Family and Species	Collection Numbers	Family and Species	Collection Numbers
EUPHORBIACEAE		*Arachis hypogaea* L.	0894
Aparisthmium cordatum Baill.	2696 2711 2730 2772 2820 2822 3003 2974	*Bowdichia nitida* Spruce ex Benth.	2236
		Coumarouna micrantha Ducke	2961
Conceveiba guianensis Aubl.	0280	*Dalbergia monetaria* L.F.	3929
Croton cajucara Benth.	2263 3036 4395	*Derris amazonica* Killip.	2789 3921
Croton cuneatus Klotesch.	3947	*Derris* sp. 1	1003
Croton matourensis Aubl.	0149 0403 0861 4405 4434	*Derris utilis* (A.C. Smith) Ducke	0857
Dodecastigma integrifolium (Lanj.) Lanj. & Sandw.	0003 0004 0160 0161 1067	*Desmodium adescendens* (Sw.) DC.	3080
Drypetes variabilis Vitt.	3009 4175 4187 4234	*Dioclea reflexa* Hook. F.	0987 3548
Euphorbia sp. 1	0818 0942	*Diplotropis purpurea* (Rich.) Amsh.	0111
Hevea guianensis Aubl.	0944	*Dipteryx odorata* L.	0972
Jatropha curcas L.	0819 0820	*Machaerium ferox* (Mart. ex Benth.) Ducke	3958
Mabea angustifolia Spruce ex Benth.	3096	*Ormosia coccinea* (Aubl.) Jacks.	0661 0662 0733
Mabea caudata Pax & Hoffm.	0041 0103 2244	*Platypodium elegans* Vog.	2168 2170 2279 3891
Mabea sp. 1	2152	*Poecilanthe effusa* (Hub.) Ducke	0178 0971 2786 2795 2809 2900
Mabea pohliana M. Arg.	3946	*Taralea oppositifolia* Aubl.	1036
Magaritaria nobilis L.F.	2165 2775	*Tephrosia sinapou* (Buchot) A.Chev.	0813
Manihot brachyloba Muell. Arg.	3063	*Vigna adnantha* (G.F. Meyer) Marechal	3099
Manihot esculenta Crantz	0848 0862 0863 0864 0865 0866 0867 0869 0870 0874 0890 0891 0892 0901 0933	**FLACOURTIACEAE**	
		Banara guianensis Aubl.	3034 3902
Manihot leptophylla Pax in Engler	0811 0812 0813 2304	*Casearia arborea* (L.C. Richard) Urban	2149 2151 2155
Manihot quinquepartita Huber ex Rog et Appan	2221	*Casearia decundra* Jacq.	4329
Phyllanthus martii Muell. Arg.	3927	*Casearia javitensis* H.B.K.	2659 3059 4411
Phyllanthus niruri L.	3085	*Laetia procera* (Poepp. et Endl.) Eichl.	0382 2942 2949 3000 3029
Phyllanthus urinaria L.	4424	*Lindackeria latifolia* Benth.	4120
Ricinus comunis L.	0930	*Lindackeria* sp. 1	0993 2214 2269
Sagotia racemosa Baill.	0031 2663 2891	*Neoptychocarpus apodanthus* Kuhlm.	0790
Sapium caspceolatum Huber	4084	*Ryania speciosa* Vahl.	0994
Sapium lanceolatum Huber	2227 2246 2260	Flacourtiaceae sp. 1	0602
FABACEAE		**GESNERIACEAE**	
Andira retusa (Lam.) H.B.K.	3893	*Drymonia coccinea* (Aubl.) Weihl.	2873
Andira sp. 1	4299		

Family and Species	Collection Numbers	Family and Species	Collection Numbers
GOUPIACEAE		*Ocotea canaliculata* (L.C.	
Goupia glabra Aubl.	1065	Rich.) Mez.	0512
		Ocotea caudata Mez.	2958
HAEMODORACEAE		*Ocotea costata* Mez.	2975
Xiphidium caeruleum Aubl.	2967	*Ocotea fasciculata* (Ness)	
		Mez	4344
HELICONIACEAE		*Ocotea glomerata* (Nees)	
Heliconia acuminata L.C.		Mez.	2662
Rich.	0615 0705 0845	*Ocotea guianensis* Aubl.	0946 0988
Heliconia bihai L.	2218 3981	*Ocotea opifera* Mart.	3539 3916
Heliconia chartacea Lane ex		*Ocotea rubra* Mez.	2946 2971
Barr.	0792	*Ocotea silvae* Vattimo-Gil	4118 4243 4239
		sp. nov.	4284 4342 4353
HUMIRIACEAE			4358
Sacoglottis amazonica Mart.	4038 4131 4145	*Ocotea* sp. 3	0330 0331
	4266 4351	Lauraceae sp. 1	0782
Sacoglottis guianensis		Lauraceae sp. 2	2992
Benth.	0368 0369 4351	Lauraceae sp. 3	2865
Sacoglottis sp. 1	0189		
		LECYTHIDACEAE	
ICACINACEAE		*Bertholletia excelsa*	
Dendrobangia boliviana		Humboldt & Bonpland	NONE
Rusby	4416	*Couratari guianensis* Aublet	0186 0463 1042
Humirianthera duckei Huber	2970	*Couratari oblongifolia* Ducke	
		& Knuth.	2242
[INDETERMINATE 1]		*Eschweilera amazonica* R.	
Indeterminate genus 2	NONE	Kunth.	4308
[INDETERMINATE 2]		*Eschweilera apiculata*	1072 4273
Indeterminate genus 3	0694	(Miers) A.C. Smith	4355 4430
[INDETERMINATE 3]		*Eschweilera coriacea* (A.P.	0010 0011 0016
Indeterminate genus 4	0638	Candolle) Mart. ex Berg	0018 0020 0023
[INDETERMINATE 5]			0024 0025 0027
Indeterminate genus 5	NONE		0032 0040 0042
[INDETERMINATE 22]			0050 0056 0057
Indeterminate genus 89	NONE		0059 0061 0067
			0068 0070 0072
LABIATEAE			0080 0083 0085
Ocimum micranthum Willd.	0931		0088 0089 0090
Ocimum sp. 1	0905		0107 0110 0126
			0132 0136 0142
LAURACEAE			0147 0150 0394
Aniba citrifolia (Nees) Mez.	3954		0410 0424 0441
Endlicheria verticellata Mez.	0404		0605 0680 0732
Licaria brasiliensis (Nees)			1030 1057 1638
Kost	3015		2760 2780 2817
Licaria debilis (Mez) Kost.	4310		2827 2833 2851
Licaria sp. 1	0454		2859 2863 2870
Ocotea amazonica (Meiss.)			2920 2927 2939
Mez.	2715		2972 3008

Family and Species	Collection Numbers	Family and Species	Collection Numbers
Eschweilera micrantha (Berg) Miers	4083	*Sida cordifolia* L.	3079
		Sida santaremensis Mont.	3054
Eschweilera obversa (Berg) Miers	0920	*Urena lobata* L.	0974 3054
Eschweilera ovata (Cambess) Miers	3031 4151 4166 4206 4271 4355 4356 4367 4401	**MARANTACEAE**	
		Calathea sp. 1	2297
		Ctenanthe sp. 1	4023
Eschweilera pedicellata (Richard) Mori	4074 4142 4323	*Ischnosiphon arouma* (Aubl.) Koern.	0691 0825
Gustavia augusta L.	0192 0193 0420 2162 2164 2167 0722 2978 2995 4325	*Ischnosiphon graciles* (Rudge) Koern.	0689
		Ischnosiphon obliquus (Rudge) Koern.	0707 0718 0786
Lecythis chartacea Berg	0019 0306	*Ischnosiphon petrolatus* (Rudge) L. And.	0791
Lecythis idatimon Aublet	0037	*Ischnosiphon* sp. 1	0607
Lecythis lurida (Miers) Mori	2269	*Ischnosiphon* sp. 2	2668
Lecythis pisonis Cambess.	1052 2248 2713	*Maranta protracta* Mig.	3071
LILIACEAE		Marantaceae sp. 1	0541 0542 0543
Allium cepa L.	NONE	Marantaceae sp. 2	0665
Allium sativum L.	NONE	Marantaceae sp. 3	0695
		Marantaceae sp. 4	0692 0696
LIMNOCHARITACEAE		**MARCGRAVIACEAE**	
Limnocharis sp. 1	0924	*Norantea guianensis* Aubl.	0838 0839
LOMARIOPSIDACEAE		**MELASTOMATACEAE**	
Lomariopsis japurensis (Mart.) J. Smith	0592 0756 0761 0763 0778	*Aciotis purpurescens* (Aubl.) Triana	1033
LORANTHACEAE		*Bellucia grossularioides* (L.) Triana	1073
Struthanthus marginatus Blume	2794	*Henrietta spruceana* Cogn.	3932
		Miconia ceramicarpa (DC.) Cogn.	0773
MALPIGHIACEAE		*Miconia* cf. *kappleri* Naud.	0105
Banisteriopsis pubipetala (Adr. Juss.) Cuatr.	4181	*Miconia nervosa* (Smith) Triana	1035
Byrsonima laevigata (Poir.) DC.	0337 1026	*Miconia serialis* DC.	4044 4124 4132 4156 4215 4307
Byrsonima stipulacea Juss.	2258	*Miconia serrulata* (DC.) Naud.	3955
Byrsonima sp. 5	0963		
Malpighia sp.1	3092	*Miconia surinamensis* Gleason	2148 2160
Malpighia sp. 2	3091	*Miconia* sp. 1	1068
Mascagnia sp. 1	2203	*Mouriri guianensis* Aubl.	4039 4051 4101 4258 4265 4040 4126
Schwannia sp. 1	3923		
Stigmaphyllon hypoleucum Miz.	0885 0886		
Tetrapterys styloptera A. Juss	4313	*Mouriri huberi* Cogn.	4222
MALVACEAE		*Mouriri sagotiana* Triana	4144 4346 4365
Gossypium barbadense L.	0850	*Mouriri trunciflora* Ducke	4072

Family and Species	Collection Numbers	Family and Species	Collection Numbers
Myriaspora egensis DC.	3028		0483 1056 2240
MELIACEAE			3002 3016
Carapa guianensis Aubl.	0279 0374 2725	*Inga cinammonea* Spruce ex Benth.	0377
	2740 2765 2821	*Inga fagifolia* (L.) Willd.	2782 2807 2812
Cedrela fissilis Vell.	0966 2250	*Inga falcistipula* Ducke	3889 4159
Guarea guidonia (L.)	2194 2195	*Inga gracilifolia* Ducke	0238 0274 2267
Sleumer	2200 2201		4134 4275 4357
Guarea kunthiana A. Juss.	2726 2796 2862	*Inga heterophylla* Willd.	2287 2936
	2899 2918 2923		2956 4428
	1066	*Inga marginata* Willd.	0443
Guarea macrophylla Vahl. ssp. *pachycarpa* (C.DC.)		*Inga miriantha* Poepp. et Endl.	2193
Penn.	1066	*Inga nobilis* Willd.	2226
Guarea sp. 1	2296	*Inga paraensis* Ducke	4402
Trichilia micrantha Benth.	2702 2732 2752	*Inga rubiginosa* (Rich.) DC.	2241 2277 2774
	2945 2963 2989		2787 2977
	2999	*Inga splendens* Willd.	0505
Trichilia quadrijuga Kunth.	0141 2211 2230	*Inga thibaudiana* DC.	0828 0975 3005
ssp. *quadrijuga*	2932 4090 4094	*Inga* sp. 5	0978
	4274 4292 4313	*Inga* sp. 6	0582
Trichilia verrocosa C. DC.	4116 4212 4226	*Inga* sp. 7	0672
	4296 4319 4320	*Inga* sp. 8	0582
	4322 4315 4326	*Inga* sp. 9	2198 2199
	4341 4391	*Inga* sp. 10	2295
		Newtonia psilostachya (DC.)	
MENDONCIACEAE		Brenan	0195 2701 4170
Mendoncia hoffmannseggiana		*Newtonia suaveolens*	
Nees	0860 3098	(Miguel) Brenan	3038
Mendoncia sp. 1	0642	*Parkia multijuga* Benth.	2245 2781 2831
MENISPERMACEAE		*Parkia paraensis* Ducke	0379
Abuta grandifolia (Mar.)		*Parkia pendula* Mig.	3547
Sandw.	0728	*Parkia ulei* (Harms) Kuhlm.	3553 4327
Caryomene foveolata		*Pithecellobium cauliflorum*	
Barneby & Krukoff	0962	Mart.	2825 3944
Cissampelos sp. 1	0641	*Pithecellobium cochleatum*	
MIMOSACEAE		(Willd.) Mart.	2834 3053 4330
Acacia multipinnata Ducke	0871 2738 2937	*Pithecellobium comunis*	
Acacia paniculata Willd.	0570	Benth.	2701 3001
Enterolobium sp. nov.	4415	*Pithecellobium foliolosum*	
Hymenolobium excelsum		Benth	3073
Ducke	3557	*Pithecellobium jupumba*	
Inga alba Willd.	0731 0986 2281	(Willd.) Urb.	2153
	2660 2661	*Pithecellobium niopoides*	
Inga auristellae Harms	2172 2213 2270	Sprucec ex Benth.	2163 2177
	2272 2291 2683	*Pithecellobium pedicellare* (DC.) Benth.	0091
Inga brevialata Ducke	2739	*Pithecellobium*	
Inga capitata Desv.	0296 0297 0376	*Pithecellobium racemosum*	2699 2868

Family and Species	Collection Numbers	Family and Species	Collection Numbers
Ducke	4080 4277	MYRSINACEAE	
Stryphnodendron guianensis (Aubl.) Benth.	2255	*Ardisia guianensis* (Aubl.) Mez	4421
Stryphnodendron polystachyum (Miq.) Kleinh.	0191 2292 2691 2769 2893 2938	MYRTACEAE	
		aff. *Myrcia*	0846 0847
MONIMIACEAE		*Calyptranthes* or *Marlierea*	2820 2835 2962 3010
Siparuna amazonica (Mart.) A.DC.	0945 2307	*Campomanesia grandiflora* (Aubl.) Sagot.	4050 4065 4106 4255 4248
Siparuna guianensis Aubl.	0834 2667	cf. *Calyptranthes* sp. 1	0999
MORACEAE		*Eugenia egensis* DC.	3882
Bagassa guianensis Aubl.	0523 2298	*Eugenia eurcheila* Berg.	4052 4285
Brosimum acutifolium ssp. *interjectum* C.C. Berg	3006	*Eugenia flavescens* DC.	2280
		Eugenia muricata DC.	2208 2271 2684
Brosimum guianense (Aubl.) Huber	0135 2921	*Eugenia omissa* McVaugh	3025 4064 4117 4204
Brosimum paclesum (S. Moore) C.C. Berg	0488 0725	*Eugenia patrissi* Vahl.	2004 4288
Brosimum rubescens Taubert	0957	*Eugenia schomburkii* Benth.	4059
		Eugenia tapecumensis Berg	0043
Clarisia racemosa Ruiz & Pavon	2305	*Eugenia* sp. 5	1071
Ficus sp. 1	0890 0888	*Eugenia* sp. 6	1047
Ficus sp. 2	2808	*Eugenia* sp. 7	2293
Ficus sp. 3	3878	*Eugenia* sp. 8	0611
Helicostylis tomentosa (P. & E.) Rusby	0333 0580 1092 2735 2792	*Myrcia paivae* Berg.	2839 4333
		Myrcia splendens (Sw.) DC.	4217
Maquira guianensis Aubl.	0608 2277 2761 2816 2861 2955	*Myrcia* sp. 1	3936
		Myrcia sp. 4	0884
		Myrcia sp. 5	0883
MUSACEAE		*Myrciaria* cf. *dubia* McVaugh	3950
Musa sp. 1	NONE	*Myrciaria flexuosa*	4061
Musa sp. 2	NONE	*Myrciaria pyriifolia* Desv. ex Hamilton	3542
Phenakospermum guianensis Peterson	3350	*Myrciaria spruceana* Berg	4069 4268
		Myrciaria tenella (DC.) Berg	0947
MYCOTA		*Psidium guajava* L.	0903
Indeterminate mycotica sp. 1	3552	*Syzygium malaccencis* (L.) Merril & Pery	3086
aff. *Rhizomorpha* sp. 1	3919		
MYRISTICACEAE		NYCTAGINACEAE	
Iryanthera juruensis Warb.	0355 0356 0371	*Neea floribunda* Poepp. & Endl.	2888
Virola carinata (Benth.) Warb.	2976 2922	*Neea glameruliflora* Heimut	4137 4157 4201
Virola michelii Heckel	0255 2881 2897 2931	*Neea oppositifolia* R. & P.	4240
		Neea ovalifolia Spruce ex J.A. Schmidt	4305
Virola sp. 1	0677 1051	*Neea* sp. 1	2947 4193 4179

Family and Species	Collection Numbers	Family and Species	Collection Numbers
	4228 4230 4376	*Piper piscatorum* Trel. &	
	4393 4293 4340	Jun.	0630
	4325 4364	POACEAE	
Neea sp. 2	2940 4082 4168	aff. *Olyra*	0690
	4173 4221 4229	*Brachiaria humidicula*	
	4232 4237 4279	(Rendle) Schweickerdt	0932
	4272 4242 4335	*Coix lacryma-jobi* L.	0928
Pisonia sp. 2	2187 2188 2274	*Cymbopogon citratus* (DC.)	
OCHNACEAE		Stapf	0955
Ouratea salicifolia (St. Hil.		*Digitaria insularis* (L.) Mez	
et Tul.) Engl.	4253	ex Ekman	3035
Ouratea sp. 2	0995	*Guadua glomerata* Munro	3463
OLACACEAE		*Gynerium sagittatum*	
Dulacia guianensis (Engl.)		(Aubl.) Beauvois	0917 1074
O. Kuntze	4049	*Hyparrhenia rufa* (Nees)	
Minquartia guianensis Aubl.	1028 2156 2941	Stapf	0967
OPILIACEAE		*Ichnanthus* sp. 1	0622 0689 0737
Agonandra brasiliensis	0395 2159	*Imperata brasiliensis* Trin.	0784
Benth. & Hook.	2183 4291	*Lasiacis ligulata*	
ORCHIDACEAE		Hitchelchase	2220
Trigonidium acuminatum		*Olyra caudata* Trin.	3064
Batem ex Lindl.	0882	*Olyra latifolia* L.	0789 0808
OXALIDACEAE		*Oryza sativa* L.	0915
Averrhoa carambola L.	3089	*Paspalum conjugatum* Berg.	3051
PASSIFLORACEAE		*Piresia goeldi* Swallen	0681 0689 0737
Passiflora aranjoi Sacco	2657	*Saccharum officinarum* L.	1572
Passiflora coccinea Aubl.	0785 0934 1048	*Zea mays* L.	NONE
Passiflora edulis L.	NONE	Poaceae sp. 1	3047
Passiflora sp. 1	0794	POLYGALACEAE	
Passiflora sp. 2	1049	*Moutabea guianensis* Aubl.	0120 0432
Passiflora sp. 3	0810	*Moutabea* sp. 1	2930
Passiflora sp. 4	0667	*Moutabea* sp. 2	0558 0610
PHYTOLACCACEAE		*Ruprechtia* sp. 1	4256
Petiveria alliacea L.	0841 0842	POLYGONACEAE	
Phytolacca rivinoides Kunth.		*Coccoloba acuminata* H.B.K.	0632
& Bouche	0896	*Coccoloba latifolia* Lam.	3876
Seguieria amazonica Huber	4053 4244	*Coccoloba paniculata* Lam.	2206
PIPERACEAE		*Coccoloba racemulosa*	
Piper anonnifolium Kunth.	0634	Meissn.	4037
Piper hostmannianum (Mig.)		*Coccoloba* sp. 1	0943
C.DC.	3538	*Coccoloba* sp. 2	4431
Piper ottonoides Jun.	0595 0872 2678	Polygonaceae sp. 1	0806
Piper pilirameum Jun.	0699	POLYPODIACEAE	
		Stenoclaena sp. 1	0940

Family and Species	Collection Numbers	Family and Species	Collection Numbers
PTEROBRYACEAE		*Psychotria poeppigiana* ssp.	
Orthostichorsis crinite (Sull)		*poeppigiana* Muell. Arg.	3100
Broter.	3077	*Psychotria racemosa* (Aubl.)	
QUIINACEAE		Baensh.	2222 3069
Lacunaria jenmani (Oliv.)		*Psychotria ulviformis*	
Ducke	0220	Steyerm.	3058
Lacunaria sp. 3	2880	*Psychotria* sp. 1	3024
Lacunaria sp. 4	0629	*Randia armata* (Sw.) Sw.	3690
Lacunaria sp. 5	3014	*Remijia* sp. 1	2871
RHAMNACEAE		*Sipanea veris* S. Moore	4407
Ampelozizyphus amazonicus		*Tocoyena foetida* Poepp. &	
Ducke	0939	Endl.	2916
Colubrina sp. 1	2180 2181	*Uncaria guianensis* (Aubl.)	
Colubrina glandulossa		Gmel	3423
Perkins var. *glandulosa*	0984	Rubiaceae sp. 1	0626
Gouania pyrifolia Reiss.	0826 0827 1043	Rubiaceae sp. 2	0616
RUBIACEAE		RUTACEAE	
Alibertia edulis (L.Rich.) A.		*Citrus aurantiifolia*	
Rick.	2306 4408	(Christm.) Swingle	1010
Bertiera guianensis Aublet	2235	*Citrus medica-acida* L.	3041
Borreria verticulata G.F.W.		*Citrus sinensis* (L.) Osbeck	0852
Mey	3094	*Galipea trifoliolata* Aubl.	3066 4427
Cephalia sp. 1	0589 0653	*Rauia resinosa* Nees &	
Chimarrhis turbinata DC.	0431 2826	Mart.	3076
Coffea arabica L.	NONE	*Zanthoxylum* cf. *juniperina*	
Coussarea ovalis Standl.	4178	Engl.	2867 2912
Coussarea paniculata (Vahl.)		*Zanthoxylum chiloperone*	
Standl.	3034	(Mart.) Engl.	4096 4272
Duroia sp. 1	4387	*Zanthoxylum rhoifolium*	
Faramea sessilifolia (H.B.K.)	3890 4073	Lam.	3900
DC.	4075 4079		
Genipa americana L.	0800	SAPINDACEAE	
Geophila repens (L.) I.M.		*Cardiospermum halicacabum*	
Johnston	0990	L.	4020
Guettarda divaricata		*Cupania scrobiculata* L.C.	0223 2286 2791
(H.B.K.) Standl.	2289	Rich.	2850 3876 4063
Malanea sp. 1	2837	*Matayba spruceana* Radlk.	0593
Posoqueria latifolia (Rudge)		*Paullinia* sp. 1	3945
R. & S. ssp. *gracilis*		*Paullinia* sp. 2	3924
(Rudge) Steyerm.	0837	*Pseudyma frutescens* (Aubl.)	
Psychotria colorata Poepp.	0752 0753 0754	Radlk.	1001 1017 2190
Psychotria hoffmannseggiana		*Serjania* aff. *lethalis* St. Hil.	0889
(Willd. ex R. & S.) M.		*Talisia acutifolia* Radlk.	3872
Arg.	3011	*Talisia carasina* (Benth.)	
Psychotria poeppigiana		Radlk.	0983
Muell. Arg.	0973	*Talisia micrantha* Radlk.	0101
		Talisia microphylla Witt.	4399

Family and Species	Collection Numbers	Family and Species	Collection Numbers
Talisia retusa Cowan	0506 2247 4078 4155 4345	*Pouteria macrophylla* (Lam.) Eyma	1004 2196 2682
		Pouteria penicillata Baehni	4388
SAPOTACEAE		*Pouteria reticulata* (Engl.)	
aff. *Pradosia*	0956	Eyma ssp. *reticulata*	4208
Chrysophyllum argenteum Jacquin ssp. *auratum* (Miquel) Pennington	4386	*Pouteria reticulata* (Engler) Eyma ssp. *reticulata*	0250
Chrysophyllum lucentifolium Crong. ssp. *pachycarpum* Pires & Penn.	3074	*Pouteria sagotiana* (Baill.) Eyma	3894
		Pouteria trigonosperma Eyma	2256 4233 4238 4324
Chrysophyllum pomiferum (Eyma) Penn.	0049 4383	*Pouteria* sect. *Franchetella* sp. 1	0449
Chrysophyllum sparsiflorum Kl. ex Miq.	1016 2207 2210 2278 4392	*Pouteria* sect. *Franchetella* sp. 2	0516
Manilkara bidentata (A.DC.) Chev. ssp. *surinamensis* (Miq) Penn.	3556	*Pouteria* sp. 2	4394 4042 4076 4100 4110 4112 4133 4135 4138 4147 4150 4152
Manilkara huberi (Ducke) Chev.	2926 4104	*Pouteria* sp. 5	0522
Micropholis guyanensis (A.DC.) Pierre ssp. *guyanensis* Penn.	0144 0435 0436 2996 4397	Sapotaceae sp. 1	0693
		Sapotaceae sp. 2	3535
		Sapotaceae sp. 3	0693
Micropholis melinoniana Pierre	4433	SCROPHULARIACEAE	
Micropholis venulosa (Mart. ex Eich.) Pierre	0060	*Scoparia dulcis* L.	3093
Pouteria bilocularis (Winkler) Baehni	4270 4219 4223 4289	SELAGINELLACEAE	
		Selaginella sp. 1	0858
Pouteria caimito (R. & P.) Radlk.	0170 0349 2890 4148 4360	SEMATOPHYLLACEAE	
		Taxithelium planum (Brid.) Mitt.	3077
Pouteria aff. *caimito* (R. & P.) Radlk.	2856	SIMARUBACEAE	
Pouteria durlandii (Standley) Baehni	0472	*Simaba* aff. *cavalcantei* Thomas	0687
Pouteria egregia Sandw.	0167 0458 0721	*Simaba cedron* Planch.	0262 2219 2233 4092
Pouteria engleri Eyma	0519 4364 4176 4200 4276		
Pouteria filipes Eyma	0412	*Simaba cuspidata* Spr. var. *typica*	4246 4338
Pouteria gongrijpii Eyma	0215 2243 2705	*Simaba guianensis* Aubl. ssp. *guianensis*	4209
Pouteria guianensis Aublet	4189	*Simaruba amara* Aubl.	0360
Pouteria hispida Eyma	0267 0526		
Pouteria jariensis Pires & Penn.	0071 0166 0172 0385 0387 0392 0469 0475 0494 2232 2771 3007	SOLANACEAE	
		Brunfelsia guianensis Benth.	4068
		Capsicum annum L.	0904
		Capsicum fructescens L.	0910
Pouteria macrocarpa (Martius) Dietrich	2171 2174 2197 2228	*Nicotiana tabacum* L.	0936 3095

Family and Species	Collection Numbers	Family and Species	Collection Numbers
Physalis angulata L.	3050	ULMACEAE	
Solanum crinitum Lam.	0874	*Ampelocera edentula*	
Solanum leucocarpon Dunal	2259 2666	Kuhlm.	0077 2707 2766
Solanum rugosum Dunal	0923 3042	*Celtis iguanea* (Jacq.)	
Solanum stramonifolium		Sargent	0996
Jacq.	1031 3040	*Trema micrantha* (L.) Blume	0788
Solanum subinerme Jacq.	0873	URTICACEAE	
Solanum vanheurckii Mulh.	2215	*Laportea aestuans* (L.) Chew	0815 0816
STERCULIACEAE			
Guazuma ulmifolia Lam.	4004 4403	VERBENACEAE	
Sterculia alata Ducke	4413	*Aegiphila* sp. 1	1007
Sterculia pruriens (Aubl.)	0148 0156 0199	*Lantana camara* L.	3030
K. Schum.	0211 0384 0697	VIOLACEAE	
	0703 2866 4086	*Paypayrola grandiflora* Tul.	2708 2762 2952
Theobroma grandiflorum	0477 0478		2953 3019
(Willd. ex Spreg.)	2300 2919	*Rinorea flavescens* (Aubl.)	
Schum.		Kuntze	4390
Theobroma speciosum Willd.	0911 1025 2229	*Rinorea pubiflora* (Benth.)	0735 1045
ex Spreng.	2261 2768 2889	Spr. & Sandw.	2679 2686
Theobroma subincanum		*Rinorea* sp. 1	0836
Mart.	4425 4432	VITACEAE	
THEOPHRASTACEAE		*Cissus cissyoides* L.	2216
Clavija lancifolia Desf.	0991	VOCHYSIACEAE	
TILIACEAE		*Erisma uncinatum* Warn.	2878 2907 2919
Apeiba aspera Aubl.	0491 0716		2998 3011
Apeiba echinata Gaertn.	2694 2944 2718	*Vochysia inundata* Ducke	4412
	2757 2898 2914	ZINGIBERACEAE	
	2944	*Costus arabicus* L.	3953
Apeiba echinata var.		*Costus scaber* R. & P.	3913
macropetala (Ducke)		*Curcuma* sp. 1	9823 0853
Gaertn.	2283 2980	*Renealmia alpinia* (Rottb.)	
Apeiba tibourbou Aubl.	2681	Maas	1011
Apeiba cf. *burchelli* Sprague	2253 2254	*Renealmia floribunda* K.	
Luehea duckeana Burret	4139	Seh.	0537 0787
Triumfetta rhomboidea Jacq.	1005	*Zingiber officinalis* L.	1075
Triumfetta semitriloba Jacq.	0898		

APPENDIX 6

Codes for Activity Contexts

PFS   (PRIMARY FOOD SOURCE)
PFS01   fruit/seed is edible
PFS02   tuber/rhizome/corm/root is edible
PFS03   leaves and/or shoots are edible

SFS   (SECONDARY FOOD SOURCE)
SFS01   fruit/seed is edible
SFS02   shoot is edible
SFS03   rhizome is edible
SFS04   used for ceremonial beverage
SFS05   fruit-boring larvae are edible
SFS06   latex is edible
SFS07   used for making tea or coffee
SFS08   stem contains potable water
SFS09   leaves are edible

SEA   (SEASONING)
SEA01   spice
SEA02   used for making vegetable salt

ANF   (ANIMAL FOOD)
ANF01   fruits and/or seeds are edible to three or more game species not counting birds
ANF02   fruits and/or seeds are edible to fewer than three game species not counting birds or edible only for certain birds
ANF03   livestock feed

PBC   (POST-AND-BEAM CONSTRUCTION)
PBC01   wood used for house post
PBC02   wood used for rafter, tie beam, or cross beam
PBC03   roofing thatch made from leaves
PBC04   ceiling slats made from wood
PBC05   vine for weaving roofing thatch, securing posts and beams
PBC06   house wall made from wood
PBC07   fence made from wood
PBC08   bench made from split wood

TLW   (TOOLS AND WEAPONS)
TLW01   rope made from fiber
TLW02   lashing material from the bark
TLW03   axe and hoe handle from the wood
TLW04   sandpaper from the leaves
TLW05   glue from latex for arrow shaft and tail feathers
TLW06   glue from latex for feather art

TLW07	fish poison
TLW08	fishing pole
TLW09	fish bait
TLW10	bow
TLW11	bow grip
TLW12	arrow shaft from stem
TLW13	arrow point from wood
TLW14	joint of composite arrow
TLW15	war club from wood
TLW16	shotgun shell stuffing
TLW17	machete handle
TLW18	string for arrow feathers
TLW19	bowstring

HHI	(HOUSEHOLD ITEMS)
HHI01	firedrill from wood
HHI02	fan from leaves
HHI03	bowl from fruit
HHI04	bottle gourd
HHI05	spoon, ladle made from wood
HHI06	grill made from wood
HHI07	skewer from wood
HHI08	manioc press
HHI09	sieve
HHI10	manioc grater from roots
HHI11	rake for manioc flour made from wood
HHI12	basket
HHI13	broom, broom handle
HHI14	hammock loom
HHI15	hammock thread
HHI16	mat from leaves
HHI17	feather chest from wood
HHI18	smoking material
HHI19	cigar paper from bark
HHI20	pipe stem
HHI21	sherd (pottery) temper from wood
HHI22	pottery slip from resin
HHI23	leaf for storing food or tobacco
HHI24	leaf for drinking toasted manioc flour soaked in water
HHI25	drying rack for animal skin made from wood
HHI26	torch material
HHI27	leaf for wrapping fish for cooking
HHI28	wood for making storage container
HHI29	wood for mortar
HHI30	wood for pestle
HHI31	trunk of tree used as trough or receptacle for processing manioc meal

PAC	(PERSONAL ADORNMENT, COSMETICS, HYGIENE, DYES)
PAC01	clothing
PAC02	necklace, bracelet, waist adornment, ring
PAC03	soap from bark or leaves

PAC04   deodorant
PAC05   perfume
PAC06   comb
PAC07   body paint
PAC08   dye for bowls and shaman's rattler
PAC09   dye for arrow shafts and points
PAC10   hair remover (resin)
PAC11   hair cream/scent
PAC12   leaf for cutting hair

MED   (MEDICINAL USES FOR PEOPLE AND PETS)
MED01   treats headache
MED02   treats fever
MED03   treats general cold symptoms
MED04   decongestant (resin is snorted)
MED05   treats sore throat
MED06   treats stomachache
MED07   anti-emetic
MED08   treats diarrhea
MED09   treats kidney stones
MED10   vermifuge
MED11   treats skin eruption
MED12   treats cuts
MED13   treats burns
MED14   cicatrizant
MED15   treats boils and pimples
MED16   used to excise undesirable moles
MED17   soothes itching
MED18   treats children's sores at the corners of the mouth ('boqueira')
MED19   kills botfly larvae under the skin
MED20   treats measles
MED21   soothes toothache
MED22   eating the larvae is prophylaxis against tooth decay
MED23   treats earache
MED24   applied to wounded or sore eyes
MED25   treats snakebite
MED26   contraceptive and/or abortifacient if taken orally
MED27   leaves rubbed on abdomen to sooth labor pains and expedite delivery
MED28   treats post-partum hemorrhage
MED29   tonic
MED30   eaten by the sick to gain weight
MED31   fed to sick dogs so that they will gain weight
MED32   treats dog mange
MED33   insect repellent
MED34   taken orally to stop excessive menstrual discharge
MED35   taken orally to make urine less yellow
MED36   treats rheumatism
MED37   taken orally to remove blood from urine

MAG   (MAGIC USES)
MAG01   love amulet
MAG02   hunting amulet

MAG03	prevents appearances of incubus
MAG04	apotropaic uses
MAG05	protects children from evil divinities
MAG06	helps children learn to speak
MAG07	helps children learn to walk
MAG08	helps children to sleep
MAG09	treats insanity, the result of taboo violations
MAG10	amulet for murder
MAG11	helps plants to grow
MAG12	prevents 'cold leg' (rheumatism)
MAG13	prevents excessive menstruation
MAG14	amulet for a boy to learn to make manioc press
MAG15	amulet to stimulate growth of children
MAG16	amulet to confer courage in war
MAG17	amulet to guarantee a long life
MAG18	amulet for a boy to learn to kill brocket deer
MAG19	bath in leaves prevents newborns from catching a cold
MAG20	seed on necklace prevents fever in children
MAG21	seed on necklace for a boy to learn to capture yellow-footed tortoises
RIT	(RITUAL USES)
RIT01	used in shamanism
RIT02	used at female initiation ceremony; fiber is used to tie ants together that sting the initiate
RIT03	ritual mat made from leaves used by mothers in infant-naming ceremony
RIT04	liquid in pod used by female sponsor to dye faces of attendees at infant-naming ceremony
RIT05	resin used to paint face of the dead at interment
MYT	(PLANTS KNOWN FROM FOLKLORE)
MYT01	plant figures in myth fragments generally
MYT02	eaten by mythical animal (e.g., the water snake **mayu** or the arboreal quadruped with blue bones, **kiki**)
MYT03	non-Indians were made from this plant
MYT04	the Ka'apor were made from this plant
MYT05	wood used to make killer darts by evil divinity
POI	(POISONOUS AND OTHER PLANTS AVOIDED BY ALL KA'APOR OR BY SPECIFIC GROUPS DEFINED BY AGE OR SEX TERMS)
POI01	avoided for nonspecific reasons
POI02	causes botfly infection if eaten
POI03	causes blindness if touched or eaten
POI04	causes baldness if eaten
POI05	inedible to children of various ages
POI06	causes death if used as firewood
POI07	causes insanity if touched or eaten
POI08	bark, stem, or leaves sting to the touch
POI09	causes itching or irritates eyes if used as firewood
POI10	stinks or crackles upon burning
POI11	causes itching if touched
POI12	causes death if eaten

POI13   causes headache if burned
POI14   causes undesirable weight loss if eaten
POI15   noxious to growing manioc tubers
POI16   forbidden to the touch (children only)
POI17   forbidden to the touch (females only)
POI18   believed to cause boils (**yaŋoyar**) if touched

FUE   (FUEL SOURCES)
FUE01   good firewood
FUE02   secondary firewood, used only when FUE01 unavailable
FUE03   resin/latex used for incandescence

COM   (COMMERCIAL USES)
COM01   cash crop
COM02   craft
COM03   boat caulking from resin
COM04   vine for wicker furniture
COM05   medicinal resin or oil

MIS   (MISCELLANEOUS USES)
MIS01   musical instrument
MIS02   toy bowl
MIS03   canoe
MIS04   donkey harness
MIS05   leaves for trail marker
MIS06   indicates rainy season when in flower
MIS07   oropendola birds make nest from this material
MIS08   toy bow
MIS09   toy arrow point
MIS10   toy arrow shaft

NONE   no activity context recorded

APPENDIX 7

Indigenous Names and Activity Contexts of Plants Known to the Ka'apor

ACANTHACEAE

Justicia pectoralis Jacq.: **pira-piŠĩ-ka'a** 'fish-?-herb'; **kumaru-ka'a** 'tonka bean tree-herb' (T); MAG01 MAG02; terrestrial herb of swamp forest and high forest.

Justicia spectabilis T. Anders. ex C.B. Clarke: **pira-piŠĩ-ka'a** 'fish-?-herb'; MAG01 MAG02; terrestrial herb of high forest and dooryard garden.

AMARANTHACEAE

Amaranthus oleraceus L.: **ka'a-memek** 'herb-weak'; SEA01; introduced cultivated pot herb of dooryard garden.

Amaranthus spinosus L.: **kururu-ka'a** 'toad-herb'; bredo de espinhos; NONE; Tiny herb of village, old swiddens.

ANACARDIACEAE

Anacardium giganteum Hancock ex Engler: **akayu-'ɨ** 'L-stem'(= **akayu-pinar-'ɨ** 'L-fishhook-stem'); **akazu-'ɨw-ate** 'L-stem-true' (T); cajuí; wild cashew; SFS01 ANF01 SFS04 FUE02; large tree of high forest.

Anacardium occidentale L.: **akayu** 'L' (**akayu-pɨraŋ, akayu-tawa**); **akayu-rána** 'cashew-false' (G); **akazu** 'L' (T); **akayu** 'L' (AR); caju; cashew; PFS01 SFS04; traditionally cultivated treelet of dooryard garden and old swidden.

Anacardium parvifolium Ducke: **akayu-mena-'ɨ** 'L-?-stem'; cajuí; wild cashew; SFS01 SFS04 FUE02 ANF01; large tree of high forest.

Astronium cf. *obliquum* Griseb.: **kirɨhu-'ɨ-pɨtaŋ** 'L-stem-red'; **ka'a-tai-'ɨwa** 'forest-spice-stem' (AS); muiraquatiara; I II II18 MAG04, large tree of old fallow.

Astronium lecointei Ducke: **yeyu-'ɨ** 'kind of fish-stem'; **wira-račĩ-čõ** 'tree-thorn-black' (G); **ka'a-tai-'ɨwa** 'forest-spice-stem' (AS); muiraquatiara; PBC02 ANF02 FUE01 HHI06 HHI29; medium tree of the old fallow.

Mangifera indica L.: **mã** 'L'(**mã-pɨraŋ, mã-te**); **mã-'ɨwa** 'some-fruit' (G); **maŋ** 'L' (T); man-

NOTE: The name in boldface immediately following each plant species' scientific name is the corresponding Ka'apor name. Hyphens in Ka'apor plant names represent morpheme boundaries. After each Ka'apor plant name, a morpheme-by-morpheme gloss in English is given. The morpheme gloss "L" indicates a literal (unanalyzable) plant term.

Indigenous names in other Tupí-Guaraní languages and glosses for the same plant species are indicated after the English gloss of the Ka'apor name. The languages to which these names pertain are indicated in parentheses following the name itself:(T) = Tembé, (AR) = Araweté, (AS) = Asurini, and (G) = Guajá. The vernacular names of the species in English or Portuguese, where available, have been indicated after the indigenous name(s) and gloss(es).

The activity context in Ka'apor (which may differ for other languages) is listed after the indigenous name(s). The codes for each activity context are given in Appendix 6.

gueira; mango tree; PFS01; introduced cultivated medium tree of the dooryard garden.

Spondias mombin L.: **taper-ɨwa-'ɨ** 'old fallow forest-fruit-stem'; **tawe-wa-'ɨw** 'old village-fruit-stem' (T); **tawa-wa-'ɨ** 'yellow-fruit-stem' (G); **kayuwa-'ɨwa** 'L-stem' (AS); **akāya-'ɨ** 'L-stem' (AR); hog plum; taperebá, cajá; PFS01 ANF01 FUE02; Medium tree found only in old fallow in the Ka'apor habitat (also seen near rapids in the Xingu River basin).

Tapirira guianensis Aublet: **tayahu-mɨra** 'white–lipped peccary-tree'; **tata-pɨrɨrɨk-'ɨw** 'fire-crackle-stem' (T); **he-pɨpɨr-ɨwa-'ɨ** 'some-bird-fruit-stem' (G); **toko-re-me'e-'a-'ɨ** '?-?-some-fruit-stem' (AR); tatapiririca; SFS01 ANF01 FUE02; large tree common both to high forest and old fallows.

Tapirira peckoltiana Engler: **tayahu-mɨra** 'white–lipped peccary-tree', tatapiririca; SFS01 ANF01 FUE02; large tree, found in both high forest and old fallows.

Thyrsodium sp. 1: **kɨrɨhu-'ɨ-pɨtaŋ** 'L-stem-red'; breu de leite; HHI18 MAG04; small tree of the high forest.

Thyrsodium spruceanum Bentham: **tatu-mɨra** 'armadillo-tree', **waruwa-'ɨrána** 'mirror-false' (AS); HHI18 MAG04; large tree of the high forest.

ANNONACEAE

Anaxagorea dolichocarpa Spr. ex Sandw.: **teremu-mɨra** 'water insect-tree'; **pira-ɨwa-pihun** 'fish-fruit-black' (T); MED06 TLW02; small tree of the swamp forest.

Annona montana Macfad.: **arašiku-ran** 'soursop-false'; **aračiko'a-'ɨ** 'L-stem' (G); **tayahə-'a-'ɨ** 'white–lipped peccary-fruit-stem' (AR); **aračiku-ran** 'soursop false' (T); graviola; PFS01; introduced, domesticated treelet of the dooryard garden.

Annona paludosa Aublet: **arašiku-'ɨ** 'L-stem'; araticum; SFS01 ANF01 FUE01; medium tree of the high forest.

Annona sericea Dunal: **arašiku-'ɨ** 'L-stem'; ata brava; SFS01 ANF01 FUE01; small, very rare tree of the old fallow.

Duguetia echinophora R.E. Fries: **pina-'ɨ** 'fishhook-stem'; pindaíba; TLW08 ANF02 FUE02 TLW02; small tree of the high forest.

Duguetia marcgraviana Mart.: **arašiku-'ɨ** 'L-stem'; **ira-ka'a-čĩ-'ɨ** 'bird-forest-white-stem' (G); SFS01 ANF01 FUE01; small tree of the old fallow.

Duguetia riparia Huber: **tata-'iran-'ɨ** 'fire-false-stem'; SFS01 TLW02 PBC02 FUE01 ANF01; medium tree of the high forest.

Duguetia surinamensis R.E. Fries: **pina-hu-'ɨ** 'fishhook-big-stem'; **mətə-'ɨ** 'collared peccary-stem' (G); pindaíba; MAG13 PAC02; uncommon small tree of the old fallow.

Duguetia yeshidah Sandw.: **makahɨ-mɨra** 'collared peccary-tree'; MAG02; treelet of the high forest.

Duguetia sp. 1: **pina-'ɨ** 'fishhook-stem'; **pina-'ɨ** 'fishhook-stem' (G); pindaíba; TLW02 ANF02 FUE02 TLW08; medium tree of the high forest.

Duguetia sp. 2: **makahɨ-mɨra** 'collared peccary-tree'; MAG02; small tree of the high forest.

Ephedranthus pisocarpus R.E. Fries: **makahɨ-mɨra** 'collared peccary-tree'; **pira-cū-'ɨ** 'fish-black-stem' (G); MAG02; medium tree of the high forest.

Fusaea longifolia (Aublet) Saff.: **taraku'ā-'ɨ** '*Camponotus* sp. ant-stem'; **pina-ɨw-hu** 'fishhook-stem-big' (T); **yakarata'a-'ɨ** '?-stem' (G); fusaia; SFS01 MED08 PBC02 FUE01; small tree of the high forest.

Guatteria elongata Benth.: **tata-'iran-'ɨ** 'fire-false-stem'; SFS01 TLW02 PBC02 FUE01 ANF01; small tree of the high forest.

Guatteria scandens Ducke: **tata-'ɨ** 'fire-stem'; **wɨpo-čə'ə** 'vine-?' (G); HHI01; large liana, treelike in appearance and common to the high forest.

Guatteria sp. 1: **tata-'iran-'ɨ** 'fire-false-stem'; **iwɨrā-amute** 'tree-other' (AR); SFS01 TLW02 PBC02 FUE01 ANF01; small tree of the high forest.

Rollinia exsucca (DC. ex Dun.) A.DC.: **ape-'ɨ-pihun** 'L-stem-black' (= **tikwer-'ɨ-pihun** 'L-stem-black'); **pira-cū-ran** 'fish-black-false' (G); envira bôbo; SFS01 TLW01 TLW02 ANF02 FUE01; small tree of the old fallow.

Rollinia mucosa Baill.: **mɨrɨma** 'L'; biribá; PFS01; introduced, domesticated treelet of the dooryard garden.

Unonopsis rufescens (Baill.) R.E. Fries: **pina-hu-'ɨ** 'fishhook-big-stem'; **pina-'ɨw-hu** 'fish-hook-stem-big' (T); envira surucucu; ANF02 TLW08 FUE01; a rare and small tree of riverbanks.

Xylopia barbata Mart.: **tata-'iran-'ɨ** 'fire-false-stem'; ANF02 FUE01 TLW02; uncommon, large tree of the old fallow.

Xylopia nitida Dunal: **yawi-'ɨ** 'L-stem' (**yawi-'ɨ-tuwɨr, yawi-'ɨ-pɨraŋ**); **pina-'ɨw-ran** 'fish-hook-stem-false' (T); **kamiča-põ-'ɨ** 'tortoise-other-stem' (G); **yawī-'ɨ** 'L-stem' (AR); **yawi-'ɨwa** 'L-stem' (AS); envira branca; PBC01 PBC07 ANF02 FUE01 TLW02; medium to large tree of the high forest.

Annonaceae sp. 1: **yaši-mukuku-'ɨ** 'tortoise-*Licania* spp.-stem'; ANF01 FUE01 MED37; small tree of the old swidden.

APIACEAE

Eryngium foetidum L.: **ka'a-piher** 'herb-aromatic'; SEA01; introduced, domesticated pot herb of the dooryard garden.

APOCYNACEAE

Allamanda cathartica L.: **sɨpo-tuwɨr** 'vine-white'; **ihipa-ratī** 'vine-white' (AR); FUE02; scandent vine of young secondary forest.

Ambelania acida Aubl.: **kuyer-'ɨ-puku** 'spoon-stem-long'; **kamiča-'ɨ** 'L-stem' (G); pau de colher; SFS01 SFS06 HHI05 MED05 MED11 FUE01; small to medium tree, facultative in old fallows and high forests.

Ambelania grandiflora Huber: **kuyer-'ɨ-ran** 'spoon-stem-false'; pepino do mato; ANF01; small tree of creek margins.

Aspidosperma cylindrocarpon Muell. Arg.: **paraku-'ɨ-te** 'fish-stem-true'; ANF02 FUE01; Very rare, small tree of the old fallow.

Aspidosperma desmanthum Bth. Cx M. Arg.: **ara-rākā-'ɨ** 'macaw-head-stem', **imemeri-'ɨwa** (AS); PBC02; large tree of the old fallow.

Aspidosperma verruculosum M. Arg.: **ara-rākā-'ɨ** 'macaw-head-stem'; araracanga; PBC02; rare medium tree of the old fallow.

aff. *Forsteronia*: **sɨpo-pihun** 'vine-black'; NONE; medium liana of the high forest.

Himatanthus articulatus (Vahl.) Woodson: **kuyer-'ɨ-ran** 'spoon-stem-false'; sucuúba; ANF02; small tree of the old fallow.

Himatanthus sucuuba (Spruce ex Muell. Arg.) Woods.: **kanaú-'ɨ** 'L-stem'; **tapi'ɨ-puhã** 'tapir-remedy' (G); **ɨwɨra-tapi-čī** 'tree-?-white' (AS); sucuúba; MED06 MED08 MED31 MED32; small tree of the old swidden

Lacmellea aculeata (Ducke) Monachino: **kuyer-'ɨ-pu'a** 'spoon-stem-round'; **wɨra-kuzer** 'tree-spoon' (T); pau de colher; SFS01 SFS06 HHI05 ANF01 MED05 MED11 FUE01; small to medium tree of the high forest.

Masechites bicornulata (Rusby) Woods.: **sɨpo-ran** 'vine-false'; **zawa-kai-zi-wɨpo** 'dog-?-?-vine' (T); NONE; small vine of river edges.

Parahancornia amapa Ducke: **apa-'ɨ-tuwɨr** 'L-stem-white'; **amapa-'ɨw-čī** 'L-stem-white' (T); amapazeiro; MED07 SFS01 SFS06 ANF01; large tree of the high forest.

Parahancornia fasciculata (Poir.) Benoist: **apa-'ɨ-tuwɨr** 'L-stem-white'; amapazeiro; MED07 SFS01 SFS06 ANF01 FUE02; large tree of the high forest.

Rauvolfia paraensis Ducke: **tayɨ-põ** 'L-other'; ANF02 FUE01; small tree of the high forest.

Tabernaemontana angulata Mart. ex Muell. Arg.: **kaŋwaruhu-mɨra** 'paca-stem'; **ameno-wai-'ɨ** '?-?-stem' (AR); **pəkə-'a-tɨpɨ** 'paca-fruit-?' (T); MED15 MED32; small tree of the old fallow.

ARACEAE

Caladium picturatum C. Koch.: **ka'a-ro-pinī** 'herb-leaf-stripe'; MAG05; terrestrial herb of the high forest, also transplanted to and cultivated in the dooryard garden.

Colocasia esculenta (L.) Schott.: **taya-ran** '*Xanthosoma* sp. 1–false'; taro; PFS02 PFS03; introduced, domesticated herb of the dooryard garden.

Dracontium sp. 1: **ani-ran** 'L-false'; tajá de cobra; MAG10 POI03; terrestrial herb of the young swidden.

Heteropsis longispathacea Engl.: **sipo-šišik** 'vine-smooth' (= **sipo-te** 'vine-true'); cipó-titica; PBC05 COM04; medium epiphytic vine of the high forest.

Heteropsis sp. 1: **sipo-šišik** 'vine-smooth' (=**sipo-te** 'vine-true'); **parakwa-čiči** '?-?' (G); cipó-titica; PBC05 COM04; small epiphytic vine of the high forest.

Monstera cf. *pertusa* (L.) Vriesia: **wa-me-sipo** 'fruit-inside-vine'; ANF02 POI08 POI11; large, succulent epiphyte of the high forest.

Monstera subpinata (Schott) Engl.: **wa-me-sipo** 'fruit-inside-vine'; **tarakwā-ama'a-kiri** '*Camponatus* sp. ant-?—?' (G); ANF02 POI08 POI11; epiphyte of the high forest.

Montrichardia linifera (Ais.) Schott.: **anī** 'L'; **aníŋa** 'L' (AS); **aniŋ** 'L' (T); NONE; large herb of Gurupi River swamps.

Philodendron grandiflorum (Jacq.) Schott.: **wa-me-sipo** 'fruit-inside-vine'; **tarakuwa-remo** '*Camponatus* sp. ant-vine' (T); cipó-ambé; ANF02 POI08 POI11; climbing vine of the high forest.

Philodendron venustum Bunt.: **wa-me-sipo** 'fruit-inside-vine'; ANF02 POI08 POI11; succulent epiphyte of the high forest.

Philodendron sp. 1: **puru'ā-rimo** 'umbilical cord-vine'; cipó-ambé; MAG17; high climbing vine of the high forest.

Xanthosoma sp. 1: **taya** 'L'; **taza** 'L' (T); cocoyam; tajá; PFS02 PFS03; rhizomatous traditional domesticate of the young swidden.

ARALIACEAE

Schefflera morototoni (Aubl.) M.S.F.: **marato-'i** 'L-stem'; **maratato-wa-'i** 'L-fruit-stem' (G); **murete-towi-'i** 'L-green-stem' (AR); morototó; ANF02 FUE02 MIS10; small to medium tree of the old fallow.

ARECACEAE

Acrocomia sp. 1: **mukaya** 'L'; mucajá; NONE; massive solitary palm, not present in the habitat of the Ka'apor today but known from folklore and oral history.

Astrocaryum javari Mart.: **yawar-'i** 'jaguar-stem'; javari; NONE; armed caespitose palm, river edge, R. Gurupi.

Astrocaryum gynacanthum Mart.: **yu-'i** 'spine-stem'; **yára-'i** '?-stem' (AR); mumbaca; SFS01 MAG07; small armed caespitose palm of the high forest and old fallow.

Astrocaryum murumuru Mart.: **muru-'i** 'L-stem'; **murumuru-'iw** 'L-stem' (T); SFS01 ANF01 PAC02; armed caespitose palm of the igapó, rare in pre-Amazonia.

Astrocaryum vulgare Mart.: **tukumā-'i** 'L-stem'; **takamā** 'L' (G); tucumā palm; ANF01 POI04 PAC02 COM02; large, armed caespitose palm of the old fallow.

Astrocaryum sp. 1: **yu-pihun** 'spine-black'; **yu-'i** 'spine-stem' (G); **maraza-wa-'iw** 'L-fruit-stem' (T); mumbaca; SFS01; small, armed caespitose palm found in old fallow.

Bactris gasipaes Mart.: **pupu** 'L'(= **tukumā-ran** '*Astrocaryum vulgare*-false'); **tukumā-ran** '*Astrocaryum vulgare*-false' (T); pupunha; peach palm; SFS01; solitary or caespitose palm of the dooryard garden (it is in the initial stages of cultivation by some Ka'apor).

Bactris humilis (Wallace) Burret: **kwere'ī** 'L'; **wɔ'ɔ** 'L' (G); marajá; TLW12; stemless palm of the high forest.

Bactris major Mart.: **piri'a-hu-'i** 'L-big-stem'; marajá; SFS01 ANF01; small armed palm of the high forest.

Bactris maraja Mart.: **maraya'i** 'L-stem'; **mariawa** (= **mariači**) 'L' (G); **maraya-'i** 'L-stem' (AR); marajá; SFS01 ANF01; caespitose palm of the seasonally inundated forest.

Bactris setosa Barb. Rodr.: **piri'a** 'L'; SFS01; small spiny palm of the old swidden.

Bactris tomentosa Mart.: **piri'a-ran** 'L-false'; **kiripirim-'i** '?-stem' (G); SFS01; small caespitose palm of the high forest.

Cocos nucifera L.: **kuk** 'L' (**kuk** and **kuk-anā**); SFS01; introduced, domesticated palm of the dooryard garden.

Desmoncus macroacanthos Mart.: **irapar-pukwa-ha** 'bow-grip-generator'; jacitara; TLW11; armed, climbing palm of the high forest.

Desmoncus polycanthus Mart.: **irapar-pukwa-ha** 'bow-grip-generator'; **yawewi-rači** 'stingray-spine' (G); **zu-iwipo** 'spine-vine' (T); jacitara; TLW11; spiny, scandent palm of swamp forests.

Euterpe oleracea Mart.: **wasaí-'i** 'L'; **pinuwa-pihun** 'L-black' (G); **ačai-'i** 'L-stem' (AR); **yuziwa** 'L-?' (AS); açaí; PFS01 SEA02 ANF01 PBC04 PBC06 PBC08 HHI12 HHI13; slender caespitose palm mostly seen in swamp forests, occasionally in the high forest.

Geonoma baculifera Kunth.: **owi** 'L'; **ka'i-yowa-'i** '*Cebus* monkey-?-stem' (G); **uwi** 'L' (T); ubim; PBC03; caespitose palm of the swamp forest.

Geonoma leptospadix Trail: **owi-ran** 'L-false'; ANF02; small palm of the high forest.

Markleya dahlgreniana Bondar: **inaya-kami** '*Maximiliana maripa*-?' (= **yetahu-kami** '*Orbignya phalerata*-?'); **parana-'i** 'big river-stem' (G); ANF02 POI01; rare solitary palm of the old swidden.

Mauritia flexuosa Mart.: **miriši-'i** 'L-stem'; buriti; moriche palm; SFS01; massive solitary palm, not present in the habitat of the Ka'apor but known from folklore and from neighboring settlers.

Maximiliana maripa Mart.: **inaya-'i** 'L-stem'; **inaya-'i** 'L-stem' (G); **náya-'i** 'L-stem' (AR); PFS01 SEA02 ANF01 PBC03 PAC06 MED14 COM02 MIS08; massive, solitary palm of the old fallow; less commonly, it is seen along river margins.

Oenocarpus distichus Mart.: **pinuwa-'i** 'L-stem; **pinuwa-'iw** 'L-stem' (T); **pinuwa-'i** 'L-stem' (G); **pinuwa-'iwa** 'L-stem' (AS); bacaba de leque; PFS01 ANF01 HHI02 HHI16 RIT03; massive solitary palm, facultative to the high forest and old fallows.

Orbignya phalerata Mart.: **yeta-hu-'i** '*Hymenaea parvifolia*-big-stem'; **iwa-hu-'iw** 'fruit-big-stem' (T); **wa-'i;wa'i'i** 'fruit-stem' (G); **marita-rána** (AS); babaçu palm; SFS01 SFS05 MED22; massive solitary palm of the old fallow (tends to be ecologically very important wherever).

Socratea exorrhiza (Mart.) H. Wendl.: **paši-'i** 'L-stem'; **pači-'i** 'L-stem' (T); HHI10; solitary palm with spiny, exposed roots; found in swamp forests.

Syagrus cf. *inajai* (Spruce) Becc.: **marari-'i** 'L-stem'; **zuta-'a-'iw** '*Hymenaea* sp.-fruit-stem' (T); pati; SFS01 HHI14 TLW07 ANF02; rare, medium, solitary palm known only from the high forest.

Syagrus sp. 1: **marari-'i** 'L-stem'; **awa-'iwa** 'person-stem' (AS); **iriri** 'L' (AR); pati; HHI14 TLW07 ANF02; rare, medium, solitary palm of the high forest.

ASCLEPIADACEAE

Asclepias curassavica L.: **ira-kiwa-ran** 'bird-comb-false'; milkweed; oficial de sala; MED11 MED16; uncultivated herb of the swidden.

Schubertia grandiflora Mart.: **ira-i-sipo** 'bird-small-vine'; MED24; uncultivated vine of the old swidden.

ASPIDIACEAE

Ctenitis pretensa (Afz.) Ching: **yakare-ka'a** 'caiman-herb'; fern; samambaia; MED27; small fern of the high forest floor.

ASTERACEAE

Clibadium sylvestre (Aubl.) Baill.: **kanamɨ** 'L'; cunumi; TLW07 POI01; traditionally culti-vated herb of the young swidden.

Conyza banariensis (L.) Crang.: **teyu-pɨtɨm** 'skink-tobacco'; **kərə-mi'i** 'divinity-?' (AR); **ka'a-kɨ** 'herb-?' (G); MED02; noncultivated herb of the old swidden.

Eupatorium macrophyllum L.: **ka'a-yuwar-ran** 'herb-itch-false'; ANF02; shrub common to trailsides.

Porophyllum ellipticum Cass.: **teyu-pɨtɨm** 'skink-tobacco'; MED02; herb of the swidden edge.

Pterocaulon vergatum (L.) DC.: **teyu-pɨtɨm** 'skink-tobacco'; MED02; herb of the young swidden.

BIGNONIACEAE

Amphilophium paniculata (L.) H.B.K. var. *moller* (Cham. & Schl.) Gentry: **sɨpo-tuwɨr** 'vine-white'; FUE02; scandent vine of the old swidden.

Anemopaegma setilobum A.Gentry: **sɨpo-tuwɨr** 'vine-white'; cipó calango; FUE02; large liana of the old fallow.

Arrabidaea cf. *florida* DC.: **sɨpo-tuwɨr** 'vine-white'; cipó cruz; FUE02; medium liana of the high forest.

Arrabidaea sp. 1: **wiriri-sɨpo** 'swift [bird]-vine'; NONE; small vine of the high forest.

Crescentia cujete L.: **kwi** 'L' (**kwi-pu'a, kwi-puku, kwi-pihun, kwi-ra'ɨr, kwi-te, kwi-tiha**); **kwi** 'L' (G); **kəi** 'L' (AR); **kuyá** 'L' (AS); **kawaw** 'L' (T); calabash tree; cuia; HHI03; traditionally cultivated treelet of the dooryard garden.

Cuspidaria sp. 1: **sɨpo-tuwɨr** 'vine-white'; FUE02; rare, large liana rare of the old fallow.

Jacaranda copaia (Aubl.) D. Don. ssp. copaia: **para-'ɨ** 'L-stem'; **yači-pərəhəm-'ɨ** 'yellow footed tortoise-food-stem' (G); **apara-'ɨwa** 'L-stem' (AS); matchbox tree; paraparaúba; ANF02 FUE02; medium tree of the old fallow.

Jacaranda copaia ssp. *spectabilis* (Mart. ex DC.) A. Gentry: **para-'ɨ** 'L-stem'; paraparaúba; ANF02 FUE02; small to medium tree of the old fallow.

Jacaranda duckei Vattimo: **para-'ɨ** 'L-stem'; paraparaúba; ANF02 FUE02; small to medium tree of the old fallow.

Jacaranda heterophylla Bur. & K. Schum.: **para-'ɨ-ran** 'L-stem-false'; **yači-pərəhəm-'ɨ** 'yellow footed tortoise-food-stem' (G); NONE; small tree of the old swidden.

Jacaranda paraensis (Huber) Vattimo: **para-'ɨ-ran** 'L-stem-false'; NONE; uncommon small tree of old swiddens.

Mansoa angustiden (DC.) Bur. & K. Schum.: **sɨpo-nem** 'vine-fetid'; NONE; uncultivated vine of the young swidden.

Memora allamandiflora Bur. ex Bur. et K. Schum.: **tatu-ka'a-ro** 'armadillo-herb-leaf'; **ihipa-ratɨ** 'vine-white' (AR); **wɨpo-ratā** 'vine-hard' (G); NONE; vine of the high forest.

Memora bracteosa (DC.) Bur. & K. Schum.: **sɨpo-ātā** 'vine-hard'; NONE; small liana of the high forest.

Memora flavida (DC.) Bur & K. Schum.: **sɨpo-piraŋ** 'vine-reddish'; **wawarew-kawaw** 'squirrel-calabash bowl' (T); garaxama; MED11; medium liana with red inner bark, found in the high forest.

aff. *Memora bracteosa* (DC.) Bur. & K. Schum.: **sɨpo-tawa** 'vine-yellow'; ANF02; medium vine of the high forest.

Rhabidalia sp. 1: **sɨpo-ātā** 'vine-hard'; NONE; medium liana of the high forest.

Styzophyllum riparium (H.B.K.) Sandw.: **musu-sɨpo** 'eel-vine'; **ihipa-pihúna** 'vine-black' (AS); MAG15; small liana of the high forest.

Tabebuia impetiginosa Standley.: **tayɨ** 'L' (**tayɨ-te, tayɨ-tawa, tayɨ-pihun**); **ɨrapa-'ɨ** 'bow-stem' (G); pau d'arco roxo; TLW10 MIS08 MIS09; large tree of the old fallow.

Tabebuia serratifolia (Vahl.) Nichols: **tayɨ-põ** 'L-other'; **ži'a** 'L' (AS); **tayɨ-pá** 'L-other' (AR); pau d'arco amarelo; ANF02 FUE01; large tree of the old fallow.

Tabebuia sp. 1: **tayɨ-põ** 'L-other'; ANF02 FUE01; medium tree of the old fallow.

Bignoniaceae sp. 1: **sɨpo-memek** 'vine-weak'; NONE; small vine of the high forest.

BIXACEAE

Bixa orellana L.: **uruku** 'L' (for cultivated variety), **uruku-ran** 'L-false' (for swamp forest variety); **araku-'ɨ** 'L-stem' (G) (for cultivated variety), **araku-rána** (for forest variety); **karuwa-náta-'ɨ** 'divinity-babaçu fruit-stem' (AR) (for forest variety), **irikə** (for cultivated variety); **uruku** for domesticated variety (AS); **uruku-'ɨw** (T) for cultivated variety, **uruku-ran** for forest variety; annatto tree; urucuzeiro; PAC07 HHI01 RIT04; one variety of this treelet is cultivated in the dooryard garden, the other occurs along creek margins.

BOMBACACEAE

Ceiba pentandra Gaertn.: **wašiɲi-'ɨ** 'L-stem'; **iwi'um-'ɨw** 'L-stem' (T); **čama'am-hu** '?-big' (G); **tarawiri-rána** 'lizard-false' (AS); kapok tree; samaumeira; ANF02 MYT03 MED30; enormous tree of swamp forest, occasional in old fallow.

Pachira aquatica Aubl.: **kɨpɨhu-ran-'ɨ** 'cupuaçu-false-stem' (= **māmā-hu-'ɨ** 'papaya-big-stem'); **māmā-ran** 'papaya-false' (T); mungubarana; TLW02; medium tree common to riverbanks except the Turiaçu River basin.

BORAGINACEAE

Cordia exaltata Lam.: **ape-'ɨ-tawa** 'L-stem-yellow'; ANF01 TLW02 FUE01; small tree of the old fallow.

Cordia lomatoloba Johnston: **ape-'ɨ-tuwɨr** 'L-stem-white'; **táiwi-'ɨ** 'L-stem' (AR); SFS01; small tree of the high forest.

Cordia multispicata Cham.: **kurupi-'ɨ-sɨpo** 'divinity-stem-vine'; ANF02 TLW02; vine of the young swidden.

Cordia polycephala (Lam.) Johnston: **kurupi-kɨ'ɨ-ran** 'divinity-chile pepper-false'; ANF02; treelet of old swiddens.

Cordia scabrida Mart.: **ape-'ɨ-te** 'L-stem-true'; **wiriri-'ɨ** 'swift (the bird)-stem' (G); TLW02 FUE01; medium tree of old fallows.

Cordia scabrifolia C.DC.: **ape-'ɨ-tuwɨr** 'L-stem-white'; **wiriri-'ɨ** 'swift-stem' (G); **apitere-wɨwa** 'bald-stem' (AS); SFS01 TLW02 ANF02 FUE01 MED27; small tree of old fallows.

Cordia sellowiana Chamb.: **ape-'ɨ-hu** 'L-stem-big'; ANF02 TLW02; medium tree of old fallows.

Cordia trachyphylla Mart.: **ape-'ɨ** 'L-stem'; TLW02; small, rare tree of old fallows.

Cordia sp. 2 **ape-'ɨ-howɨ** 'L-stem-blue'; grão de galo; SFS01; fairly common small tree of old swiddens.

BROMELIACEAE

Aechmea brevicellis L.B. Smith: **karawata** 'L'; ANF01 MED10; arboreal bromeliad common in old swiddens.

Aechmea bromeliifolia Baker: **karawata** 'L'; **nana-'ĩ** 'L-not' (G); ANF01 MED10; epiphyte common in the high forest.

Ananas comosus L.: **nana** 'L' (**nana-te** and **nana-yɨkɨr**); **nana** 'L' (T); **nana-'ĩ** 'L-not' (G); **nanĩ** 'L' (AR); **parawa-wɨ'a** 'Mealy parrot-?' (AS); pineapple; abacaxi; PFS01; traditionally cultivated terrestrial herb of the dooryard garden.

Ananas nanas (L.B. Smith) L.B.Smith: **nana-ran** 'pineapple-false' (= **kurupi-nana** 'divinity-pineapple'); abacaxi bravo; SFS01 ANF02; terrestrial herb with tiny, pineapple-like fruits, found in marshland.

Bromelia goeldiana L.B. Smith: **karawata** 'L'; **karawata-ran** 'L-false' (T); ANF01 MED10; terrestrial bromeliad of marshland.

Guzmania lingulata (L.) Mez.: **karawata** 'L'; **tuparapa** 'L' (T); ANF01 MED10; arboreal bromeliad of riverbanks.

Neoglaziovia variegata (Arr. Cam.) Mez.: **kɨrawa** 'L'(**kɨrawa-howɨ** and **kɨrawa-pɨrãŋ**); **kɨrawã** 'L' (AR); **kurawa** 'L' (T); caroá; TLW01 TLW19 RIT02; traditionally cultivated terrestrial bromeliad of the dooryard garden, also seen in the young swidden.

Tillandsia usneoides (L.) L.: **ɨra-hu-ra-wi** 'bird-big-teather-fine'; **wɨra-mutaw** 'tree-beard' (T); MIS07; arboreal bromeliad of the old swidden.

Bromeliaceae sp. 1: **karawata** 'L'; ANF01 MED10; arboreal bromeliad of the high forest.

BURSERACEAE

Protium altsoni Sandw.: **ara-kanei-'ɨ** 'macaw-resin-stem' (=**kanei-'ɨ-pɨtaŋ** 'resin-stem-red'); breu; ANF02 FUE01 FUE03; medium tree of the high forest.

Protium aracouchini (Aubl.) Marchand: **yawa-mɨra** 'jaguar-tree' (=**sē-kãtãi-iran-'ɨ** 'honeycreeper-Brazil nut-false-stem'); **yawar-'a-'ɨ** 'jaguar-fruit-stem' (G); breu; SFS01 ANF02 FUE01; small tree of the swamp forest.

Protium decandrum (Aublet) Marchand: **ara-kanei-'ɨ** 'macaw-resin-stem' (=**kanei-'ɨ-pɨtaŋ** 'resin-stem-red'); **hikətə-'ɨw-pihun** 'resin-stem-black' (T); **yawar-'a-'ɨ** 'jaguar-fruit-stem' (G); breu; SFS01 ANF02 FUE01 FUE03; medium tree of the high forest.

Protium giganteum Engl. var. *giganteum*: **kanei-'ɨ-tuwɨr** 'resin-stem-white'; **waruwa-'ɨwa** 'mirror-stem' (AS); **hikətə-'ɨw-čī** 'resin-stem-white' (T); breu branco; SFS01 ANF02 MED21 FUE01 FUE03 HHI18 COM03; medium tree of the high forest.

Protium heptaphyllum ssp. *heptaphyllum* (Aublet) Marchand: **ara-kanei-'ɨ** 'macaw-resin-stem' (=**kanei-'ɨ-pɨtaŋ** 'resin-stem-red'); **yawar-'a-'ɨ** 'jaguar-fruit-stem' (G); **hikətə-'ɨw** 'resin-stem' (T); breu branco verdadeiro; ANF02 FUE01 FUE03; small tree of the old fallow.

Protium pallidum Cuatrec.: **kanei-'ɨ-tuwɨr** 'resin-white-stem'; **hikətə-'ɨw-čī** 'resin-white-stem' (T); **yawara-'a-čū-'ɨ** 'jaguar-fruit-black-stem' (G); breu branco; SFS01 ANF02 MED21 FUE01 FUE03 HHI18 COM03; small to medium tree of the high forest.

Protium polybotryum (Turcz.) Engl.: **kanei-a-pe-'ɨ** 'resin-fruit-flat-stem'; **hikətə-iran-'ɨw** 'resin-false-stem' (T); **yawar-'a-'ɨ** 'jaguar-fruit-stem' (G); breu; ANF02 SFS01 MED04 FUE01 FUE03 PAC04; medium tree of the high forest.

Protium sagotianum Engl.: **kanei-aka-'ɨ** 'resin-fat-stem'; **yawar-'a-čū-'ɨ** 'jaguar-fruit-black-stem' (G); breu; SFS01 ANF02 FUE01 FUE03; small tree of the high forest.

Protium aracouchini (Aubl.) Marchand: **yawa-mɨra** 'jaguar-tree' (=**sē-kãtãi-iran-'ɨ** 'honeycreeper-Brazil nut-false-stem'); **yawar-'a-'ɨ** 'jaguar-fruit-stem' (G); breu; SFS01 ANF02 FUE01; small tree of the swamp forest.

Protium tenuifolium (Engl.) Engl.: **kanei-a-pe-'ɨ** 'resin-fruit-flat-stem'; **yawar-'a-'ɨ** 'jaguar-fruit-stem' (G); breu; ANF02 SFS01 MED04 FUE01 FUE03 PAC04; small to medium tree of the high forest.

Protium trifoliolatum Engl.: **sē-kãtãi-'ɨ** 'honeycreeper-Brazil nut-stem'; **hikətə-'ɨw-pihun** 'resin-stem-black' (T); **yawar-'a-'ɨ** 'jaguar-fruit-stem' (G); breu almecega; SFS01 ANF02 PBC02 FUE01; small tree of the high forest.

Tetragastris altissima (Aublet) Swart.: **waruwa-'ɨ** 'mirror-stem'; **ɨwa-po-pe-pɨraŋ-'ɨw** 'fruit-?-flat-reddish-stem' (T); **wa-pɨrã-'ɨ** 'fruit-reddish-stem' (G); **piñi-'ɨ-hə** 'L-stem-big' (AR); breu manga; SFS01 SFS04 ANF01 HHI25 FUE01; medium tree facultative to high forests and fallows.

Tetragastris panamensis (Engl.) O.K.: **waruwa-ɨwa-pɨtaŋ-'ɨ** 'mirror-fruit-red-stem'; **papara-'ɨ** 'L-not' (G); breu preto; SFS01 ANF01 FUE01; medium tree facultative to high forests and fallows.

Trattinickia burserifolia Mart.: **kɨrɨhu-'ɨ** 'L-stem'; **kɨrɨhɨ-'ɨ** 'L-stem' (G); ANF02 PAC10 HHI18 MAG05 RIT01 FUE01 FUE03 MED06; large tree of the high forest.

Trattinickia rhoifolia Willd.: **kirihu-'i** 'L-stem'; **kiriwa-'iw** 'L-stem' (T); **ičiri-'i** 'resin-stem' (AR); **kirihi-'i** 'L-stem' (G); **ihiki-riwa** 'resin-stem' (AS); breu mescla; ANF02 PAC10 HHI18 MAG05 RIT01 FUE01 FUE03 MED06; uncommon large tree facultative to fallow and high forest.

Trattinickia sp. 1: **kirihu-'i** 'L-stem'; breu mescla; ANF02 PAC10 HHI18 MAG05 RIT01 FUE01 FUE03 MED06; medium to large tree of the high forest.

CACTACEAE

Rhipsalis baccifera (J. Miller) Stearn.: **ira-hu-ka'a-ran** 'bird-big-herb-false' (= **mira-hu-reha** 'tree-big-eye'); NONE; epiphytic cactus of the high forest.

Rhipsalis mijosurus K. Schum.: **ira-hu-ra-wi** 'bird-big-down-fine'; MIS07; small creeper of the old fallow.

CAESALPINIACEAE

Adenanthera pavonina L.: **pu'i-piraŋ-'i-karaí-ma'e** 'bead-reddish-stem-nonIndian-possess-ive'; PAC02 COM02; introduced, domesticated treelet of the dooryard garden.

Apuleia leiocarpa (Vogel) Spr. var. *molaris* (Spruce ex Benth.) Koeppen: **kumaru-'i-ší** 'L-stem-white'; **wira-tawa** 'tree-yellow' (T); **yapu-'i** 'oropendola-stem' (G); **i-hadn-'i** 'its-child-stem' (AR); pau amarelo; FUE01; large tree of the old fallow.

Bauhinia rubiginosa Bong.: **yaši-sipo-pe** 'tortoise-vine-flat'; monkey ladder vine; escada de jaboti; MED29 MED08 FUE08 MIS05; large tree of the high forest.

Bauhinia viridiflorens Ducke: **ainumir-mira** 'hummingbird-tree'; **tami-'i** 'L-stem' (G); TLW10 TLW13; treelet of the young swidden.

Bauhinia sp. 1: **yaši-sipo-pe** 'tortoise-vine-flat'; **ihipa-pe-pe** 'vine-flat-trail' (AR); **wipo-pe** 'vine-trail' (G); **iwipo-péw** 'vine-trail' (T); monkey ladder vine; escada de jaboti; MED29 MED08 FUE08 MIS05; medium to large liana of the high forest.

Bauhinia sp. 2: **akuši-yu-'i** 'agouti-spine-stem'; TLW02; very rare, medium tree of the high forest.

Cassia fastuosa Willd.: **amaŋa-putir-'i** 'time measure-flower-stem'; **ačiči-rapā** '?-?' (AR); MIS06 ANF01; treelet of the old fallow.

Copaifera duckei Dwyer: **kupa-'i** 'L-stem'; **kapowa-'i** 'L-stem' (G); copaíba; MED14 MED23; medium tree of the high forest.

Copaifera reticulata Ducke: **kupa-'i** 'L-stem'; copaíba; MED14 MED23; uncommon medium tree of the old fallow.

Copaifera sp. 1: **kupa-'i-ran** 'L-stem-false'; copaíba; FUE02; medium tree of the high forest.

Crudia parivoa DC.: **kumaru-'i-iapo-rupi-har** 'L-stem-swamp forest-through-place'; **kupa-'iw-ran** 'L-stem false' (T); FUE01; small tree of river margins.

Dialium guianense (Aubl.) Sandw.: **yuru-pe-pe-'i** 'mouth-flat-flat-stem'; **iwa-popok-'iw** 'fruit-dehiscent-stem' (T); **yeta-'i-pipiru-'i** 'L-stem-crackle-stem' (G); **ayuru-mope-'iwa** 'Mealy parrot-food-stem' (AS); **yanita-'i** 'L-stem' (AR); jutaípororoca; SFS01 FUE01 ANF02; small tree of the old fallow.

Hymenaea courbaril L.: **tarapa-i-pitaŋ** 'L-stem-red'; **tawa-'i**; 'yellow-stem' (G); **yuta-'i** 'L-stem' (AR); **zuta-'iwa-'iw** 'L-fruit-stem' (T); copal tree; jatobá; SFS01 HHI22 HHI31 FUE01 FUE03; large tree of the old fallow.

Hymenaea parvifolia Huber: **yeta-'i** 'L-stem'; **wita-i-'i** 'rock-small-stem' (G); **zuta-'i-'iw** 'L-small-stem'; copal tree; jutaí; SFS01 ANF02 HHI22 MED10 MED24 MED34 FUE01 FUE03; medium tree of the old fallow.

Hymenaea reticulata Ducke: **tarapai-'i** 'L-stem'; jatobá; SFS01 HHI22 HHI31 FUE01 FUE03; large tree of the old fallow.

Macrolobium acaciaefolium Bth.: **arapari-'i** 'L-stem'; **arapari-'i** 'L-stem' (G); **arapari-'iw** 'L-stem' (T); **mari-rupe-'iwa** '-?-through-stem' (AS); ANF01 FUE01; abundant, small tree of river margins.

Sclerolobium albiflorum R. Ben.: **ka'a-meri-'i-tuwɨr** 'herb-?-stem-white'; tachi branco; ANF02 FUE02; medium tree of the high forest.

Sclerolobium guianense Aubl.: **ka'a-meri-'i** 'herb-female pers. name-stem'; tachirana; ANF01 FUE02; medium to large tree of the high forest.

Sclerolobium paraense Huber: **ka'a-meri-'i** 'herb-female pers. name-stem'; tachirana; ANF02 FUE01; small to medium tree of the high forest.

Sclerolobium sp. 16: **taši-'i-ātā-ran** '*Pseudomyrmax* sp. ant-stem-hard-false'; tachi branco; NONE; very large tree of the high forest.

Sclerolobium sp. 20: **taši-'iran-'i** '*Pseudomyrmax* sp. ant-false-stem'; tachirana; MED08 FUE01; large tree of the high forest.

Senna chrysocarpa (Desv.) Irwin & Barneby: **amaŋa-tiri-sɨpo** 'time measure-?-stem'; **wɨpo-pe** 'vine-flat' (G); MIS06; vine of the old fallow.

Senna pendula (Lamk.) Irwin & Barneby: **amaŋa-putɨr-'i** 'time measure-flower-stem'; MIS06 ANF01; treelet of the old fallow.

Senna sylvestris (Vell.) Irwin & Barneby: **šimo-'i-ran** 'L-stem-false' (=**maširawa** 'L'); **ɨ-tata-'ɨ** 'its-fire-stem' (G); ANF02 FUE01; small tree of old swiddens.

Swartzia brachyrachis var. Harms: **parei'a** 'L'; **mai-towa-'i-ran** 'divinity-?-stem-false' (G); jacarandá de veado; ANF01 FUE01; medium tree of the high forest.

Swartzia brachyrachis var. *snethlageae* (Ducke) Ducke: **parei'a-ran-'i** 'L-false-stem'; jacarandá de veado; ANF01 FUE01; medium tree of the high forest.

Tachigali macrostachya Huber: **taši-'i-ātā** '*Pseudomyrmax* sp. ant-stem-hard'; **tači-'ɨw-ete** 'L-stem-true' (T); tachizeiro; NONE; medium tree of the high forest.

Tachigali myrmecophila Ducke: **taši-'ɨ** '*Pseudomyrmax* sp. ant-stem'; **tači-'ɨ** '*Pseudomyrmax* sp. ant-stem' (G); **tači-'ɨwa** '*Pseudomyrmax* ant-stem' (AS); **tači-'i** '*Pseudomyrmax* sp. ant-stem' (AR); **tači-'ɨw-pihun** '*Pseudomyrmax* sp. ant-stem-black' (T); tachi preto; MED02 MED08 MED31 POI10; medium tree of the high forest.

Tachigali paniculata Aublet: **taši-'i-ātā** '*Pseudomyrmax* sp. ant-stem-hard'; **tači-'ɨ** '*Pseudomyrmax* sp. ant-stem' (G); tachi vermelho; NONE; medium tree of the high forest.

Tachigali sp. 1: **taši-'ɨ-hu** '*Pseudomyrmax* sp. ant-stem-big'; tachizeiro; NONE; small tree of the high forest.

Zollernia paraensis Huber: **tamaran-'ɨ** 'war club-stem'; **mai-towa-'ɨ** 'divinity-?-stem' (G); **yapém-i** 'war club-stem' (AR); **iyapem-'ɨwa** 'war club-stem' (AS); pau santo; TLW15 COM02 POI17; medium tree facultative in high forest and old fallows; heaviest wood of any Amazonian species.

CANNABIDACEAE

Cannabis sativa L.: **mara-ran** 'malaria-false'; marijuana; maconha; NONE; not present in habitat of the Ka'apor, but known to them because nearby settlers once cultivated it.

CANNACEAE

Canna indica L.: **awa-i** 'person-small'; Indian shot; PAC02 COM02; introduced domesticate of the swidden and dooryard garden.

CAPPARIDACEAE

Capparis sola Macbr.: **sawɨya-mɨra** 'rat-tree'; **ɨwɨra-tai** 'tree-spice' (AS); ANF02 FUE01; very rare treelet of the old fallow.

CARICACEAE

Carica papaya L.: **māmā** 'L'; **mɔ̃mɔ̃** 'L' (G); **mɔ̃mɔ̃** 'L' (AR); **yarakači'a-ú** '*Jacaratia spinosa*-big' (AS); **zarakači'a** 'L' (T); papaya; mamão; PFS01 PAC03; traditionally cultivated treelet found in swiddens and dooryard gardens.

Jacaratia spinosa (Aubl.) A.DC.: **māmā-ran-'ɨ** '*Carica papaya*-false-stem'; **zarakači'a-'ɨw** 'L-stem' (T); **arakači'a** 'L' (G); **arakači-'i** 'L-stem' (AR); mamão da mata; SFS01 ANF01; small tree, extremely common in old fallows.

CARYOCARACEAE

Caryocar glabrum (Aubl.) Pers.: **piki'a-ran-'i** 'L-false-stem'; **piki'a-ran-'iw** 'L-false-stem' (T); piquiarana; SFS01 ANF01 POI09; medium to large tree of the high forest.

Caryocar villosum Aubl.: **piki'a-'i** 'L-stem'; **piki'a-'iw** 'L-stem' (T); piquiá; PFS01 ANF01 POI09; large tree of the high forest.

CECROPIACEAE

Cecropia concolor Willd.: **ama-'i-tuwir** 'L-stem-white'; imbaúba; ANF02; medium tree of the old fallow.

Cecropia obtusa Trecue: **ama-'i-tuwir** 'L-stem-white'; **ama-'i-hêtê** 'L-stem-true' (AR); imbaúba; ANF02; medium tree of the high forest, commonly seen in light gaps.

Cecropia palmata Willd.: **ama-'i-puku** 'L-stem'long'; **ama-'i** 'L-stem' (G); **ama-'iwa** 'L-stem' (AS); imbaúba; SFSO1; small tree of the old swidden.

Cecropia purpurescens C.C. Berg: **ama-'i-te** 'L-stem-true'; imbaúba; ANF01 FUE01 MIS01 PBC07; uncommon small tree of the old fallow.

Cecropia sciadophylla Mart.: **ama-'i-te** 'L-stem-true'; **ama-'i-ate** 'L-stem-true' (G); imbaúba; ANF01 FUE01 MIS01 PBC07; medium to large tree, found in light gaps in the high forest.

Coussapoa sp. 1: **apo-'i-pihun** 'L-stem-black'; NONE; small tree of the old fallow.

Pourouma guianensis Aubl. ssp. *guianensis*: **ka'a-me-'i** 'herb-inside-stem'; **ama-'i-ča'ā-'i** 'Cecropia sp.-stem-?-stem' (G); SFS01 ANF02 TLW04; small tree of the high forest.

Pourouma minor Benoist: **ama-'i-rari-tuwir** 'Cecropia sp.-stem-rachis-white'; SFS01 ANF01; small tree of the high forest.

Pourouma mollis Trec. ssp. *mollis*: **ama-'i-rari** 'Cecropia sp. -stem-rachis'; SFS01 ANF01; medium tree found both in high forest and in old fallows.

CELASTRACEAE

Cheiloclinium cognatum (Miers) A.C.: **maka-wa'ē-sipo** 'Cebus monkey-fruit-vine', **wipo-'aru-'i** 'vine-?-stem' (G); ANF02 FUE01 SFS01 MAG02; treelike liana of the high forest.

aff. *Cheiloclinium* or *Salacia* sp. 1: **maka-wa'ē-sipo** 'Cebus monkey-fruit-vine'; ANF02 FUE01 SFS01 MAG02; large liana of the high forest.

Hippocratea ovata Lam.: **kumari-yu-'i** 'L-yellow-stem'; **zuri-wipo** 'L-vine' (T); TLW02; liana of river edges.

Maytenus sp. 10: **maka-waē-'i** 'Cebus monkey-fruit-stem'; chichuasca; ANF01; small tree of the old fallow.

Maytenus sp. nov.: **paraku-'i-ran** 'kind of fish-stem-false'; chichuasca; NONE; small tree of the old fallow.

Peritassa huanuclara (Loes.) A.C.: **sipo-piraŋ** 'vine-red'; NONE; large liana of the high forest.

Pristimera tenuifolia (Mart.) A.C.: **sipo-piraŋ** 'vine-reddish'; ANF02 MED03 MED11 FUE02; large liana of the high forest.

Salacia insignis A.C. Smith: **maka-wa'ē-sipo** 'Cebus monkey-fruit-vine'; **iwipo-pihun** (T); chichuasca; ANF02 FUE01 SFS01 MAG02; large liana of the high forest.

Salacia multiflora (Lam.) DC.: **sipo-piraŋ** 'vine-reddish'; chichuasca; ANF02 MED03 MED11 FUE02; large liana of the high forest.

CHENOPODIACEAE

Chenopodium ambrosioides L.: **motoroi** 'L'; wormseed; mastruço; MED06 MED08 MAG03; introduced, domesticated herb of the dooryard garden.

CHRYSOBALANACEAE

Couepia guianensis Aubl. ssp. *divaricata* (Hub.) Prance: **payu'ā-'i-howi** 'L-stem-green' (=**anira-wiši** 'bat-feces'); **iwa-pin-'iw** 'fruit-stripe-stem' (T); **wa-pini-hu-'i** 'fruit-stripe-big-stem' (G); SFS01 PBC02 ANF02 FUE01; large tree of the high forest.

Couepia guianensis Aubl. ssp. *guianensis*: **payu'ã-'ɨ** 'L-stem', **makuku-'ɨw** 'L-stem' (T); SFS01 PBC02 ANF01 FUE01; large tree of the high forest.

Couepia guianensis ssp. *glandulosa* (Mig.) Prance: **payu'ã-'ɨ-tawa** 'L-stem-yellow'; SFS01 ANF01 PBC02 FUE01; large tree of the high forest.

Exellodendron barbatum Prance: **inamu-mɨra** 'curassow-tree', **yakami-akərə-'ɨ** 'trumpeter-?-stem' (G); SFS01 ANF02 FUE01; unusual, small to medium tree of the high forest.

Hirtella bicornis Mart. ex Zucc.: **payu'ã-'ɨ-howɨ** 'L-stem-green' (= **anira-wiši** 'bat-feces'); SFS01 ANF01 PBC02 FUE01; rare, small tree of the old fallow.

Hirtella cf. *paraensis* Prance: **akuši-mɨra** 'agouti-tree'; ANF02 FUE01; treelet of the high forest.

Hirtella eriandra Benth.: **akuši-iŋa** 'agouti-*Inga* sp.'; ANF01 FUE01; rare treelet of the old fallow.

Hirtella racemosa Lam. var. *racemosa*: **mukuku-'ɨ-wi** 'L-stem-thin'; **ira-pirã-'ɨ** 'tree-reddish-stem' (G); ANF01 HHI06 HHI07; small tree of old swiddens and fallows.

Hirtella racemosa var. *hexandra* (Willd. ex R.G.S.) Prance: **akuši-mɨra** 'agouti-tree'; ANF02 FUE01; small tree of old swiddens..

Licania apetala (E. Mey.) Fritsch: **karaí-pe-'ɨ** 'non– Indian-trail-stem'; **arar-'i-tã-'ɨ** 'macaw-small-hard-stem' (G); **takipe-'iwa** 'L-stem' (AS); **tačipe-'i** 'L-stem' (AR); caraipé; ANF02 HHI21 FUE02; enormous tree exclusive to the old fallow.

Licania canescens R. Ben.: **wa-pini-'i-tuwɨr** 'fruit-stripe-stem-white'; **ɨwa-pin-'iw** 'fruit-stripe-stem' (T); **wa-pini-hu-'ɨ** 'fruit-stripe-big-stem' (G); macucu; SFS01 ANF02 PBC02 TLW13 HHI30 FUE01 MIS09; large tree of the high forest.

Licania glabriflora Prance: **mukuku-'ɨ-te** 'L-stem-true'; macucu; SFS01 PBC02 PAC07 PAC08 ANF02 TLW13 HHI30 FUE01 MED18 RIT05 MIS09; large tree of the high forest.

Licania heteromorpha Benth. var. *heteromorpha*: **mukuku-'ɨ-te** 'L-stem-true'; **tatə-papé-nani-'i** 'armadillo-claw-?-stem' (AR); **ira-pirã** 'wood-red' (G); macucu; SFS01 ANF02 PBC02 PAC07 PAC08 FUE01 TLW13 HHI30 MED18 RIT05 MIS09; medium tree of the high forest.

Licania sp. 1: **mukuku-'ɨ-hu** 'L-stem-big'; macacu; NONE; medium tree of the high forest.

Licania kunthiana Hook. F.: **wa-pini-hu-'ɨ** 'fruit-stripe-big-stem'; **takipe-ran-'ɨw** 'L-false-stem' (T); **wa-pini-hu-'ɨ** 'tree-stripe-big-stem' (G); pajurazinho; SFS01 PBC02 TLW13 HHI30 FUE01 MIS09; large tree of the high forest.

Licania latifolia Benth.: **mukuku-'ɨ-te** 'L-stem-true'; macucu de sangue; SFS01 ANF02 PBC02 PAC07 PAC08 FUE01 TLW13 HHI30 MED18 RIT05 MIS09; medium tree of the high forest.

Licania macrophylla Benth.: **anã-mɨra** 'kinsman-tree'; **anã-wira-'ɨw** 'kinsman-tree-stem' (T); macucu terra; ANF01 FUE01; uncommon large tree of the edge of the Gurupiuna stream.

Licania membranacea Sagot. ex Laness.: **karaí-pe-'ɨ** 'nonIndian-trail-stem'; caraipé; ANF02 HHI21 FUE02; large tree of the high forest.

Licania octandra (Hoffmgg. ex R. & S.) Kuntze: **karaí-pe-'ɨ** 'nonIndian-trail-stem'; caraipé; ANF02 HHI21 FUE02; small tree of the high forest.

Parinari excelsa Sabine: **payu'ã-'ɨ** 'L-stem'; pajurazinho; SFS01 PBC02 ANF01 FUE01; large tree of the high forest.

CLUSIACEAE

Caraipa grandiflora Mart.: **aman-'ɨ** 'rain-stem'; **tamakware-'ɨw** 'L-stem' (T); tamanquaré; ANF02; small tree of seasonally inundated forest.

Clusia sp. 1: **apo-'ɨ** 'L-stem'; **apu-'iw-pihun** 'L-stem-black' (T); **añĩ-adapeká** 'divinity-?' (AR); apuí; MAG10; large, treelike liana of the high forest.

Platonia insignis Mart.: **pakuri-'ɨ** 'L-stem' (includes **pakuri-te, pakuri-pu'a, pakuri-hãšĩ, pakuri-pihun**); **mukur-'ɨ** 'opossum-stem' (G); PFS01 ANF01; enormous tree of the old fallow.

Rheedia acuminata Pl. et Tr.: **pakuri-sōsō-'i** 'L-?-stem'; SFS01; medium tree of the high forest.

Rheedia brasiliensis (Mart.) Pl. et Tr.: **pakuri-sōsō-'i** 'L-?-stem'; bacurizinho; PFS01 FUE01 ANF01; small tree of the old fallow.

Symphonia globulifera L.: **irati-'i** 'wax-stem' (includes **irati-ātā-'i** and **irati-'i-te**) ; **irati-'i** 'wax-stem' (G); uanani; TLW05 PAC09 MED26 FUE01 FUE03 ANF02 POI12 POI16; medium tree usually of the swamp forest (one variety occurs in the high forest).

Tovomita brasiliensis (Mart.) Walp.: **yapu-mira-ran** 'oropendola-tree-false'; **maŋ-'iw-ran** 'mango-stem-false' (T); TLW14; small tree of the inundated forest along small creeks.

Tovomita brevistaminea Engl.: **yapu-mira** 'oropendola-tree'; **maŋ-'iw-ran** 'mango-stem-false' (T); ANF02 FUE01; small tree of inundated forests.

Vismia guianensis (Aubl.) Choisy: **wayaŋ-i** 'L-stem' (= **āyaŋ-ruku** 'divinity-annatto'); **kiriŋ-iwa-'iw** '?-fruit-stem' (T); **wira-ra'i** 'tree-small' (G); lacre; PAC07; small tree of the old swidden.

COCHLOSPERMACEAE

Cochlospermum orinocense (H.B.K.) Steud.: **samo'ā-'i** 'fem.pers.name-stem'; **arapa^ha-'i** 'red brocket deer-stem' (G); **mo'i-rape-'i** 'bead-trail-stem' (AR); periquiteira; TLW01 TLW02 ANF02; decidedly rare, small tree of the old swidden.

COMBRETACEAE

Buchenavia cf. *tetraphylla* (Aubl.) Howard: **tukur-iwa-'i** 'grasshopper-fruit-stem'; tanim-buca; ANF01 FUE01; small tree of the high forest.

Combretum laxum Jacq.: **yuti'āi-rimo** '?-?-vine'; NONE; fairly common, small vine of the high forest.

Terminalia amazonia (J. Gmel.) Exell.: **tukur-iwa-'i** 'grasshopper-fruit-stem'; ANF01 FUE01; small tree of the old fallow.

Terminalia dichotoma G. Meyer: **tukur-iwa-'i** 'grasshopper-fruit-stem'; ANF01 FUE01; small tree of the old fallow.

Terminalia lucida Hoffmgg.: **tukur-iwa-'i** 'grasshopper-fruit-stem'; ANF01 FUE01; small tree of the old fallow.

COMMELINACEAE

Dichorisandra affinis Mart. ex Roem. & Schultes: **yamir-ran-ka'a** 'L-false-herb'; NONE; vine of the high forest.

CONNARACEAE

Connarus favosus Planch. **sipo-hu** 'vine-big;' **wipo-hu** 'vine-big' (G); NONE; large liana of the high forest.

Connaraceae sp. 1: **sipo-ran** 'vine-false'; NONE; large liana of the high forest.

Connaraceae sp. 2: **maha-sipo** 'grey brocket deer (*Mazama guazoubira*)-vine'; POI12; large liana of the high forest.

CONVOLVULACEAE

Calyobolus sp. 1: **sipo-ran** 'vine-false'; cipo tuíra; NONE; large liana of the high forest.

Ipomoea batatas Lam.: **yitik** 'L' (includes **yitik-howi**, **yitik-marū**, **yitik-marū-ši**, **yitik-pihun**, **yitik-tawa**, **yitik-tuwir**, **yitik-'i**, **yitik-piraŋ**); **matak** 'L' (= **mačitu** 'L') (G); **yitíka** 'L' (AS); **de'ti** 'L' (AR); **zitik** 'L' (T); sweet potato; batata doce; PFS02; traditionally cultivated prostrate vine found in swiddens and dooryard gardens.

Ipomoea phyllomega (Velloso) House **yitik-ran** 'L-false'; batatarana; morning glory; ANF01; ruderal creeper of young swidden edges and old swiddens.

Ipomea setofera Oir.: **āyaŋ-nami** 'divinity-year'; ANF02; vine of the old swidden.

Ipomoea aff. *squamosa* Choisy: **yitik-ran** 'sweet potato-false'; **zitik-ran** 'sweet potato-false' (T); morning glory; batatarana; ANF02; creeping vine of young and old swiddens.

Ipomoea sp. 1: **tayahu-sɨpo** 'white lipped peccary-vine'; ANF02; small vine of the high forest.

Maripa sp. 1: **sɨpo-ātā** 'vine-hard'; NONE; large liana of the high forest.

CRASSULACEAE

Bryophyllum sp. 1: **kure-nami** 'domestic-pig-ear'; kalanchöe; MED03; introduced, domesticated herb of the dooryard garden.

CUCURBITACEAE

Cayaponia sp. 1: **kawasu-ran** 'L-false'; MED24; vine of the old swidden.

Citrullus lanatus (Thunb.) Matsumi & Nakai: **waraši** 'L' (includes **waraši-te** and **waraši-pinī**); **yanači** 'L' (G); **yeú-arána** 'L-false' (AS); watermelon; melancia; SFS01; introduced, domesticated vine of dooryard gardens and swiddens.

Cucumis anguria L.: **waraši-ran** 'L-false'; West Indian gherkin; maxixe; SFS01; introduced, domesticated vine of the dooryard garden.

Cucurbita moschata Duch.: **yurumū** 'L' (includes **yurumū-pe, yurumū-pu'a,** and **yurumū-puku**); **urumū** 'L' (G); **yerimu** 'L' (AS); **zurumu** 'L' (T); crookneck or winter squash; jerimum; PFS01; traditionally cultivated, domesticated vine of the young swidden and dooryard gardens.

Gurania cissoides (Benth.) Cogn.: **kawasu-ran-puku** 'bottle gourd-false-long'; **ape-čiti-'i** '?-?-stem' (AR); HHI27 ANF02; scandent vine of the old fallow.

Gurania eriantha (Poepp. & Endl.) Cogn.: **kawasu-ran** 'L-false'; MED24; scandent vine of the old swidden.

Lagenaria siceraria (Molina) Standl.: **kawasu** 'L' (includes **kawasu-tiha** and **kawasu-ra'ɨr**); bottle gourd; cabaça; HHI04 PAC02 (this latter for **kawasura'ɨr** only); traditionally cultivated vine of the swidden.

Luffa cylindrica (L.) M.J. Roem.: **u'ɨ-hu-ruwɨ** 'arrow-big-blood'; **kumana-'iran** 'bean-false' (T); bath sponge; bucha; TLW16; introduced, domesticated, and scandent vine of the young swidden and dooryard garden.

Posadea sphaerocarpa Cogn.: **yere'a** 'L'; HHI03; vine of the old swidden, occasionally also seen in dooryard gardens.

Cucurbitaceae sp. 1: **murukuya-ran** 'passion fruit-false'; ANF02; vine of river banks.

CYCLANTHACEAE

Evodianthus funifer (Poit.) Lindm.: **sɨpo-hu** 'vine-big'; cipó timbó-açu; HHI08 HHI12 MED28 MAG12; scandent vine of the swamp forest.

CYPERACEAE

Cyperus corymbosus L.F. **pipir-ɨwa** 'tanager-fruit' (includes **pipir-ɨwa-hu**); PAC05 MAG19; introduced, cultivated, medium sedge of the dooryard garden and young swidden.

Diplasia karataefolia Rich.: **karaya-hɨ** 'enemy Indian tribe-hurt (?)'; **arira-kɨ** 'bat-?' (G); PAC12; terrestrial herb of the old swidden.

Scleria secans (L.) Urb.: **tɨrɨrɨ** 'rip'; **tamakisa** 'L' (T); capim navalha; ANF02; adhesive sedge of old swiddens.

Scleria sp. 1: **karaya-hɨ** 'enemy Indian tribe-hurt (?)'; PAC12; sedge of the high forest.

Scleria sp. 2: **karaya-pihun** 'enemy Indian tribe-black'; PAC12; sedge of the high forest.

Cyperaceae sp. 1: **kāpī-ran** 'grass-false'; NONE; sedge of the high forest.

DENNSTAEDTIACEAE

Pteridium aquilinium (L.) Kuhn.: **yakare-ka'a** 'jaguar herb'; bracken; samambaia; MED27; large and very rare fern of trailsides.

DILLENIACEAE

Davilla kunthii St. Hil.: **tɨrɨrɨ-sɨpo** 'rip-vine'; cipó de fogo; SFS08 ANF02 FUE02 MED29; large liana of the high forest.

Davilla nitida (Vahl.) Kubitzki: **tɨrɨrɨ-sɨpo** 'rip-vine'; cipó de fogo; SFS08 ANF02 FUE02 MED29; large liana of the high forest.

Doliocarpus sp. 1: **tɨrɨrɨ-sɨpo** 'rip-vine'; **muruči-ti-hipa** '*Byrsonima* spp.-grove-vine' (AR); **zapekuromoŋ-wɨpo** 'L-vine' (T); cipó d'água; SFS08 ANF02 FUE02 MED29; large liana of the high forest.

Tetracera volubilis L. ssp. *volubilis*: **tɨrɨrɨ-sɨpo** 'rip-vine'; cipó de fogo; SFS08 ANF02 FUE02 MED29; large liana of the high forest.

Tetracera willdenowiana; Schlechtd. ssp. *willdenowiana*; **tɨrɨrɨ-sɨpo** 'rip-vine'; SFS08 ANF02 FUE02 MED29; large liana of the high forest.

DIOSCOREACEAE

Dioscorea trifida L.: **kara** 'L' (includes **kara-māmā, kara-pe, kara-pihun**); **karã** (AR); **kara** (AS); **kara** (T); **kara** 'L' (G); yam; inhame; PFS02; traditionally cultivated scandent vine of young and old swiddens (the only *traditional* domesticate in old fallows, persisting there for forty or more years).

Dioscorea sp. 1: **kara-ran** 'yam-false'; **maratō** 'L' (G); wild yam; scandent vine of the high forest with an enormous, hard tuber.

Dioscorea sp. 2: **kara-hu** 'L-big'; **yowoɨ-kara** 'boa constrictor-L' (T); PFS02; traditionally cultivated, scandent vine of the old swidden.

Dioscorea sp. 3: **yakamī-tɨma-kaŋwer** 'trumpeter (bird)-leg-bone'; NONE; small terrestrial herb of the high forest.

EBENACEAE

Diospyros artanthifolia Mart.: **tamari-mɨra** 'tortoise shelled tamarind monkey-tree'; **wariw-apɨ'a-ran-'ɨw** 'howler monkey-female-false-stem' (T); **inamu-'ɨw-úna** 'curassow-stem-black' (AS); ANF02 PBC02 HHI11 TLW03 FUE01; small tree of river edges.

Diospyros duckei Sandw.: **tamari-mɨra** 'tortoise shelled tamarind monkey-tree'; **wɨra-kē'ē** 'tree-?' (G); ANF02 PBC02 HHI11 TLW03; small tree of the high forest.

Diospyros melinoni (Hiern.) A.C. Smith: **yanu-kiha-wɨ-'ɨ** 'spider-hammock-thin-stem'; PBC07 ANF02 FUE01; small tree of the old fallow.

ELAEOCARPACEAE

Sloanea grandiflora Smith: **murukuya-'ɨ-wɨ** 'passion fruit-stem-thin'; ANF02 FUE01; small tree of the high forest.

Sloanea grandis Ducke: **murukuya-'ɨ** 'passion fruit-stem'; **wɨra-i-čĩ-'ɨ** 'tree-small-white-stem' (G); ANF02 FUE01; small tree of the high forest.

Sloanea guianensis (Aubl.) Benth.: **murukuya-'ɨ** 'passion fruit-stem'; **uruku-ran-'ɨw** 'annatto-false-stem' (T); **wɨra-i-čĩ-'ɨ** 'tree-small-white-stem' (G); ANF02 FUE01; small tree of the high forest.

Sloanea porphyrocarpa Ducke: **murukuya-'ɨ-wɨ** 'passion fruit-stem-fine'; ANF02 FUE01; medium tree of the old fallow.

ERYTHROXYLACEAE

Erythroxylum cf. *leptoneurum* O.E. Schulz: **tapiša-'ɨ** 'broom-stem'; ANF02 FUE01; treelet of old swiddens.

Erythroxylum citrifolium A. St. H.L.: **mɨra-ātā** 'tree-hard' (= **mɨra-tawa-ran** 'tree-yellow-false'); ANF02; treelet of the old swidden.

EUPHORBIACEAE

Aparisthmium cordatum Baill.: **ara-kɨ'ī** 'macaw-chile pepper'; **uruku-ran-'ɨw** 'annatto-false-stem' (T); mameleiro; ANF02 FUE01; small tree of light gaps in the high forest, also on rock outcroppings.

Conceveiba guianensis Aubl.: **arapuha-mɨra** 'brocket deer-tree'; **wɨra-uhu** (?) 'tree-big' (G);

wɨra-ki'ɨw-ran 'tree-chile pepper-false' (T); iŋa-yúna 'inga-black' (AS); ANF02; small tree of the high forest.

Croton cajucara Benth.: **kurupi-yi-'ɨ** 'divinity-axe-stem'; **tata-kaya-'ɨ** 'fire-?-stem' (G); sacaca; ANF01 FUE01; small tree of the old fallow.

Croton cuneatus Klotesch.: **kara-yuru-'ɨ** 'yam-mouth-stem' (= **karapatu-'ɨ** 'L-stem'); **ka-hapa-čĩ** 'L-white' (T); sacaca; ANF02; POI12; common small tree of river margins.

Croton matourensis Aubl.: **kurupi-ši-'ɨ** 'divinity-feces-stem' (= **paku-mi'u-'ɨ** 'characin fish-food-stem'); **wɨra-kurum-hu-'ɨ** 'tree-?-big-stem' (G); sacaca; PBC04 FUE02 MED06; small tree of the swamp forest.

Dodecastigma integrifolium (Lanj.) Lanj. & Sandw.: **pa'i-mɨra** 'priest-tree'; **uha-ka'a-ran-'ɨw** 'crab-herb-false-stem' (T); **mɨrikə -'ɨ** 'woman-stem' (G); ANF02 FUE01; very common, small tree of the high forest.

Drypetes variabilis Vitt.: **kururu-'iran-'ɨ** 'toad-false-stem'; **wɨra-him-hu-'ɨ** 'tree-slippery-much-stem' (G); MED29; small tree of the high forest.

Euphorbia sp. 1: **kupi'i-pu'ã** 'termite-erect'; spurge; assacurana; MED11; tiny, uncultivated herb of dooryard gardens.

Hevea guianensis Aubl.: **širiŋi-'ɨ** 'L-stem'; rubber tree; seringa itaúba; NONE; decidedly rare, medium tree of swamp forest.

Jatropha curcas L.: **piyã-'ɨ** 'L-stem'; purge nut; pião branco; MED12 MED13 MED31; introduced, domesticated shrub only in dooryard gardens.

Jatropha gossypifolia L.: **piyã-'ɨ-pihun** 'L-stem-black'; purge nut; pião vermelho; MED12 MED13 MED31; Introduced, domesticated shrub found only in dooryard gardens.

Mabea angustifolia Spruce ex Benth.: **kašima-'ɨ** 'pipe-stem'; ANF02 HHI20 FUE01; treelet of the old swidden.

Mabea caudata Pax & Hoffm.: **kašima-'ɨ** 'pipe-stem'; **wəkəwə-'ɨ** 'Laughing falcon-stem' (G); **kačimi-'ɨw** 'pipe-stem' (T); pau de cachimbo; ANF02 HHI20 FUE01; small, fairly common tree of the high forest.

Mabea pohliana M. Arg.: **sɨpo-ran** 'vine-false'; **čimo-'ɨw** '*Derris utilis* (a fish poison)-stem' (T); NONE; small vine of river edges.

Mabea sp. 1: **kašima-'ɨ** 'pipe-stem'; pau de cachimbo; ANF02 HHI20 FUE01; small tree of the old fallow.

Magaritaria nobilis L.F.: **taŋwa-'ɨ** 'L-stem'; ANF02 FUE01; small, rare tree of the high forest.

Manihot brachyloba Muell. Arg.: **arapuha-mani'ɨ** 'red brocket deer-manioc'; **mačã** 'L' (AR); ANF02; nondomesticated shrub of old swiddens.

Manihot esculenta Crantz: **mani'ɨ** 'L' (includes **ara-rũ-mani'ɨ, mani'ɨ-hu, mani'ɨ-paitɨ, mani'ɨ-põ, mani'ɨ-puku, mani'ɨ-tuwɨr-ran, tapayũ-mani'ɨ, mani'ɨ-howɨ, mani'ɨ-pihun, mani'ɨ-pɨtaŋ, mani'ɨ-se, mani'ɨ-tawa** (= **mani'ɨ-te**), **mani'ɨ-tuwɨr, sarakur-mani'ɨ, šimo-kape-mani'ɨ, tikuwi, yararak-mani'ɨ, yaši-mani'ɨ**); **makaser** 'L' (includes **makaser-te, makaser-pɨraŋ, makaser-ran, makaser-tuwɨr**); **maniaka** 'L'; **mani'ɨw** (T) [bitter varieties only]; **mani'o** (AR) [bitter varieties only]; **tərəmõ** 'flour' (G) [bitter and sweet varieties]; **mani'ak** 'L' (AS) [bitter varieties only]; PFS02 (all varieties) SFS04 (**ara-rũ-mani'ɨ, mani'ɨ-tawa** only) MED36 (**mani'ɨ-se** only) MED27 (**mani'ɨ-se, mani'ɨ-tuwɨr** only); manioc or cassava; mandioca, macaxeira; traditionally cultivated herb of young swiddens, old swiddens, and occasionally dooryard gardens.

Manihot leptophylla Pax in Engler: **arapuha-mani'ɨ** 'red brocket deer-manioc'; **arapaha-mani'ɨ-wa** 'red brocket deer-manioc-fruit' (T); **arapʰa-mani'ɨ-i-'ɨ** 'red brocket deer-manioc-little-stem' (G); wild manioc; maniva de veado; ANF02; noncultivated shrub of young and old swiddens.

Manihot quinquepartita Huber ex Rogers et Appan: **arapuha-mani-'ɨ** 'brocket deer-manioc-stem'; **arapʰa-mani-'ɨ** 'red brocket deer-manioc-stem' (G); **mayi-'irĩ** 'manioc-stem-false'

(AR); wild manioc; maniva de veado; ANF01; very woody, treelet-like nondomesticated species of manioc found on young swidden edges and in old swiddens.

Phyllanthus martii Muell. Arg.: **pišū-ran** 'L-false'; **taŋara-wira** 'manakin-tree' (T); FUE02; treelet of river banks.

Phyllanthus niruri L.: **teyu-pitim-ran** 'skink-tobacco-false'; quebra pedra; NONE; uncultivated herb of dooryard gardens.

Phyllanthus urinaria L.: **ita-mira** 'stone-tree'; quebra pedra; MED09; small shrub of trailsides in old swiddens.

Ricinus comunis L.: **karapatu** 'L'; **kazapat** 'L' (T); **məmə-ra'ī** 'papaya-seed' (G); castor bean; mamona; TLW09 PAC11 ; introduced, cultivated treelet of dooryard gardens.

Sagotia racemosa Baill.: **mira-wawak** 'tree-spin'; **mirikə-'i** 'woman-stem' (G); **ka'a-'iw-pihun** 'forest-stem-black' (T); ANF02 MED11 MED17FUE01; small tree, very common in the high forest.

Sapium caspceolatum Huber: **wakura-mira-hu** 'nighthawk-tree-big'; ANF02 FUE02; very rare, medium tree of the old fallow.

Sapium lanceolatum Huber: **wakura-mira-hu** 'nighthawk-tree-big'; **kiripi-kawa-'i** 'divinity-calabash-stem' (G); **yuwa-'iwa** (AS); ANF02 FUE02; small tree of the old fallow.

FABACEAE

Andira retusa (Lam.) H.B.K.: **pu'i-piraŋ-pihun-'i** 'bead-reddish-black-stem'; **ira-pihun** 'tree-black' (G); andira-uchi; PAC02 COM02; medium tree of the old fallow.

Andira sp. 1: **yeyu-'i** 'kind of fish-stem'; PBC02 HHI06 HHI29 FUE01; very large tree of old fallows.

Arachis hypogaea L.: **manuwi** 'L'(includes **manuwi-tuwir** and **manuwi-piraŋ**); peanut; amendoim; SFS01; traditionally cultivated, terrestrial herb of swiddens and dooryard gardens.

Bowdichia nitida Spruce ex Benth.: **kumaru-'i-ran** 'L-stem-false'; sucupira da terra firme; FUE01; medium tree of the old fallow.

Coumarouna micrantha Ducke: **yupara-mira** 'kinkajou-tree'; ANF02 FUE02; very large tree of the high forest.

Dalbergia monetaria L.F.: **tayahu-iŋa** 'white–lipped peccary-*Inga* spp.' (=**iŋa-ran-'i** '*Inga* spp.-false-stem'); **tariri-iwipo** '?-vine' (T); Verônica branca; FUE02 ANF02; treelike liana of river banks.

Derris amazonica Killip.: **šimo-ran-sipo** 'L-false-vine'; **čimo-ran** 'L-false' (T); **ačī-ipa** 'head-vine' (AR); **a'i-mō-wipo** 'sloth-?-vine' (G); ANF02; large liana, facultative in the high forest and in swamp forests.

Derris utilis (A.C. Smith) Ducke: **šimo** 'L'; **kururu-čimo** 'toad-fish poison' (G); timbó; TLW07; large llana of the high forest.

Derris sp. 1: **šimo-ran** 'L-false'; ANF02; large liana of swamp forests and old fallow.

Desmodium adescendens (Sw.) DC.: **ka'a-pe** 'herb-flat'; carrapicho; ANF03; introduced, cultivated, prostrate herb of dooryard gardens in Gurupiuna only.

Dioclea reflexa Hook. F.: **yahi-sipo** 'moon-vine'; **karuwa-'i-rimo** 'divinity-little-vine' (G); orelha de veado; POI12 ANF02; scandent, fairly common vine of old fallows.

Diplotropis purpurea (Rich.) Amsh.: **yeyu-'i-ran** 'kind of fish-stem-false'; sucupira; PBC01 HHI06 HHI29 FUE01; large tree of the high forest.

Dipteryx odorata L.: **kumaru-'i** 'L-tree'; **kumaru-'iw** 'L-stem' (T); cumaruzeiro; PAC05 MED23 POI12; large tree of the high forest.

Machaerium ferox (Mart. ex Benth.) Ducke: **yuti'āi-rimo** '?-stem'; **a'i-wipo** 'sloth-vine' (T); **u'i-ruwa-pahā** 'flour-face-?' (AR); NONE; uncommon, small vine of river edges.

Ormosia coccinea (Aubl.) Jacks.: **pu'i-piraŋ-'i** 'bead-reddish-stem'; tento; PAC02 COM02; enormous tree of the high forest.

Phaseolus lunatus L.: **kamana** 'L'; **kamána-'i** 'L-little' (AR); **kumana** 'L' (T); PFS01; traditionally cultivated, uncommon vine of young swiddens.

Platypodium elegans Vog.: **panā-ka'a-mira** 'butterfly-herb-tree'; **yami-'i** 'squeeze-stem'; **mani'o'o-'i** 'manioc-stem' (G); ANF02 FUE01; medium tree of the old fallow.

Poecilanthe effusa (Hub.) Ducke: **moi-mira** 'snake-tree'; **ara-kowā-'i** 'macaw-vessel-stem' (G); gema de ovo; MED25; medium tree facultative to high forests and old fallows.

Taralea oppositifolia Aubl.: **kururu-'i** 'toad-stem'; **kumaru-ran-'iw** 'L-false-stem' (T); **ira-pō** 'tree-other' (G); cumarurana; ANF01 FUE01; medium tree of the swamp forest.

Tephrosia sinapou (Buchot) A. Chev.: **šimo-'i** '*Derris utilis* (nondomesticated fish poison)-little'; TLW07; introduced domesticated herb of young swiddens.

Vigna adnantha (G.F. Meyer) Marechal: **kamana-'i-tuwir** 'L-small-white'; **komana** (AS); PFS01; uncommon, introduced, cultivated vine of young swiddens.

FLACOURTIACEAE

Banara guianensis Aubl.: **mitū-eha-'i** 'curassow-eye-stem'; **haíra-maka-'i** 'Blue headed parrot-?-stem' (G); ANF02 FUE01; small tree of old swiddens.

Casearia arborea (L.C. Richard) Urban: **inamu-kiha-wi-'i** 'curassow-hammock-thin-stem'; **wənənə-rána** 'L-false' (G); **awači-tu'awī-rána** 'maize-?-false' (AS); ANF02 FUE01; small tree of the old fallow.

Casearia decandra Jacq.: **karaí-pe-ran-'i** 'nonIndian-trail-false-stem'; NONE; medium tree of the old fallow.

Casearia javitensis H.B.K.: **kase-ran-'i** 'coffee-false-stem'; **čai-pin-čī-'i** '?-stripe-white-stem' (G); ANF02; small tree of old fallows.

Laetia procera (Poepp. et Endl.) Eichl.: **parani-'i** 'L-stem'; pau jacaré; ANF02 HHI29 MED29 FUE01; large tree of the swamp forest.

Lindackeria latifolia Benth.: **ka'a-ro-tiapu-'i** 'herb-leaf-noise-stem' (= **kupapa-'iran-'i** 'L-false-stem'); **wari-rána-ka'aw** 'howler monkey-false-calabash' (G); **aniŋa-kiwawa-'iwa** 'divinity-comb-stem' (AS); **ara-ra-i** 'macaw-down-stem' (AR); farinha seca; ANF02 MAG06 FUE02; small tree, very common in old fallows.

Lindackeria sp. 1: **ka'a-ro-tiapu-i** 'herb-leaf-noise-stem' (= **kupapa-'iran-'i** 'L-false-stem'); MAG06 FUE02 ANF02; very uncommon, small tree of light gaps in the high forest.

Neoptychocarpus apodanthus Kuhlm.: **yaši-mukuku-'i** 'yellow footed tortoise-*Licania* spp.-tree'; ANF01 MED37 FUE01; uncommon shrub of old swiddens.

Ryania speciosa Vahl.: **parani-ran-'i** 'L-false-stem'; mata calado; FUE02; rare treelet of old fallows.

Flacourtiaceae sp. 1: **yaši-mukuku-'i** 'yellow footed tortoise-*Licania* spp.-stem'; ANF01 MED37 FUE01; small tree of old fallows.

GESNERIACEAE

Drymonia coccinea (Aubl.) Weihl.: **akuši-sipo** 'agouti-vine' (= **akuši-nami** 'agouti-ear'); ANF01; epiphyte of the high forest.

GOUPIACEAE

Goupia glabra Aubl.: **kupi'i-'i** 'termite-stem'; **yaku-wi-hu-'i** 'guan-thin-big-stem' (G); cupiúba; SFS01 ANF01 HHI31 MIS03 FUE02; large tree found in light gaps of the high forest.

HAEMODORACEAE

Xiphidium caeruleum Aubl.: **irakahu-ka'a** 'weasel-herb'; **tučā-na-'i-ha** 'toucan-?-small-down' (AR); **yawa-waya** 'jaguar-guava' (G); **marakatai-ran** 'ginger root-false' (T); POI18; terrestrial herb of the old swidden.

HELICONIACEAE

Heliconia acuminata L.C. Rich.: **tayahu-pako-ro** 'white lipped peccary-banana-leaf'; **pai-didə-čĩ** 'L-white' (AR); bico de tucano; ANF02; terrestrial herb of the high forest.

Heliconia bihai L.: **tayahu-pako-ro** 'white lipped peccary-banana-leaf'; **paku-ran-'ɨw** 'banana-false-stem' (T); **paididə** 'L' (AR); bico de tucano; ANF02; terrestrial herb of the old swidden.

Heliconia chartacea Lane ex Barr.: **tayahu-pako-ro** 'white lipped peccary-banana-leaf'; **paididə-čɜ̃** 'L-black' (AR); bico de tucano; ANF02; terrestrial herb of the high forest.

HUMIRIACEAE

Sacoglottis amazonica Mart.: **paruru-'ɨ** 'kind of fish-stem'; uchirana; SFS01 ANF02 FUE01; medium tree of the old fallow, occasionally seen in the high forest.

Sacoglottis guianensis Benth.: **paruru-'ɨ** 'kind of fish-stem'; **paruru-'ɨw** 'kind of fish-stem' (T); **čaperai-čĩ-'ɨ** 'L-white-stem' (G); uchirana; SFS01 ANF02 FUE01; medium tree of the old fallow, occasionally seen in the high forest.

Sacoglottis sp. 1: **yaši-wa-'ɨ** 'yellow footed tortoise-fruit-stem'; uchirana; SFS01 ANF02 FUE01; medium tree of the high forest.

ICACINACEAE

Dendrobangia boliviana Rusby: **yanaí-mɨra** 'kind of turtle-tree'; ANF02; small tree of the high forest.

Humirianthera duckei Huber: **sɨpo-tawa** 'vine-yellow'; **ihipa-péwa** 'vine-flat' (AS); ANF02; large liana of the high forest.

[INDETERMINATE 1]

Indeterminate genus 2: **waya-mɨra-'a** 'guava-tree-fruit'; NONE; small tree of the high forest.

[INDETERMINATE 2]

Indeterminate genus 3: **yahɨ-sɨpo-ran** 'moon-vine-false'; NONE; small liana of the high forest.

[INDETERMINATE 3]

Indeterminate genus 4: **maŋwa-šape-'ɨ** *Leptodactylus* sp. frog-back-stem'; NONE; small liana of the high forest.

[INDETERMINATE 7]

Indeterminate genus 5: **anã-sɨpo** 'kinsman-vine'; NONE; liana of the high forest.

[INDETERMINATE 22]

Indeterminate genus 09. **sɨpo-tal** 'vine-spicy'; POI12; liana of the high forest.

LABIATEAE

Ocimum sp. 1: **arapawak** 'L'; basil; SEA01; introduced, cultivated pot herb of dooryard gardens.

Ocimum micranthum Willd.: **arapawak** 'L'; SEA01; introduced, cultivated pot herb of dooryard gardens.

LAURACEAE

Aniba citrifolia (Nees) Mez.: **ayu-'ɨ** 'L-stem'; **azu-'ɨw** 'L-stem' (T); PBC02 ANF01 FUE01; rare, medium tree of river margins.

Endlicheria verticellata Mez.: **ayu-'ɨ** 'L-stem'; louro; PBC02 ANF01 FUE01; medium tree of the high forest.

Licaria brasiliensis (Nees) Kost.: **ayu-wa-hu-'ɨ** 'L-fruit-big-stem'; **ayu-'ɨ** 'L-stem' (AR); louro roxo; ANF01 FUE01; medium tree of the high forest.

Licaria debilis (Mez) Kost.: **ayu-'iran-'ɨ** 'L-false-stem'; PAC02 ANF01 FUE01; very rare, small tree of the old fallow.

Licaria sp. 1: **ayu-'ɨ** 'L-stem'; louro; PBC02 ANF01 FUE01; medium tree of the high forest.

Ocotea amazonica (Meiss.) Mez.: **ayu-'ɨ-pihun** 'L-stem-black'; louro; PBC02 ANF01 FUE01; medium tree of the high forest.

Ocotea canaliculata (L.C. Rich.) Mez.: **ayu-'ɨ-ran** 'L-stem-false'; louro; ANF02 FUE01 MIS04; medium tree of the high forest.

Ocotea caudata Mez.: **ayu-'ɨ-pihun** 'L-stem-black'; **azu-ɨw-pihun** 'L-stem-black' (T); **wa-yuwa-'ɨ** 'L-stem' (G); **ayu-'ɨwa** 'L-stem' (AS); **ayu-'ɨ** 'L-stem' (AR); louro; PBC02 ANF01 FUE01; large tree of the high forest.

Ocotea costata Mez.: **ayu-wa-puku-'ɨ** 'L-fruit-long-stem'; louro; PBC02 ANF01 FUE01; small tree of the high forest.

Ocotea fasciculata (Ness) Mez.: **ayu-'ɨ-tuwɨr** 'L-stem-white'; **wayuwa-'ɨ** 'L-stem' (G); louro; ANF01 FUE01; very rare, medium tree of the old fallow.

Ocotea glomerata (Nees) Mez.: **maha-yu-wa-'ɨ** 'grey brocket deer (*Mazama guazoubira*)-L-fruit-stem' (= **maha-mɨra** 'grey brocket deer-tree'); ANF02 FUE01 MAG02; small tree of old swiddens, also seen in high forest.

Ocotea guianensis Aubl.: **maha-yu-wa-'ɨ** 'grey brocket deer (*Mazama guazoubira*)-L-fruit-stem' (= **maha-mɨra** 'grey brocket deer-tree'); louro prata; MAG02 ANF01 FUE01; medium tree of the high forest.

Ocotea opifera Mart.: **maha-yu-wa-'ɨ** 'grey brocket deer-L-fruit-stem' (= **maha-mɨra** 'grey brocket deer-tree'); **təkənə-hə-'ɨ** 'toucan-big-stem' (G); **ayu-'ɨ** 'L-stem' (AR); louro alca-trão; ANF01 MAG02 FUE01; treelet of the high forest.

Ocotea rubra Mez.: **ayu-wa-tā-'ɨ** 'L-fruit-hard-stem'; **azu-ɨw-pɨraŋ** 'L-stem-reddish' (T); louro vermelho; ANF01 FUE01; very large tree of the high forest.

Ocotea silvae Vattimo-Gil sp. nov.: **ayu-wa-puku-'ɨ** 'L-fruit-long-stem'; ANF01 FUE01; large tree of the old fallow.

Ocotea sp. 3: **ayu-'ɨ-ran** 'L-stem-false'; **wayuwa-'ɨ** 'L-stem' (G); louro; ANF02 FUE01 MIS04; small tree of the high forest.

Persea americana L.: **awakaši** 'L'; SFS01; treelet of dooryard gardens ('lost' or **kāyī** in most of the Turiaçu basin).

Lauraceae sp. 1: **ayu-'ɨ-pɨtaŋ** 'L-stem-red'; louro; PBC02 ANF01 FUE01; medium tree of the high forest.

Lauraceae sp. 2: **ayu-'ɨ-te** 'L-stem-true'; louro; PBC02 ANF01 FUE01; small tree of the high forest.

Lauraceae sp. 3: **ayu-wa-puku-'ɨ** 'L-fruit-long-stem'; louro; ANF02 FUE01; small tree of the high forest.

LECYTHIDACEAE

Bertholletia excelsa Humboldt & Bonpland: **kātāi-'ɨ** 'L-stem'; **ya-'ɨ** 'L-stem' (AR); **ya-'ɨwa** 'L-stem' (AS); Brazil nut tree; castanheira; MYT01; enormous tree of old fallows west of the Ka'apor habitat, known from folklore.

Couratari guianensis Aublet: **pɨtɨm-'ɨ** 'tobacco-stem'; **mətəm-hu;** 'curassow-big' (= **pɨtɨm-hu** 'tobacco-big') (G); toari; ANF02 TLW02 HHI19 FUE02; large tree of the high forest.

Couratari oblongifolia Ducke & Kunth.: **pɨtɨm-inem-'ɨ** 'tobacco-rancid-stem'; **pɨtɨm-ɨ** 'tobacco-stem' (G); **petɨm-'ɨ** 'tobacco-stem' (AR); toari; ANF01 TLW02 FUE01; very large tree of old fallows.

Eschweilera amazonica R. Kunth.: **parawa-'ɨ-wi** 'Mealy parrot-stem-fine'; **wɨri-'ɨ** 'lashing material-stem' (G); ANF01 PBC01 PBC02 PBC07 TLW02 HHI26 MED08 MED14 FUE01; rare, small tree of old fallows.

Eschweilera apiculata (Miers) A.C. Smith: **parawa-'ɨ-pihun** 'Mealy parrot-stem-black';

wɨri-'ɨ 'lashing material-stem' (G); matámatá; ANF01 PBC01 PBC02 PBC07 TLW02 HHI26 MED08 MED14 FUE01; medium tree of the high forest.

Eschweilera coriacea (A.P. Candolle) Mart. ex. Berg: **parawa-'ɨ** 'Mealy parrot-stem'; **awa-wiha-'ɨw-pihun** 'person-scalp-stem-black' (T); **wɨri-'ɨ** 'lashing material-stem' (G); **iwɨri-tiri-'ɨwa** 'lashing material-stem' (AS); **iwitir-'ɨ** 'lashing material-stem' (AR); matámatá; ANF01 PBC01 PBC02 PBC07 TLW02 HHI26 MED08 MED14 FUE02; extremely common medium tree in the high forest, somewhat less common in old fallow.

Eschweilera micrantha (Berg) Miers: **parawa-'ɨ-wi** 'Mealy parrot-tree-thin'; **wɨri-'ɨ** 'lashing material-stem' (G); matámatá; ANF01 PBC01 PBC02 PBC07 TLW02 HHI26 MED08 MED14 FUE01; very rare, small tree of old fallows.

Eschweilera obversa (Berg) Miers: **arar-uhu-kãtãi-'ɨ** 'macaw-big-Brazil nut-stem'; ANF01 TLW02 PBC01 FUE01; large tree of the high forest.

Eschweilera ovata (Cambess) Miers: **pɨtɨm-inem-ran-'ɨ** 'tobacco-fetid-false-stem'; **iwɨr-ɨ-pihun** 'lashing material-stem-black' (T); **wɨri-'ɨ** 'lashing material-stem' (G); **iwitiri-'ɨwa** 'lashing material-stem' (AS); TLW02 PBC01; small tree of river margins and old fallows.

Eschweilera pedicellata (Richard) Mori: **parawa-'ɨ** 'Mealy parrot-stem'; **wai-ha-'ɨ** 'fruit-hair-stem' (G); matámatá; ANF01 PBC01 PBC02 PBC07 TLW02 HHI26 MED08 MED14 FUE01; very rare, small tree of old fallows.

Gustavia augusta L.: **mɨtũ-pusu-'ɨ** 'curassow-crop-stem'; **zanipa-ran-'ɨw** 'genipapo-false-stem' (T); **tərəkəmtərə-'ɨ** '?-stem' (G); **yanipa-ran-'ɨwa** 'genipapo-falso-stem' (AS); **yan-ipa-ran-'ɨ** 'genipapo-false-stem' (AR); geniparana; ANF02 MED14 MED25 MIS04 FUE01 TLW17; small tree of old fallows and swamp forests.

Lecythis chartacea Berg: **iwɨr-'ɨ-pɨtaŋ** 'lashing material-stem-red'; **iwɨr-'ɨw-pɨraŋ** 'lashing-material-stem-reddish' (T); **wai-ha-ɨ** 'fruit-hair-stem' (G); ANF01 PBC07 TLW02 FUE01 PBC02; uncommon, medium tree of the high forest.

Lecythis idatimon Aublet: **yaši-amɨr** 'yellow footed tortoise-deceased'; **pɨtata-'ɨ** 'L-stem' (G); caçador; ANF02 HHI12 TLW02 FUE01; very common, medium tree of the high forest.

Lecythis lurida (Miers) Mori: **iwɨr-'ɨ-ran** 'lashing material-stem-false'; **ya'ɨ-rána** 'Brazil nut tree-false' (AS); ANF02 TLW02 FUE01; medium tree of the high forest.

Lecythis pisonis Cambess.: **ya-pukai-'ɨ** 'we-scream-stem'; **ya-pukəi** 'we-scream' (AR); **ya-məkai-'ɨ** 'we-scream-stem' (G); **za-pukai-'ɨw** 'we-scream-stem' (T); sapucaia; PFS01 ANF01 TLW02 FUE01 PBC02; enormous tree predominantly of old fallows.

LILIACEAE

Allium cepa L.: **ma'e-wa-nem** 'some-fruit-fetid'; onion; cebola; SFS02; Pot herb, not present in the Ka'apor habitat (known from visits by outsiders, most Ka'apor have eaten this species).

Allium sativum L.: **ma'e-wa-nem** 'some-fruit-fetid'; garlic; alho; SFS02; Pot herb, not present in the Ka'apor habitat (known from visits by outsiders, most Ka'apor have eaten it).

LIMNOCHARITACEAE

Limnocharis sp. 1: **mayu-ka'a-ro** 'divinity-herb-leaf'; MYT02; small, terrestrial herb of swamp forests.

LOMARIOPSIDACEAE

Lomariopsis japurensis (Mart.) J. Smith: **wari-ruwai** 'howler monkey-tail'; samambaia; MAG09 MAG10; terrestrial fern of the high forest, especially light gaps.

LORANTHACEAE

Struthanthus marginatus Blume: **ma'e-wɨra-puši** 'some-bird-feces'; ANF02; parasitic epiphyte of old swiddens.

MALPIGHIACEAE

Banisteriopsis pubipetala (Adr. Juss.) Cuatr.: **sɨpo-pɨraŋ** 'vine-reddish'; **kururu-'i** 'toad-stem' (AR); ANF02 MED03 MED11 FUE02; very rare, large liana of old fallows.

Byrsonima laevigata (Poir.) DC.: **meraí-pɨraŋ-'ɨ** 'anteater-reddish-stem'; muruçi do mato; PFS01 ANF01 FUE01; small tree of the high forest.

Byrsonima stipulacea Juss.: **meraí-ātā-'ɨ** 'anteater-hard-stem'(=**meraí-tawa-'ɨ-ran** 'anteater-yellow-stem-false' ; muruçi do mato; SFS01; small tree of the high forest.

Byrsonima sp. 5: **meraí-tawa-'ɨ** 'anteater-yellow-stem'; muruçi do mato; SFS01; small tree of the high forest.

Malpighia sp.1: **ma'e-ɨwa-pɨraŋ-ran** 'some-fruit-reddish-false'; West Indian cherry; acerola; PFS01; introduced, cultivated shrub of dooryard gardens.

Malpighia sp.2: **ma'e-ɨwa-pɨraŋ-ran** 'some-fruit-reddish-false'; West Indian cherry; acerola; PFS01; introduced, cultivated shrub of dooryard gardens.

Mascagnia sp. 1 **sɨpo-tuwɨr** 'vine-white'; NONE; large liana of old fallows.

Schwannia sp. 1: **mišik-sɨpo** 'roast-vine'; **wɨpo-hu** 'vine-big' (G); MED11; vine of the high forest.

Stigmaphyllon hypoleucum Miz.: **so'o-ran-sɨpo** 'game-false-vine'; MED06; fairly common, medium liana of the high forest.

Tetrapterys styloptera A. Juss.: **sɨpo-pɨraŋ** 'vine-reddish'; ANF02 MED03 MED11 FUE02; large liana of old fallows.

MALVACEAE

Gossypium barbadense L.: **maneyu** 'L' (includes **maneyu-pɨraŋ** and **maneyu-tuwɨr**); **manizu** 'L' (T); **miniyu** 'L' (AR); **aminiyu** (AS); sea island cotton; algodão; TLW18 HHI15 PAC01; traditionally cultivated shrub normally found only in old swiddens.

Hibiscus rosa-sinensis L.: **tupā-ka'a** 'thunder-herb'; hibiscus; papola; NONE; recently introduced domesticate, once planted at the now extinct FUNAI post on the Turiaçu River.

Sida cordifolia L.: **mɨra-kɨrawa-ran** 'tree-caroá (*Neoglaziovia variegata*)-false'; malva branca; TLW01; terrestrial, uncultivated herb of young swiddens.

Sida santaremensis Mont.: **mɨra-kɨrawa-ran** 'tree-caroá (*Neoglaziovia variegata*)-false'; erva relógio; TLW01; shrub of old swiddens.

Urena lobata L.: **mɨra-kɨrawa** 'tree-caroá (*Neoglaziovia variegata*)'; **wɨra-kurawa** 'tree-caroá (T); **iwɨr** 'lashing material' (G); aramina fiber; malva; TLW01 TLW02; shrub of old swiddens.

MARANTACEAE

Calathea sp. 1: **suruwi-ka'a** 'surubim fish-herb'; **ka'a-pō** 'herb-other' (T); **ka'a-ta'ī** 'herb-small' (G); **tuwa-čiŋo** (AS); ANF02 HHI27; terrestrial herb of the old fallow.

Ctenanthe sp. 1: **tayahu-manuwi** 'white—lipped peccary-peanut'; **uru-wɨw-ran** 'ground pheasant (*Odontophorus*)-stem-false' (T) SFS03 ANF02; fairly common, rhizomatous herb of the high forest.

Ischnosiphon arouma (Aubl.) Koern.: **warumā-te** 'L-true'; **uru-wɨw-wa-'ɨ** 'ground pheasant (*Odontophorus*)-stem-fruit-small' (T); guarumā; SFS02 HHI12 HHI08 HHI09 ; terrestrial herb of the swamp forest.

Ischnosiphon graciles (Rudge) Koern.: **warumā-širik** 'L-dried out'; arumā canela; POI14; terrestrial herb of the high forest.

Ischnosiphon obliquus (Rudge) Koern.: **ka'a-ro-tuwɨr** 'herb-leaf-white'; **ariwɨ** 'surubim fish' (G); **uru-'ɨ** 'L-stem' (AR); HHI12; terrestrial herb of the swamp forest.

Ischnosiphon petrolatus (rudge) L. And.: **warumā-ra'i** 'L-little'; arumā; ANF02; terrestrial herb of the transitional forest between swamp and *terra firme*.

Ischnosiphon sp. 1: **warumā-hu** 'L-big'; HHI12; terrestrial herb of the high forest, especially in light gaps.

Ischnosiphon sp. 2: **ka'a-ro-hātā** 'herb-leaf-hard'; HHI23; terrestrial herb of the high forest.

Maranta protracta Mig.: **warumā-šī** 'L-white'; ANF02; shrub of the high forest.

Marantaceae sp. 1: **tayahu-manuwi** 'white—lipped peccary-peanut'; SFS03 ANF02; tiny terrestrial herb of the high forest.

Marantaceae sp. 2: **tayahu-manuwi-ran** 'white—lipped peccary-peanut-false'; NONE; terrestrial herb of the high forest.

Marantaceae sp. 3: **ka'a-ro-šī** 'herb-leaf-white'; HHI12; terrestrial herb of the high forest.

Marantaceae sp. 4: **yanaí-ka'a-ro** 'kind of turtle-herb-leaf'; HHI24; terrestrial herb of the high forest.

MARCGRAVIACEAE

Norantea guianensis Aubl.: **arar-uhu-wai-rimo** 'macaw-big-tail-vine'; **tarakwa-wĩpo-ran** '*Camponatus* spp. ant-vine-tail' (T); rabo de arara; MAG11; arboreal, epiphytic herb of the high forest.

MELASTOMATACEAE

Aciotis purpurescens (Aubl.) Triana: **yu'i-ka'a** 'kind of frog-herb'; folha azeda; ANF02; terrestrial herb of the old swidden.

Bellucia grossularioides (L.) Triana: **mu-'ĩ** 'agentive prefix-stem'; **məkərə-to-wĩra** 'opossum-?-tree' (G); **papater** 'L' (T); goiabinha de anta; SFS01 ANF02 PAC08; large tree facultative to swamp forests and old swiddens.

Henrietta spruceana Cogn.: **taŋara-tai-ran-'ĩ** 'manakin-spice-false-stem'; **taŋara-wĩra-iapo-pe-har** 'manakin-tree-swamp-at-place' (T); ANF02 FUE02; treelet of river margins.

Miconia ceramicarpa (DC.) Cogn.: **yu'i-ka'a** 'kind of frog-herb'; canela de velho; ANF02; shrub of the high forest.

Miconia cf. *kappleri* Naud.: **taŋara-tai-'ĩ** 'manakin-spice-stem'; ANF01 FUE01; medium to large tree of the high forest.

Miconia nervosa (Smith) Triana: **taŋara-tai-'ĩ** 'manakin-spice-stem'; ANF01 FUE01; medium tree of the high forest.

Miconia serialis DC.: **taŋara-tai-'ĩ** 'manakin-spice-stem'; ANF01 FUE01; small tree of the old fallow.

Miconia serrulata (DC.) Naud.: **taŋara-tai-'ĩ** 'manakin-spice-stem'; **uru-wĩra** '*Odontophorus* (pheasant)-tree' (T); ANF01 FUE01; small tree of river margins.

Miconia surinamensis Gleason: **taŋara-tai-'ĩ** 'manakin-spice-stem'; ANF01 FUE01; medium tree of the high forest.

Miconia sp. 1: **taŋara-tai-'ĩ** 'manakin-spice-stem'; ANF01 FUE01; medium tree of the high forest.

Mouriri guianensis Aubl.: **mĩra-'i-'a-'ĩ** 'tree-small-fruit-stem'; SFS01 ANF01 FUE01; small tree of the old fallow.

Mouriri huberi Cogn.: **mĩra-'i-'a-'ĩ** 'tree-small-fruit-stem'; mamãozinho; SFS01; rare, medium tree of the old fallow.

Mouriri sagotiana Triana: **mĩra-'i-'a-'ĩ** 'tree-small-fruit-stem'; **irar-'i-'ĩ** 'L-small-stem' (G); SFS01; uncommon, small tree of old fallows.

Mouriri trunciflora Ducke: **mĩra-'i-'a-'ĩ** 'tree-small-fruit-stem'; mamãozinho; SFS01 ANF01 FUE01; small tree of old fallows.

Myriaspora egensis DC.: **mu-'ĩ-ran** 'agentive prefix-stem-false'; ANF02; small tree of old swiddens.

MELIACEAE

Carapa guianensis Aubl.: **yani-ro-'ĩ** 'oil-bitter-stem' (includes **yani-ro-'ĩ-pĩraŋ** and **yani-ro-'ĩ-tuwĩr**); **zani-ro-'ĩw** 'oil-bitter-stem' (T); **yare-ro-'ĩ** 'oil-bitter-stem' (G); crabwood tree; andiroba; ANF01 PAC09 MED33 COM05; large tree—one variety found in high forest, the other more commonly found in swamp forest.

Cedrela fissilis Vell.: **irar-ĩ** 'L-tree' (= **patuwa-'ĩ** 'feather box-stem'); **yare-ro-ran-'ĩ** 'oil-bitter-false-stem' (G); **tata-kačǐ-'ɨwa** 'fire-smoke-stem' (AS); tropical cedar; cedro; HHI17 HHI28 MYT05; medium tree of the high forest.

Guarea guidonia (L.) Sleumer: **yaku-širi-'ɨ-pɨtaŋ** 'Spix's guan-?-stem-red'(= **waruwa-'iran-'ɨ** 'mirror-false-stem'); **wəkəwə-'ɨw** 'Laughing falcon-stem' (T); **ira-pərəhəm-'ɨ** 'tree-?-stem' (G); **pia-'i-rī** 'L-stem-false' (AR); jitó; ANF02 FUE01; large tree of old fallows.

Guarea sp. 1: **yaku-širi-'iran-'ɨ** 'Spix's guan-?-false-stem'; (= **waruwa-'iran-'ɨ** 'mirror-false-stem'); jitó; ANF02 FUE01; large tree of old fallows.

Guarea kunthiana A. Juss.: **yaku-širi-hu-'ɨ** 'Spix's guan-?-big-stem'; (= **waruwa-'iran-'ɨ** 'mirror-false-stem'); **təkənə-haíra-'ɨ** 'toucan-blue headed parrot-stem' (G); jitó; ANF01 FUE01; large tree of old fallows.

Guarea macrophylla Vahl. ssp. *pachycarpa* (C.DC.) Penn.: **yaku-širi-hu-'ɨ** 'Spix's guan-?-big-stem' (= **waruwa-'iran-'ɨ** 'mirror-false-stem'); **yare-ro-ran-'ɨ** 'oil-bitter-false-stem' (G); jitó; ANF01 FUE01; very large tree of the old fallow.

Trichilia micrantha Benth.: **yaku-širi-'ɨ-tuwɨr** 'Spix's guan-?-stem-white' (= **waruwa-'iran-'ɨ** 'mirror-false-stem'); **zani-ro-ran-'ɨw** 'oil-bitter-false-stem' (T); **ka'a-čowa-'ɨ** '*Cebus* monkey-calabash bowl-stem' (G); caxuá; ANF01 FUE01; small tree of the old fallow.

Trichilia quadrijuga Kunth. ssp. *quadrijuga*: **yaku-širi-'ɨ-tawa** 'Spix's guan-?-stem-yellow' (= **waruwa-'iran-'ɨ** 'mirror-false-stem'); **ka'i-čowa-'ɨ** '*Cebus* monkey-calabash bowl-stem' (G); **pia-'i** 'L-stem' (AR); **ɨwɨra-pɨtik-'ɨwa** 'tree-?-stem' (AS); **zani-ro-ran-'ɨw** 'oil-bitter-false-stem'; caxuá; ANF02 HHI06 FUE01; small tree of the old fallow.

Trichilia verrocosa C. DC.: **yaku-širi-hu-'ɨ** 'Spix's guan-hide-big-stem' (= **waruwa-'iran-'ɨ** 'mirror-false-stem'); caxuá; ANF01 FUE01; uncommon, medium tree of old fallows.

MENDONCIACEAE

Mendoncia hoffmannseggiana Nees: **pikahu-āmpūi-rimo** 'pigeon-bill-vine'; MAG20; fairly common, small vine of the high forest.

Mendoncia sp. 1: **pikahu-āmpūi-rimo** 'pigeon-bill-vine'; MAG20; small vine of the high forest.

MENISPERMACEAE

Abuta grandifolia (Mar.) Sandw.: **aputa** 'L'; **wari-makwa-'ɨ** 'howler monkey-?-stem' (G); cipó abuta; MED26 MED34; large liana of the high forest, sometimes appearing shrub-like.

Caryomene foveolata Barneby & Krukoff: **tapamɨr** 'L'; **ihipa-kaúwa** 'vine-insanity' (AS); SFS04; large liana of the high forest.

Cissampelos sp. 1: **āyaŋ-āmpūi-putɨr** 'divinity-nose-flower'; NONE; large liana of the high forest.

MIMOSACEAE

Acacia multipinnata Ducke: **yikiri-'ɨ** 'L-stem' (= **yikiri-sɨpo** 'L-vine'); **yami-'ɨ** '?-stem' (G); **yu-'me'e** 'spine-something' (AS); juquirí; MAG08 FUE01; fairly common treelet of the high forest.

Acacia paniculata Willd.: **šiŋ-i** 'L-stem'; rabo de camaleão; NONE; small tree of the high forest.

Enterolobium sp. nov.: **šimo-'ɨ** '*Derris utilis* (a fish poison)-stem'; orelha de negro; PAC03 FUE02; rare, medium tree of the old swidden.

Hymenolobium excelsum Ducke: **pani-'ɨ** 'L-stem'; **warara-'ɨ** 'crab-stem' (G); angelim da mata; ANF02 MIS03; enormous tree of the high forest.

Inga alba Willd.: **iŋa-šiši-'ɨ** 'L-smooth-stem'; **iŋa-pi-'ɨw** 'L-thin-stem' (T); **čičipe-'ɨ** 'L-stem' (G); **iŋa-i-yúna** 'L-little-black' (AS); **čiči-'i** 'smooth-stem' (AR); ingatitica; SFS01 ANF01 PAC08; fairly common, medium tree of old fallows.

Inga auristellae Harms: iŋa-pērē-'ɨ 'L-?-stem' (=iŋa-tawa-'ɨ 'L-yellow-stem'); čičipe-'ɨ 'L-stem' (G); ña-pəkə-'i 'L-long-stem' (AR); iŋa-pi'ɨw-him 'L-thin-slippery' (T); ingá; SFS01 ANF01 FUE01; fairly common, small tree of old fallows.

Inga brevialata Ducke: iŋa-šiši-'ɨ; 'L-smooth-stem'; ingá; SFS01 ANF01 PAC08 FUE01; small tree of old fallow forest.

Inga capitata Desv.: iŋa-hu-'ɨ 'L-big-stem'; tapi'i-riŋa 'tapir-L' (T); čičipe-'ɨ 'L-stem' (G); kururu-iŋa 'toad-L' (AS); ña-pəkə-'i 'L-long-stem' (AR); ingá-açu; SFS01 ANF01 FUE01; small tree of the high forest.

Inga cinammonea Spruce ex Benth.: iŋa-hu-'ɨ 'L-big-stem'; ingá canela; SFS01 ANF01 FUE01; small tree of the high forest.

Inga fagifolia (L.) Willd.: kaŋwaruhu-iŋa 'paca-L'; čičipe-'ɨ 'L-stem' (G); ingá; SFS01 ANF01 FUE01; small tree of the high forest.

Inga falcistipula Ducke: kaŋwaruhu-iŋa 'paca-L'; čičipe-'ɨ 'L-stem' (G); ingatitica; SFS01 ANF01 FUE01; small tree of old fallows.

Inga gracilifolia Ducke: taír-iŋa-'ɨ 'Blue headed parrot-L-stem'; iŋa-pi-'ɨw-ran 'L-thin-stem-false' (T); angelim; ANF02 FUE01; small to medium tree facultative to high forest and old fallows.

Inga heterophylla Willd.: iŋa-pērē-'ɨ 'L-?-stem' (=iŋa-tawa-'ɨ 'L-yellow-stem'); čičipe-'ɨ 'L-stem' (G); ingatitica; SFS01 ANF01 FUE01; small tree of the old fallow.

Inga marginata Willd.: iŋa-pērē-'ɨ 'L-?-stem' (=iŋa-tawa-'ɨ 'L-yellow-stem'); čičipe-'ɨ 'L-stem' (G); iŋa-'i-una-'iwa 'L-small-black-stem' (AS); ingá; SFS01 ANF01 FUE01; small tree of the high forest.

Inga miriantha Poepp. et Endl.: iŋa-pērē-'ɨ 'L-?-stem' (=iŋa-tawa-'ɨ 'L-yellow-stem'); čičipe-'ɨ 'L-stem' (G); ingá; SFS01 ANF01 FUE01; small tree of old fallows.

Inga nobilis Willd.: iŋa-howɨ-'ɨ 'L-blue/green-stem'; ña-pəkə-'i 'L-long-stem' (AR); yurupi-rána-iŋa 'throat-false-L' (AS); ingá; SFS01 ANF01 FUE01; small tree of old fallows.

Inga paraensis Ducke: kaŋwaruhu-iŋa 'paca-L'; SFS01 ANF01 FUE01; small tree of the old fallow.

Inga rubiginosa (Rich.) DC.: tapi'ir-iŋa-'ɨ 'tapir-L-stem'; čičipe-'ɨ 'L-stem' (G); iŋa-pe-iwi 'L-flat-thin' (AS); ña-pəkə-'i 'L-long-stem' (AR); ingá; SFS01 ANF01 FUE01; small tree of the old fallow.

Inga splendens Willd.: tapi'ir-namir-iŋa-'ɨ 'tapir-ear-L-stem'; ingá; SFS01 ANF01 FUE01; small tree of the high forest.

Inga thibaudiana DC.: iŋa-pitaŋ-'ɨ 'L—red-stem'; iŋa-wi-zu-'ɨw 'L-thin-yellow-stem' (T); muruwi-yawa-iŋa '?-jaguar-L' (AS); ña-pəkə-'i 'L-long-stem' (AR); čičipe-'ɨ 'L-stem' (G); ingá; SFS01 ANF01 FUE01; small tree of the high forest.

Inga sp. 5: iŋa-ha-'ɨ 'L-hair-stem'; yurupɨr-iŋa 'throat-L' (AS); ingá; SFS01 ANF01 FUE01; small tree of the swamp forest.

Inga sp. 6: tapi'ir-nami-iŋa-'ɨ 'tapir-ear-L-stem'; ingá; ANF01 SFS01 FUE01; treelet of the high forest.

Inga sp. 7: iŋa-howɨ-'ɨ 'L-blue/green-stem'; ingá; ANF01 SFS01 FUE01; small tree of the high forest.

Inga sp. 8: tapi'ir-namir-iŋa-'ɨ 'tapir-ear-L-stem'; ingá; SFS01 ANF01 FUE01; small tree of the high forest.

Inga sp. 9: iŋa-howɨ-'ɨ 'L-green-stem'; ingá; SFS01 ANF01 FUE01; small tree of the high forest.

Inga sp. 10: musu-iŋa 'eel-L' (=iŋa-pō-'ɨ 'L-other-stem'); ingá; SFS01 ANF01 FUE01; small tree of old fallow.

Newtonia psilostachya (DC.) Brenan: kɨkɨ-'ɨ 'monster-stem'; čimo-'ɨ '*Derris utilis* (a fish poison)-stem' (G); čimo-'ɨw '*Derris utilis*-stem' (T); faveira folha fina; ANF02 HHI31 FUE01; fairly common, large tree of the high forest.

Newtonia suaveolens (Miguel) Brenan: **kɨkɨ-'ɨ** 'monster-stem'; **čimo-'ɨw-ran** '*Derris utilis* (a fish poison)-stem-false' (T); **čimo-'ɨ** '*Derris utilis*-stem' (G); faveira folha fina; ANF02 HHI31 FUE01; very large tree of the old fallow.

Parkia multijuga Benth.: **tayahu-mi'u-rupe-'ɨ** 'white lipped peccary-food-pod-stem'; **warara-'ɨw** 'crab-stem' (T); faveira bengue; ANF01; large tree of the old fallow.

Parkia paraensis Ducke: **šimo-'ɨ** '*Derris utilis*-stem'; PAC03 FUE02; large tree of the high forest.

Parkia pendula Miq.: **yupu-'ɨ** 'L-stem'; **azɨrɨ-'ɨw** 'L-stem' (T); **warara-'ɨ** 'crab-stem' (G); **yəpə-'ɨ** 'L-stem' (AR); ANF01 MYT02; enormous tree of the high forest and also seen on river banks.

Parkia ulei (Harms) Kuhlm.: **ɨra-muru-'ɨ** 'bird-?-stem'; **warara-'ɨ** 'crab-stem' (G); ANF01 FUE01; small tree of old swiddens.

Pithecellobium cauliflorum Mart.: **taír-iŋa-'ɨ** 'Blue headed parrot-*Inga* spp.-stem'; **iŋa-pi-'ɨw-ran** 'L-thin-stem-false' (T); **pira-iŋa-'ɨwa** 'fish-inga-stem' (AS); **arapai-yi-i-te** 'L-axe-stem-true' (AR); angelim; ANF02 FUE01; medium tree of the high forest.

Pithecellobium cochleatum (Willd.) Mart.: **tayahu-iŋa** 'white lipped peccary-*Inga* spp.' (=**iŋa-ran-'ɨ** '*Inga* spp.-false-stem'); ingarana; FUE02 ANF02; large tree of the high forest.

Pithecellobium comunis Benth.: **kɨkɨ-'ɨ** 'monster-stem'; ANF02 HHI31 FUE01; large tree of the high forest.

Pithecellobium foliolosum Benth: **suruku-yu-rāšĩ** 'bushmaster-yellow-spiny'; ANF01; small tree of the old fallow.

Pithecellobium jupumba (Willd.) Urb.: **mɨra-wewi-atu-'ɨ** 'tree-light-good-stem'; **wɨra-ro-'ɨ** 'tree-bitter-stem' (G); angelim falso; ANF01; large tree of the old fallow.

Pithecellobium niopoides Spruce ex Benth.: **šiŋ-ɨ** 'L-stem'; FUE01; large tree of the old fallow.

Pithecellobium pedicellare (DC.) Benth.: **arapasu-mɨra** 'woodpecker-tree'; FUE02; large tree of the high forest.

Pithecellobium racemosum Ducke: **taír-iŋa-'ɨ** 'Blue headed parrot-*Inga* spp.-stem'; **warara-'ɨ** 'crab-stem' (G); **iŋa-ran-'ɨw** '*Inga* spp.-false-stem' (T); angelim rajado; ANF02 FUE01; large tree of the old fallow.

Stryphnodendron guianensis (Aubl.) Benth.: **ɨra-i-rupe-atu-'ɨ** 'bird-small-pod-good-stem'; faveira camuzê; ANF02 FUE01; large tree of the old fallow.

Stryphnodendron polystachyum (Miq.) Kleinh.: **mani-'iran-'ɨ** 'manioc-false-stem'; **tači-'iran-'ɨw** '*Pseudomyrmax* spp. ant-false-stem' (T); **arapa-pi-apo-'ɨ** 'kind of bird-?-?-stem' (G); faveira camuzê; ANF01 FUE01; large tree facultative to old fallows and high forests.

MONIMIACEAE

Siparuna amazonica (Mart.) A.DC.: **ka'uwa-pusan** 'insanity-remedy' (=**ka'a-piši'u-ran** 'herb-fetid-false'); **wɨra-him-hu** 'tree-slippery-much' (G); capitiú; MAG09 MED02; fairly common treelet of the high forest especially in light gaps and other openings.

Siparuna guianensis Aubl.: **ka'a-piši'u** 'herb-fetid'; capitiú; MED02; small tree of the high forest—not common.

MORACEAE

Bagassa guianensis Aubl.: **tareka-'ɨ** 'L-stem'; **taraka-'ɨ** 'L-stem' (G); **tarawire-'ɨwa** 'L-stem' (AS); **taraikə-'ɨ** 'L-stem' (AR); tatajuba; SFS01 ANF01 HHI29 POI02 POI04 POI05 FUE02; enormous tree of the high forest, occasional in old fallow.

Brosimum acutifolium ssp. *interjectum* C.C. Berg: **murure-'ɨ** 'L-stem'; **merere-'ɨ** 'L-stem' (G); **murure-ete** 'L-true' (AS); mururé; SFS01 ANF02 FUE02; medium tree of the high forest.

Brosimum guianense (Aubl.) Huber: **mɨra-pɨtaŋ-ran** 'tree-red-false'; **arara-kaŋ-tapiwa-'ɨ** 'macaw-head-nail-stem' (G); ANF02 FUE02; large tree of the high forest.

Brosimum paclesum (S. Moore) C.C. Berg: **mɨra-pɨtaŋ-ran** 'tree-red-false'; **ita-čɨ-'i** 'stone-white-stem' (AR); **tamamari-'ɨ** 'L-stem' (G); ANF02 FUE02; large tree of the high forest.

Brosimum rubescens Taubert: **mɨra-pɨtaŋ** 'tree-red'; **uruwa-pirirun-'ɨwa** 'snail-?-stem' (AS); pau brasil; TLW10 TLW13 HHI30 MYT04 MIS08; medium tree of the high forest.

Clarisia racemosa Ruiz & Pavon: **wari-mɨra** 'howler monkey-tree'; **wari-wa-'ɨ** 'howler monkey-fruit-stem' (G); guariúba; ANF01 FUE01 MED29; medium tree of the high forest.

Ficus sp. 1: **parawa-ka'a** 'Mealy parrot-herb'; caxinguba; MED06; treelet of the high forest.

Ficus sp. 2: **apo-'ɨ-pini** 'L-stem-stripe'; caxinguba; ANF02 FUE02; strangler fig of the high forest.

Ficus sp. 3: **wari-'ɨ** 'howler monkey-stem'; caxinguba; ANF01; medium tree of the old fallow.

Helicostylis tomentosa (P. & E.) Rusby: **akaú-'ɨ** 'L-stem'; **murure-ran-'ɨw** 'L-false-stem' (T); **taraka-i-hu** 'L-stem-big' (G); inharé; SFS01 PBC02 HHI29 ANF01 FUE01; large tree of the high forest.

Maquira guianensis Aubl.: **mɨra-pɨtaŋ-ran** 'tree-red-false'; **ɨwira-piru-'ɨwa** 'tree-?-stem' (AS); ANF02 FUE02; medium tree of the high forest.

MUSACEAE

Musa sp. 1: **pako** 'L' (includes **pako-pɨraŋ**, **pako-tawa** [=**pako-ran**], **pako-nerɨ** [=**pako-kururu**], **katumē** [var. **katumē-howɨ**], **pako-parawa** and **yakare-rãi**); **pátsitsi** 'L' (AR); **paku** 'L' (G); **paku** 'L' (AS); **paku** 'L' (T); banana; PFS01 SFS04; traditionally cultivated, large herb seen of young and old swiddens.

Musa sp. 2: **pako** 'L' (includes **pako-te** and **pako-hu**); **paku** 'L' (G); **paku** 'L' (T); **pátsitsi** 'L' (AR); **paku** 'L' (AS); plantain; banana; PFS01 SFS04; traditionally cultivated herb of young and old swiddens.

Phenakospermum guianensis Peterson: **pako-sororo** 'banana-hungry'; **yawar-ka'a-hu** 'jaguar-herb-big' (G); **paku-ran** 'banana-false' (T); pacosororoca; NONE; enormous herb, most often seen in light gaps of the high forest.

MYCOTA

Indeterminate mycotica sp.1: **urupe-pɨraŋ** 'L-reddish'; **urupe** 'L' (G); NONE; mushroom species commonly seen on fallen logs in the high forest.

aff. *Rhizomorpha* sp. 1: **ka'a-tuwɨr** 'herb-white'; **ka'a-čū** 'herb-white' (G); **wararu-'ɨ-raitɨ** 'kind of frog-stem-nest' (T); NONE; luminescent fungus, very common on the high forest floor.

MYRISTICACEAE

Iryanthera juruensis Warb.: **mɨkur-pɨ'a-'ɨ** 'opossum-female-stem'; **koko-'iran-'ɨw** 'tiger heron-false-stem' (T); FUE02; large tree of the high forest.

Virola carinata (Benth.) Warb.: **tukwā-mi'u-'ɨ** 'toucan-food-stem' (=**tukwā-mɨra** 'toucan-tree'); ANF02 MED05 MED17; medium tree of the high forest.

Virola michelii Heckel: **tukwā-mi'u-'ɨ** 'toucan-food-stem'; (=**tukwā-mɨra** 'toucan-tree'); **kɨwa-'ɨ** 'comb-stem' (G); virola; ANF02 MED05 MED17; medium tree of the swamp forest.

Virola sp. 1: **uru-kɨwa-'ɨ** '*Odontophorus* (pheasant)-comb-stem'; ANF01 MED17 FUE01; medium tree of the high forest.

MYRSINACEAE

Ardisia guianensis (Aubl.) Mez: **yukuna-ka'a** *Crenicichla* (fish)-herb'; MAG01 HHI18; tiny herb of the high forest.

MYRTACEAE

aff. *Myrcia* sp. 1: **u'i-tima** 'arrow point-leg'; SFS01 TLW14 HHI12 MIS09; small tree of the high forest.

Calyptranthes or *Marlierea*: **arapuha-tima-kaŋwer-'i** 'red brocket deer-leg-bone-stem'; **peye'i'L-stem'** (AR); **wira-'i-ku** 'tree-small-?' (G); SFS01 ANF01 TLW13; small tree of the high forest.

cf. *Calyptranthes* sp. 1: **waya-taper-rupi-har** 'guava-old fallow-through-place'; SFS01 ANF01; very rare treelet of the old fallow.

Campomanesia grandiflora (Aubl.) Sagot.: **paraku-'i-ran** 'kind of fish-stem-false'; **ayā-piri-ran** 'divinity-crackle-false' (G); ANF02 FUE02; small tree with a deeply grooved trunk; found in the old fallow.

Eugenia egensis DC.: **kwi-ra'i-mira** '*Crescentia cujete*-small-tree'; **ayā-piri-ate** 'divinity-crackle-true' (G); NONE; medium tree of the old fallow.

Eugenia eurcheila Berg.: **u'i-tima-ran** 'arrow point-leg-false'; SFS01 ANF01 FUE02; rare, small tree of the old fallow.

Eugenia flavescens DC.: **tayi-ran** '*Tabebuia impetiginosa* (pau d'arco)-false'; **atarā-'i** 'L-stem' (G); **me'e-'a-'i** 'some-fruit-stem' (AR); FUE01; very large tree (unusual for a myrtaceous species) of the old fallow.

Eugenia muricata DC.: **arākwā-mira** 'Little chachalacha (*Ortalis motmot*)-tree' (= **arākwā-mi'u-'i** 'Little chachalacha (*Ortalis motmot*)-food-stem'; **waya-ran** 'guava-false' (T); **itai-wira** 'smile-tree' (G); SFS01 ANF02 FUE01; small tree of the old fallow.

Eugenia omissa McVaugh: **yanu-mira** 'spider-tree'; **yanu-'i** 'spider-stem' (G); **piči-reni-'iwa** 'small-sister-stem' (AS); POI11 POI03; small tree of the old fallow.

Eugenia patrissi Vahl.: **ma'e-iwa-pitaŋ-'i** 'some-fruit-red-stem' (= **ira-pitaŋ-'i** 'bird-red-stem'); **yanu-'i** 'spider-stem' (G); SFS01; treelet of the old swidden.

Eugenia schomburkii Benth.: **u'i-tima-piriri-ran-'i** 'arrow point-leg-crackle-false-stem'; **atarā'i** 'L-stem' (G); ANF02 SFS01 FUE02; rare, medium tree of the old fallow.

Eugenia tapecumensis Berg: **arākwā-mira** 'Little chachalacha (*Ortalis motmot*)-tree' (= **arākwā-mi'u-'i** 'Little chachalacha-food-stem'); SFS01 ANF01 FUE01; small tree of the high forest.

Eugenia sp. 5: **arapuha-tima-kaŋwer-'i** 'red brocket deer-leg-bone-stem'; SFS01 ANF01 TLW14; treelet of the high forest.

Eugenia sp. 6: **ma'e-iwa-pitaŋ-'i** 'some-fruit-red-stem'; (= **ira-pitaŋ-'i** 'bird-red-stem'); pitanga; SFS01 ANF01 FUE01; medium tree of the high forest.

Eugenia sp. 7: **arākwā-mira** 'Little chachalacha (*Ortalis motmot*)-tree'(= **arākwā-mi'u-'i** 'Little chachalacha-food-stem'); SFS01 ANF02 FUE01; treelet of the old fallow.

Eugenia sp. 8: **u'i-tima-piriri-'i** 'arrow-leg-crackle-stem'; SFS01 ANF02 TLW14 FUE01 MIS09; small tree of the high forest.

Myrcia paivae Berg: **u'i-tima-ran-'i** 'arrow point-leg-false-stem'; **yapu-čī-ran-'i** 'oropendola-white-false-stem' (G); SFS01; treelet of old swiddens.

Myrcia splendens (Sw.) DC.: **mira-piraŋ-'i** 'tree-reddish-stem'; SFS01; rare, medium tree of old fallows.

Myrcia sp. 1: **mayaŋ-i** 'L-stem'; ANF02; small tree of river margins.

Myrcia sp. 4: **u'i-tima-ran-'i** 'arrow point-leg-false-stem'; SFS01; treelet of the high forest.

Myrcia sp. 5: **arapuha-tima-kaŋwer-'i** 'red brocket deer-leg-bone-stem'; SFS01; treelet of the high forest.

Myrciaria cf. *dubia* McVaugh: **mira-šī-'i** 'tree-white-stem'; **muka** 'L' (T); **pečiči-'i** '?-stem' (AR); PFS01 FUE02 ANF02; fairly common small tree of river margins.

Myrciaria flexuosa: **ira-šī-'i** 'bird-white-stem'; SFS01; small tree of the old fallow.

Myrciaria pyriifolia Desv. ex Hamilton: **yanu-mira** 'spider-tree'; POI03 POI11; small tree of old fallows.

Myrciaria spruceana Berg: ɨra-ši-'ɨ 'bird-white-stem'; SFS01; small tree of old fallows.

Myrciaria tenella (DC.) Berg: mɨra-kɨ'ɨ 'tree-chile pepper'(= ita-piša-'ɨ 'stone-?-stem'); ɨra-ro-ɨ 'tree-bitter-stem' (G); MED28; small tree of the old fallow.

Psidium guajava L.: **waya** 'L'; **waya** 'L' (G); **gwaya** 'L' (T); **kuyawá** 'L' (AR); guava tree; goiabeira; PFS01; traditionally cultivated treelet of dooryard gardens, also seen in old swiddens.

Syzygium malaccencis (L.) Merril & Pery: **yam** 'L'; rose apple; jambo; PFS01; introduced, domesticated small tree of dooryard gardens.

NYCTAGINACEAE

Neea floribunda Poepp. & Endl.: **tapi'i-sikiri-'ɨ** 'tapir-rough hide-stem'; **tepeči-kuruw-'ɨwa** 'L-?-stem' (AS); **depečiri-'i** 'L-stem' (AR); ANF02 FUE02; small tree of the high forest.

Neea glameruliflora Heimut: **tapi'i-sikiri-'ɨ** 'tapir-rough hide-stem'; **depečiri-'i** 'L-stem' (AR); wəkəwə-'ɨw 'Laughing falcon-stem' (T); joão mole;; ANF02 FUE02; small tree of the old fallow.

Neea oppositifolia R. & P.: **tapi'i-sikiri-'ɨ** 'tapir-rough hide-stem'; **wɨra-tata-'ī** 'tree-fire-not' (G); **tepeči-kuruw-'ɨwa** 'L-?-stem' (AS); **depečiri-'i** '?-stem' (AR); joão mole; ANF02 FUE02; uncommon, small tree of old fallows.

Neea ovalifolia Spruce ex J.A. Schmidt: **tapi'i-sikiri-'ɨ** 'tapir-rough hide-stem'; **wɨra-tata-'ī** 'tree-fire-not' (G); ANF02 FUE02; medium tree of old fallows.

Neea sp. 1: **tapi'i-sikiri-'ɨ** 'tapir-rough hide-stem'; **depečiri-'i** 'L-stem' (AR); **wɨra-tata-'ī** 'tree-fire-not' (G); **tepeči-kuruw-'ɨwa** 'L-?-stem' (AS); wəkəwə-'ɨw 'crab-stem' (T); ANF02 FUE02; small tree of the old fallow.

Neea sp. 2: **tapi'i-sikiri-'ɨ** 'tapir-rough hide-stem'; ANF02 FUE02; small tree of the high forest.

Pisonia sp. 2: **tapi'i-sikiri-'ɨ** 'tapir-rough hide-stem'; **wɨra-tata-'ī** 'tree-fire-not' (G); joão mole; ANF02 FUE02; small tree of old fallows.

OCHNACEAE

Ouratea sp. 2: **mukuku-'ɨ-ran** '*Licania* spp.-stem-false'; HHI07; treelet of old fallows.

Ouratea salicifolia (St. Hil. et Tul.) Engl.: **mukuku-'ɨ-tuwɨr** '*Licania* spp.-stem-white'; NONE; rare, small tree of the old fallow.

OLACACEAE

Dulacia guianensis (Engl.) O. Kuntze: **pina-hu-ran-'ɨ** 'fishhook-big-false-stem'; ANF02 FUE01; rare, medium tree of old fallows.

Minquartia guianensis Aubl.: **yowoi-ɨ** 'boa constrictor-stem'; ɨra-tama-kamakama-'ɨ 'wood-leg-?-stem' (G); **wakari-'ɨw** 'kind of fish-stem' (T); acariquara; PBC01 POI06; uncommon, medium tree of the high forest.

OPILIACEAE

Agonandra brasiliensis Benth. & Hook.: **kaŋwaruhu-mɨra** 'paca-tree'; žɨpɨ'o-'ɨw 'L-stem' (T); pau marfim; MED15 MED32; medium tree common to both old fallows and high forests.

ORCHIDACEAE

Trigonidium acuminatum Batem ex Lindl.: **a'ɨhu-pako** 'two–toed sloth-banana'; MED11 ANF02; arboreal orchid of the high forest.

OXALIDACEAE

Averrhoa carambola L.: ɨwa-me-'ɨ 'fruit-inside-stem'; **tai-ran** 'spice-false' (G); akara-wɨra 'kind of fish-tree' (T); carambola; SFS01; introduced, domesticated small tree of dooryard gardens.

PASSIFLORACEAE

Passiflora aranjoi Sacco: **murukuya-ran** 'L-false'; ANF02; vine of old swiddens.

Passiflora coccinea Aubl.: **murukuya-ran** 'passion fruit-false'; **marikuya** 'L' (AS); passion fruit; maracujá bravo; ANF02; scandent vine in old swiddens.

Passiflora edulis L.: **murukuya** 'L'; **murukuza** 'L' (T); SFS01; passion fruit; maracujá; traditional domesticate, occasionally cultivated in dooryard gardens.

Passiflora sp. 1: **murukuya-tawa-rimo** 'passion fruit-yellow-vine'; maracujá bravo; passion fruit; SFS01; scandent vine of the high forest.

Passiflora sp. 2: **murukuya-ran** 'L-false'; ANF02; vine of old swiddens.

Passiflora sp. 3: **murukuya-pinī** 'L-stripe'; passion fruit; maracujá bravo; SFS01 ANF02; small vine of high forest.

Passiflora sp. 4: **murukuya-ran-sipo** 'L-false-vine'; passion fruit; maracujá bravo; ANF02; scandent vine of the high forest.

PHYTOLACCACEAE

Petiveria alliacea L.: **mikur-ka'a** 'opossum-herb'; guinea hen weed; mucuracáa; MED01 MED02 MAG05 MAG04; terrestrial herb cultivated in dooryard gardens.

Phytolacca rivinoides Kunth. & Bouche: **ka'a-riru** 'herb-vessel'; **ka'a-piw** 'herb-thin' (T); **wira-kerem-ti** 'tree-?-?' (G); PAC09 MAG09; terrestrial herb, agrestic and very common in young swiddens.

Seguieria amazonica Huber: **surukuku-yu-rāšī** 'bushmaster-spine-thorn'; ANF01; large liana, rare except locally, in old fallows near the village of Gurupiuna.

PIPERACEAE

Piper anonnifolium Kunth.: **yamir-hu** 'L-big'; NONE; terrestrial herb of the high forest, also agrestic in young swiddens.

Piper hostmannianum (Mig.) C.DC.: **yamir-hu** 'L-big'; ANF02; terrestrial herb of the high forest.

Piper ottonoides Jun.: **yamir** 'L'; **zamir** 'L' (T); **yamir** 'L' (AS); MED21; terrestrial herb of the high forest.

Piper pilirameum Jun.: **yamir-hu** 'L-big'; **yai-ka'a** 'menstruation-herb' (AR); **yakamī-ta-makana-hu** 'trumpeter-?-big' (G); NONE; terrestrial herb of the high forest.

Piper piscatorum Trel. & Jun.: **yamir** 'L'; MED21; terrestrial herb of the high forest.

POACEAE

Brachiaria humidicula (Rendle) Schweickerdt: **kāpī-ran** 'L-false'; **ka'a-pi'i** 'herb-thin' (G); pasture grass; braquiária; ANF03; introduced, domesticated grass of one or two dooryard gardens only in the Turiaçu basin (for donkey forage).

Coix lacryma-jobi L.: **pu'i-risa** 'bead-cold'; Job's tears; lágrimas de Nossa Senhora; PAC02 COM02; probably introduced, domesticated grass, common in young swiddens and dooryard gardens.

Cymbopogon citratus (DC.) Stapf: **kāpī-piher** 'grass-aromatic'; lemon grass; capim santo; SFS07; introduced, domesticated grass of dooryard gardens.

Digitaria insularis (L.) Mez ex Ekman: **u'iwa-ran** 'arrow cane (*Gynerium sagittatum*)-false'; NONE; grass of young swiddens.

Guadua glomerata Munro: **takwar** 'arrow point'; **takwar** 'arrow point' (G); bamboo; TLW13; large caespitose bamboo of old fallows and light gaps in high forest.

Gynerium sagittatum (Aubl.) Beauvois: **u'iwa** 'L'; arrow cane; tacana, flecheira; TLW12; traditionally cultivated, very tall grass of both young swiddens and old swiddens.

Hyparrhenia rufa (Nees) Stapf: **kāpī-ran** 'grass-false'; jaraguá grass; lagiado; NONE; introduced, cultivated grass, found on pastures near the Ka'apor frontier but not used by the Ka'apor.

Ichnanthus sp. 1: **takwar-ɨ-ran** 'arrow point-water-false'; NONE; grass found in light gaps in the high forest.

Imperata brasiliensis Trin.: **u'ɨwa-ran** 'arrow cane-false'; capim furão, sapé; POI15; grass found in old swiddens, especially those of nearby settlers (it appears to thrive, therefore, in intensively degraded pastures).

Lasiacis ligulata Hitchelchase: **takwar-i** 'arrow point-small'; **takwar-i-čï** 'arrow point-small-white' (T); ANF02; grass of the high forest, found in light gaps.

Olyra caudata Trin.: **kere-wi** 'L-narrow'; **takwar-i-kənə** 'arrow point-water-sugar cane (G); ANF02; grass found on edges of old swiddens.

Olyra latifolia L.: **takwar-ɨ-pinï** 'arrow point-water-stripe'; **takwar-i** 'arrow point-small' (T); **takwar-ɨ-pinimü** 'arrow point-water-stripe (AS); **ta'akačï** 'L' (AR); taboquinha; HHI18 MIS01; small grass seen in light gaps of the high forest.

aff. *Olyra*: **takwar-ɨ-ran** 'arrow point-water-false'; NONE; grass found in light gaps in the high forest.

Oryza sativa L.: **ahúi** 'L' (=**awaši-'i** 'maize-little') [includes **ahúi-puku** and **ahúi-pu'a**); **takwari-ake** 'bamboo-?' (G); **awači-apo** 'maize-now' (T); COM01 SFS01; introduced, cultivated grass of young swiddens, most commonly grown by Indians in the Turiaçu basin, having assumed commercial importance only by the early 1980s.

Paspalum conjugatum Berg: **kupi'i-pe** 'termite-flat'; ANF02; grass found along edges of young swiddens.

Piresia goeldi Swallen: **takwar-i-pinï** 'arrow point-little-stripe'; NONE; fairly common grass seen in light gaps of the high forest.

Saccharum officinarum L.: **kã** (includes **kã-te, kã-howɨ, kã-memek, kã-pinï, kã-tuwɨr, kã-tã** [=**kã-hu**]); **kã** (T); sugar cane; cana de açúcar; SFS02; introduced, domesticated grass, frequently grown by the Ka'apor in diverse villages.

Zea mays L.: **awaši** 'L'; **wači** or ᵏ**wači** 'L' (G); **awači** 'L' (AS); **awači** 'L' (AR); **awači** (T); maize; milho; PFS01 ANF01; traditionally cultivated grass of the young swidden.

Poaceae sp. 1: **kãpï-puku** 'L-long'; ANF02; grass of the young swidden.

POLYGALACEAE

Moutabea guianensis Aubl.: **mišik-sɨpo** 'roast-vine'; MED11; large liana of the high forest.

Moutabea sp. 1: **sɨpo-tawa** 'vine-yellow'; ANF02; large liana of the high forest.

Moutabea sp. 2: **ka'a-ro-yẽ'ẽ-hãtã** 'herb-leaf-speak-strong'; MAG06; liana of the high forest.

Ruprechtia sp. 1: **ma'e-ɨra-pɨraŋ** 'some-bird-reddish'; NONE; small tree of the old fallow.

POLYGONACEAE

Coccoloba acuminata H.B.K.: **yawa-pɨtaŋ-sɨpo** 'jaguar-red-vine'; FUE02; large liana of the high forest.

Coccoloba latifolia Lam.: **sɨpo-mɨra** 'vine-tree'; **yawa-čü-wɨra** 'jaguar-black-tree' (G); **kwi-piaw-'ɨwa** 'calabash (*Crescentia cujete*)-?-stem' (AS); NONE; small tree of the old fallow.

Coccoloba paniculata Lam.: **takwar-i-mɨra** 'arrow point-little-tree'; FUE01; small tree with hollow stem of the old fallow.

Coccoloba racemulosa Meissn.: **sɨpo-mɨra** 'vine-tree'; NONE; very rare, small tree of the old fallow (it has a hollow stem.)

Coccoloba sp. 1: **arapuha-sɨpo** 'brocket deer-vine'; MAG02 MAG18; large liana of the high forest.

Coccoloba sp. 2: **yaši-mɨra** 'yellow footed tortoise-tree'; MAG21; treelet of the transitional forest between swamp forest and *terra firme*.

Polygonaceae sp. 1: **tatu-ruwai** 'armadillo-tail'; MED11 MED23; shrub of old swiddens.

POLYPODIACEAE

Stenoclaena sp. 1: **ɨra-hu-ka'a** 'bird-big-herb'; MED01; epiphytic herb of the high forest.

PTEROBRYACEAE

Orthostichorsis crinite (Sull) Broter.: **mɨra-meyu** 'tree-manioc bread'; NONE; bryophyte common on tree trunks in old fallows and high forests.

QUIINACEAE

Lacunaria jenmani (Oliv.) Ducke **ɨra-tawa-'ɨ-hu** 'bird-yellow-stem-big'; **ɨwa-kaw-'ɨwa** 'fruit-?-stem' (AS); **ɨwa-pe-di** 'fruit-flat-stem' (AR); moela de mutum; PBC02 ANF02 FUE01; fairly common, very large tree of the high forest.

Lacunaria sp. 3: **ɨra-tawa-'ɨ-hu** 'bird-yellow-stem-big'; moela de mutum; FUE02; small tree of the high forest.

Lacunaria sp. 4: **arapasu-ātā-'ɨ** 'woodpecker-hard-stem'; moela de mutum; ANF02 FUE01; small tree of the high forest.

Lacunaria sp. 5: **arapasu-ātā-'ɨ** 'woodpecker-hard-stem'; moela de mutum; ANF02 FUE01; medium tree of the high forest.

RHAMNACEAE

Ampelozizyphus amazonicus Ducke: **eremenɨ** 'L'; MED11 MED20; medium liana of the high forest.

Colubrina glandulosa Perkins var. *glandulosa*: **pu'ɨ-pihun-'ɨ** 'bead-black-stem'; PAC02; uncommon, large tree of the high forest.

Colubrina sp. 1: **pu'ɨ-pihun-'ɨ** 'bead-black-stem'; PAC02; large tree of old fallows.

Gouania pyrifolia Reiss.: **tapuru-ka'a** 'grub-herb'; PAC03; scandent vine of old swiddens.

RUBIACEAE

Alibertia edulis (L.Rich.) A. Rick.: **kiraru** 'L'; **tamari-kawaw** 'tortoise shelled tamarind monkey-calabash bowl' (T); puruizinho; ANF01 MIS02 FUE02; small tree of swamp forest.

Bertiera guianensis Aublet: **kuyuí-ka'a** 'Blue throated piping guan-herb'; **kunamɨ-ran** '*Clibadium sylvestre* (a fish poison)-false' (T); ANF02; treelet of the old fallow.

Borreria verticulata G.F.W. Mey: **kururu-ka'a** 'toad-herb'; **ami'e-rɨ** '?-false' (AR); NONE; small herb, agrestic in dooryard gardens.

Cephalia sp. 1: **tapeši-'ɨ** 'manioc press-stem'; SFS01 MAG14; herb of the high forest.

Chimarrhis turbinata DC.: **paraku-'ɨ** 'L-stem'; pau de remo; ANF02 FUE01; enormous tree of the high forest.

Coffea arabica L.: **kase** 'L'; **kafe** 'L' (T); **kače** (AR); coffee; café; SFS07; uncommon, introduced, cultivated treelet, typically seen in dooryard gardens in Gurupiuna.

Coussarea ovalis Standl.: **yakamɨ-mɨra** 'trumpeter-tree' (= **tapi'i-ka'a-mɨra** 'tapir-herb-tree'); ANF02; very rare, small tree of the old fallow.

Coussarea paniculata (Vahl.) Standl.: **yakamɨ-mɨra** 'trumpeter-tree' (= **tapi'i-ka'a-mɨra** 'tapir-herb-tree'); ANF02; small tree found in old swiddens.

Duroia sp. 1: **wa-'ɨ-'ɨ-ran** 'fruit-small-stem-false'; jenipapinho; NONE; small tree of the old fallow.

Faramea sessilifolia (H.B.K.) DC.: **yakamɨ-mɨra** 'trumpeter-tree' (= **tapi'i-ka'a-mɨra** 'tapir-herb-tree'); **təkənə-haíha-ra-'ɨ**; 'toucan-eyelash-hair-stem' (G); ANF02 FUE01; small tree of old fallows.

Genipa americana L.: **yenipa-'ɨ** 'L-stem'; **yanupa-'ɨ** 'L-stem' (G); genipapo; PAC07 MAG16 MYT01; traditionally cultivated, small tree of dooryard gardens.

Geophila repens (L.) I.M. Johnston: **yuruši-kɨ'ĩ** 'ruddy ground dove-chile pepper'; **ɨwa-pe** 'fruit-flat' (T); **ka'a-mururu** 'herb-viscera' (G); **točĩ-ka'a** 'pigeon-herb' (AR); ANF02; prostrate, pantropical herb of the old fallow.

Guettarda divaricata (H.B.K.) Standl.: **tukur-ɨ** 'grasshopper-stem'; ANF02 FUE02; very rare, small tree only seen in the high forest.

Malanea sp. 1: **paraku-'iran-'ɨ** 'kind of fish-false-stem'; ANF02 FUE02; large tree of the high forest.

Posoqueria latifolia (Rudge) R. & S. ssp. *gracilis* (Rudge) Steyerm.: **pakurɨ-sōsō-ran-'ɨ** 'bacuri (*Platonia insignis*)-?-false-stem'; FUE02 PAC02; rare, small tree of the high forest.

Psychotria colorata Poepp.: **tapi-i-ka'a** 'tapir-herb'; perpétua do mato; ANF02 TLW09 MAG14; terrestrial herb of the high forest.

Psychotria hoffmannseggiana (Willd. ex R. & S.) M. Arg. **yu'i-ka'a** 'frog-herb'; **warara-wɨra** 'crab-tree' (G); ANF01; terrestrial herb of the old fallow.

Psychotria poeppigiana Muell. Arg.: **tapi'i-kanamɨ** 'tapir-*Clibadium sylvestre* (a fish poison)' (= **yawarũ-ka'a** 'black jaguar-herb'); **ka'a-rahɨ** 'herb-poison' (T); MAG02 ANF02; terrestrial herb of the high forest.

Psychotria poeppigiana ssp. *poeppigiana* Muell. Arg.: **tapi'i-ka'a** 'tapir-herb'; ANF02; herb of the high forest.

Psychotria racemosa (Aubl.) Baensh.: **tapi'i-ka'a** 'tapir-herb'; **ka'a-rahɨ** 'herb-poison' (T); **warara-piya-ran** 'crab-?-false' (G); **yawči-iŋa** 'yellow footed tortoise-*Inga* spp.' (AS); **akuči-wɨrã** 'agouti-tree' (AR); ANF02; small herb of the high forest.

Psychotria ulviformis Steyerm.: **ipe-ka'a** 'flat-herb'; ANF02 MAG02; terrestrial herb of seasonally inundated forest.

Psychotria sp. 1: **tapi'i-ka'a** 'tapir-herb'; ANF02; herb of old swiddens.

Randia armata (Sw.) Sw.: **akuši-yu-'ɨ** 'agouti-spine-stem'; **akuči-ri-wɨra** 'agouti-?-tree' (T); **akuči-wɨra** 'agouti-tree' (G); espinho de porco; ANF01; small tree of the old fallow.

Remijia sp. 1: **paraku-'ɨ-ran** 'kind of fish-stem-false'; ANF02 FUE01; small tree of the high forest.

Sipanea veris S. Moore: **ipe-ka'a** 'flat-herb'; NONE; tiny herb of deep forest creek margins.

Tocoyena foetida Poepp. & Endl.: **tapi'i-ka'a** 'tapir-herb'; mata cavalo; ANF02; very tall, terrestrial herb of light gaps in the high forest.

Uncaria guianensis (Aubl.) Gmel: **parawa-sɨpo** 'Mealy parrot-vine'; **su-pina-pina-'iw** 'spine-fishhook-fishhook-stem' (T); **wɨpo-rači** 'vine-thorn' (G); **iwɨrã-atɨ** 'tree-spine' (AR); MED24; common creeper with very sharp, thorny tendrils, found in swamp forests and river banks.

Rubiaceae sp. 1: **kururu-ka'a** 'toad-herd'; NONE; terrestrial herb of the high forest.

Rubiaceae sp. 2: **arari-ãkã-'ɨ** 'kind of fish-head-stem'; POI13; treelet of the high forest.

RUTACEAE

Citrus aurantiifolia (Christm.) Swingle: **irimã-te** 'L-true'; **ɨra-tai** 'tree-spice' (G); **límə** 'L' (T); lime tree; limão; SFS01; introduced, domesticated treelet of dooryard gardens.

Citrus medica-acida L.: **irimã-hu** 'L-true'; lime; limão galego; SFS01; introduced domesticated treelet of old swiddens, less common than *C. aurantifolia*.

Citrus sinensis (L.) Osbeck: **narãi-'ɨ** 'L-stem'; **tai-hamãi** 'spice-much' (G); **narãi** 'L' (T); orange tree; laranjeira; SFS01; introduced cultivated treelet of the dooryard garden.

Galipea trifoliolata Aubl.: **teyu-ka'a** 'skink-herb'; MED17; tall shrub of old swiddens.

Rauia resinosa Nees & Mart.: **mɨra-tawa-tuwɨr** 'tree-yellow-white'; **ara-ka'a-'ɨ** 'macaw-herb-stem' (G); **uha-ka'a-'ɨw** 'crab-herb-stem' (T); ANF02; treelet of old fallows.

Zanthoxylum cf. *juniperina* Engl.: **wa-šɨ-šɨ-hu-'ɨ** 'fruit-white-white-big-stem'; limãozinho; ANF02 MED21 FUE01; medium to large tree of the old fallow.

Zanthoxylum chiloperone (Mart.) Engl.: **wa-šɨ-šɨ-hu-'ɨ** 'fruit-white-white-big-stem'; limãozinho; ANF02 FUE02; uncommon, large tree of the old fallow.

Zanthoxylum rhoifolium Lam.: **wa-šɨ-šɨ-'ɨ** 'fruit-white-white-stem'; **yakare-kəməta-'ɨ** 'caiman-?-not' (G); **yawči-mu-'ɨ-wi** 'tortosie-food-stem-thin' (AS); **i-wa** 'stem-fruit' (AR); limãozinho; ANF02 FUE02; small tree of the old fallow.

SAPINDACEAE

Cardiospermum halicacabum L.: **awa-í-ran** 'person-little-false'; **pariri-ran** 'L-false' (T); **činā** 'L' (AR); PAC02 COM02; introduced domesticate (although appears to be native to the Araweté), prostrate vine of swiddens and dooryard gardens.

Cupania scrobiculata L.C. Rich.: **tarara-mɨra** 'manioc griddle rake-tree'; **yakamī-təməkənə-'ɨ** 'trumpeter-leg-stem' (G); **pərərəne-'ɨ** '?-stem' (AR); ANF02 HHI11 FUE01; medium tree of the high forest, occasional in old fallows.

Matayba spruceana Radlk.: **tarara-mɨra** 'manioc griddle rake-tree'; **yakamī-təməkənə-'ɨ** 'trumpeter-leg-stem' (G); breu de tucano; ANF02 HHI11 FUE01; small tree of the high forest.

Paullinia sp. 1: **taŋara-'ɨ** 'manakin-stem'; FUE02; treelet of river margins.

Paullinia sp. 2: **sɨpo-ātā** 'vine-hard'; NONE; small vine of the high forest.

Pseudyma frutescens (Aubl.) Radlk.: **aŋwa-yar-mɨra** 'drum-owner-tree'; **wɨra-te** 'tree-true' (G); **arapoa-rena-'ɨwa** 'red brocket deer-sitting place-stem' (AS); ANF02 POI12 MED32 MED15 MAG10; treelet of the old fallow.

Serjania aff. *lethalis* St. Hil.: **kururu-šimo** 'toad-*Derris utilis* (a fish poison)'; TLW07; large liana of the high forest.

Talisia acutifolia Radlk.: **kɨrɨhu-'ɨ-ran** 'L-stem-false'; **čiči-apo** '?-?' (G); SFS01 ANF02; small tree of the old fallow.

Talisia carasina (Benth.) Radlk.: **tupiyarɨma-mɨra** 'Long tailed tyrant (a bird)-tree'; SFS01 TLW13; small tree of the high forest.

Talisia micrantha Radlk.: **tupiyarɨma-mɨra** 'Long tailed tyrant (a bird)-tree'; SFS01 TLW13 ANF02 FUE01; small tree of the old fallow.

Talisia microphylla Witt.: **yaši-pɨta-'ɨ** 'tortoise-staying place-stem'; SFS01 ANF02 FUE02; rare, small tree of the old fallow.

Talisia retusa Cowan: **yaši-pɨta-'ɨ** 'tortoise-staying place-stem'; **mitū-wɨra** 'curassow-tree' (G); SFS01 ANF02 FUE02; small tree of the old fallow.

SAPOTACEAE

Chrysophyllum argenteum Jacquin ssp. *auratum* (Miquel) Pennington: **parawa-mi'u-'ɨ** 'Mealy parrot-food-stem'; SFS01 ANF02 FUE01; medium tree of the old fallow.

Chrysophyllum lucentifolium Crong. ssp. *pachycarpum* Pires & Pennington: **wari-kawī-'ɨ** 'howler monkey-beer-stem'; SFS01 ANF01 FUE01; large tree of the old fallow.

Chrysophyllum pomiferum (Eyma) Pennington: **ɨra-tawa-'ɨ-ran** 'bird-yellow-stem-false'; SFS01 ANF02 FUE01; very rare, small tree, facultative in high forests and old fallows.

Chrysophyllum sparsiflorum Kl. ex Miq.: **ara-kā-šɨ-'a-ran** 'macaw-head-white-fruit-false'; **wa-yu-'ɨ** 'fruit-yellow-stem' (G); SFS01 ANF01 POI08 FUE01; treelet of the old swidden.

Manilkara bidentata (A.DC.) Chev. ssp. *surinamensis* (Miq) Pennington: **ɨrɨkɨwa-yu-'ɨ** 'L-yellow-stem'; maçaranduba da folha miúda; SFS01 ANF01 FUE02; uncommon, enormous tree of the high forest.

Manilkara huberi (Ducke) Chev.: **ɨrɨkɨwa-'ɨ** 'L-stem'; **masaranu-'ɨw** 'L-stem' (T); **apari-**

hu-'ɨ 'L-big-stem' (G); maçaranduba; SFS01 SFS06 ANF01 TLW06; fairly common, very large tree of both in high forests and old fallows.

Micropholis guyanensis (A.DC.) Pierre ssp. *guyanensis* Pennington: **kara-miri-'ɨ-tuwɨr** 'yam-small-stem-white' (= **wari-kawī-'ɨ** 'howler monkey-beer-stem' in Gurupiuna only); **ɨwa-zu-ran-'ɨ** 'fruit-yellow-false-stem' (T); SFS01 ANF02 FUE01; large tree of the high forest.

Micropholis melinoniana Pierre: **ɨwa-hu-'ɨ** 'fruit-big-stem'; **arakwa-re-ro-'ɨw** 'Little chachal-acha (*Ortalis motmot*)-?-leaf-stem' (T); abiu; SFS01; uncommon, enormous tree of the high forest.

Micropholis venulosa (Mart. ex Eich.) Pierre: **tapiša-'ɨ** 'broom-stem'; **wɨra-mani'akaw-ran-'ɨw** 'tree-sweet manioc (variety)-false-stem' (T); **aparata-'ɨ-ran** 'L-stem-false' (G); abiu folha fina; ANF02 FUE01; small tree of the high forest.

Pouteria bilocularis (Winkler) Baehni: **akaú-'ɨ** 'L-stem'; **arakwa-re-ro-'ɨ** 'Little chachalacha-?-leaf-stem' (G); SFS01 HHI29; small tree of the high forest.

Pouteria caimito (R. & P.) Radlk. **ɨra-tawa-'ɨ** 'bird-yellow-stem' (= **ɨwa-tawa-'ɨ** 'fruit-yellow-stem' in Gurupiuna only) **wɨra-mani'akaw-ran-'ɨw** 'tree-sweet manioc (variety)-false-stem' (T); **wa-yu-'ɨ** 'fruit-yellow-stem' (G); star apple; caimito; SFS01 ANF01 FUE01; very large tree of the high forest.

Pouteria aff. *caimito* (R. & P.) Radlk.: **kupapa-'ɨ** 'L-stem'; SFS01 SFS04 TLW03; small tree of the high forest.

Pouteria durlandii (Standley) Baehni: **kanawaru-mɨra** 'kind of frog-tree'; FUE02; large tree of the high forest.

Pouteria egregia Sandw.: **yapu-rɨwa-'ɨ** 'oropendola-fruit-stem'; SFS01 ANF02 PBC02 FUE01; large tree of the high forest.

Pouteria engleri Eyma: **kara-mirī-hu-'ɨ** 'yam-small-big-stem' (= **wari-kawī-'ɨ** 'howler mon-key-beer-stem' in Gurupiuna only); **ɨra-ruhu-'ɨ** 'bird-big-stem' (G); SFS01 ANF01 FUE01; large tree facultative to high forests and old fallows.

Pouteria filipes Eyma: **wiriri-mi'u-'ɨ** 'swift-food-stem'; PBC02 TLW03 FUE01; small tree of the high forest.

Pouteria gongrijpii Eyma: **ara-kã-šī-'ɨ** 'macaw-head-white-stem'; **ɨwa-zu-'ɨw-ete** 'fruit-yel-low-stem-true' (T); **wa-yu-'ɨ** 'fruit-yellow-stem' (G); PBC02 ANF01; small tree of the high forest.

Pouteria guianensis Aublet: **mɨra-hãtã-'ɨ-ran** 'tree-hard-stem-false'; **wiči-ran-'ɨw** 'L-false-stem' (T); abiurana; NONE; large tree of the old fallow.

Pouteria hispida Eyma: **kupapa-'ɨ** 'L-stem'; **wa-yu-'ɨ** 'fruit-yellow-stem' (G); abiu peludo; FUE01; small tree of the high forest.

Pouteria jariensis Pires & Pennington: **kupapa-'ɨ** 'L-stem'; **wa-yu-'ɨ** 'fruit-yellow-stem' (G); SFS01 SFS04 TLW03 FUE01 ANF01; small tree of the high forest.

Pouteria macrocarpa (Martius) Dietrich: **akuši-tɨr-ɨwa-hu-'ɨ** 'agouti-place-fruit-big-stem'; **akuči-tɨr-ɨwa-'ɨw** 'agouti-place-fruit-stem' (T); **akuči-ter-ewa-hu-'ɨ** 'agouti-place-fruit-big-stem' (G); **ara-hə-'ɨ** 'macaw-big-stem' (AR); abiu cutite; SFS01 PBC02 ANF01 FUE01; large tree of the old fallow.

Pouteria macrophylla (Lam.) Eyma: **akuši-tɨr-ɨwa-'ɨ** 'agouti-place-fruit-stem'; **ɨwa-zu-ete** 'fruit-yellow-true' (T); **akuči-ter-ewa-'ɨ** 'agouti-place-fruit-stem' G); **akuči-tɨr-ɨwa-'ɨwa** 'agouti-place-fruit-stem' (AS); abiu cutite; PFS01 ANF01 PBC02 FUE01; medium tree of the old fallow.

Pouteria penicillata Baehni: **wa'-ɨ-'ɨ-ran** 'fruit-small-stem-false'; SFS01; small tree of the old fallow.

Pouteria reticulata (Engl.) Eyma ssp. *reticulata*: **wa-'ɨ-'ɨ-te** 'fruit-small-stem-true'; SFS01; small tree of the old fallow.

Pouteria sagotiana (Baill.) Eyma: **kara-miri-ran-'i** 'yam-small-false-stem'; **iwa-zu-ran-'iw** 'fruit-yellow-false-stem' (T); **wa-yu-'i** 'fruit-yellow-stem' (G); ANF01 PBC02; medium tree of the old fallow.

Pouteria trigonosperma Eyma: **ira-tawa-'i** 'bird-yellow-stem'; **akuči-ter-ewa-'i** 'agouti-place-fruit-stem' (G); SFS01 ANF01 PBC02 FUE01; large tree found in both high forests and old fallows.

Pouteria sect. *Franchetella* sp. 1: **akuši-tir-iwa-hu-'i** 'agouti-place-fruit-big-stem'; SFS01 PBC02 ANF01 FUE01; large tree of the high forest.

Pouteria sect. *Franchetella* sp. 2: **ira-tawa-'i** 'bird-yellow-stem' (= **iwa-tawa-'i** 'fruit-yellow-stem' in Gurupiuna only); SFS01 PBC02 ANF01 FUE01; large tree of the high forest.

Pouteria sp. 2: **kupapa-'i** 'L-stem'; SFS01 FUE01 ANF01; small tree of the old fallow.

Pouteria sp. 5: **kara-miri-ran-pitaŋ-'i** 'yam-small-false-red-stem'; SFS01 ANF01 FUE01; small tree of the high forest.

aff. *Pradosia* sp. 1: **mira-pirer-hě'ě-'i** 'tree-bark-sweet-stem'; SEA02 MAG04; very rare, large tree of the high forest.

Sapotaceae sp. 1: **mira-howi-'i** 'tree-green-stem'; NONE; small tree of the high forest.

Sapotaceae sp. 2: **ira-tawa-'i** 'bird-yellow-stem' (= **iwa-tawa-'i** 'fruit-yellow-stem' in Gurupiuna only); SFS01 PBC02 ANF01 FUE01; medium tree of the high forest.

Sapotaceae sp. 3: **mira-howi-'i** 'tree-blue/green-stem'; NONE; treelet of the high forest.

SCROPHULARIACEAE

Scoparia dulcis L.: **tapiša** 'broom'; vassourinha; HHI13 MED02; small, uncultivated herb of dooryard gardens.

SELAGINELLACEAE

Selaginella sp. 1: **yaŋwate-ka'a** 'spotted jaguar-herb'; MED11; terrestrial herb of the high forest.

SEMATOPHYLLACEAE

Taxithelium planum (Brid.) Mitt.: **mira-meyu** 'tree-manioc bread'; NONE; common bryophyte found on tree trunks in old fallows.

SIMARUBACEAE

Simaba aff. *cavalcantei* Thomas: **pere-pusan-'i** 'skin eruption-remedy-stem' (= **kuru-pusan-'i** 'skin eruption-remedy-stem'); serve p'rá tudo; MED11; small tree of the high forest.

Simaba cedron Planch.: **pere-pusan-'i** 'skin eruption-remedy-stem' (= **kuru-pusan-'i** 'skin eruption-remedy-stem'); **akuči-akarã-'i** 'agouti-?-stem' (G); **ari-kara-'iwa** '?-*Dioscorea* (yam)-stem' (AS); serve p'rá tudo; MED11; small to medium tree of old fallows.

Simaba cuspidata Spr. var. *typica*: **maka-wa'ě-'i** '*Cebus* monkey-fruit-stem'; maruparana; ANF01; rare, small tree of old fallows.

Simaba guianensis Aubl. ssp. *guianensis*: **maka-wa'ě-'i** '*Cebus* monkey-fruit-stem'; cajurana; ANF01; rare, small tree of old fallows.

Simaruba amara Aubl.: **iwise-'i** 'manioc grater-stem'; **hãwã-'i** '?-stem' (G); **tukuri-'iwa** 'grasshopper-stem' (AS); **marupa-'iw** 'L-stem' (T); marupazeiro; ANF02 FUE01 HHI29; rare, small tree of the high forest.

SOLANACEAE

Brunfelsia guianensis Benth.: **irimã-'i-ran** 'lime-stem-false'; manaca; NONE; rare, medium to large tree of the old fallow.

Capsicum annum L.: **ki'ï** 'L' (includes **ki'ï-hu** and **ki'ï-te**); chile pepper; pimenta de cheiro; SEA01; traditionally cultivated small shrub of the dooryard garden.

Capsicum fructescens L.: **ki'ï-awi** 'L-needle'; **tai-pičika'e** 'spice-tiny' (G); **iki'ïya** 'L' (AS); **ki'iy** 'L' (T); **kĩ'ï** 'L' (AR); chile pepper; pimenta malagueta; SEA01; traditionally cultivated terrestrial herb of the dooryard garden.

Nicotiana tabacum L.: **pɨtɨm** 'L'; **petɨm** 'L' (AS); **petɨm** 'L' (T); **pétī** 'L' (AR); HHI18 MED19 RIT01; traditionally cultivated herb of dooryard gardens.

Physalis angulata L.: **kamamu** 'L'; **kanapu** 'L' (AR); **kəməmə-uhu** 'L-big' (T); SFS01; terrestrial herb of the young swiden.

Solanum crinitum Lam.: **yu-ruwe-te** 'spine-?-true'; **zu-ruwe** 'spine-?' (T); **wɨra-račĭ** 'tree-thorn' (G); jurubeba; SFS01 ANF02; tall shrub of the old swidden.

Solanum leucocarpon Dunal: **āyaŋ-ara-mɨra** 'divinity-?-tree' (= **āyaŋ-ara-ka'a** 'divinity-?-herb'); POI07 MAG09; treelet of the old fallow.

Solanum rugosum Dunal: **ka'a-yuwar** 'herb-itch'; **zuhar-'ɨw** 'itch-stem' (T); **me-e'-ra-'tī-'i** 'some-fruit-white-stem' (AR); amor de cunhã; ANF02 POI03 MED15; treelet of the old swidden.

Solanum stramonifolium Jacq.: **yu-ruwe-pɨraŋ** 'spine-?-reddish'; **wɨra-račĭ** 'tree-thorn' (G); **zu-ruwe-čĭ** 'spine-?-white' (T); jurubeba; SFS01; treelet of the old swidden.

Solanum subinerme Jacq.: **yu-ruwe-tawa** 'spine-?-yellow'; jurubeba; ANF02; small shrub of the old swidden.

Solanum vanheurckii Mulh.: **yu-ruwe-hu** 'spine-?-big'; jurubeba; ANF02; large, woody shrub of old fallows.

STERCULIACEAE

Guazuma ulmifolia Lam.: **apɨ'a-'ɨ** 'L-stem'; **kiri-wɨra-'ɨw** '?-tree-stem' (T); **yane-'i** 'we-stem' (AR); mutamba; SFS01 ANF01 FUE01; uncommon, small tree of old swiddens.

Sterculia alata Ducke: **šimo-hu-'ɨ** '*Derris utilis* (a fish poison)-big-stem'; **wa-čiči-'ɨ** 'fruit-?-stem' (G); tacacazeiro da várzea; ANF02; rare, large tree of old swiddens.

Sterculia pruriens (Aubl.) K. Schum.: **tapi'i-pamɨr-'ɨ**; 'tapir-?-stem' (= **kiha-pirita-'ɨ** 'hammock-loom-stem'); **tapi'ira-pawmi-'ɨw** 'tapir-?-stem' (T); **tapi-ir-pami-'ɨ** 'tapir-?-stem' (G); **tapi-do-páimi-'i** 'tapir-?-?-stem' (AR); tacacazeiro; TLW02 HHI14 FUE01; medium tree of high forests.

Theobroma cacao L.: **kaka** 'L'; **aka-'i** 'L-stem' (AR); **ako'o-'ɨ** 'L-stem' (G); **aka-'ɨw-ete** 'L-stem-true' (T); **aka-'ɨwa** 'L-stem' (AS); cacao; SFS01; small tree of dooryard gardens.

Theobroma grandiflorum (Willd. ex Spreg.) Schum.: **kɨpɨ-hu-'ɨ** 'L-big-stem'; **kupɨ'a-'ɨw** 'L-stem' (T); **kɨpɨ-'ɨ** 'L-stem' (G); cupuaçu; PFS01 ANF01; fairly common, small to medium tree of the high forest.

Theobroma speciosum Willd. ex Spreng.: **kaka-ran-'ɨ** 'L-false-tree'; **aka'u-'ɨw** 'L-stem' (T); **ako'o-'ɨ** 'L-stem' (G); **aka-'ɨwa** 'L-stem' (AS); **aka'á-wi-'i** 'L-fine-stem' (AR); SFS01 ANF01 FUE02; medium tree, very common in old fallows.

Theobroma subincanum Mart.: **kɨpɨ-'a-'ɨ** 'L-fruit-stem'(= **nu-kɨpɨ-'ɨ** '?-L-stem'); cupuí; SFS01 ANF01; very rare, small tree of high forests.

THEOPHRASTACEAE

Clavija lancifolia Desf.: **karume-pɨta-'ɨ** 'red footed tortoise-staying place-stem'; **zawiči-pɨta** 'tortoise-staying place' (T); **kamiča-apini'a-'ɨ** 'yellow footed tortoise-?-stem' (G); **yawči-rupi'a-rena** 'yellow footed tortoise-egg-place' (AS); **yani-dópiā** '?-egg' (AR); SFS01 ANF02; treelet of the high forest.

TILIACEAE

Apeiba aspera Aubl.: **āyaŋ-kɨwa-'i** 'divinity-comb-stem'; **ka'i-kɨwa-čĭ-'iw** 'capuchin monkey-comb-white-stem' (T); pente de macaco; ANF02 TLW02 PAC06 FUE01; large tree of the high forest.

Apeiba cf. *burchelli* Sprague: **āyaŋ-kɨwa-'ɨ** 'divinity-comb-stem'; pente de macaco; ANF02 TLW02 PAC06 FUE01; large tree of the high forest.

Apeiba echinata Gaertn.: **āyaŋ-kɨwa-'ɨ** 'divinity-comb-stem'; **kičʰu-kowa-'ɨ** 'Bearded saki-calabash bowl-stem' (G); pente de macaco; ANF02 TLW02 PAC06 FUE01; large tree of the high forest.

Apeiba echinata var. *macropetala* (Ducke) Gaertn.: **āyaŋ-kɨwa-'ɨ** 'divinity-comb-stem'; **wir-iri-'ɨ** 'swift-stem' (G); pente de macaco; ANF02 TLW02 PAC06 FUE01; large tree of the high forest.

Apeiba tibourbou Aubl.: **kaŋw-ɨ** '?-stem'; **kičʰu-kowa-'ɨ** 'Bearded saki-calabash bowl-stem' (G); pente de macaco; TLW02 FUE01; rare, very large tree of the high forest.

Luehea duckeana Burret: **parawa-mi'u-'ɨ** 'Mealy parrot-food-stem'; **kɨpɨ-ran-o'o-'ɨ** '*Theobroma* sp.-false-?-stem' (G); **mukape-'ɨwa** 'L-stem' (AS); ANF01 TLW02; rare, enormous tree of the old fallow.

Triumfetta rhomboidea Jacq.: **mɨra-kɨrawa** 'tree-caroá (*Neoglaziovia variegata*)'; malva; TLW01; tall, uncultivated shrub: spontaneous in young swiddens.

Triumfetta semitriloba Jacq.: **mɨra-kɨrawa** 'tree-caroá (*Neoglaziovia variegata*)'; TLW01; tall, uncultivated shrub, commonly found along edges of young and old swiddens.

ULMACEAE

Ampelocera edentula Kuhlm.: **tapi'i-rākwāi-pe-'ɨ** 'tapir-penis-flat-stem'; **a'ɨ-hu-'ɨ** 'sloth-big-stem' (G); **ɨwɨ-paye** 'earth-shaman' (AS); **yači-pape-'ɨ** 'tortoise-claw-stem' (AR); ANF02 TLW02 MED03 MED04 FUE01; very large tree of the high forest.

Celtis iguanea (Jacq.) Sargent: **akuši-ka'a** 'agouti-herb'; **mata-'ɨ** '?-stem' (AR); MAG02; treelet of the old fallow.

Trema micrantha (L.) Blume: **kuru-mi'u-'ɨ** 'divinity-food-stem'; (= **sa'ɨ-mi'u-'ɨ** 'honeycreeper-food-stem'); **wɨra-yu** 'tree-yellow' (G); **kuru-mi'i** 'divinity-food' (AR); trema; ANF02 SFS01; small tree, fairly common in old swiddens.

URTICACEAE

Laportea aestuans (L.) Chew: **purake-ka'a** 'electric eel-herb'; **pinu** 'L' (AR); **pinū** 'L' (AS); stinging nettles; urtiga vermelha; MED01 MED02 POI08; terrestrial herb of the old swidden.

VERBENACEAE

Aegiphila sp. 1: **sɨpo-ātā** 'vine-hard'; **pikəhu-rimi'u** 'pigeon-?' (T); NONE; small liana of the old fallow.

Lantana camara L.: **kanamɨ-ran** '*Clibadium sylvestre* (a fish poison)-false'; **ərəkwə-wɨra-'ɨw** 'Little chachalacha (*Ortalis motmot*)-tree-stem' (T); **tai-rána** 'spice-false' (G); **ñe-me'e** 'nothing-thing' (AR); lantana; chumbinho; erva cidreira; NONE; terrestrial herb of old swiddens.

VIOLACEAE

Paypayrola grandiflora Tul.: **sawɨya-mɨra** 'rat-tree' **ka'a-'ɨw-ran** 'herb-stem-false' (T); ANF02 FUE01; small tree of the high forest.

Rinorea flavescens (Aubl.) Kuntze: **pɨwa-'ɨ-hu** 'blunt arrow point (for killing birds)-stem-big'; **pɨwa-'ɨw** 'blunt arrow point (for killing birds)-stem' (T); canela de velha; ANF02 TLW13 HHI07 FUE01; uncommon, small tree uncommon of old fallows.

Rinorea pubiflora (Benth.) Spr. & Sandw.: **pɨwa-'ɨ** 'blunt arrow point (for killing birds)-stem'; **yanu-'ɨ** 'spider-stem' (AR); **yaí-'ɨ** 'menstruation-stem' (G); **wayaw-'a-ɨy** 'guava-fruit-seed' (AS); canela de velha; ANF02 TLW13 HHI07 FUE01 MIS08 MIS09; treelet of the high forest.

Rinorea sp. 1: **pɨwa-'ɨ-hu** 'kind of arrow point-stem-big'; canela de velha; ANF02 TLW13 HHI07 FUE01; small tree of the high forest.

VITACEAE

Cissus cissyoides L.: **sɨpo-ran** 'vine-false'; **tukur-ɨwɨpo** 'grasshopper-vine' (G); NONE; scandent vine of the old fallow.

VOCHYSIACEAE

Erisma uncinatum Warn.: **mitū-mɨra** 'curassow-tree'; **ɨra-tā-'ɨ** 'tree-hard-stem' (G); **tarapai-'i-namo** 'L-stem-?' (AR); ANF02 FUE01; large tree of the high forest.

Vochysia inundata Ducke: **serēserē-'ɨ** 'parakeet-stem'; ANF02 FUE02; common, medium tree of river edges.

ZINGIBERACEAE

Costus arabicus L.: **tayahu-pako-ro** 'white—lipped peccary-leaf-butter'; **taza-ran** 'cocoyam-false' (T); **ka'i-kəmə-'ɨ** '*Cebus* monkey-?-stem' (G); ANF02; terrestrial herb of the swamp forest.

Costus scaber R. & P.: **āyaŋ-rākwāi** 'divinity-penis'; **ka'a-tamkɨ** 'herb-?' (G); NONE; terrestrial herb found in light gaps of the high forest.

Curcuma sp. 1: **tawa** 'yellow'; turmeric; MED20 MED35; Introduced, domesticated herb of young swiddens.

Renealmia alpinia (Rottb.) Maas: **kurupi-kā** 'divinity-sugar cane'; **kənə-ran** 'sugar cane-false' (T); **kani-əhə** 'sugar cane-big' (AR); canarana; PAC07; terrestrial herb of the old swidden.

Renealmia floribunda K. Seh.: **kurupi-pɨtɨm** 'divinity-tobacco'; **ka'a-kacə** 'herb-?' (G); canarana; HHI18; terrestrial herb of the high forest, occasional in old swiddens.

Zingiber officinalis L.: **marakatai** 'L'; **marakatai** 'L' (T); **žiži** 'L' (AR); ginger root; gengibre; MED03 MED05; introduced, domesticated, rhizomatous herb of the young swidden.

Specific Activities and Associated Plants in Ka'apor Culture

PFS (PRIMARY FOOD SOURCE)

PFS01
  *Anacardium occidentale L.
  *Ananas comosus L.
  Annona montana Macfad.
  Byrsonima laevigata (Poir.) DC.
  *Carica papaya L.
  Caryocar villosum Aubl.
  *Cucurbita moschata Duch.
  Euterpe oleracea Mart.
  Lecythis pisonis Cambess.
  *Malpighia sp. 1
  *Malpighia sp. 2
  *Mangifera indica L.
  Maximiliana maripa Mart.
  *Musa sp. 1
  *Musa sp. 2
  Myrciaria cf. dubia McVaugh
  Oenocarpus distichus Mart.
  Platonia insignis Mart.
  Pouteria macrophylla (Lam.) Eyma
  *Psidium guajava L.
  Rheedia brasiliensis (Mart.) Pl. et Tr.
  *Rollinia mucosa Baill.
  Spondias mombin L.
  *Syzygium malaccencis (L.) Merril & Pery
  Theobroma grandiflorum (Willd. ex Spreg.)
    Schum.
  *Vigna adnantha (G.F. Meyer) Marechal
  *Zea mays L.

PFS02
  *Colocasia esculenta (L.) Schott.
  *Dioscorea sp. 2

NOTE: See Appendix 6 for definition of codes for each specific activity. Each species denoted with an asterisk is a domesticate.

  *Dioscorea trifida L.
  *Ipomoea batatas Lam.
  *Manihot esculenta Crantz
  *Xanthosoma sp. 1

PFS03
  *Colocasia esculenta (L.) Schott.
  *Xanthosoma sp. 1

SFS (SECONDARY FOOD SOURCE)

SFS01
  Ambelania acida Aubl.
  Anacardium giganteum Hancock ex
    Engler
  Anacardium parvifolium Ducke
  Ananas nanas (L.B. Smith) L.B. Smith
  Annona paludosa Aublet
  Annona sericea Dunal
  *Arachis hypogaea L.
  *Averrhoa carambola L.
  Astrocaryum gynacanthum Mart.
  Astrocaryum sp. 1
  Astrocaryum murumuru Mart.
  *Bactris gasipaes Mart.
  Bactris major Mart.
  Bactris maraja Mart.
  Bactris setosa Barb. Rodr.
  Bactris tomentosa Mart.
  Bagassa guianensis Aubl.
  Bellucia grossularioides (L.) Triana
  Brosimum acutifolium ssp. interjectum
    C.C. Berg
  Byrsonima stipulacea Juss.
  Byrsonima sp. 5
  Calyptranthes or Marlierea
  cf. Calyptranthes sp. 1
  Caryocar glabrum (Aubl.) Pers.

Cephalia sp. 1
Cheiloclinium cognatum (Miers) A.C.
aff. *Cheiloclinium* or *Salacia*
Chrysophyllum argenteum Jacquin ssp.
  auratum (Miquel) Pennington
Chrysophyllum lucentifolium Crong. ssp.
  pachycarpum Pires & Penn.
Chrysophyllum pomiferum (Eyma) Penn.
Chrysophyllum sparsiflorum Kl. ex Miq.
Citrullus lanatus (Thunb.) Matsumi &
  Nakai
Citrus aurantiifolia (Christm.) Swingle
Citrus medica-acida L.
Citrus sinensis (L.) Osbeck
Clavija lancifolia Desf.
Cocos nucifera L.
Cordia lomatoloba Johnst.
Cordia scabrifolia C.DC.
Cordia sp. 2
Couepia guianensis Aubl. ssp. *divaricata*
  (Hub.) Prance
Couepia guianensis Aubl. ssp. *guianensis*
Couepia guianensis ssp. *glandulosa* (Mig.)
  Prance
Cucumis anguria L.
Dialium guianense (Aubl.) Sandw.
Duguetia marcgraviana Mart.
Duguetia riparia Huber
Eugenia eurcheila Berg.
Eugenia muricata DC.
Eugenia patrissi Vahl.
Eugenia schomburkii Benth.
Eugenia tapecumensis Berg
Eugenia sp. 5
Eugenia sp. 6
Eugenia sp. 7
Eugenia sp. 8
Exellodendron barbatum Prance
Fusaea longifolia (Aublet) Saff.
Goupia glabra Aubl.
Guatteria elongata Benth.
Guatteria sp. 1
Guazuma ulmifolia Lam.
Helicostylis tomentosa (P. & E.) Rusby
Hirtella bicornis Mart. ex Zucc.
Hymenaea courbaril L.
Hymenaea parvifolia Huber
Hymenaea reticulata Ducke
Inga alba Willd.
Inga auristellae Harms
Inga brevialata Ducke

Inga capitata Desv.
Inga cinammomea Spruce ex Benth.
Inga fagifolia (L.) Willd.
Inga falcistipula Ducke
Inga heterophylla Willd.
Inga marginata Willd.
Inga miriantha Poepp. et Endl.
Inga nobilis Willd.
Inga paraensis Ducke
Inga rubiginosa (Rich.) DC.
Inga splendens Willd.
Inga thibaudiana DC.
Inga sp. 5
Inga sp. 6
Inga sp. 7
Inga sp. 8
Inga sp. 9
Inga sp. 10
Jacaratia spinosa (Aubl.) A.DC.
Lacmellea aculeata (Ducke) Monachino
Licania canescens R. Ben.
Licania glabriflora Prance
Licania heteromorpha Benth. var. *hetero-
  morpha*
Licania kunthiana Hook. F.
Licania latifolia Benth.
Manilkara bidentata (A.DC.) Chev. ssp.
  surinamensis (Miq) Penn.
Manilkara huberi (Ducke) Chev.
Mauritia flexuosa Mart.
Micropholis guyanensis (A.DC.) Pierre
  ssp. *guyanensis* Penn.
Micropholis melinoniana Pierre
Mouriri guianensis Aubl.
Mouriri huberi Cogn.
Mouriri sagotiana Triana
Mouriri trunciflora Ducke
Myrcia paivae Berg.
Myrcia splendens (Sw.) DC.
aff. *Myrcia* sp. 1
Myrcia sp. 4
Myrcia sp. 5
Myrciaria flexuosa
Myrciaria spruceana Berg
Orbignya phalerata Mart.
Oryza sativa L.
Parahancornia amapa Ducke
Parahancornia fasciculata (Poir.) Benoist
Parinari excelsa Sabine
Passiflora sp. 1
Passiflora sp. 3

Physalis angulata L.
Pourouma guianensis Aubl. ssp.
  guianensis
Pourouma minor Benoist
Pourouma mollis Trec. ssp. *mollis*
Pouteria bilocularis (Winkler) Baehni
Pouteria caimito (R. & P.) Radlk.
Pouteria aff. *caimito* (R. & P.) Radlk.
Pouteria egregia Sandw.
Pouteria engleri Eyma
Pouteria jariensis Pires & Penn.
Pouteria macrocarpa (Martius) Dietrich
Pouteria penicillata Baehni
Pouteria reticulata (Engl.) Eyma ssp.
  reticulata
Pouteria trigonosperma Eyma
Pouteria sect. *Franchetella* sp. 1
Pouteria sect. *Franchetella* sp. 2
Pouteria sp. 2
Pouteria sp. 5
Protium aracouchini (Aubl.) March.
Protium decandrum (Aublet) Marchand
Protium giganteum Engl. var. *giganteum*
Protium pallidum Cuatrec.
Protium polybotryum (Turcz.) Engl.
Protium sagotianum Engl.
Protium spruceanum (Benth.) Engl.
Protium tenuifolium (Engl.) Engl.
Protium trifoliolatum Engl.
Rheedia acuminata Pl. et Tr.
Rollinia exsucca (DC. ex Dun.) A.DC.
Saccharum officinarum L.
Sacoglottis amazonica Mart.
Sacoglottis guianensis Benth.
Sacoglottis sp. 1
Salacia insignis A.C. Smith
Sapotaceae sp. 2
Solanum crinitum Lam.
Solanum stramonifolium Jacq.
Talisia acutifolia Radlk.
Talisia carasina (Benth.) Radlk.
Talisia micrantha Radlk.
Talisia microphylla Witt.
Talisia retusa Cowan
Tapirira guianensis Aublet
Tapirira peckoltiana Engler
Tetragastris altissima (Aublet) Swart.
Tetragastris panamensis (Engl.) O.K.
Theobroma speciosum Willd. ex Spreng.
Theobroma subincanum Mart.
Trema micrantha (L.) Blume

SFS02
  *Allium cepa L.
  *Allium sativum L.
  Ischnosiphon arouma (Aubl.) Koern.

SFS03
  Ctenanthe sp. 1
  Marantaceae sp. 1

SFS04
  Anacardium giganteum Hancock ex En-
    gler
  *Anacardium occidentale L.
  Anacardium parvifolium Ducke
  Caryomene foveolata Barneby & Krukoff
  *Manihot esculenta Crantz
  *Musa sp. 1
  *Musa sp. 2
  Pouteria aff. *caimito* (R. & P.) Radlk.
  Pouteria jariensis Pires & Penn.
  Tetragastris altissima (Aublet) Swart.

SFS05
  Orbignya phalerata Mart.

SFS06
  Ambelania acida Aubl.
  Lacmellea aculeata (Ducke) Monachino
  Manilkara huberi (Ducke) Chev.
  Parahancornia amapa Ducke
  Parahancornia fasciculata (Poir.) Benoist

SFS07
  *Coffea arabica L.
  *Cymbopogon citratus (DC.) Stapf.

SFS08
  Davilla kunthii St. Hil.
  Davilla nitida (Vahl.) Kubitzki
  Doliocarpus sp. 1
  Tetracera volubilis L. ssp. *volubilis*
  Tetracera willdenowiana Schlechtd. ssp.
    willldenowiana

SEA (SEASONING)
SEA01
  *Amaranthus oleraceus L.
  *Capsicum annum L.
  *Capsicum fructescens L.
  *Eryngium foetidum L.

Ocimum micranthum Willd.
Ocimum sp. 1

SEA02

Euterpe oleracea Mart.
Maximiliana maripa Mart.
aff. *Pradosia* sp. 1

ANF (ANIMAL FOOD)

ANF01

Aechmea brevicellis L.B. Smith
Aechmea bromeliifolia Baker
Alibertia edulis (L.Rich.) A. Rick.
Ambelania grandiflora Hubar
Anacardium giganteum Hancock ex En-
  gler
Anacardium parvifolium Ducke
Aniba citrifolia (Nees) Mez.
Annona paludosa Aublet
Annona sericea Dunal
Annonaceae sp. 1
Aspidosperma cylindrocarpon Muell. Arg.
Astrocaryum murumuru Mart.
Astrocaryum vulgare Mart.
Bactris major Mart.
Bactris maraja Mart.
Bagassa guianensis Aubl.
Bromelia goeldiana L.B. Smith
Bromeliaceae sp. 1
Buchenavia cf. *tetraphylla* (Aubl.) How-
  ard
Byrsonima laevigata (Poir.) DC.
Calyptranthes or Marlierea
cf. *Calyptranthes* sp. 1
Carapa guianensis Aubl.
Caryocar glabrum (Aubl.) Pers
Caryocar villosum Aubl.
Cassia fastuosa Willd.
Cecropia purpurescens C.C. Berg
Cecropia sciadophylla Mart.
Chrysophyllum lucentifolium Crong. ssp.
  pachycarpum Pires & Penn.
Chrysophyllum sparsiflorum Kl. ex Miq.
Clarisia racemosa Ruiz & Pavon
Cordia exaltata Lam.
Couepia guianensis Aubl. ssp. *guianensis*
Couepia guianensis ssp. *glandulosa* (Mig.)
  Prance
Couratari oblongifolia Ducke & Knuth.
Croton cajucara Benth.

Drymonia coccinea (Aubl.) Weihl.
Duguetia riparia Huber
Endlicheria verticellata Mez.
Eschweilera amazonica R. Kunth.
Eschweilera apiculata (Miers) A.C. Smith
Eschweilera coriacea (A.P. Candolle) Mart.
  ex Berg
Eschweilera micrantha (Berg) Miers
Eschweilera obversa (Berg) Miers
Eschweilera pedicellata (Richard) Mori
Eugenia eurcheila Berg.
Eugenia sp. 5
Eugenia sp. 6
Eugenia tapecumensis Berg
Euterpe oleracea Mart.
Ficus sp. 3
Flacourtiaceae sp. 1
Goupia glabra Aubl.
Guarea kunthiana A. Juss.
Guarea macrophylla Vahl. ssp. *pachy-
  carpa* (C.DC.) Penn.
Guatteria elongata Benth.
Guatteria sp. 1
Guazuma ulmifolia Lam.
Guzmania lingulata (L.) Mez.
Helicostylis tomentosa (P. & E.) Rusby
Hirtella bicornis Mart. ex Zucc.
Hirtella eriandra Benth.
Hirtella racemosa Lam. var. *racemosa*
Inga alba Willd.
Inga auristellae Harms
Inga brevialata Ducke
Inga capitata Desv.
Inga cinammomea Spruce ex Benth.
Inga fagifolia (L.) Willd.
Inga falcistipula Ducke
Inga heterophylla Willd.
Inga marginata Willd.
Inga miriantha Poepp. et Endl.
Inga nobilis Willd.
Inga paraensis Ducke
Inga rubiginosa (Rich.) DC.
Inga splendens Willd.
Inga thibaudiana DC.
Inga sp. 5
Inga sp. 6
Inga sp. 7
Inga sp. 8
Inga sp. 9
Inga sp. 10
Ipomoea phyllomega (Velloso) House

Jacaratia spinosa (Aubl.) A.DC.
Lacmellea aculeata (Ducke) Monachino
Lauraceae sp. 1
Lauraceae sp. 2
Lecythis chartacea Berg
Lecythis pisonis Cambess.
Licania macrophylla Benth.
Licaria brasiliensis (Nees) Kost
Licaria debilis (Mez) Kost.
Licaria sp. 1
Luehea duckeana Burret
Macrolobium acaciaefolium Benth.
Manihot quinquepartita Huber ex Rog et
  Appan
Manilkara bidentata (A.DC.) Chev. ssp.
  surinamensis (Miq) Penn.
Manilkara huberi (Ducke) Chev.
Maximiliana maripa Mart.
Maytenus sp. 10
Miconia cf. *kappleri* Naud.
Miconia nervosa (Smith) Triana
Miconia serialis DC.
Miconia serrulata (DC.) Naud.
Miconia surinamensis Gleason
Miconia sp. 1
Mouriri guianensis Aubl.
Mouriri trunciflora Ducke
Neoptychocarpus apodanthus Kuhlm.
Ocotea amazonica (Meiss.) Mez.
Ocotea caudata Mez.
Ocotea costata Mez.
Ocotea fasciculata (Ness) Mez
Ocotea guianensis Aubl.
Ocotea opifera Mart.
Ocotea rubra Mez.
Ocotea silvae Vattimo-Gil sp. nov.
Oenocarpus distichus Mart.
Parahancornia amapa Ducke
Parahancornia fasciculata (Poir.) Benoist
Parinari excelsa Sabine
Parkia multijuga Benth.
Parkia pendula Mig.
Parkia ulei (Harms) Kuhlm.
Pithecellobium foliolosum Benth
Pithecellobium jupumba (Willd.) Urb.
Platonia insignis Mart.
Pourouma minor Benoist
Pourouma mollis Trec. ssp. *mollis*
Pouteria caimito (R. & P.) Radlk.
Pouteria engleri Eyma
Pouteria gongrijpii Eyma

Pouteria jariensis Pires & Penn.
Pouteria macrocarpa (Martius) Dietrich
Pouteria macrophylla (Lam.) Eyma
Pouteria sagotiana (Baill.) Eyma
Pouteria sect. *Franchetella* sp. 1
Pouteria sect. *Franchetella* sp. 2
Pouteria sp. 2
Pouteria sp. 5
Pouteria trigonosperma Eyma
Psychotria hoffmannseggiana (Willd. ex R.
  & S.) M. Arg.
Randia armata (Sw.) Sw.
Rheedia brasiliensis (Mart.) Pl. et Tr.
Sapotaceae sp. 2
Sclerolobium guianense Aubl.
Senna pendula (Lamk.) Irwin & Barneby
Seguieria amazonica Huber
Simaba cuspidata Spr. var. *typica*
Simaba guianensis Aubl. ssp. *guianensis*
Spondias mombin L.
Stryphnodendron   polystachyum   (Miq.)
  Kleinh.
Swartzia brachyrachis var. Harms
Swartzia   brachyrachis   var.   *snethlageae*
  (Ducke) Ducke
Tapirira guianensis Aublet
Tapirira peckoltiana Engler
Taralea oppositifolia Aubl.
Terminalia amazonia (J. Gmel.) Exell.
Terminalia lucida Hoffmgg.
Tetragastris altissima (Aublet) Swart.
Tetragastris panamensis (Engl.) O.K.
Theobroma grandiflorum (Willd. ex Spreg.)
  Schum.
Theobroma speciosum Willd. ex Spreng.
Theobroma subincanum Mart.
Trichilia micrantha Benth.
Trichilia verrocosa C. DC.
Virola sp. 1
*Zea mays L.

ANF02
Aciotis purpurescens (Aubl.) Triana
Ampelocera edentula Kuhlm.
Ananas nanas (L.B. Smith) L.B.Smith
Aparisthmium cordatum Baill.
Apeiba aspera Aubl.
Apeiba cf. *burchelli* Sprague
Apeiba echinata Gaertn.
Apeiba echinata var. *macropetala* (Ducke)
  Gaertn.

Aspidosperma cylindrocapon Muell. Arg.
Astronium lecointei Ducke
Banara guianensis Aubl.
Banisteriopsis pubipetala (Adr. Juss.)
  Cuatr.
Bellucia grossularioides (L.) Triana
Bertiera guianensis Aublet
Brosimum acutifolium ssp. *interjectum*
  C.C. Berg
Brosimum guianense (Aubl.) Huber
Brosimum paclesum (S. Moore) C.C. Berg
Calathea sp. 1
Campomanesia grandiflora (Aubl.) Sagot.
Capparis sola Macbr.
Caraipa grandiflora Mart.
Casearia arborea (L.C. Richard) Urban
Casearia javitensis H.B.K.
Cecropia concolor Willd.
Cecropia obtusa Trecue
Ceiba pentandra Gaertn.
Cheiloclinium cognatum (Miers) A.C.
aff. *Cheiloclinium* or *Salacia* sp. 1
Chimarrhis turbinata DC.
Chrysophyllum argenteum Jacquin ssp.
  auratum (Miquel) Pennington
Chrysophyllum pomiferum (Eyma) Penn.
Clavija lancifolia Desf.
Cochlospermum orinocense (H.B.K.) Steud.
Conceveiba guianensis Aubl.
Cordia multispicata Cham.
Cordia polycephala (Lam.) Johnston
Cordia scabrifolia C.DC.
Cordia sellowiana Chamb.
Costus arabicus L.
Couepia guianensis Aubl. ssp. *divaricata*
  (Hub.) Prance
Couratari guianensis Aublet
Coumarouna micrantha Ducke
Coussarea ovalis Standl.
Coussarea paniculata (Vahl.) Standl.
Croton cuneatus Klotesch.
Ctenanthe sp. 1
Cucurbitaceae sp. 1
Cupania scrobiculata L.C. Rich.
Dalbergia monetaria L.F.
Davilla kunthii St. Hil.
Davilla nitida (Vahl.) Kubitzki
Dendrobangia boliviana Rusby
Derris amazonica Killip.
Derris sp. 1
Dialium guianense (Aubl.) Sandw.

Dioclea reflexa Hook. F.
Diospyros artanthifolia Mart.
Diospyros duckei Sandw.
Diospyros melinoni (Hiern.) A.C. Smith
Dodecastigma integrifolium (Lanj.) Lanj.
  & Sandw.
Doliocarpus sp. 1
Duguetia echinophora R.E. Fries
Duguetia sp. 1
Dulacia guianensis (Engl.) O. Kuntze
Erisma uncinatum Warn.
Erythroxylum cf. *leptoneurum* O.E. Schulz
Erythroxylum citrifolium A. St. H.L.
Eugenia muricata DC.
Eugenia schomburkii Benth.
Eugenia sp. 7
Eugenia sp. 8
Eupatorium macrophyllum L.
Exellodendron barbatum Prance
Faramea sessilifolia (H.B.K.) DC.
Ficus sp. 2
Geonoma leptospadix Trail
Geophila repens (L.) I.M. Johnston
Guarea guidonia (L.) Sleumer
Guarea sp. 1
Guettarda divaricata (H.B.K.) Standl.
Gurania cissoides (Benth.) Cogn.
Gustavia augusta L.
Heliconia acuminata L.C. Rich.
Heliconia bihai L.
Heliconia chartacea Lane ex Barr.
Henrietta spruceana Cogn.
Himatanthus articulatus (Vahl.) Wood-
son
Hirtella cf. *paraensis* Prance
Hirtella racemosa var. *hexandra* (Willd.
  ex R.G.S.) Prance
Humirianthera duckei Huber
Hymenaea parvifolia Huber
Hymenolobium excelsum Ducke
Inga gracilifolia Ducke
Ipomea setofera Oir.
Ipomoea aff. *squamosa* Choisy
Ipomoea sp. 1
Ischnosiphon petrolatus (rudge) L. And.
Jacaranda copaia (Aubl.) D. Don. ssp.
  copaia
Jacaranda copaia ssp. *spectabilis* (Mart. ex
  DC.) A. Gentry
Jacaranda duckei Vattimo
Lacunaria jenmani (Oliv.) Ducke

Lacunaria sp. 4
Lacunaria sp. 5
Laetia procera (Poepp. et Endl.) Eichl.
Lasiacis ligulata Hitchelchase
Lauraceae sp. 1
Lecythis idatimon Aublet
Lecythis lurida (Miers) Mori
Licania apetala (E. Mey.) Fritsch
Licania canescens R. Ben.
Licania glabriflora Prance
Licania heteromorpha Benth. var. *hetero-morpha*
Licania latifolia Benth.
Licania membranacea Sagot. ex Laness.
Licania octandra (Hoffmgg. ex R. & S.) Kuntze
Lindackeria latifolia Benth.
Lindackeria sp. 1
Mabea angustifolia Spruce ex Benth.
Mabea caudata Pax & Hoffm.
Mabea sp. 1
Magaritaria nobilis L.F.
Malanea
Manihot brachyloba Muell. Arg.
Manihot leptophylla Pax in Engler
Maquira guianensis Aubl.
Maranta protracta Mig.
Marantaceae sp. 1
Markleya dahlgreniana Bondar
Matayba spruceana Radlk.
aff. *Memora bracteosa* (DC.) Bur. & K. Schum.
Miconia ceramicarpa (DC.) Cogn.
Micropholis guyanensis (A.DC.) Pierre ssp. *guyanensis* Penn.
Micropholis venulosa (Mart. ex Eich.) Pierre
Monstera cf. *pertusa* (L.) Vriesia
Monstera subpinata (Schott) Engl.
Moutabea sp. 1
Myrcia sp. 1
Myrciaria cf. *dubia* McVaugh
Myriaspora egensis DC.
Neea floribunda Poepp. & Endl.
Neea glameruliflora Heimut
Neea oppositifolia R. & P.
Neea ovalifolia Spruce ex J.A. Schmidt
Neea sp. 1
Neea sp. 2
Newtonia psilostachya (DC.) Brenan
Newtonia suaveolens (Miguel) Brenan

Ocotea canaliculata (L.C. Rich.) Mez.
Ocotea glomerata (Nees) Mez.
Ocotea sp. 3
Olyra caudata Trin.
Paspalum conjugatum Berg.
Passiflora aranjoi Sacco
Passiflora coccinea Aubl.
Passiflora sp. 2
Passiflora sp. 3
Passiflora sp. 4.
Paypayrola grandiflora Tul.
Philodendron grandiflorum (Jacq.) Schott.
Philodendron venustum Bunt.
Piper hostmannianum (Mig.) C.DC.
Pisonia sp. 2
Pithecellobium cauliflorum Mart.
Pithecellobium cochleatum (Willd.) Mart.
Pithecellobium comunis Benth.
Pithecellobium racemosum Ducke
Platypodium elegans Vog.
Poaceae sp. 1
Pourouma guianensis Aubl. ssp. *guianensis*
Pouteria egregia Sandw.
Pristimera tenuifolia (Mart.) A.C.
Protium altsoni Sandw.
Protium aracouchini (Aubl.) March.
Protium decandrum (Aublet) Marchand
Protium giganteum Engl. var. *giganteum*
Protium heptaphyllum ssp. *heptaphyllum* (Aublet) Marchand
Protium pallidum Cuatrec.
Protium polybotryum (Turcz.) Engl.
Protium sagotianum Engl.
Protium spruceanum (Benth.) Engl.
Protium tenuifolium (Engl.) Engl.
Protium trifoliolatum Engl.
Pseudyma frutescens (Aubl.) Radlk.
Psychotria colorata Poepp.
Psychotria poeppigiana Muell. Arg.
Psychotria poeppigiana ssp. *poeppigiana* Muell. Arg.
Psychotria racemosa (Aubl.) Baensh.
Psychotria ulviformis Steyerm.
Psychotria sp. 1
Rauia resinosa Nees & Mart.
Rauvolfia paraensis Ducke
Remijia sp. 1
Rinorea flavescens (Aubl.) kuntze
Rinorea pubiflora (Benth.) Spr. & Sandw.
Rinorea sp. 1

Rollinia exsucca (DC. ex Dun.) A.DC.
Sacoglottis amazonica Mart.
Sacoglottis guianensis Benth.
Sacoglottis sp. 1
Sagotia racemosa Baill.
Salacia insignis A.C. Smith
Salacia multiflora (Lam.) DC.
Sapium caspceolatum Huber
Sapium lanceolatum Huber
Schefflera morototoni (Aubl.) M.S.F.
Scleria secans (L.) Urb.
Sclerolobium albiflorum R. Ben.
Sclerolobium paraense Huber
Senna sylvestris (Vell.) Irwin & Barneby
Simaruba amara Aubl.
Sloanea grandiflora Smith
Sloanea grandis Ducke
Sloanea guianensis (Aubl.) Benth.
Sloanea porphyrocarpa Ducke
Solanum crinitum Lam.
Solanum rugosum Dunal
Solanum subinerme Jacq.
Solanum vanheurckii Mulh.
Sterculia alata Ducke
Struthanthus marginatus Blume
Stryphnodendron guianensis (Aubl.)
 Benth.
Syagrus cf. *inajai* (Spruce) Becc.
Syagrus sp. 1
Symphonia globulifera L.
Tabebuia sp. 1
Tabebuia serratifolia (Vahl.) Nichols
Talisia acutifolia Radlk.
Talisia micrantha Radlk.
Talisia microphylla Witt.
Talisia retusa Cowan
Tetracera volubilis L. ssp. *volubilis*
Tetracera willdenowiana Schlechtd. ssp.
 willdenowiana
Tetrapterys styloptera A. Juss.
Tocoyena foetida Poepp. & Endl.
Tovomita brevistaminea Engl.
Trattinickia burserifolia Mart.
Trattinickia rhoifolia Willd.
Trattinickia sp. 1
Trema micrantha (L.) Blume
Trichilia quadrijuga Kunth. ssp. *quadri-
juga*
Trigonidium acuminatum Batem ex Lindl.
Unonopsis rufescens (Baill.) R.E. Fries
Virola carinata (Benth. Warb.)

Virola michelii Heckel
Vochysia inundata Ducke
Xylopia barbata Mart.
Xylopia nitida Dunal
Zanthoxylum cf. *juniperina* Engl.
Zanthoxylum chiloperone (Mart.) Engl.
Zanthoxylum rhoifolium Lam.

ANF03
 Brachiaria humidicula (Rendle)
  Schweickerdt
 Desmodium adescendens (Sw.) DC.

PBC (POST-AND-BEAM CONSTRUCTION)

PBC01
 Diplotropis purpurea (Rich.) Amsh.
 Eschweilera amazonica R. Kunth.
 Eschweilera apiculata (Miers) A.C. Smith
 Eschweilera coriacea (A.P. Candolle) Mart.
  ex Berg
 Eschweilera micrantha (Berg) Miers
 Eschweilera obversa (Berg) Miers
 Eschweilera ovata (Cambess) Miers
 Eschweilera pedicellata (Richard) Mori
 Minquartia guianensis Aubl.
 Xylopia nitida Dunal

PBC02
 Andira sp. 1
 Aniba citrifolia (Nees) Mez.
 Aspidosperma desmanthum Bth. ex M.
  Arg.
 Aspidosperma verruculosum M. Arg.
 Astronium lecointei Ducke
 Couepia guianensis Aubl. ssp. *divaricata*
  (Hub.) Prance
 Couepia guianensis Aubl. ssp. *guianensis*
 Couepia guianensis ssp. *glandulosa* (Mig.)
  Prance
 Diospyros artanthifolia Mart.
 Diospyros duckei Sandw.
 Duguetia riparia Huber
 Endlicheria verticellata Mez.
 Eschweilera amazonica R. Kunth.
 Eschweilera apiculata (Miers) A.C. Smith
 Eschweilera coriacea (A.P. Candolle) Mart.
  ex Berg
 Eschweilera micrantha (Berg) Miers
 Eschweilera pedicellata (Richard) Mori
 Fusaea longifolia (Aublet) Saff.

Guatteria elongata Benth.
Guatteria sp. 1
Helicostylis tomentosa (P. & E.) Rusby
Hirtella bicornis Mart. ex Zucc.
Lacunaria jenmani (Oliv.) Ducke
Lauraceae sp. 1
Lauraceae sp. 2
Lecythis chartacea Berg
Lecythis pisonis Cambess.
Licania canescens R. Ben.
Licania glabriflora Prance
Licania heteromorpha Benth. var. *hetero-morpha*
Licania kunthiana Hook. F.
Licania latifolia Benth.
Licaria sp. 1
Ocotea amazonica (Meiss.) Mez.
Ocotea caudata Mez.
Ocotea costata Mez.
Parinari excelsa Sabine
Pouteria egregia Sandw.
Pouteria filipes Eyma
Pouteria gongrijpii Eyma
Pouteria macrocarpa (Martius) Dietrich
Pouteria macrophylla (Lam.) Eyma
Pouteria sagotiana (Baill.) Eyma
Pouteria trigonosperma Eyma
Pouteria sect. *Franchetella* sp. 1
Pouteria sect. *Franchetella* sp. 2
Protium trifoliolatum Engl.
Sapotaceae sp. 2

PBC03
Geonoma baculifera Kunth.
Maximiliana maripa Mart.

PBC04
Croton matourensis Aubl.
Euterpe oleracea Mart.

PBC05
Heteropsis longispathacea Engl.
Heteropsis sp. 1

PBC06
Euterpe oleracea Mart.

PBC07
Cecropia purpurescens C.C. Berg
Cecropia sciadophylla Mart.

Diospyros melinoni (Hiern.) A.C. Smith
Eschweilera amazonica R. Kunth.
Eschweilera apiculata (Miers) A.C. Smith
Eschweilera coriacea (A.P. Candolle) Mart. ex Berg
Eschweilera micrantha (Berg) Miers
Eschweilera pedicellata (Richard) Mori
Lecythis chartacea Berg
Xylopia nitida Dunal

PBC08
Euterpe oleracea Mart.

TLW (TOOLS AND WEAPONS)

TLW01
Cochlospermum orinocense (H.B.K.) Steud.
*Neoglaziovia variegata (Arr. Cam.) Mez.
Rollinia exsucca (DC. ex Dun.) A.DC.
Sida cordifolia L.
Sida santaremensis Mont.
Triumfetta rhomboidea Jacq.
Triumfetta semitriloba Jacq.
Urena lobata L.

TLW02
Ampelocera edentula Kuhlm.
Anaxagorea dolichocarpa Spr. ex Sandw.
Apeiba aspera Aubl.
Apeiba cf. *burchelli* Sprague
Apeiba echinata Gaertn.
Apeiba echinata var. *macropetala* (Ducke) Gaertn.
Apeiba tibourbou Aubl.
Bauhinia sp. 2
Cochlospermum orinocense (H.B.K.) Steud.
Cordia exaltata Lam.
Cordia multispicata Cham.
Cordia scabrida Mart.
Cordia scabrifolia C.DC.
Cordia sellowiana Chamb.
Cordia trachyphylla mart.
Couratari guianensis Aublet
Couratari oblongifolia Ducke & Knuth.
Duguetia echinophora R.E. Fries
Duguetia riparia Huber
Duguetia sp. 1
Eschweilera amazonica R. Kunth.
Eschweilera apiculata (Miers) A.C. Smith
Eschweilera coriacea (A.P. Candolle) Mart. ex Berg

Eschweilera micrantha (Berg) Miers
Eschweilera obversa (Berg) Miers
Eschweilera ovata (Cambess) Miers
Eschweilera pedicellata (Richard) Mori
Guatteria elongata Benth.
Guatteria sp. 1
Hippocratea ovata Lam.
Lecythis chartacea Berg
Lecythis idatimon Aublet
Lecythis lurida (Miers) Mori
Lecythis pisonis Cambess.
Luehea duckeana Burret
Pachira aquatica Aubl.
Rollinia exsucca (DC. ex Dun.) A.DC.
Sterculia pruriens (Aubl.) K. Schum.
Urena lobata L.
Xylopia barbata Mart.
Xylopia nitida Dunal

TLW03
   Diospyros artanthifolia Mart.
   Diospyros duckei Sandw.
   Pouteria aff. *caimito* (R. & P.) Radlk.
   Pouteria filipes Eyma
   Pouteria jariensis Pires & Penn.

TLW04
   Pourouma guianensis Aubl. ssp. *guianen-sis*

TLW05
   Symphonia globulifera L.

TLW06
   Manilkara huberi (Ducke) Chev.

TLW07
   *Clibadium sylvestre (Aubl.) Baill.
   Derris utilis (A.C. Smith) Ducke
   Serjania aff. *lethalis* St. Hil.
   Syagrus cf. *inajai* (Spruce) Becc.
   Syagrus sp. 1
   *Tephrosia sinapou (Buchot) A. Chev.

TLW08
   Duguetia echinophora R.E. Fries
   Duguetia sp. 1
   Duguetia sp. 2
   Unonopsis rufescens (Baill.) R.E. Fries

TLW09
   Psychotria colorata Poepp.
   Ricinus comunis L.

TLW10
   Bauhinia viridiflorens Ducke
   Brosimum rubescens Taubert
   Tabebuia impetiginosa Standley.

TLW11
   Desmoncus macroacanthos Mart.
   Desmoncus polycanthus Mart.

TLW12
   Bactris humilis (Wallace) Burret
   *Gynerium sagittatum (Aubl.) Beauvois

TLW13
   Bauhinia viridiflorens Ducke
   Brosimum rubescens Taubert
   Calyptranthes or *Marlierea*
   Guadua glomerata Munro
   Licania canescens R. Ben.
   Licania glabriflora Prance
   Licania heteromorpha Benth. var. *heteromorpha*
   Licania kunthiana Hook. F.
   Licania latifolia Benth.
   Rinorea flavescens (Aubl.) Kuntze
   Rinorea pubiflora (Benth.) Spr. & Sandw.
   Rinorea sp. 1
   Talisia carasina (Benth.) Radlk.
   Talisia micrantha Radlk.

TLW14
   aff. *Myrcia*
   Eugenia sp. 5
   Eugenia sp. 8
   Tovomita brasiliensis (Mart.) Walp.

TLW15
   Zollernia paraensis Hub.

TLW16
   *Luffa cylindrica (L.) M.J. Roem.

TLW17
   Gustavia augusta L.

TLW18
   *Gossypium barbadense L.

TLW19
*Neoglaziovia variegata (Arr. Cam.) Mez.

HHI (HOUSEHOLD ITEMS)

HHI01
*Bixa orellana L.
Guatteria scandens Ducke

HHI02
Oenocarpus distichus Mart.

HHI03
*Crescentia cujete L.
Posadea sphaerocarpa Cogn.

HHI04
*Lagenaria siceraria (Molina) Standl.

HHI05
Ambelania acida Aubl.
Lacmellea aculeata (Ducke) Monachino

HHI06
Andira sp. 1
Astronium lecointei Ducke
Diplotropis purpurea (Rich.) Amsh.
Hirtella racemosa Lam. var. racemosa
Trichilia quadrijuga Kunth. ssp. quadrijuga

HHI07
Hirtella racemosa Lam. var. racemosa
Ouratea sp. 2
Rinorea flavescens (Aubl.) Kuntze
Rinorea pubiflora (Benth.) Spr. & Sandw.
Rinorea sp. 1

HHI08
Evodianthus funifer (Poit.) Lindm.
Ischnosiphon arouma (Aubl.) Koern.

HHI09
Ischnosiphon arouma (Aubl.) Koern.

HHI10
Socratea exorrhiza (Mart.) H. Wendl.

HHI11
Cupania scrobiculata L.C. Rich.

Diospyros artanthifolia Mart.
Diospyros duckei Sandw.
Matayba spruceana Radlk.

HHI12
aff. Myrcia sp. 1
Euterpe oleracea Mart.
Evodianthus funifer (Poit.) Lindm.
Ischnosiphon arouma (Aubl.) Koern.
Ischnosiphon obliquus (Rudge) Koern.
Ischnosiphon sp. 1
Lecythis idatimon Aublet
Marantaceae sp. 3

HHI13
Euterpe oleracea Mart.
Scoparia dulcis L.

HHI14
Sterculia pruriens (Aubl.) K. Schum.
Syagrus cf. inajai (Spruce) Becc.
Syagrus sp. 1

HHI15
*Gossypium barbadense L.

HHI16
Oenocarpus distichus Mart.

HHI17
Cedrela fissilis Vell.

HHI18
Ardisia guianensis (Aubl.) Mez
Astronium cf. obliquum Griseb.
*Nicotiana tabacum L.
Olyra latifolia L.
Protium giganteum Engl. var. giganteum
Protium pallidum Cuatrec.
Protium spruceanum (Benth.) Engl.
Renealmia floribunda K. Seh.
Thyrsodium sp. 1
Thyrsodium spruceanum Bentham
Trattinickia burserifolia Mart.
Trattinickia rhoifolia Willd.
Trattinickia sp. 1

HHI19
Couratari guianensis Aublet

HHI20
>   *Mabea angustifolia* Spruce ex Benth.
>   *Mabea caudata* Pax & Hoffm.
>   *Mabea* sp. 1

HHI21
>   *Licania apetala* (E. Mey.) Fritsch
>   *Licania membranacea* Sagot. ex Laness.
>   *Licania octandra* (Hoffmgg. ex R. & S.)
>   Kuntze

HHI22
>   *Hymenaea courbaril* L.
>   *Hymenaea parvifolia* Huber
>   *Hymenaea reticulata* Ducke

HHI23
>   *Ischnosiphon* sp. 2

HHI24
>   *Marantaceae* sp. 4

HHI25
>   *Tetragastris altissima* (Aublet) Swart.

HHI26
>   *Eschweilera amazonica* R. Kunth.
>   *Eschweilera apiculata* (Miers) A.C. Smith
>   *Eschweilera coriacea* (A.P. Candolle) Mart.
>   ex Berg
>   *Eschweilera micrantha* (Berg) Miers
>   *Eschweilera pedicellata* (Richard) Mori

HHI27
>   *Calathea* sp. 1
>   *Gurania cissoides* (Benth.) Cogn.

HHI28
>   *Cedrela fissilis* Vell.

HHI29
>   *Andira* sp. 1
>   *Astronium lecointei* Ducke
>   *Bagassa guianensis* Aubl.
>   *Diplotropis purpurea* (Rich.) Amsh.
>   *Helicostylis tomentosa* (P. & E.) Rusby
>   *Laetia procera* (Poepp. et Endl.) Eichl.
>   *Pouteria bilocularis* (Winkler) Baehni
>   *Simaruba amara* Aubl.

HHI30
>   *Brosimum rubescens* Taubert
>   *Licania canescens* R. Ben.
>   *Licania glabriflora* Prance
>   *Licania heteromorpha* Benth. var. *hetero-*
>   *morpha*
>   *Licania latifolia* Benth.
>   *Licania kunthiana* Hook. F.

HHI31
>   *Goupia glabra* Aubl.
>   *Hymenaea courbaril* L.
>   *Hymenaea reticulata* Ducke
>   *Newtonia psilostachya* (DC.) Brenan
>   *Newtonia suaveolens* (Miguel) Brenan
>   *Pithecellobium comunis* Benth.

PAC (PERSONAL ADORNMENT, COSMETICS, HYGIENE, DYES)

PAC01
>   **Gossypium barbadense* L.

PAC02
>   **Adenanthera pavonina* L.
>   *Andira retusa* (Lam.) H.B.K.
>   *Astrocaryum murumuru* Mart.
>   *Astrocaryum vulgare* Mart.
>   **Canna indica* L.
>   **Cardiospermum halicacabum* L.
>   **Coix lacryma-jobi* L.
>   *Colubrina* sp. 1
>   *Colubrina glandulossa* Perkins var. *glan-*
>   *dulosa*
>   *Duguetia surinamensis* R. E. Fries
>   **Lagenaria siceraria* (Molina) Standl.
>   *Licaria debilis* (Mez) Kost.
>   *Ormosia coccinea* (Aubl.) Jacks.
>   *Posoqueria latifolia* (Rudge) R. & S. ssp.
>   *gracilis* (Rudge) Steyerm.

PAC03
>   **Carica papaya* L.
>   *Enterolobium* sp. nov.
>   *Gouania pyrifolia* Reiss.
>   *Parkia paraensis* Ducke

PAC04
>   *Protium polybotryum* (Turcz.) Engl.
>   *Protium tenuifolium* (Engl.) Engl.

PAC05
Cyperus corymbosus L.F.
Dipteryx odorata L.

PAC06
Apeiba aspera Aubl.
Apeiba cf. *burchelli* Sprague
Apeiba echinata Gaertn.
Apeiba echinata var. *macropetala* (Ducke)
   Gaertn.
Maximiliana maripa Mart.

PAC07
*Bixa orellana L.
Genipa americana L.
Licania heteromorpha Benth. var. *hetero-morpha*
Licania glabriflora Prance
Licania latifolia Benth.
Renealmia alpinia (Rottb.) Maas
Vismia guianensis (Aubl.) Choisy

PAC08
Bellucia grossularioides (L.) Triana
Inga alba Willd.
Inga brevialata Ducke
Licania glabriflora Prance
Licania heteromorpha Benth. var. *hetero-morpha*
Licania latifolia Benth.

PAC09
Carapa guianensis Aubl.
Phytolacca rivinoides Kunth. & Bouche
Symphonia globulifera L.

PAC10
Trattinickia burserifolia Mart.
Trattinickia rhoifolia Willd.
Trattinickia sp. 1

PAC11
*Ricinus comunis L.

PAC12
Diplasia karataefolia Rich.
Scleria sp. 1
Scleria sp. 2

MED (MEDICINAL USES FOR PEOPLE AND PETS)

MED01
Laportea aestuans (L.) Chew
Petiveria alliacea L.

MED02
Conyza banariensis (L.) Crang.
Laportea aestuans (L.) Chew
Petiveria alliacea L.
Porophyllum ellipticum Cass.
Pterocaulon vergatum (L.) DC.
Scoparia dulcis L.
Siparuna amazonica (Mart.) A.DC.
Siparuna guianensis Aubl.
Tachigali myrmecophila Ducke
Stenoclaena sp. 1

MED03
Ampelocera edentula Kuhlm.
Banisteriopsis pubipetala (Adr. Juss.)
   Cuatr.
Bryophyllum sp. 1
Pristimera tenuifolia (Mart.) A.C.
Salacia multiflora (Lam.) DC.
Tetrapterys styloptera A. Juss.
*Zingiber officinalis L.

MED04
Ampelocera edentula Kuhlm.
Protium polybotryum (Turcz.) Engl.
Protium tenuifolium (Engl.) Engl.

MED05
Ambelania acida Aubl.
Lacmellea aculeata (Ducke) Monachino
Virola carinata (Benth. Warb.)
Virola michelii Heckel
*Zingiber officinalis L.

MED06
Anaxagorea dolichocarpa Spr. ex Sandw.
*Chenopodium ambrosioides L.
Croton matourensis Aubl.
Ficus sp. 1
Himatanthus sucuuba (Spruce ex Muell.
   Arg.) Woods.
Stigmaphyllon hypoleucum Miz.
Trattinickia burserifolia Mart.
Trattinickia rhoifolia Willd.
Trattinickia sp. 1

MED07
   Parahancornia amapa Ducke
   Parahancornia fasciculata (Poir.) Benoist

MED08
   Bauhinia rubiginosa Bong.
   Bauhinia sp. 1
   *Chenopodium ambrosioides L.
   Eschweilera amazonica R. Kunth.
   Eschweilera apiculata (Miers) A.C. Smith
   Eschweilera coriacea (A.P. Candolle) Mart.
      ex Berg
   Eschweilera micrantha (Berg) Miers
   Eschweilera pedicellata (Richard) Mori
   Fusaea longifolia (Aublet) Saff.
   Himatanthus sucuuba (Spruce ex Muell.
      Arg.) Woods.
   Sclerolobium sp. 20
   Tachigali myrmecophila Ducke

MED09
   Phyllanthus urinaria L.

MED10
   Aechmea brevicellis L.B. Smith
   Aechmea bromeliifolia Baker
   Bromelia goeldiana L.B. Smith
   Guzmania lingulata (L.) Mez.
   Hymenaea parvifolia Huber
   Bromeliaceae sp. 1

MED11
   Ambelania acida Aubl.
   Ampelozizyphus amazonicus Ducke
   Asclepias curassavica L.
   Banisteriopsis pubipetala (Adr. Juss.)
      Cuatr.
   Conyza banariensis (L.) Crang.
   Euphorbia sp. 1
   Lacmellea aculeata (Ducke) Monachino
   Memora flavida (DC.) Bur & K. Schum.
   Moutabea guianensis Aubl.
   Polygonaceae sp. 1
   Pristimera tenuifolia (Mart.) A.C.
   Sagotia racemosa Baill.
   Salacia multiflora (Lam.) DC.
   Schwannia sp. 1
   Selaginella sp. 1
   Simaba aff. *cavalcantei* Thomas
   Simaba cedron Planch.

Tetrapterys styloptera A. Juss.
Trigonidium acuminatum Batem ex Lindl.

MED12
   *Jatropha curcas L.
   *Jatropha gossypifolia L.

MED13
   *Jatropha curcas L.
   *Jatropha gossypifolia L.

MED14
   Copaifera duckei Dwyer
   Copaifera reticulata Ducke
   Eschweilera amazonica R. Kunth.
   Eschweilera apiculata (Miers) A.C. Smith
   Eschweilera coriacea (A.P. Candolle) Mart.
      ex Berg
   Eschweilera micrantha (Berg) Miers
   Eschweilera pedicellata (Richard) Mori
   Gustavia augusta L.
   Maximiliana maripa Mart.

MED15
   Agonandra brasiliensis Benth. & Hook.
   Pseudyma frutescens (Aubl.) Radlk.
   Solanum rugosum Dunal
   Tabernaemontana angulata Mart. ex Muell.
      Arg.

MED16
   Asclepias curassavica L.

MED17
   Galipea trifoliolata Aubl.
   Sagotia racemosa Daill.

MED18
   Licania glabriflora Prance
   Licania heteromorpha Benth. var. *hetero-
      morpha*
   Licania latifolia Benth.
   Virola carinata (Benth. Warb.)
   Virola michelii Heckel
   Virola sp. 1

MED19
   *Nicotiana tabacum L.

MED20
> *Ampelozizyphus amazonicus* Ducke
> **Curcuma* sp. 1

MED21
> *Piper ottonoides* Jun.
> *Piper piscatorum* Trel. & Jun.
> *Protium giganteum* Engl. var. *giganteum*
> *Protium pallidum* Cuatrec.
> *Protium spruceanum* (Benth.) Engl.
> *Zanthoxylum* cf. *juniperina* Engl.

MED22
> *Orbignya phalerata* Mart.

MED23
> *Copaifera duckei* Dwyer
> *Copaifera reticulata* Ducke
> *Dipteryx odorata* L.
> Polygonaceae sp. 1

MED24
> *Cayaponia* sp. 1
> *Gurania eriantha* (Poepp. & Endl.) Cogn.
> *Hymenaea parvifolia* Huber
> *Schubertia grandiflora* Mart.
> *Uncaria guianensis* (Aubl.) Gmel.

MED25
> *Gustavia augusta* L.
> *Poecilanthe effusa* (Hub.) Ducke

MED26
> *Abuta grandifolia* (Mar.) Sandw.
> *Symphonia globulifera* L.

MED27
> *Cordia scabrifolia* C.DC.
> *Ctenitis pretensa* (Afz.) Ching
> **Manihot esculenta* Crantz
> *Pteridium aquilinium* (L.) Kuhn.

MED28
> *Evodianthus funifer* (Poit.) Lindm.
> *Myrciaria tenella* (DC.) Berg.

MED29
> *Bauhinia rubiginosa* Bong.
> *Bauhinia* sp. 1
> *Clarisia racemosa* Ruiz & Pavon
> *Davilla kunthii* St. Hil.

> *Davilla nitida* (Vahl.) Kubitzki
> *Doliocarpus* sp. 1
> *Drypetes variabilis* Vitt.
> *Laetia procera* (Poepp. et Endl.) Eichl.
> *Tetracera volubilis* L. ssp. *volubilis*
> *Tetracera willdenowiana* Schlechtd. ssp. *willdenowiana*

MED30
> *Ceiba pentandra* Gaertn.

MED31
> *Himatanthus sucuuba* (Spruce ex Muell. Arg.) Woods.
> **Jatropha curcas* L.
> *Tachigali myrmecophila* Ducke

MED32
> *Agonandra brasiliensis* Benth. & Hook.
> *Himatanthus sucuuba* (Spruce ex Muell. Arg.) Woods.
> *Pseudyma frutescens* (Aubl.) Radlk.
> *Tabernaemontana angulata* Mart. ex Muell. Arg.

MED33
> *Carapa guianensis* Aubl.

MED34
> *Abuta grandifolia* (Mar.) Sandw.
> *Hymenaea parvifolia* Huber

MED35
> **Curcuma* sp. 1

MED36
> **Manihot esculenta* Crantz

MED37
> Annonaceae sp. 1
> Flacourtiaceae sp. 1
> *Neoptychocarpus apodanthus* Kuhlm.

MAG (MAGIC USES)

MAG01
> *Ardisia guianensis* (Aubl.) Mez
> *Justicia pectoralis* Jacq.
> *Justicia spectabilis* T. Anders. ex C.B. Clarke

MAG02
aff. *Cheiloclinium* or *Salacia* sp. 1
Celtis iguanea (Jacq.) Sargent
Cheiloclinium cognatum (Miers) A.C.
Coccoloba sp. 1
Duguetia sp. 2
Duguetia yeshidah Sandw.
Ephedranthus pisocarpus R.E. Fries
Justicia pectoralis Jacq.
Justicia spectabilis T. Anders. ex C.B.
  Clarke
Ocotea glomerata (Nees) Mez.
Ocotea guianensis Aubl.
Ocotea opifera Mart.
Psychotria poeppigiana Muell. Arg.
Psychotria ulviformis Steyerm.
Salacia insignis A.C. Smith

MAG03
Chenopodium ambrosioides L.

MAG04
aff. *Pradosia* sp. 1
Astronium cf. *obliquum* Griseb.
Petiveria alliacea L.
Thyrsodium sp. 1
Thyrsodium spruceanum Bentham

MAG05
Caladium picturatum C. Koch.
Petiveria alliacea L.
Trattinickia burserifolia Mart.
Trattinickia rhoifolia Willd.
Trattinickia sp. 1

MAG06
Lindackeria latifolia Benth.
Lindackeria sp. 1
Moutabea sp. 2

MAG07
Astrocaryum gynacanthum Mart.

MAG08
Acacia multipinnata Ducke

MAG09
Lomariopsis japurensis (Mart.) J. Smith
Phytolacca rivinoides Kunth. & Bouche
Siparuna amazonica (Mart.) A.DC.
Solanum leucocarpon Dunal

MAG10
Clusia sp. 1
Dracontium sp. 1
Pseudyma frutescens (Aubl.) Radlk.
Lomariopsis japurensis (Mart.) J. Smith

MAG11
Norantea guianensis Aubl.

MAG12
Evodianthus funifer (Poit.) Lindm.

MAG13
Duguetia surinamensis R.E. Fries

MAG14
Cephalia sp. 1
Psychotria colorata Poepp.

MAG15
Styzophyllum riparium (H.B.K.) Sandw.

MAG16
Genipa americana L.

MAG17
Philodendron sp. 1

MAG18
Coccoloba sp. 1

MAG19
Cyperus corymbosus L.F.

MAG20
Mendoncia hoffmannseggiana Nees
Mendoncia sp. 1

MAG21
Coccoloba sp. 2

RIT (RITUAL USES)

RIT01
Nicotiana tabacum L.
Trattinickia burserifolia Mart.
Trattinickia rhoifolia Willd.
Trattinickia sp. 1

RIT02
Neoglaziovia variegata (Arr. Cam.)
  Mez.

RIT03
Oenocarpus distichus Mart.

RIT04
**Bixa orellana* L.

RIT05
Licania glabriflora Prance
Licania heteromorpha Benth. var. *hetero-morpha*
Licania latifolia Benth.

MYT (MYTHICAL CONNOTATIONS)

MYT01
Bertholletia excelsa Humboldt & Bonpland
Genipa americana L.

MYT02
Limnocharis sp. 1
Parkia pendula Mig.

MYT03
Ceiba pentandra Gaertn.

MYT04
Brosimum rubescens Taubert

MYT05
Cedrela fissilis Vell.

POI (POISONOUS PLANTS AND OTHER PLANTS HABITUALLY AVOIDED BY ALL KA'APOR OR BY SPECIFIC GROUPS DEFINED IN AGE/ SEX TERMS)

POI01
**Clibadium sylvestre* (Aubl.) Baill.
Markleya dahlgreniana Bondar

POI02
Bagassa guianensis Aubl.

POI03
Dracontium sp. 1
Eugenia omissa McVaugh
Myrciaria pyriifolia Desv. ex hamilton
Solanum rugosum Dunal

POI04
Astrocaryum vulgare Mart.
Bagassa guianensis Aubl.

POI05
Bagassa guianensis Aubl.

POI06
Minquartia guianensis Aubl.

POI07
Solanum leucocarpon Dunal

POI08
Chrysophyllum sparsiflorum Kl. ex Miq.
Laportea aestuans (L.) Chew
Monstera cf. *pertusa* (L.) Vriesia
Monstera subpinata (Schott) Engl.
Philodendron grandiflorum (Jacq.) Schott.
Philodendron venustum Bunt.

POI09
Caryocar glabrum (Aubl.) Pers.
Caryocar villosum Aubl.

POI10
Tachigali myrmecophila Ducke

POI11
Eugenia omissa McVaugh
Monstera cf. *pertusa* (L.) Vriesia
Monstera subpinata (Schott) Engl.
Myrciaria pyriifolia Desv. ex Hamilton
Philodendron grandiflorum (Jacq.) Schott
Philodendron venustum Bunt.

POI12
Connaraceae sp. 2
Dioclea reflexa Hook. F.
Pseudyma frutescens (Aubl.) Radlk.
Indeterminate genus 89

POI13
Rubiaceae sp. 2

POI14
Ischnosiphon graciles (Rudge) Koern.

POI15
Imperata brasiliensis Trin.

POI16
  Symphonia globulifera L.

POI17
  Zollernia paraensis Hub.

POI18
  Xiphidium caeruleum Aubl.

FUE (FUEL SOURCES)

FUE01
  Acacia multipinnata Ducke
  aff. *Cheiloclinium* or *Salacia* sp. 1
  Ambelania acida Aubl.
  Ampelocera edentula Kuhlm.
  Andira sp. 1
  Aniba citrifolia (Nees) Mez.
  Annona paludosa Aublet
  Annona sericea Dunal
  Annonaceae sp. 1
  Aparisthmium cordatum Baill.
  Apeiba aspera Aubl.
  Apeiba cf. *burchelli* Sprague
  Apeiba echinata Gaertn.
  Apeiba echinata var. *macropetala* (Ducke)
    Gaertn.
  Apeiba tibourbou Aubl.
  Apuleia leiocarpa (Vogel) Spr. var. *mo-laris* (Spruce ex Benth.) Koeppen
  Aspidosperma cylindrocapon Muell. Arg.
  Astronium lecointei Ducke
  Banara guianensis Aubl.
  Bowdichia nitida Spruce ex Benth.
  Buchenavia cf. *tetraphylla* (Aubl.) Howard
  Byrsonima laevigata (Poir.) DC.
  Capparis sola Macbr.
  Casearia arborea (L.C. Richard) Urban
  Cecropia purpurescens C.C. Berg
  Cecropia sciadophylla Mart.
  Cheiloclinium cognatum (Miers) A.C.
  Chimarrhis turbinata DC.
  Chrysophyllum argenteum Jacquin ssp.
    auratum (Miquel) Pennington
  Chrysophyllum lucentifolium Crong. ssp.
    pachycarpum Pires & Penn.
  Chrysophyllum pomiferum (Eyma) Penn.
  Chrysophyllum sparsiflorum Kl. ex Miq.

Clarisia racemosa Ruiz & Pavon
Coccoloba paniculata Lam.
Cordia exaltata Lam.
Cordia scabrida Mart.
Cordia scabrifolia C.DC.
Couepia guianensis Aubl. ssp. *divaricata*
  (Hub.) Prance
Couepia guianensis Aubl. ssp. *guianensis*
Couepia guianensis ssp. *glandulosa* (Mig.)
  Prance
Couratari oblongifolia Ducke & Knuth.
Croton cajucara Benth.
Crudia parivoa DC.
Cupania scrobiculata L.C. Rich.
Dialium guianense (Aubl.) Sandw.
Diospyros artanthifolia Mart.
Diospyros melinoni (Hiern.) A.C. Smith
Diplotropis purpurea (Rich.) Amsh.
Dodecastigma integrifolium (Lanj.) Lanj.
  & Sandw.
Duguetia riparia Huber
Dulacia guianensis (Engl.) O. Kuntze
Endlicheria verticellata Mez.
Erisma uncinatum Warn.
Erythroxylum cf. *leptoneurum* O.E. Schulz
Eschweilera amazonica R. Kunth.
Eschweilera apiculata (Miers) A.C. Smith
Eschweilera micrantha (Berg) Miers
Eschweilera obversa (Berg) Miers
Eschweilera pedicellata (Richard) Mori
Eugenia flavescens DC.
Eugenia muricata DC.
Eugenia sp. 6
Eugenia sp. 7
Eugenia sp. 8
Eugenia tapecumensis Berg
Exellodendron barbatum Prance
Faramea sessilifolia (H.B.K.) DC.
Flacourtiaceae sp. 1
Fusaea longifolia (Aublet) Saff.
Guarea guidonia (L.) Sleumer
Guarea sp. 1
Guarea kunthiana A. Juss.
Guarea macrophylla Vahl. ssp. *pachy-carpa* (C.DC.) Penn.
Guatteria elongata Benth.
Guatteria sp. 1
Guazuma ulmifolia Lam.
Gustavia augusta L.
Helicostylis tomentosa (P. & E.) Rusby

Hirtella bicornis Mart. ex Zucc.
Hirtella cf. *paraensis* Prance
Hirtella eriandra Benth.
Hirtella racemosa var. *hexandra* (Willd. ex R.G.S.) Prance
Hymenaea courbaril L.
Hymenaea parvifolia Huber
Hymenaea reticulata Ducke
Inga auristellae Harms
Inga brevialata Ducke
Inga capitata Desv.
Inga cinammonea Spruce ex Benth.
Inga fagifolia (L.) Willd.
Inga falcistipula Ducke
Inga gracilifolia Ducke
Inga heterophylla Willd.
Inga marginata Willd.
Inga miriantha Poepp. et Endl.
Inga nobilis Willd.
Inga paraensis Ducke
Inga rubiginosa (Rich.) DC.
Inga splendens Willd.
Inga thibaudiana DC.
Inga sp. 5
Inga sp. 6
Inga sp. 7
Inga sp. 8
Inga sp. 9
Inga sp. 10
Lacmellea aculeata (Ducke) Monachino
Lacunaria jenmani (Oliv.) Ducke
Lacunaria sp. 4
Lacunaria sp. 5
Laetia procera (Poepp. et Endl.) Eichl.
Lauraceae sp. 1
Lauraceae sp. 2
Lauraceae sp. 3
Lecythis chartacea Berg
Lecythis idatimon Aublet
Lecythis lurida (Miers) Mori
Lecythis pisonis Cambess.
Licania canescens R. Ben.
Licania glabriflora Prance
Licania heteromorpha Benth. var. *heteromorpha*
Licania kunthiana Hook. F.
Licania latifolia Benth.
Licania macrophylla Benth.
Licaria brasiliensis (Nees) Kost
Licaria debilis (Mez) Kost.
Licaria sp. 1

Mabea angustifolia Spruce ex Benth.
Mabea caudata Pax & Hoffm.
Mabea sp. 1
Macrolobium acaciaefolium Bth.
Magaritaria nobilis L.F.
Matayba spruceana Radlk.
Miconia cf. *kappleri* Naud.
Miconia nervosa (Smith) Triana
Miconia serialis DC.
Miconia serrulata (DC.) Naud.
Miconia surinamensis Gleason
Miconia sp. 1
Micropholis guyanensis (A.DC.) Pierre ssp. *guyanensis* Penn.
Micropholis venulosa (Mart. ex Eich.) Pierre
Mouriri guianensis Aubl.
Mouriri trunciflora Ducke
Neoptychocarpus apodanthus Kuhlm.
Newtonia psilostachya (DC.) Brenan
Newtonia suaveolens (Miguel) Brenan
Ocotea amazonica (Meiss.) Mez.
Ocotea canaliculata (L.C. Rich.) Mez.
Ocotea costata Mez.
Ocotea fasciculata (Ness) Mez
Ocotea glomerata (Nees) Mez.
Ocotea guianensis Aubl.
Ocotea rubra Mez.
Ocotea silvae Vattimo-Gil sp. nov.
Ocotea sp. 3
Parinari excelsa Sabine
Parkia ulei (Harms) Kuhlm.
Paypayrola grandiflora Tul.
Pithecellobium cauliflorum Mart.
Pithecellobium comunis Benth.
Pithecellobium niopoides Spruce ex Benth.
Pithecellobium racemosum Ducke
Platypodium elegans Vog.
Pouteria caimito (R. & P.) Radlk.
Pouteria egregia Sandw.
Pouteria engleri Eyma
Pouteria filipes Eyma
Pouteria hispida Eyma
Pouteria jariensis Pires & Penn.
Pouteria macrocarpa (Martius) Dietrich
Pouteria macrophylla (Lam.) Eyma
Pouteria reticulata (Engler) Eyma ssp. *reticulata*
Pouteria trigonosperma Eyma
Pouteria sect. *Franchetella* sp. 1
Pouteria sect. *Franchetella* sp. 2

Pouteria sp. 2
Pouteria sp. 5
Protium altsoni Sandw.
Protium aracouchini (Aubl.) March.
Protium decandrum (Aublet) Marchand
Protium giganteum Engl. var. *giganteum*
Protium heptaphyllum ssp. *heptaphyllum*
  (Aublet) Marchand
Protium pallidum Cuatrec.
Protium polybotryum (Turcz.) Engl.
Protium sagotianum Engl.
Protium spruceanum (Benth.) Engl.
Protium tenuifolium (Engl.) Engl.
Protium trifoliolatum Engl.
Rauvolfia paraensis Ducke
Remijia sp. 1
Rheedia brasiliensis (Mart.) Pl. et Tr.
Rinorea flavescens (Aubl.) kuntze
Rinorea pubiflora (Benth.) Spr. & Sandw.
Rinorea sp. 1
Rollinia exsucca (DC. ex Dun.) A.DC.
Sacoglottis amazonica Mart.
Sacoglottis guianensis Benth.
Sacoglottis sp. 1
Sagotia racemosa Baill.
Salacia insignis A.C. Smith
Sapotaceae sp. 2
Sclerolobium paraense Huber
Sclerolobium sp. 20
Senna sylvestris (Vell.) Irwin & Barneby
Simaruba amara Aubl.
Sloanea grandiflora Smith
Sloanea grandis Ducke
Sloanea guianensis (Aubl.) Benth.
Sloanea porphyrocarpa Ducke
Sterculia pruriens (Aubl.) K. Schum.
Stryphnodendron guianensis (Aubl.)
  Benth.
Stryphnodendron polystachyum (Miq.)
  Kleinh.
Swartzia brachyrachis var. Harms
Swartzia brachyrachis var. *snethlageae*
  (Ducke) Ducke
Symphonia globulifera L.
Tabebuia serratifolia (Vahl.) Nichols
Tabebuia sp. 1
Talisia micrantha Radlk.
Taralea oppositifolia Aubl.
Terminalia amazonia (J. Gmel.) Exell.
Terminalia dichotoma G. Meyer
Tetragastris altissima (Aublet) Swart.

Tetragastris panamensis (Engl.) O.K.
Tovomita brevistaminea Engl.
Trattinickia burserifolia Mart.
Trattinickia rhoifolia Willd.
Trattinickia sp. 1
Trichilia micrantha Benth.
Trichilia quadrijuga Kunth. ssp. *quadri-
juga*
Trichilia verrocosa C. DC.
Unonopsis rufescens (Baill.) R.E. Fries
Virola sp. 1
Xylopia barbata Mart.
Xylopia nitida Dunal
Zanthoxylum cf. *juniperina* Engl.

FUE02
Alibertia edulis (L.Rich.) A. Rick.
Allamanda carthartica L.
Amphilophium paniculata (L.) H.B.K. var.
  moller (Cham. & Schl.) Gentry
Anacardium giganteum Hancock ex En-
gler
Anacardium parvifolium Ducke
Anemopaegma setilobum A.Gentry
Arrabidaea cf. *florida* DC.
Bagassa guianensis Aubl.
Banisteriopsis pubipetala (Adr. Juss.)
  Cuatr.
Brosimum acutifolium ssp. *interjectum*
  C.C. Berg
Brosimum guianense (Aubl.) Huber
Brosimum paclesum (S. Moore) C.C. Berg
Campomanesia grandiflora (Aubl.) Sagot.
Coccoloba acuminata H.B.K.
Copaifera sp. 1
Couratari guianensis Aublet
Courmarouna micrantha Ducke
Croton matourensis Aubl.
Cuspidaria sp. 1
Dalbergia monetaria L.F.
Davilla kunthii St. Hil.
Davilla nitida (Vahl.) Kubitzki
Doliocarpus sp. 1
Duguetia echinophora R.E. Fries
Duguetia sp. 1
Enterolobium sp. nov.
Eschweilera coriacea (A.P. Candolle) Mart.
  ex Berg
Eugenia eurcheila Berg
Eugenia schomburkii Benth.
Ficus sp. 2

Goupia glabra Aubl.
Guettarda divaricata (H.B.K.) Standl.
Henrietta spruceana Cogn.
Iryanthera juruensis Warb.
Jacaranda copaia (Aubl.) D. Don. ssp.
  copaia
Jacaranda copaia ssp. *spectabilis* (Mart. ex
  DC.) A. Gentry
Jacaranda duckei Vattimo
Lacunaria sp. 3
Licania apetala (E. Mey.) Fritsch
Licania membranacea Sagot. ex Laness.
Licania octandra (Hoffmgg. ex R. & S.)
  Kuntze
Lindackeria latifolia Benth.
Lindackeria sp. 1
Malanea sp. 1
Manilkara bidentata (A.DC.) Chev. ssp.
  surinamensis (Miq) Penn.
Maquira guianensis Aubl.
Myrciaria cf. *dubia* McVaugh
Neea floribunda Poepp. & Endl.
Neea glameruliflora Heimut
Neea oppositifolia R. & P.
Neea ovalifolia Spruce ex J.A. Schmidt
Neea sp. 1
Neea sp. 2
Parahancornia fasciculata (Poir.) Benoist
Parkia paraensis Ducke
Paullinia sp. 1
Phyllanthus martii Muell. Arg.
Pisonia sp. 2
Pithecellobium cochleatum (Willd.) Mart.
Pithecellobium pedicellare (DC.) Benth.
Posoqueria latifolia (Rudge) R. & S. ssp.
  gracilis (Rudge) Steyerm.
Pouteria durlandii (Standley) Baehni
Pristimera tenuifolia (Mart.) A.C.
Ryania speciosa Vahl.
Salacia multiflora (Lam.) DC.
Sapium caspceolatum Huber
Sapium lanceolatum Huber
Schefflera morototoni (Aubl.) M.S.F.
Sclerolobium albiflorum R. Ben.
Sclerolobium guianense Aubl.
Spondias mombin L.
Talisia microphylla Witt.
Talisia retusa Cowan
Tapirira guianensis Aublet
Tapirira peckoltiana Engler
Tetracera volubilis L. ssp. *volubilis*

Tetracera willdenowiana Schlechtd. ssp.
  willdenowiana
Tetrapterys styloptera A. Juss.
Theobroma speciosum Willd. ex Spreng.
Vochysia inundata Ducke
Zanthoxylum chiloperone (Mart.) Engl.
Zanthoxylum rhoifolium Lam.

FUE03
Hymenaea courbaril L.
Hymenaea parvifolia Huber
Hymenaea reticulata Ducke
Protium altsoni Sandw.
Protium decandrum (Aublet) Marchand
Protium giganteum Engl. var. *giganteum*
Protium heptaphyllum ssp. *heptaphyllum*
  (Aublet) Marchand
Protium pallidum Cuatrec.
Protium polybotryum (Turcz.) Engl.
Protium sagotianum Engl.
Protium spruceanum (Benth.) Engl.
Protium tenuifolium (Engl.) Engl.
Symphonia globulifera L.
Trattinickia burserifolia Mart.
Trattinickia rhoifolia Willd.
Trattinickia sp. 1

COM (COMMERCIAL USES)

COM01
  *Oryza sativa L.

COM02
  *Adenanthera pavonina L.
  Andira retusa (Lam.) H.B.K.
  Astrocaryum vulgare Mart.
  *Canna indica L.
  *Cardiospermum halicacabum L.
  *Coix lacryma-jobi L.
  Maximiliana maripa Mart.
  Ormosia coccinea (Aubl.) Jacks.
  Zollernia paraensis Hub.

COM03
  Protium giganteum Engl. var. *giganteum*
  Protium pallidum Cuatrec.

COM04
  Heteropsis longispathacea Engl.
  Heteropsis sp. 1

COM05
Carapa guianensis Aubl.

MIS (MISCELLANEOUS USES)

MIS01
Cecropia purpurescens C.C. Berg
Cecropia sciadophylla Mart.
Olyra latifolia L.

MIS02
Alibertia edulis (L.Rich.) A. Rick.

MIS03
Goupia glabra Aubl.
Hymenolobium excelsum Ducke

MIS04
Gustavia augusta L.
Ocotea canaliculata (L.C. Rich.) Mez.
Ocotea sp. 3

MIS05
Bauhinia rubiginosa Bong.
Bauhinia sp. 1

MIS06
Cassia fastuosa Willd.
Senna chrysocarpa (Desv.) Irwin & Barneby
Senna pendula (Lamk.) Irwin & Barneby

MIS07
Rhipsalis mijosurus K. Schum.
Tillandsia usneoides (L.) L.

MIS08
Brosimum rubescens Taubert
Maximiliana maripa (Corr. Serr.) Drude
Rinorea pubiflora (Benth.) Spr.
Tabebuia impegitinosa Standley

MIS09
Eugenia sp. 8
Licania canescens R. Ben.
Licania glabriflora Prance
Licania heteromorpha Benth. var. *heteromorpha*
Licania kunthiana Hook. F.
Licania latifolia Benth.
aff. *Myrcia* sp. 1

Rinorea pubiflora (Benth.) Spr.
Tabebuia impegitinosa Standley.

MIS10
Schefflera morototoni (Aubl.) M.S.F.

NONE (NO KNOWN USE)

Acacia paniculata Willd.
Acrocomia sp. 1
Aegiphila sp. 1
Amaranthus spinosus L.
Arrabidaea sp. 1
Astrocaryum javari Mart.
Bignoniaceae sp. 1
Borreria verticulata G.F.W. Mey
Brunfelsia guianensis Benth.
Calyobolus sp. 1
Cannabis sativa L.
Casearia decandra Jacq.
Cissampelos sp. 1
Cissus cissyoides L.
Coccoloba latifolia Lam.
Coccoloba racemulosa Meissn.
Combretum laxum Jacq.
Connaraceae sp. 1
Connarus favosus Planch.
Costus scaber R. & P.
Coussapoa sp. 1
Cyperaceae sp. 1
Dichorisandra affinis Mart. ex Roem. & Schultes
Digitaria insularis (L.) Mez ex Ekman
Dioscorea sp. 1
Duroia sp. 1
Eugenia egensis DC.
aff. *Forsteronia* sp. 1
Hevea guianensis Aubl.
Hyparrhenia rufa (Nees) Stapf
Ichnanthus sp. 1
Indeterminate genus 2
Indeterminate genus 3
Indeterminate genus 4
Indeterminate genus 5
Indeterminate mycotica sp. 1
Jacaranda heterophylla Bur. & K. Schum.
Jacaranda paraensis (Huber) Vattimo
Lantana camara L.
Licania sp. 1
Mabea pohliana M. Arg.

Machaerium ferox (Mart. ex Benth.) Ducke

Mansoa angustiden (DC.) Bur. & K. Schum.

Marantaceae sp. 2

Maripa sp. 1

Mascagnia sp. 1

Masechites bicornulata (Rusby) Woods.

Maytenus sp. nov.

Memora allamandiflora Bur ex Bur et K. Schum.

Memora bracteosa (DC.) Bur. & K. Schum.

aff. *Olyra* sp. 1

Orthostichorsis crinite (Sull) Broter.

Ouratea salicifolia (St. Hil. et Tul.) Engl.

Paullinia sp. 2

Peritassa huanuclara (Loes.) A.C.

Phenakospermum guianensis Peterson

Phyllanthus niruri L.

Piper anonnifolium Kunth.

Piper pilirameum Jun.

Piresia goeldi Swallen

Pouteria guianensis Aublet

Rhabidalia sp. 1

Rhipsalis baccifera (J. Miller) Stearn.

aff. *Rhizomorpha* sp. 1

Rubiaceae sp. 1

Ruprechtia sp. 1

Sapotaceae sp. 1

Sapotaceae sp. 3

Sclerolobium sp. 16

Sipanea veris S. Moore

Tachigali sp. 1

Tachigali macrostachya Huber

Tachigali paniculata Aublet

Taxithelium planum (Brid.) Mitt.

Concordance of Folk and Botanical Nomenclature

'Trees' (Mɨra)			
Folk Genus	Folk Species	Botanical Species	Family
aŋwayarmɨra	aŋwa-yar-mɨra	*Pseudyma frutescens*	SAPINDACEAE
ainumɨrmɨra	ainumɨr-mɨra	*Bauhinia viridiflorens*	CAESALPINIACEAE
akaú'ɨ	akaú-'ɨ	*Helicostylis tomentosa*	MORACEAE
		Pouteria bilocularis	SAPOTACEAE
akayu'ɨ	akayu-pinar-'ɨ	*Anacardium giganteum*	ANACARDIACEAE
	akayu-mena-'ɨ	*Anacardium parvifolium*	ANACARDIACEAE
akušiŋa	akuši-iŋa	*Hirtella eriandra*	CHRYSOBALANACEAE
akušika'a	akuši-ka'a	*Celtis iguanea*	ULMACEAE
akušimɨra	akuši-mɨra	*Hirtella* cf. *paraensis*	CHRYSOBALANACEAE
		Hirtella racemosa var. *hexandra*	CHRYSOBALANACEAE
akušitɨriwa'ɨ	akuši-tɨr-iwa-'ɨ	*Pouteria macrophylla*	SAPOTACEAE
	akuši-tɨr-iwa-hu-'ɨ	*Pouteria macrocarpa*	SAPOTACEAE
		Pouteria sect. *Franchetella* sp. 1	SAPOTACEAE
akušiyu'ɨ	akuši-yu-'ɨ	*Bauhinia* sp. 2	CAESALPINIACEAE
		Randia armata	RUBIACEAE
ama'ɨ	ama-'ɨ-te	*Cecropia purpurescens*	CECROPIACEAE
		Cecropia sciadophylla	CECROPIACEAE
	ama-'ɨ-puku	*Cecropia palmata*	CECROPIACEAE
	ama-'ɨ-tuwɨr	*Cecropia concolor*	CECROPIACEAE
		Cecropia obtusa	CECROPIACEAE
ama'ɨrarɨ	ama-'ɨ-rarɨ	*Pourouma mollis* ssp. *mollis*	CECROPIACEAE
	ama-'ɨ-rarɨ-tuwɨr	*Pourouma minor*	CECROPIACEAE
amaŋaputɨr'ɨ	amaŋa-putɨr-'ɨ	*Cassia fastuosa*	CAESALPINIACEAE
		Senna pendula	CAESALPINIACEAE

'Trees' (**Mɨra**) (*Continued*)

Folk Genus	Folk Species	Botanical Species	Family
aman'ɨ	aman-'ɨ	*Caraipa grandiflora*	CLUSIACEAE
anãmɨra	anã-mɨra	*Licania macrophylla*	CHRYSOBALANACEAE
apa'ɨ	apa-'ɨ-tuwɨr	*Parahancornia amapa*	APOCYNACEAE
		Parahancornia fasciculata	APOCYNACEAE
ape'ɨ	ape-'ɨ-tuwɨr	*Cordia scabrifolia*	BORAGINACEAE
	(= ape-'ɨ-te)	*Cordia lomatoloba*	BORAGINACEAE
		Cordia scabrida	BORAGINACEAE
		Cordia trachyphylla	BORAGINACEAE
(= tikwer'ɨ)	ape-'ɨ-howɨ	*Cordia* sp. 2	BORAGINACEAE
	ape-'ɨ-hu	*Cordia sellowiana*	BORAGINACEAE
	ape-'ɨ-pihun;	*Rollinia exsucca*	ANNONACEAE
	(= tikwer-'ɨ-pihun)		
	ape-'ɨ-tawa	*Cordia exaltata*	BORAGINACEAE
	kaŋw-ɨ	*Apeiba tibourbou*	TILIACEAE
	(= ape-'ɨ-'ɨ)		
apɨ'a'ɨ	apɨ'a-'ɨ	*Guazuma ulmifolia*	STERCULIACEAE
arakãšɨ'ɨ	ara-kã-šɨ-'ɨ	*Pouteria gongrijpii*	SAPOTACEAE
arakãšɨ'iran'ɨ	ara-kã-šɨ-'iran-'ɨ	*Chrysophyllum sparsiflorum*	SAPOTACEAE
arakɨ'ĩ	ara-kɨ'ĩ	*Aparisthmium cordatum*	EUPHORBIACEAE
arãkwãmɨra	arãkwã-mɨra	*Eugenia tapecumensis*	MYRTACEAE
(= arãkwãmi'u'ɨ)		*Eugenia muricata*	MYRTACEAE
		Eugenia sp. 7	MYRTACEAE
	u'ɨ-tɨma-ran	*Myrcia paivae*	MYRTACEAE
arapari'ɨ	arapari-'ɨ	*Macrolobium acaciaefolium*	CAESALPINIACEAE
arapasuãtã'ɨ	arapasu-ãtã-'ɨ	*Lacunaria* sp. 4	QUIINACEAE
		Lacunaria sp. 5	QUIINACEAE
arapasumɨra	arapasu-mɨra	*Pithecellobium pedicellare*	MIMOSACEAE
arapuhamɨra	arapuha-mɨra	*Conceveiba guianensis*	EUPHORBIACEAE
arapuhatɨmaka-ŋwer'ɨ	arapuha-tɨma-kaŋwer-'ɨ	*Calyptranthes* or *Marlierea*	MYRTACEAE
		Eugenia sp. 5	MYRTACEAE
		Myrcia sp. 5	MYRTACEAE
ararãkã'ɨ	ara-rãkã-'ɨ	*Aspidosperma desmanthum*	APOCYNACEAE
		Aspidosperma verruculosum	APOCYNACEAE
arariãkã'ɨ	arari-ãkã-'ɨ	Rubiaceae sp. 2	RUBIACEAE
araruhukãtãi'ɨ	ara-uhu-kãtãi-'ɨ	*Eschweilera obversa*	LECYTHIDACEAE
arašiku'ɨ	arašiku-'ɨ	*Annona paludosa*	ANNONACEAE
		Annona sericea	ANNONACEAE
		Duguetia marcgraviana	ANNONACEAE
ãyaŋkɨwa'ɨ	ãyaŋ-kɨwa-'ɨ	*Apeiba aspera*	TILIACEAE

'Trees' (**Mɨra**) (*Continued*)

Folk Genus	Folk Species	Botanical Species	Family
		Apeiba cf. *burchelli*	TILIACEAE
		Apeiba echinata	TILIACEAE
		Apeiba echinata var. *macropetala*	TILIACEAE
ayu'ɨ	**ayu-'ɨ**	*Aniba citrifolia*	LAURACEAE
		Endlicheria verticellata	LAURACEAE
(free variants =		*Licaria* sp. 1	LAURACEAE
ayuwa'ɨ; wayuwa'ɨ)		Lauraceae sp. 2	LAURACEAE
	ayu-'ɨ-pihun	*Ocotea amazonica*	LAURACEAE
		Ocotea caudata	LAURACEAE
	ayu-'ɨ-pɨtaŋ	Lauraceae sp. 1	LAURACEAE
	ayu-wãtã-'ɨ	*Ocotea rubra*	LAURACEAE
	ayu-'ɨ-tuwɨr	*Ocotea fasciculata*	LAURACEAE
	ayu-wa-hu-'ɨ	*Licaria brasiliensis*	LAURACEAE
	ayu-wa-puku-'ɨ	Lauraceae sp. 3	LAURACEAE
		Ocotea silvae	LAURACEAE
		Ocotea costata	LAURACEAE
	maha-yu-wa-'ɨ	*Ocotea glomerata*	LAURACEAE
	(**= maha-mɨra**)	*Ocotea guianensis*	LAURACEAE
		Ocotea opifera	LAURACEAE
ayu'ɨran	**ayu-'ɨ-ran**	*Ocotea canaliculata*	LAURACEAE
(free variants =		*Ocotea* sp. 3	LAURACEAE
ayuwa'ɨran;		*Licaria debilis*	LAURACEAE
wayuwa'ɨran)			
ɨŋa	**ɨŋa-ha-'ɨ**	*Inga* sp. 5	MIMOSACEAE
	ɨŋa-howɨ	*Inga* sp. 7	MIMOSACEAE
		Inga nobilis	MIMOSACEAE
		Inga sp. 9	MIMOSACEAE
	ɨŋa-hu-'ɨ	*Inga capitata*	MIMOSACEAE
		Inga cinammomea	MIMOSACEAE
	ɨŋa-pẽrẽ-'ɨ	*Inga marginata*	MIMOSACEAE
	(**= ɨŋa-tawa-'ɨ**)	*Inga miriantha*	MIMOSACEAE
		Inga auristellae	MIMOSACEAE
		Inga heterophylla	MIMOSACEAE
	ɨŋa-pɨtaŋ-'ɨ	*Inga thibaudiana*	MIMOSACEAE
	ɨŋa-šiši-'ɨ	*Inga alba*	MIMOSACEAE
		Inga brevialata	MIMOSACEAE
	kaŋwaruhu-ɨŋa	*Inga falcistipula*	MIMOSACEAE
		Inga paraensis	MIMOSACEAE
		Inga fagifolia	MIMOSACEAE
	musu-ɨŋa	*Inga* sp. 10	MIMOSACEAE
	(**= ɨŋa-põ-'ɨ**)		
	tapi'ir-namir-ɨŋa-'ɨ	*Inga* sp. 6	MIMOSACEAE
		Inga sp. 8	MIMOSACEAE
		Inga splendens	MIMOSACEAE

'Trees' (**Mɨra**) (*Continued*)

Folk Genus	Folk Species	Botanical Species	Family
	tapi'ir-ɨŋa-'ɨ	*Inga rubiginosa*	MIMOSACEAE
ɨŋaran	ɨŋa-ran-'ɨ	*Pithecellobium cochleatum*	MIMOSACEAE
(= tayahuɨŋa)			
		Dalbergia monetaria	FABACEAE
inamukihawi'ɨ	inamu-kiha-wi-'ɨ	*Casearia arborea*	FLACOURTIACEAE
inamumɨra	inamu-mɨra	*Exellodendron barbatum*	CHRYSOBALANACEAE
iraírupeatu'ɨ	ira-i-rupe-atu-'ɨ	*Stryphnodendron guianensis*	MIMOSACEAE
iramuru'ɨ	ira-muru-'ɨ	*Parkia ulei*	MIMOSACEAE
ɨrarɨ	ɨrar-ɨ	*Cedrela fissilis*	MELIACEAE
(= patuwa'ɨ)			
ɨraší'ɨ	ira-šī-'ɨ	*Myrciaria flexuosa*	MYRTACEAE
		Myrciaria spruceana	MYRTACEAE
ɨratawa'ɨ	ira-tawa-'ɨ	*Pouteria* sect.	SAPOTACEAE
(free variant =		*Franchetella* sp. 2	SAPOTACEAE
ɨwa-tawa-'ɨ; also		*Pouteria caimito*	SAPOTACEAE
called ma'eɨwatawa'ɨ		*Pouteria trigonosperma*	SAPOTACEAE
in Gurupiuna)			
		Sapotaceae sp. 2	SAPOTACEAE
ɨratawa'ɨ'hu	ira-tawa-'ɨ-hu	*Lacunaria jenmani*	QUIINACEAE
		Lacunaria sp. 3	QUIINACEAE
ɨratawa'ɨran	ira-tawa-'ɨ-ran	*Chrysophyllum pomiferum*	SAPOTACEAE
ɨratɨ'ɨ	irati-i-te	*Symphonia lobulifera*	CLUSIACEAE
	irati-ātā-'ɨ		
ɨrɨkɨwa'ɨ	irɨkɨwa-'ɨ	*Manilkara huberi*	SAPOTACEAE
	irɨkɨwa-yu-'ɨ	*Manilkara bidentata* ssp. *surinamensis*	SAPOTACEAE
irimã'ɨ-ran	irimã-'ɨ-ran	*Brunfelsia guianensis*	SOLANACEAE
ɨwahu'ɨ	iwa-hu-'ɨ	*Micropholis melinoniana*	SAPOTACEAE
iwir'ɨ	iwir-'ɨ-pɨtaŋ	*Lecythis chartacea*	LECYTHIDACEAE
iwɨr-'ɨ-ran	iwɨr-'ɨ-ran	*Lecythis lurida*	LECYTHIDACEAE
ɨwise'ɨ	iwise-'ɨ	*Simaruba amara*	SIMARUBACEAE
ka'ame'ɨ	ka'a-me-'ɨ	*Pourouma guianensis* ssp. *guianensis*	CECROPIACEAE
ka'ameri'ɨ	ka'a-meri-'ɨ	*Sclerolobium guianense*	CAESALPINIACEAE
		Sclerolobium paraense	CAESALPINIACEAE
	ka'a-meri-'ɨ-tuwɨr	*Sclerolobium albiflorum*	CAESALPINIACEAE
ka'arotɨapu'ɨ	ka'a-ro-tɨapu-'ɨ	*Lindackeria* sp. 1	FLACOURTIACEAE
(= kupapa'ɨran-'ɨ)		*Lindackeria latifolia*	FLACOURTIACEAE

'Trees' (**Mɨra**) (*Continued*)

Folk Genus	Folk Species	Botanical Species	Family
kaŋwaruhumɨra	kaŋwaruhu-mɨra	*Agonandra brasiliensis*	OPILIACEAE
		Tabernaemontana angulata	APOCYNACEAE
kakaraní'ɨ	kaka-ran-'ɨ	*Theobroma speciosum*	STERCULIACEAE
kanaú'ɨ	kanaú-'ɨ	*Himatanthus sucuuba*	APOCYNACEAE
kanawarumɨra	kanawaru-mɨra	*Pouteria durlandii*	SAPOTACEAE
kaneí'ɨ	ara-kanei-'ɨ (= kanei-'ɨ-pɨtaŋ)	*Protium altsoni*	BURSERACEAE
		Protium heptaphyllum ssp. *heptaphyllum*	BURSERACEAE
		Protium decandrum	BURSERACEAE
	kanei-'ɨ-tuwɨr	*Protium giganteum* var. *giganteum*	BURSERACEAE
		Protium pallidum	BURSERACEAE
		Protium spruceanum	BURSERACEAE
	kanei-a-pe-'ɨ	*Protium polybotryum*	BURSERACEAE
		Protium tenuifolium	BURSERACEAE
	kanei-aka-'ɨ	*Protium sagotianum*	BURSERACEAE
karaípe'ɨ	karaí-pe-'ɨ	*Licania apetala*	CHRYSOBALANACEAE
		Licania membranacea	CHRYSOBALANACEAE
		Licania octandra	CHRYSOBALANACEAE
karaíperan'ɨ	karaí-pe-ran-'ɨ	*Casearia decandra*	FLACOURTIACEAE
karamiri'ɨ (= warikawī'ɨ in Gurupiuna)	kara-miri-'ɨ-tuwɨr	*Micropholis guyanensis* ssp. *guyanensis*	SAPOTACEAE
	kara-miri-hu-'ɨ	*Pouteria engleri*	SAPOTACEAE
	kara-miri-'ɨ-te	*Chrysophyllum lucentifolium* ssp. *pachycarpum*	SAPOTACEAE
karamiri'iran'ɨ	kara-miri-ran-'ɨ-te	*Pouteria sagotiana*	SAPOTACEAE
	kara-miri-ran-pɨtaŋ-'ɨ	*Pouteria* sp. 5	SAPOTACEAE
karayuru'ɨ (= karapatu'ɨ)	kara-yuru-'ɨ	*Croton cuneatus*	EUPHORBIACEAE
karumepɨta'ɨ	karume-pɨta-'ɨ	*Clavija lancifolia*	THEOPHRASTACEAE
kaseran'ɨ	kase-ran-'ɨ	*Casearia javitensis*	FLACOURTIACEAE
kašima'ɨ	kašima-'ɨ	*Mabea angustifolia*	EUPHORBIACEAE
		Mabea caudata	EUPHORBIACEAE
		Mabea sp. 1	EUPHORBIACEAE
kātāi'ɨ	kātāi-'ɨ	*Bertholletia excelsa*	LECYTHIDACEAE
kɨkɨ'ɨ	kɨkɨ-'ɨ	*Newtonia psilostachya*	MIMOSACEAE
		Newtonia suaveolens	MIMOSACEAE
		Pithecellobium comunis	MIMOSACEAE

'Trees' (**Mɨra**) (*Continued*)

Folk Genus	Folk Species	Botanical Species	Family
kɨpɨ	kɨpɨ-'a-'ɨ (=nukupu'ɨ)	*Theobroma subincanum*	STERCULIACEAE
	kɨpɨ-hu-'ɨ	*Theobroma grandiflorum*	STERCULIACEAE
kɨpɨhuran'ɨ (=mãmãhu'ɨ)	kɨpɨhu-ran-'ɨ	*Pachira aquatica*	BOMBACACEAE
kiraru	kiraru	*Alibertia edulis*	RUBIACEAE
kɨrɨhu'ɨ	kɨrɨhu-'ɨ	*Trattinickia burserifolia*	BURSERACEAE
		Trattinickia rhoifolia	BURSERACEAE
		Trattinickia sp. 1	BURSERACEAE
	kɨrɨhu-'ɨ-pɨtaŋ	*Astronium* cf. *obliquum*	ANACARDIACEAE
		Thyrsodium sp. 1	ANACARDIACEAE
kɨrɨhu'ɨran	kɨrɨhu-'ɨ-ran	*Talisia acutifolia*	SAPINDACEAE
kumaru'ɨ	kumaru-'ɨ	*Dipteryx odorata*	FABACEAE
kumaru'ɨšĩ	kumaru-'ɨ-šĩ	*Apuleia leiocarpa* var. *molaris*	CAESALPINIACEAE
kumaru'ɨran	kumaru-'ɨ-ran	*Bowdichia nitida*	FABACEAE
kumaruɨaporupi-har	kumaru-ɨapo-rupi-har	*Crudia parivoa*	CAESALPINIACEAE
kupa'ɨ	kupa-'ɨ	*Copaifera duckei*	CAESALPINIACEAE
		Copaifera reticulata	CAESALPINIACEAE
kupa'ɨran	kupa-'ɨ-ran	*Copaifera* sp. 1	CAESALPINIACEAE
kupapa'ɨ	kupapa-'ɨ	*Pouteria jariensis*	SAPOTACEAE
		Pouteria sp. 2	SAPOTACEAE
		Pouteria aff. *caimito*	SAPOTACEAE
		Pouteria hispida	SAPOTACEAE
kupi'ɨ'ɨ	kupi'ɨ-'ɨ	*Goupia glabra*	GOUPIACEAE
kurumi'u'ɨ (=sa'imi'u'ɨ)	kuru-mi'u-'ɨ	*Trema micrantha*	ULMACEAE
kurupi'ɨ	kurupi-šɨ-'ɨ (=pakumi'u'ɨ)	*Croton matourensis*	EUPHORBIACEAE
	kurupi-yɨ-'ɨ	*Croton cajucara*	EUPHORBIACEAE
kurupikɨ'ĩran	kurupi-kɨ'ĩ-ran	*Cordia polycephala*	BORAGINACEAE
kururu'iran'ɨ	kururu-'iran-'ɨ	*Drypetes variabilis*	EUPHORBIACEAE
kururu'ɨ	kururu-'ɨ	*Taralea oppositifolia*	FABACEAE
kuyer'ɨ	kuyer-'ɨ-pu'a	*Lacmellea aculeata*	APOCYNACEAE
	kuyer-'ɨ-puku	*Ambelania acida*	APOCYNACEAE
kuyer'ɨran	kuyer-'ɨ-ran	*Himatanthus articulatus*	APOCYNACEAE
		Ambelania grandiflora	APOCYNACEAE
kwira'imɨra	kwɨ-ra'ɨ-mɨra	*Eugenia egensis*	MYRTACEAE

'Trees' (**Mɨra**) (*Continued*)

Folk Genus	Folk Species	Botanical Species	Family
ma'eɨrapɨraŋ	ma'e-ɨra-pɨraŋ	*Ruprechtia* sp. 1	POLYGALACEAE
ma'eɨwapɨtaŋ (=ɨrapɨtaŋ'ɨ; free variant = wapɨtaŋ'ɨ)	ma'e-ɨwa-pɨtaŋ	*Eugenia patrissi* *Eugenia* sp. 6	MYRTACEAE
makahɨmɨra	makahɨ-mɨra	*Duguetia* sp. 2	ANNONACEAE
		Duguetia yeshidah	ANNONACEAE
		Ephedranthus pisocarpus	ANNONACEAE
makawa'ē'ɨ	maka-wa'ē-'ɨ	*Simaba cuspidata* var. *typica*	SIMARUBACEAE
		Simaba guianensis ssp. *guianensis*	SIMARUBACEAE
		Maytenus sp. 10	CELASTRACEAE
māmāran'ɨ	māmā-ran-'ɨ	*Jacaratia spinosa*	CARICACEAE
mani'iran'ɨ	mani-'iran-'ɨ	*Stryphnodendron polystachyum*	MIMOSACEAE
marato'ɨ	marato-'ɨ	*Schefflera morototoni*	ARALIACEAE
mayaŋɨ	mayaŋ-ɨ	*Myrcia* sp. 1	MYRTACEAE
meraí'ɨ	merai-ātā-'ɨ; (=meraí-tawa-ran)	*Byrsonima stipulacea*	MALPIGHIACEAE
	meraí-pɨraŋ-'ɨ	*Byrsonima laevigata*	MALPIGHIACEAE
	meraí-tawa	*Byrsonima* sp. 5	MALPIGHIACEAE
mɨkupɨ'a'ɨ	mɨku-pɨ'a-'ɨ	*Iryanthera juruensis*	MYRISTICACEAE
mɨra'i'a'ɨ	mɨra-'i-'a-'ɨ	*Mouriri guianensis*	MELASTOMATACEAE
		Mouriri huberi	MELASTOMATACEAE
		Mouriri sagotiana	MELASTOMATACEAE
		Mouriri trunciflora	MELASTOMATACEAE
mɨra-ātā (=mɨra-tawa-ran)	mɨra-ātā	*Erythroxylum citrifolium*	ERYTHROXYLACEAE
mɨrahātā'ɨran	mɨra-hātā-'ɨ-ran	*Pouteria guianensis*	SAPOTACEAE
mɨrahowɨ'ɨ	mɨra-howɨ-'ɨ	Sapotaceae sp. 1	SAPOTACEAE
		Sapotaceae sp. 3	SAPOTACEAE
mɨrakɨ'ī (=itapiša'ɨ)	mɨra-kɨ'ī	*Myrciaria tenella*	MYRTACEAE
mɨrapɨraŋ'ɨ	mɨra-pɨraŋ-'ɨ	*Myrcia splendens*	MYRTACEAE
mɨrapirerhē'ē'ɨ	mɨra-pirer-hē'ē-'ɨ	aff. *Pradosia* sp. 1	SAPOTACEAE
mɨrapɨtaŋ	mɨra-pɨtaŋ	*Brosimum rubescens*	MORACEAE
mɨrapɨtaŋran	mɨra-pɨtaŋ-ran	*Brosimum guianense*	MORACEAE
		Brosimum paclesum	MORACEAE
		Maquira guianensis	MORACEAE
mɨrašī'ɨ	mɨra-šī-'ɨ	*Myrciaria* cf. *dubia*	MYRTACEAE
mɨratawa	mɨra-tawa-tuwɨr	*Rauia resinosa*	RUTACEAE
mɨrawawak	mɨra-wawak	*Sagotia racemosa*	EUPHORBIACEAE
mɨrawewiatu'ɨ	mɨra-wewi-atu-'ɨ	*Pithecellobium jupumba*	MIMOSACEAE

'Trees' (**Mɨra**) (*Continued*)

Folk Genus	Folk Species	Botanical Species	Family
mɨtũeha'ɨ	mɨtũ-eha-'ɨ	*Banara guianensis*	FLACOURTIACEAE
mɨtũmɨra	mɨtũ-mɨra	*Erisma uncinatum*	VOCHYSIACEAE
mɨtũpusu'ɨ	mɨtũ-pusu-'ɨ	*Gustavia augusta*	LECYTHIDACEAE
moimɨra	moi-mɨra	*Poecilanthe effusa*	FABACEAE
mu'ɨ	mu-'ɨ	*Bellucia grossularioides*	MELASTOMATACEAE
mu'ɨran	mu-'ɨ-ran	*Myriaspora egensis*	MELASTOMATACEAE
mukuku'ɨ	mukuku-'ɨ-te	*Licania glabriflora*	CHRYSOBALANACEAE
		Licania heteromorpha var. *heteromorpha*	CHRYSOBALANACEAE
		Licania latifolia	CHRYSOBALANACEAE
	mukuku-'ɨ-hu	*Licania* sp. 1	CHRYSOBALANACEAE
	mukuku-'ɨ-tuwɨr	*Ouratea salicifolia*	OCHNACEAE
		Hirtella racemosa var. *racemosa*	CHRYSOBALANACEAE
mukuku-'ɨ-wi mukuku'ɨran	mukuku-ɨ-ran	*Ouratea* sp. 2	OCHNACEAE
murukuya'ɨ	murukuya-'ɨ	*Sloanea grandis*	ELAEOCARPACEAE
		Sloanea guianensis	ELAEOCARPACEAE
	murukuya-'ɨ-wi	*Sloanea grandiflora*	ELAEOCARPACEAE
		Sloanea porphyrocarpa	ELAEOCARPACEAE
murure'ɨ	murure-'ɨ	*Brosimum acutifolium* ssp. *interjectum*	MORACEAE
pa'imɨra	pa'i-mɨra	*Dodecastigma integrifolium*	EUPHORBIACEAE
pakurɨ'ɨ	pakurɨ-'ɨ-te (=pakurɨ-tawa) pakurɨ-pu'a pakurɨ-hãšĩ pakurɨ-pihun	*Platonia insignis*	CLUSIACEAE
pakurɨsoso'ɨ	pakurɨ-soso-'ɨ	*Rheedia acuminata*	CLUSIACEAE
		Rheedia brasiliensis	CLUSIACEAE
pakurɨsõsõran	pakurɨ-sõsõ-ran	*Posoqueria latifolia* ssp. *gracilis*	RUBIACEAE
panaka'amɨra	pana-ka'a-mɨra	*Platypodium elegans*	FABACEAE
pani'ɨ	pani-'ɨ	*Hymenolobium excelsum*	MIMOSACEAE
para'ɨ	para-'ɨ	*Jacaranda copaia* ssp. *copaia*	BIGNONIACEAE
		Jacaranda copaia ssp. *spectabilis*	BIGNONIACEAE
		Jacaranda duckei	BIGNONIACEAE

'Trees' (**Mɨra**) (*Continued*)

Folk Genus	Folk Species	Botanical Species	Family
para'ɨran	para-'ɨ-ran	*Jacaranda paraensis*	BIGNONIACEAE
		Jacaranda heterophylla	BIGNONIACEAE
paraku'ɨ	paraku-'ɨ	*Chimarrhis turbinata*	RUBIACEAE
		Aspidosperma cylindrocarpon	APOCYNACEAE
paraku'ɨran	paraku-'ɨ-ran	*Campomanesia grandiflora*	MYRTACEAE
		Maytenus sp. nov.	CELASTRACEAE
		Remijia sp. 1	RUBIACEAE
		Malanea sp. 1	RUBIACEAE
parani'ɨ	parani-'ɨ	*Laetia procera*	FLACOURTIACEAE
paraniran'ɨ	parani-ran-'ɨ	*Ryania speciosa*	FLACOURTIACEAE
parawa'ɨ	parawa-'ɨ	*Eschweilera coriacea*	LECYTHIDACEAE
		Eschweilera pedicellata	LECYTHIDACEAE
	parawa-'ɨ-pihun	*Eschweilera apiculata*	LECYTHIDACEAE
	parawa-'ɨ-wi	*Eschweilera amazonica*	LECYTHIDACEAE
		Eschweilera micrantha	LECYTHIDACEAE
parawami'u'ɨ	parawa-mi'u-'ɨ	*Chrysophyllum argenteum* ssp. *auratum*	SAPOTACEAE
		Luehea duckeana	TILIACEAE
parei'a'ɨ	parei'a-'ɨ	*Swartzia brachyrachis* var.	CAESALPINIACEAE
parei'a'ran'ɨ	parei'a-ran-'ɨ	*Swartzia brachyrachis* var. *snethlageae*	CAESALPINIACEAE
paruru'ɨ	paruru-'ɨ	*Sacoglottis amazonica*	HUMIRIACEAE
		Sacoglottis guianensis	HUMIRIACEAE
payu'ã'ɨ	payu'ã-'ɨ	*Couepia guianensis* ssp. *guianensis*	CHRYSOBALANACEAE
		Parinari excelsa	CHRYSOBALANACEAE
	payu'ã-'ɨ-howɨ	*Hirtella bicornis*	CHRYSOBALANACEAE
	(= anirawiši)	*Couepia guianensis* ssp. *divaricata*	CHRYSOBALANACEAE
	payu'ã-'ɨ-tawa	*Couepia guianensis* ssp. *glandulosa*	CHRYSOBALANACEAE
perepusan'ɨ (= kurupusan'ɨ)	pere-pusan-'ɨ	*Simaba* aff. *cavalcantei*	SIMARUBACEAE

'Trees' (**Mɨra**) (*Continued*)

Folk Genus	Folk Species	Botanical Species	Family
		Simaba cedron	SIMARUBACEAE
pɨkɨ'a'ɨ	pɨkɨ'a-'ɨ	*Caryocar villosum*	CARYOCARACEAE
pɨkɨ'aran'ɨ	pɨkɨ'a-ran-'ɨ	*Caryocar glabrum*	CARYOCARACEAE
pina'ɨ	pina-'ɨ	*Duguetia echinophora*	ANNONACEAE
		Duguetia sp. 1	ANNONACEAE
	pina-hu-'ɨ	*Unonopsis rufescens*	ANNONACEAE
		Duguetia surinamensis	ANNONACEAE
pinahuran'ɨ	pina-hu-ran-'ɨ	*Dulacia guianensis*	OLACACEAE
pɨtɨm'ɨ	pɨtɨm-'ɨ	*Couratari guianensis*	LECYTHIDACEAE
	pɨtɨm-inem-'ɨ	*Couratari oblongifolia*	LECYTHIDACEAE
pɨtɨm-inem-ran-'ɨ	pɨtɨm-inem-ran-'ɨ	*Eschweilera ovata*	LECYTHIDACEAE
pɨwa'ɨ	pɨwa-'ɨ	*Rinorea pubiflora*	VIOLACEAE
	pɨwa-'ɨ-hu	*Rinorea flavescens*	VIOLACEAE
		Rinorea sp. 1	VIOLACEAE
pu'ɨpihun'ɨ	pu'ɨ-pihun-'ɨ	*Colubrina glandulosa* var. *glandulosa*	RHAMNACEAE
		Colubrina sp. 1	RHAMNACEAE
pu'ɨpɨraŋ'ɨ	pu'ɨ-pɨraŋ-'ɨ-te	*Ormosia coccinea*	FABACEAE
	pu'ɨ-pɨraŋ-pihun-'ɨ	*Andira retusa*	FABACEAE
samo'ã'ɨ	samo'ã-'ɨ	*Cochlospermum orinocense*	COCHLOSPERMACEAE
sawɨyamɨra	sawɨya-mɨra	*Capparis sola*	CAPPARIDACEAE
		Paypayrola grandiflora	VIOLACEAE
sekãtãi'ɨ	se-kãtãi-'ɨ	*Protium trifoliolatum*	BURSERACEAE
serẽserẽ'ɨ	serẽserẽ-'ɨ	*Vochysia inundata*	VOCHYSIACEAE
šiŋɨ	šiŋ-ɨ	*Pithecellobium niopoides*	MIMOSACEAE
		Acacia paniculata	MIMOSACEAE
šimo'ɨ	šimo-'ɨ	*Enterolobium* sp. nov.	MIMOSACEAE
		Parkia paraensis	MIMOSACEAE
šimo'hu'ɨ	šimo-hu-'ɨ	*Sterculia alata*	STERCULIACEAE
šimo'ɨran (= mašɨrawa)	šimo-'ɨ-ran	*Senna sylvestris*	CAESALPINIACEAE
sɨpomɨra	sɨpo-mɨra	*Coccoloba latifolia*	POLYGONACEAE
		Coccoloba racemulosa	POLYGONACEAE
širiŋi'ɨ	širiŋi-'ɨ	*Hevea guianensis*	EUPHORBIACEAE

'Trees' (**Mɨra**) (*Continued*)

Folk Genus	Folk Species	Botanical Species	Family
surukuyurãšĩ'ɨ	**suruku-yu-rãšĩ-'ɨ**	*Pithecellobium foliolosum*	MIMOSACEAE
taŋara'ɨ	**taŋara-'ɨ**	*Paullinia* sp. 1	SAPINDACEAE
taŋaratai'ɨ	**taŋara-tai-'ɨ**	*Miconia* cf. *kappleri*	MELASTOMATACEAE
		Miconia sp. 1	MELASTOMATACEAE
		Miconia nervosa	MELASTOMATACEAE
		Miconia serialis	MELASTOMATACEAE
		Miconia serrulata	MELASTOMATACEAE
		Miconia surinamensis	MELASTOMATACEAE
taŋaratairan'ɨ	**taŋara-tai-ran-'ɨ**	*Henrietta spruceana*	MELASTOMATACEAE
taŋwa-'ɨ	**taŋwa-'ɨ**	*Magaritaria nobilis*	EUPHORBIACEAE
tahayumɨra	**tayahu-mɨra**	*Tapirira guianensis*	ANACARDIACEAE
taíriŋa'ɨ	**taír-iŋa-'ɨ**	*Inga gracilifolia*	MIMOSACEAE
		Pithecellobium cauliflorum	MIMOSACEAE
			MIMOSACEAE
		Pithecellobium racemosum	MIMOSACEAE
takwarɨmɨra	**takwar-ɨ-mɨra**	*Coccoloba paniculata*	POLYGONACEAE
tamaran'ɨ	**tamaran-'ɨ**	*Zollernia paraensis*	CAESALPINIACEAE
tamarimɨra	**tamari-mɨra**	*Diospyros artanthifolia*	EBENACEAE
		Diospyros duckei	EBENACEAE
taperɨwa'ɨ	**taper-ɨwa-'ɨ**	*Spondias mombin*	ANACARDIACEAE
tapeši'ɨ	**tapeši'ɨ**	*Cephalia* sp. 1	RUBIACEAE
tapi'ipamɨr (=**kihapirita'ɨ**)	**tapi'i-pamɨr-'ɨ**	*Sterculia pruriens*	STERCULIACEAE
tapi'irãkwãipe'ɨ	**tapi'i-rãkwãi-pe-'ɨ**	*Ampelocera edentula*	ULMACEAE
tapi'isikiri'ɨ	**tapi'i-sikiri-'ɨ**	*Neea floribunda*	NYCTAGINACEAE
		Neea glamerulɨflora	NYCTAGINACEAE
		Neea oppositifolia	NYCTAGINACEAE
		Neea ovalifolia	NYCTAGINACEAE
		Neea sp. 1	NYCTAGINACEAE
		Neea sp. 2	NYCTAGINACEAE
		Pisonia sp. 2	NYCTAGINACEAE
tapiša'ɨ	**tapiša-'ɨ**	*Erythroxylum* cf. *leptoneurum*	ERYTHROXYLACEAE
		Micropholis venulosa	SAPOTACEAE
taraku'ã'ɨ	**taraku'ã-'ɨ**	*Fusaea longifolia*	ANNONACEAE
tarapa'ɨ	**tarapa-'ɨ-pɨtaŋ**	*Hymenaea courbaril*	CAESALPINIACEAE
	tarapa-'ɨ-te	*Hymenaea reticulata*	CAESALPINIACEAE
tararamɨra	**tarara-mɨra**	*Cupania scrobiculata*	SAPINDACEAE

'Trees' (**Mɨra**) (*Continued*)

Folk Genus	Folk Species	Botanical Species	Family
		Matayba spruceana	SAPINDACEAE
tareka'ɨ	tareka-'ɨ	*Bagassa guianensis*	MORACEAE
taši'ɨ	taši-'ɨ	*Tachigali myrmecophila*	CAESALPINIACEAE
	taši-'ɨ-ātā	*Tachigali macrostachya*	CAESALPINIACEAE
		Tachigali paniculata	CAESALPINIACEAE
	taši-'ɨ-hu	*Tachigali* sp. 1	CAESALPINIACEAE
taši-'iran-'ɨ	taši-'iran-'ɨ	*Sclerolobium* sp. 20	CAESALPINIACEAE
	taši-'ɨ-ātā-ran	*Sclerolobium* sp. 16	CAESALPINIACEAE
	tata'iran'ɨ	*Xylopia barbata*	ANNONACEAE
	tata-'iran-'ɨ	*Duguetia riparia*	ANNONACEAE
		Guatteria elongata	ANNONACEAE
		Guatteria sp. 1	ANNONACEAE
tatumɨra	tatu-mɨra	*Thyrsodium spruceanum*	ANACARDIACEAE
tayahumi'urupe'ɨ	tayahu-mi'u-rupe-'ɨ	*Parkia multijuga*	MIMOSACEAE
tayahumɨra	tayahu-mɨra	*Tapirira guianensis*	ANACARDIACEAE
		Tapirira peckoltiana	ANACARDIACEAE
tayɨ	tayɨ-te	*Tabebuia impetiginosa*	BIGNONIACEAE
	tayɨ-tawa		
	tayɨ-pihun		
tayɨpō	tayɨ-pō	*Tabebuia* sp. 1	BIGNONIACEAE
		Tabebuia serratifolia	BIGNONIACEAE
tayɨran	tayɨ-ran	*Eugenia flavescens*	MYRTACEAE
		Rauvolfia paraensis	APOCYNACEAE
teremu-mɨra	teremu-mɨra	*Anaxagorea dolichocarpa*	ANNONACEAE
tukwãmi'u'ɨ	tukwã-mi'u-'ɨ	*Virola carinata*	MYRISTICACEAE
(= tukwãmɨra)		*Virola michelii*	MYRISTICACEAE
tukurɨ	tukur-ɨ	*Guettarda divaricata*	RUBIACEAE
tukurɨwa'ɨ	tukur-ɨwa-'ɨ	*Buchenavia* cf. *tetraphylla*	COMBRETACEAE
		Terminalia amazonia	COMBRETACEAE
		Terminalia dichotoma	COMBRETACEAE
		Terminalia lucida	COMBRETACEAE
tupiyarɨmamɨra	tupiyarɨma-mɨra	*Talisia micrantha*	SAPINDACEAE
		Talisia carasina	SAPINDACEAE
u'ɨtɨma'ɨ	u'ɨ-tɨma-'ɨ	aff. *Myrcia* sp. 1	MYRTACEAE
	u'ɨ-tɨma-piriri-'ɨ	*Eugenia* sp. 8	MYRTACEAE
u'ɨtɨmaran'ɨ	u'ɨ-tɨma-ran-'ɨ	*Myrcia* sp. 4	MYRTACEAE
		Eugenia eurcheila	MYRTACEAE

'Trees' (**Mɨra**) (*Continued*)

Folk Genus	Folk Species	Botanical Species	Family
	u'ɨ-tɨma-piriri-ran-'ɨ	*Eugenia schomburkii*	MYRTACEAE
urukɨwa'ɨ	uru-kɨwa-'ɨ	*Virola* sp. 1	MYRISTICACEAE
urukuran'ɨ	uruku-ran-'ɨ	*Bixa orellana*	BIXACEAE
wa'ɨ'ɨ	wa-'ɨ-'ɨ-te	*Pouteria reticulata* ssp. *reticulata*	SAPOTACEAE
wa'ɨ'ɨran	wa'-ɨ-'ɨ-ran	*Pouteria penicillata*	SAPOTACEAE
		Duroia sp. 1	RUBIACEAE
wakuramɨra	wakura-mɨra-hu	*Sapium caspceolatum*	EUPHORBIACEAE
		Sapium lanceolatum	EUPHORBIACEAE
wapini'ɨ	wa-pini-'ɨ-tuwɨr	*Licania canescens*	CHRYSOBALANACEAE
	wa-pini-hu-'ɨ	*Licania kunthiana*	CHRYSOBALANACEAE
wari'ɨ	wari-'ɨ	*Ficus* sp. 3	MORACEAE
warimɨra	wari-mɨra	*Clarisia racemosa*	MORACEAE
waruwa'ɨ	waruwa-'ɨ	*Tetragastris altissima*	BURSERACEAE
	waruwa-ɨwa-pɨtaŋ-'ɨ	*Tetragastris panamensis*	BURSERACEAE
wašiŋi'ɨ	wašiŋi-'ɨ	*Ceiba pentandra*	BOMBACACEAE
wašɨšɨ'ɨ	wa-šɨ-šɨ-hu-'ɨ	*Zanthoxylum* cf. *juniperina*	RUTACEAE
		Zanthoxylum chiloperone	RUTACEAE
	wa-šɨ-šɨ-'ɨ	*Zanthoxylum rhoifolium*	RUTACEAE
wayamɨra	waya-mɨra-	Indeterminate genus 2	[INDETER-MINATE 1]
wayaŋɨ (= āyaŋruku)	wayaŋ-ɨ	*Vismia guianensis*	CLUSIACEAE
wayataper-rupɨhar	waya-taper rupɨ-har	cf. *Calyptranthes* sp. 1	MYRTACEAE
wiririmi'u'ɨ	wiriri-mi'u-'ɨ	*Pouteria filipes*	SAPOTACEAE
yakamīmɨra (= tapi'ika'amɨra)	yakamī-mɨra	*Coussarea ovalis*	RUBIACEAE
		Coussarea paniculata	RUBIACEAE
		Faramea sessilifolia	RUBIACEAE
yakuširi'ɨ (= waruwa-'iran'ɨ)	pɨtaŋ yaku-širi-'ɨ-	*Guarea guidonia*	MELIACEAE
	yaku-širi-hu-'ɨ	*Guarea macrophylla* ssp. *pachycarpa*	MELIACEAE
		Guarea kunthiana	MELIACEAE
		Trichilia verrocosa	MELIACEAE
	yaku-širi-'ɨ tuwɨr	*Trichilia micrantha*	MELIACEAE
	yaku-širi-'ɨ-	*Trichilia*	MELIACEAE

'Trees' (**Mɨra**) (*Continued*)

Folk Genus	Folk Species	Botanical Species	Family
	tawa	*quadrijuga*	
yakuširi'iran'ɨ	yaku-širi-'iran-'ɨ	*Guarea* sp. 1	MELIACEAE
yanaímɨra	yanaí-mɨra	*Dendrobangia boliviana*	ICACINACEAE
yaniro'ɨ (free variants = aniro'ɨ; yinɨro'ɨ)	yani-ro-'ɨ- tuwɨr yani-ro-'ɨ-pɨraŋ	*Carapa guianensis*	MELIACEAE
yanukihawi'ɨ	yanu-kiha-wi-'ɨ	*Diospyros melinoni*	EBENACEAE
yanumɨra	yanu-mɨra	*Eugenia omissa*	MYRTACEAE
		Myrciaria pyriifolia	MYRTACEAE
yapukai'ɨ	ya-pukai-'ɨ	*Lecythis pisonis*	LECYTHIDACEAE
yapumɨra	yapu-mɨra	*Tovomita brevistaminea*	CLUSIACEAE
yapuri'a'ran'ɨ	yapu-ri'a- 'ran-'ɨ	*Tovomita brasiliensis*	CLUSIACEAE
yapurɨwa'ɨ	yapu-rɨwa-'ɨ	*Pouteria egregia*	SAPOTACEAE
yašiamɨr	yaši-amɨr	*Lecythis idatimon*	LECYTHIDACEAE
yašimɨra	yaši-mɨra	*Coccoloba* sp. 2	POLYGONACEAE
yašimukuku'ɨ	yaši-mukuku-'ɨ	Annonaceae sp. 1	ANNONACEAE
		Flacourtiaceae sp. 1	FLACOURTIACEAE
		Neoptychocarpus apodanthus	FLACOURTIACEAE
yašipɨta'ɨ	yaši-pɨta-'ɨ	*Talisia retusa*	SAPINDACEAE
		Talisia microphylla	SAPINDACEAE
yašɨwa'ɨ	yaši-wa-'ɨ	*Sacoglottis* sp. 1	HUMIRIACEAE
yawamɨra (= sekãtãi- iran'ɨ)	yawa-mɨra	*Protium aracouchini*	BURSERACEAE
yawi'ɨ	yawi-'ɨ-tuwɨr yawi-'ɨ-pɨraŋ	*Xylopia nitida*	ANNONACEAE
yenipa'ɨ	yenipa-'ɨ	*Genipa americana*	RUBIACEAE
yeta'ɨ	yeta-'ɨ	*Hymenaea parvifolia*	CAESALPINIACEAE
yeyu'ɨ	yeyu-'ɨ	*Andira* sp. 1	FABACEAE
		Astronium lecointei	ANACARDIACEAE
yeyu'ɨran	yeyu-'ɨ-ran	*Diplotropis purpurea*	FABACEAE
yowoiɨ	yowoi-ɨ	*Minquartia guianensis*	OLACACEAE
yuparamɨra	yupara-mɨra	*Coumarouna micrantha*	FABACEAE
yupu'ɨ	yupu-'ɨ	*Parkia pendula*	MIMOSACEAE
yurupepe'ɨ	yuru-pe-pe-'ɨ	*Dialium guianense*	CAESALPINIACEAE

'Herbs' **(Ka'a)**

Folk Genus	Folk Species	Botanical Species	Family
anī	anī	*Montrichardia linifera*	ARACEAE
anī-ran	anī-ran	*Dracontium* sp. 1	ARACEAE
arapuhamani'ɨ	arapuha-mani'ɨ	*Manihot brachyloba*	EUPHORBIACEAE
		Manihot leptophylla	EUPHORBIACEAE
		Manihot quinquepartita	EUPHORBIACEAE
āyaŋrākwāi	āyaŋ-rākwāi	*Costus scaber*	ZINGIBERACEAE
ipeka'a	ipe-ka'a	*Psychotria ulviformis*	RUBIACEAE
		Sipanea veris	RUBIACEAE
ɨrahuka'a	ɨra-hu-ka'a	*Stenoclaena* sp. 1	POLYPODIACEAE
ɨrahuka'aran (=mɨrahureha)	ɨra-hu-ka'a-ran	*Rhipsalis baccifera*	CACTACEAE
ɨrakahuka'a	ɨrakahu-ka'a	*Xiphidium caeruleum*	HAEMODORACEAE
ɨrakɨwaran	ɨra-kɨwa-ran	*Asclepias curassavica*	ASCLEPIADACEAE
itamɨra	ita-mɨra	*Phyllanthus urinaria*	EUPHORBIACEAE
ka'apiši'u	ka'a-piši'u	*Siparuna guianensis*	MONIMIACEAE
ka'ariru	ka'a-riru	*Phytolacca rivinoides*	PHYTOLACCACEAE
ka'aro	ka'a-ro-hātā	*Ischnosiphon* sp. 2.	MARANTACEAE
	ka'a-ro-pinī	*Caladium picturatum*	ARACEAE
	ka'a-ro-šī	Marantaceae sp. 3.	MARANTACEAE
	ka'a-ro-tuwɨr	*Ischnosiphon obliquus*	MARANTACEAE
	yanaí-ka'a-ro	Marantaceae sp. 4	MARANTACEAE
	ka'a-ro-yē'ē-hātā	*Moutabea* sp. 2	POLYGALACEAE
ka'atuwɨr	ka'a-tuwɨr	aff. *Rhizomorpha* sp. 1	MYCOTA
ka'ayuwar	ka'a-yuwar	*Solanum rugosum*	SOLANACEAE
ka'ayuwaran	ka'a-yuwar-ran	*Eupatorium macrophyllum*	ASTERACEAE
ka'uwapusan (=ka'apiši'uran)	ka'uwa-pusan	*Siparuna amazonica*	MONIMIACEAE
kamamu	kamamu	*Physalis angulata*	SOLANACEAE
kanamɨran	kanamɨ-ran	*Lantana camara*	VERBENACEAE
kararan	kara-ran	*Dioscorea* sp. 2	DIOSCOREACEAE
kupi'ipe	kupi'i-pe	*Paspalum conjugatum*	POACEAE
kupi'ipu'ā	kupi'i-pu'ā	*Euphorbia* sp. 1	EUPHORBIACEAE
kurupikā	kurupir-kā	*Renealmia alpinia*	ZINGIBERACEAE
kurupipɨtɨm	kurupi-pɨtɨm	*Renealmia floribunda*	ZINGIBERACEAE
kururuka'a	kururu-ka'a	*Amaranthus spinosus*	AMARANTHACEAE
		Borreria verticulata	RUBIACEAE
		Rubiaceae sp. 1	RUBIACEAE

'Herbs' **(Ka'a)** (*Continued*)

Folk Genus	Folk Species	Botanical Species	Family
kuyuíka'a	kuyuí-ka'a	*Bertiera guianensis*	RUBIACEAE
mayuka'aro	mayu-ka'a-ro	*Limnocharis* sp. 1	LIMNOCHARITACEAE
pakosororo	pako-sororo	*Phenakospermum guianensis*	MUSACEAE
parawaka'a	parawa-ka'a	*Ficus* sp. 1	MORACEAE
pirapišīka'a	pira-piší-ka'a	*Justicia pectoralis*	ACANTHACEAE
		Justicia spectabilis	ACANTHACEAE
pišuran	pišu-ran	*Phyllanthus martii*	EUPHORBIACEAE
purakeka'a	purake-ka'a	*Laportea aestuans*	URTICACEAE
suruwika'a	suruwi-ka'a	*Calathea* sp. 1	MARANTACEAE
tapi'ika'a	tapi'ika'a	*Psychotria* sp. 1	RUBIACEAE
		Psychotria poeppigiana ssp. *poeppigiana*	RUBIACEAE
		Psychotria racemosa	RUBIACEAE
		Tocoyena foetida	RUBIACEAE
		Psychotria colorata	RUBIACEAE
tapi'ikanamɨ (=yawarūka'a)	tapi'i-kanamɨ	*Psychotria poeppigiana*	RUBIACEAE
tapiša	tapiša	*Scoparia dulcis*	SCROPHULARIACEAE
tatuka'aro	tatu-ka'a-ro	*Memora allamandiflora*	BIGNONIACEAE
taturuwai	tatu-ruwai	Polygonaceae sp. 1	POLYGONACEAE
tayahumanuwi	tayahu-manuwi	*Ctenanthe* sp. 1	MARANTACEAE
		Marantaceae sp. 1	MARANTACEAE
tayahumanuwiran	tayahu-manuwiran	Marantaceae sp. 2	MARANTACEAE
tayahupakoro	tayahu-pako-ro	*Costus arabicus*	ZINGIBERACEAE
		Heliconia acuminata	HELICONIACEAE
		Heliconia bihai	HELICONIACEAE
		Heliconia chartacea	HELICONIACEAE
teyuka'a	teyu-ka'a	*Galipea trifoliolata*	RUTACEAE
teyupɨtɨm	teyu-pɨtɨm	*Conyza banariensis*	ASTERACEAE
		Porophyllum ellipticum	ASTERACEAE
		Pterocaulon vergatum	ASTERACEAE
teyupɨtɨmran	teyu-pɨtɨm-ran	*Phyllanthus niruri*	EUPHORBIACEAE
wariruwai	wari-ruwai	*Lomariopsis japurensis*	LOMARIOPSIDACEAE
warumā	warumā-hu	*Ischnosiphon* sp. 1	MARANTACEAE
	warumā-ra'ɨr	*Ischnosiphon petrolatus*	MARANTACEAE
	warumā-šī	*Maranta protracta*	MARANTACEAE
	warumā-širik	*Ischnosiphon graciles*	MARANTACEAE
	warumā-te	*Ischnosiphon arouma*	MARANTACEAE

'Herbs' **(Ka'a)** (*Continued*)

Folk Genus	Folk Species	Botanical Species	Family
yaŋwateka'a	yaŋwate-ka'a	*Selaginella* sp. 1	SELAGINELLACEAE
yakamɨtɨma-kaŋwer	yakami-tɨma-kaŋwer	*Dioscorea* sp. 1	DIOSCOREACEAE
yakareka'a	yakare-ka'a	*Ctenitis pretensa*	ASPIDIACEAE
		Pteridium aquilinium	DENNSTAEDTIACEAE
yamɨr	yamɨr	*Piper ottonoides*	PIPERACEAE
		Piper piscatorum	PIPERACEAE
	yamɨr-hu	*Piper anonnifolium*	PIPERACEAE
		Piper hostmannianum	PIPERACEAE
		Piper pilirameum	PIPERACEAE
yamɨran	yamɨr-ran-ka'a	*Dichorisandra affinis*	COMMELIACEAE
yu'ika'a	yu'i-ka'a	*Aciotis purpurescens*	MELASTOMATACEAE
		Miconia ceramicarpa	MELASTOMATACEAE
		Psychotria hoffmannseggiana	RUBIACEAE
yukunaka'a	yukuna-ka'a	*Ardisia guianensis*	MYRSINACEAE
yurušikɨ'ĩ	yuruši-kɨ'ĩ	*Geophila repens*	RUBIACEAE
yuruwe	yu-ruwe-hu	*Solanum vanheurckii*	SOLANACEAE
	yu-ruwe-pɨraŋ	*Solanum stramonifolium*	SOLANACEAE
	yu-ruwe-tawa	*Solanum subinerme*	SOLANACEAE
	yu-ruwe-te	*Solanum crinitum*	SOLANACEAE

'Vines' **(Sɨpo)**

Folk Genus	Folk Species	Botanical Species	Family
a'ɨhupako	a'ɨhu-pako	*Trigonidium acuminatum*	ORCHIDACEAE
akušisɨpo (= akušinamɨ)	akuši-sɨpo	*Drymonia coccinea*	GESNERIACEAE
amaŋatirisɨpo	amaŋa-tiri-sɨpo	*Senna chrysocarpa*	CAESALPINIACEAE
anãsɨpo	anã-sɨpo	Indeterminate genus 5	[INDETERMINATE 4]
apo'ɨ	apo-'ɨ	*Clusia* sp. 1	CLUSIACEAE
	apo-'ɨ-pihun	*Coussapoa* sp. 1	CECROPIACEAE
	apo-'ɨ-pinĩ	*Ficus* sp. 2	MORACEAE
aputa	aputa	*Abuta grandifolia*	MENISPERMACEAE
arapuhasɨpo	arapuha-sɨpo	*Coccoloba* sp. 1	POLYGONACEAE
araruhuwairimo	arar-uhu-wai-rimo	*Norantea guianensis*	MARCGRAVIACEAE
ãyaŋrampũiputɨr	ãyaŋ-rampũi-putɨr	*Cissampelos* sp. 1	MENISPERMACEAE
ãyaŋnami	ãyaŋ-nami	*Ipomea setofera*	CONVOLVULACEAE
eremenɨ	eremenɨ	*Ampelozizyphus mamazonicus*	RHAMNACEAE
ɨraísɨpo	ɨra-i-sɨpo	*Schubertia grandiflora*	ASCLEPIADACEAE

'Vines' **(Sɨpo)** *(Continued)*

Folk Genus	Folk Species	Botanical Species	Family
kawasuran	**kawasu-ran**	*Cayaponia*	CUCURBITACEAE
		Gurania eriantha	CUCURBITACEAE
	kawasu-ran-puku	*Gurania cissoides*	CUCURBITACEAE
kumariyu′ɨ	**kumari-yu-′ɨ**	*Hippocratea ovata*	CELASTRACEAE
kurupi′isɨpo	**kurupi-′i-sɨpo**	*Cordia multispicata*	BORAGINACEAE
maŋwasape′ɨ	**maŋwa-sape-′ɨ**	Indeterminate genus 4	[INDETERMINATE 3]
mahasɨpo	**maha-sɨpo**	Connaraceae sp. 2	CONNARACEAE
makawa′ēsɨpo	**maka-wa′ē-sɨpo**	aff. *Cheiloclinium* or *Salacia* sp. 1	CELASTRACEAE
		Cheiloclinium cognatum	CELASTRACEAE
		Salacia insignis	CELASTRACEAE
mišiksɨpo	**mišik-sɨpo**	*Moutabea guianensis*	POLYGALACEAE
		Schwannia sp. 1	MALPIGHIACEAE
murukuyapinī	**murukuya-pinī**	*Passiflora* sp. 3	PASSIFLORACEAE
murukuyaran	**murukuya-ran**	Cucurbitaceae sp. 1	CUCURBITACEAE
		Passiflora sp. 2	PASSIFLORACEAE
		Passiflora aranjoi	PASSIFLORACEAE
		Passiflora coccinea	PASSIFLORACEAE
	murukuya-ran-sɨpo	*Passiflora* sp. 4	PASSIFLORACEAE
murukuyatawa	**murukuya-tawa-rimo**	*Passiflora* sp. 1	PASSIFLORACEAE
mususɨpo	**musu-sɨpo**	*Styzophyllum riparium*	BIGNONIACEAE
parawasɨpo	**parawa-sɨpo**	*Uncaria guianensis*	RUBIACEAE
pikahuampūirimo	**pikahu-ampūi-rimo**	*Mendoncia hoffmannseggiana*	MENDONCIACEAE
		Mendoncia sp. 1	MENDONCIACEAE
puru′ārimo	**puru′ā-rimo**	*Philodendron* sp. 1	ARACEAE
šimo	**kururu-šimo**	*Serjania* aff. *lethalis*	SAPINDACEAE
	šimo-rimo	*Derris utilis*	FABACEAE
šimoran	**šimo-ran**	*Derris* sp. 1	FABACEAE
		Derris amazonica	FABACEAE
sɨpo-memek	**sɨpo-memek**	Bignoniaceae sp. 1	BIGNONIACEAE
sɨpo-te (=sɨpošišik)	**sɨpo-te**	*Heteropsis longispathacea*	ARACEAE
		Heteropsis sp. 1	
sɨpoātā	**sɨpo-ātā**	*Aegiphila* sp. 1	VERBENACEAE
		Maripa sp. 1	CONVOLVULACEAE
		Memora bracteosa	BIGNONIACEAE
		Paullinia sp. 2	SAPINDACEAE
		Rhabidalia sp. 1	BIGNONIACEAE
sɨpohu	**sɨpo-hu**	*Connarus favosus*	CONNARACEAE
		Evodianthus funifer	CYCLANTHACEAE
sɨponem	**sɨpo-nem**	*Mansoa angustiden*	BIGNONIACEAE

'Vines' **(Sɨpo)** (*Continued*)

Folk Genus	Folk Species	Botanical Species	Family
sɨpopihun	sɨpo-pihun	aff. *Forsteronia* sp. 1	APOCYNACEAE
sɨpopɨraŋ	sɨpo-pɨraŋ	*Banisteriopsis pubipetala*	MALPIGHIACEAE
		Memora flavida	BIGNONIACEAE
		Pristimera tenuifolia	CELASTRACEAE
		Salacia multiflora	CELASTRACEAE
		Tetrapterys styloptera	MALPIGHIACEAE
sɨpopɨtaŋ	sɨpo-pɨtaŋ	*Peritassa huanuclara*	CELASTRACEAE
sɨporan	sɨpo-ran	*Calyobolus* sp. 1	CONVOLVULACEAE
		Cissus cissyoides	VITACEAE
		Connaraceae sp. 1	CONNARACEAE
		Mabea pohliana	EUPHORBIACEAE
		Masechites bicornulata	APOCYNACEAE
sɨpotai	sɨpo-tai	Indeterminate genus 89	[INDETERMINATE 22]
sɨpotawa	sɨpo-tawa	aff. *Memora bracteosa*	BIGNONIACEAE
		Humirianthera duckei	ICACINACEAE
		Moutabea sp. 1	POLYGALACEAE
sɨpotuwɨr	sɨpo-tuwɨr	*Allamanda cathartica*	APOCYNACEAE
		Amphilophium paniculata var. *moller*	BIGNONIACEAE
		Anemopaegma setilobum	BIGNONIACEAE
		Arrabidaea cf. *florida*	BIGNONIACEAE
		Cuspidaria sp. 1	BIGNONIACEAE
		Mascagnia sp. 1	MALPIGHIACEAE
so'oransɨpo	so'o-ran-sɨpo	*Stigmaphyllon hypoleucum*	MALPIGHIACEAE
surukukuyurãšĩ	surukuku-yu-rãšĩ	*Seguieria amazonica*	PHYTOLACCACEAE
tapamɨr	tapamɨr	*Caryomene foveolata*	MENISPERMACEAE
tapuruka'a	tapuru-ka'a	*Gouania pyrifolia*	RHAMNACEAE
tata'ɨ	tata-'ɨ	*Guattteria scandens*	ANNONACEAE
tayahusɨpo	tayahu-sɨpo	*Ipomoea* sp. 1	CONVOLVULACEAE
tɨrɨrɨsɨpo	tɨrɨrɨ-sɨpo	*Davilla kunthii*	DILLENIACEAE
		Davilla nitida	DILLENIACEAE
		Doliocarpus sp. 1	DILLENIACEAE
		Tetracera volubilis ssp. *volubilis*	DILLENIACEAE

'Vines' **(Sɨpo)** (*Continued*)

Folk Genus	Folk Species	Botanical Species	Family
		Tetracera willdenowiana ssp. *willdenowiana*	DILLENIACEAE
wamēsɨpo	wa-mē-sɨpo	*Monstera* cf. *pertusa*	ARACEAE
		Monstera subpinata	ARACEAE
		Philodendron grandiflorum	ARACEAE
		Philodendron venustum	ARACEAE
wiririsɨpo	wiriri-sɨpo	*Arrabidaea* sp. 1	BIGNONIACEAE
yahɨsɨpo	yahɨ-sɨpo	*Dioclea reflexa*	FABACEAE
yahɨsɨporan	yahɨ-sɨpo-ran	Indeterminate genus 3	[INDETERMINATE 2]
yašisɨpope	yaši-sɨpo-pe	*Bauhinia rubiginosa*	CAESALPINIACEAE
		Bauhinia sp. 1	CAESALPINIACEAE
yawapɨtaŋsɨpo	yawa-pɨtaŋ-sɨpo	*Coccoloba acuminata*	POLYGONACEAE
yere'a	yere'a	*Posadea sphaerocarpa*	CUCURBITACEAE
yikiri'ɨ (=yikiri-sɨpo)	yikiri-'ɨ	*Acacia multipinnata*	MIMOSACEAE
yɨtɨkran	yɨtɨk-ran	*Ipomoea* aff. *squamosa*	CONVOLVULACEAE
		Ipomoea phyllomega	CONVOLVULACEAE
yuti'āirimo	yuti'āi-rimo	*Combretum laxum*	COMBRETACEAE
		Machaerium ferox	FABACEAE

Domesticates

Folk Genus	Folk Species	Botanical Species	Family
ahúi (=awaši'i)	ahúi-puku ahúi-pu'a	*Oryza sativa*	POACEAE
akayu	akayu-pɨraŋ akayu-pu'a akayu-pihun akayu-tawa	*Anacardium occidentale*	ANACARDIACEAE
awakaši	awakaši	*Persea americana*	LAURACEAE
arapawak	arapawak	*Ocimum micranthum* *Ocimum* sp. 1	LABIATEAE
arašikuran	arašiku-ran	*Annona montana*	ANNONACEAE
awaí	awa-í	*Canna indica*	CANNACEAE
awaíran	awa-í-ran	*Cardiospermum halicacabum*	SAPINDACEAE
awaši	awaši	*Zea mays*	POACEAE
irimā	irimā-te irimā-hu	*Citrus aurantiifolia*	RUTACEAE
		Citrus medica-acida	RUTACEAE
ɨwamē'ɨ	ɨwa-mē-'ɨ	*Averrhoa carambola*	OXALIDACEAE
kā	kā-te	*Saccharum*	POACEAE

Domesticates (*Continued*)

Folk Genus	Folk Species	Botanical Species	Family
	kā-howɨ	*officinarum*	
	kā-hu (= kā-tā)		
	kā-memek		
	kā-pinī		
	kā-tuwɨr		
ka'amemek	ka'a-memek	*Amaranthus oleraceus*	AMARANTHACEAE
ka'ape	ka'a-pe	*Desmodium adescendens*	FABACEAE
ka'apiher	ka'a-piher	*Eryngium foetidum*	APIACEAE
kaka	kaka	*Theobroma cacao*	STERCULIACEAE
kamana	kamana-'i-tuwɨr	*Vigna adnantha*	FABACEAE
	kamana	*Phaseolus lunatus*	FABACEAE
kanamɨ	kanamɨ	*Clibadium sylvestre*	ASTERACEAE
kāpīpiher	kāpī-piher	*Cymbopogon citratus*	POACEAE
kara	kara-hu	*Dioscorea* sp. 2	DIOSCOREACEAE
	kara-māmā	*Dioscorea trifida*	DIOSCOREACEAE
	kara-pe		
	kara-pihun		
karapatu	karapatu	*Ricinus comunis*	EUPHORBIACEAE
kase	kase	*Coffea arabica*	RUBIACEAE
kawasu	kawasu-tiha	*Lagenaria siceraria*	CUCURBITACEAE
	kawasu-ra'ɨr		
kɨ'ī	kɨ'ī-awi	*Capsicum fructescens*	SOLANACEAE
	kɨ'ī-hu	*Capsicum annum*	SOLANACEAE
	kɨ'ī-te		
kɨrawa	kɨrawa-howɨ	*Neoglaziovia variegata*	BROMELIACEAE
	kɨrawa-pɨraŋ		
kuk	kuk	*Cocos nucifera*	ARECACEAE
	kuk-anā		
kurenami	kure-nami	*Bryophyllum* sp. 1	CRASSULACEAE
kwi	kwi-te	*Crescentia cujete*	BIGNONIACEAE
	kwɨ-pu'a		
	kwi-puku		
	kwi-pihun		
	kwi-ra'ɨr		
	kwi-tiha		
mā	mā-pɨraŋ	*Mangifera indica*	ANACARDIACEAE
	ma-te		
ma'ewanem	ma'e-wa-nem	*Allium cepa*	LILIACEAE
		Allium sativum	LILIACEAE
ma'eɨwapiragran	ma'e-ɨwa-pɨraŋ-ran	*Malpighia* sp. 1	MALPIGHIACEAE
		Malpighia sp. 2	MALPIGHIACEAE
makaser	makaser-te	*Manihot esculenta*	EUPHORBIACEAE
	makaser-pɨraŋ		
	makaser-tuwɨr		

Domesticates (*Continued*)

Folk Genus	Folk Species	Botanical Species	Family
	makaser-ran		
māmā	māmā-te	*Carica papaya*	CARICACEAE
	māmā-howɨ		
	māmā-hu		
	māmā-pinī		
maneyu	maneyu-pɨraŋ	*Gossypium*	MALVACEAE
	maneyu-te	*barbadense*	
mani'ɨ	ara-rū-mani'ɨ	*Manihot esculenta*	EUPHORBIACEAE
	mani'ɨ-hu		
	mani'ɨ-paitɨ		
	mani'ɨ-pō		
	mani'ɨ-puku		
	mani'ɨ-tuwɨr-ran		
	tapayū-mani'ɨ		
	mani'ɨ-howɨ		
	mani'ɨ-pihun		
	mani'ɨ-pɨtaŋ		
	mani'ɨ-se		
	mani'ɨ-tawa		
	(= mani'ɨ-te)		
	mani'ɨ-tuwɨr		
	sarakur-mani'ɨ		
	šimokape-mani'ɨ		
	tikuwi		
	yararak-mani'ɨ		
	yaši-mani'ɨ		
maniaka	maniaka	*Manihot esculenta*	EUPHORBIACEAE
manuwi	manuwi-tuwɨr	*Arachis hypogaea*	FABACEAE
	manuwi-pɨraŋ		
marakatai	marakatai	*Zingiber*	ZINGIBERACEAE
		officinalis	
mararan	mara-ran	*Cannabis sativa*	CANNABIDACEAE
mɨkur-ka'a	mɨkur-ka'a	*Petiveria alliacea*	PHYTOLACCACEAE
mirimā	mirimā	*Rollinia mucosa*	ANNONACEAE
motoroi	motoroi	*Chenopodium*	CHENOPODIACEAE
		ambrosioides	
murukuya	murukuya	*Passiflora edulis*	PASSIFLORACEAE
nana	nana-te	*Ananas comosus*	BROMELIACEAE
	nana-yɨkɨr		
narāi'ɨ	narāi-'ɨ	*Citrus sinensis*	RUTACEAE
pako	pako-te	*Musa* sp. 2	MUSACEAE
	pako-hu		
	pako-pɨraŋ	*Musa* sp. 1	MUSACEAE
	pako-tawa		
	(= pako-ran)		
	pako-nerī		
	(= pako-kururu)		
	katumē		

Domesticates (*Continued*)

Folk Genus	Folk Species	Botanical Species	Family
	katumē-howɨ		
	pako-parawa		
	yakare-rāi		
pipirɨwa	pipir-ɨwa-te	*Cyperus corymbosus*	CYPERACEAE
	pipir-ɨwa-hu		
pɨtɨm	pɨtɨm	*Nicotiana tabacum*	SOLANACEAE
piyā'ɨ	piyā-'ɨ	*Jatropha curcas*	EUPHORBIACEAE
	piyā-'ɨ-pihun	*Jatropha gossypifolia*	EUPHORBIACEAE
pu'ɨpɨraŋ'ɨ-karaí-ma'e	pu'ɨ-pɨraŋ-'ɨ	*Adenanthera pavonina*	CAESALPINIACEAE
pu'ɨrisa	pu'ɨ-risa	*Coix lacryma-jobi*	POACEAE
pupu (= tukumā-ran)	pupu	*Bactris gasipaes*	ARECACEAE
šimo'i	šimo-'i	*Tephrosia sinapou*	FABACEAE
tawa	tawa	*Curcuma* sp. 1	ZINGIBERACEAE
taya	taya	*Xanthosoma* sp. 1	ARACEAE
tayaran	taya-ran	*Colocasia esculenta*	ARACEAE
tupāka'a	tupā-ka'a	*Hibiscus sinensis*	MALVACEAE
u'ɨhuruwɨ	u'ɨ-hu-ruwɨ	*Luffa cylindrica*	CUCURBITACEAE
u'ɨwa	u'ɨwa	*Gynerium sagittatum*	POACEAE
uruku	uruku	*Bixa orellana*	BIXACEAE
waraši	waraši-te	*Citrullus lanatus*	CUCURBITACEAE
	waraši-pinī		
waraši-ran	waraši-ran	*Cucumis anguria*	CUCURBITACEAE
waya	waya	*Psidium guajava*	MYRTACEAE
yam	yam	*Syzygium malaccencis*	MYRTACEAE
yɨtɨk	yɨtɨk-howɨ	*Ipomoea batatas*	CONVOLVULACEAE
	yɨtɨk-marū		
	yɨtɨk-marū-šī		
	yɨtɨk-pihun		
	yɨtɨk-tawa		
	yɨtɨk-tuwɨr		
	yɨtɨk-'i		
	yɨtɨk-pɨraŋ		
yurumū	yurumū-pe	*Cucurbita moschata*	CUCURBITACEAE
	yurumū-pu'a		
	yurumū-puku		

Other Unclassified Plants

Folk Genus	Folk Species	Botanical Species	Family
āyaŋaramɨra (=āyaŋaraka'a)	āyaŋ-ara-mɨra; āyaŋ-ara-ka'a	*Solanum leucocarpon*	SOLANACEAE
inaya'ɨ	inaya-'ɨ	*Maximiliana maripa*	ARECACEAE
inayakamɨ (=yetahu-kamɨ)	inaya-kamɨ	*Markleya dahlgreniana*	ARECACEAE
ɨrahurawi	ɨra-hu-ra-wi	*Rhipsalis mijosurus*	CACTACEAE
		Tillandsia usneoides	BROMELIACEAE
ɨrapapukwaha	ɨrapar-pukwa-ha	*Desmoncus polycanthus*	ARECACEAE
		Desmoncus macroacanthos	ARECACEAE
kãpĩ	kãpĩ	*Brachiaria humidicula*	POACEAE
		Hyparrhenia rufa	POACEAE
	kãpĩ-puku	Poaceae sp. 1	POACEAE
kãpĩran	kãpĩ-ran	Cyperaceae sp. 1	CYPERACEAE
karaya	karaya-pihun	*Scleria* sp. 2	CYPERACEAE
	karaya-hɨ	*Scleria* sp. 1	CYPERACEAE
		Diplasia karataefolia	CYPERACEAE
karawata	karawata	*Aechmea brevicellis*	BROMELIACEAE
		Aechmea bromeliifolia	BROMELIACEAE
		Bromelia goeldiana	BROMELIACEAE
		Guzmania lingulata	BROMELIACEAE
		Bromeliaceae sp. 1	BROMELIACEAE
kerewi (free variant = kereí)	kere-wi	*Olyra caudata*	POACEAE
kwere'ĩ	kwere'ĩ	*Bactris humilis*	ARECACEAE
ma'ewɨrapuši	ma'e-wɨra-puši	*Struthanthus marginatus*	LORANTHACEAE
marari'ɨ	marari-'ɨ	*Syagrus* sp. 1	ARECACEAE
		Syagrus cf. *inajai*	ARECACEAE
maraya'ɨ	maraya-'ɨ	*Bactris maraja*	ARECACEAE
mɨrakɨrawa	mɨra-kɨrawa	*Triumfetta rhomboidea*	TILIACEAE
		Triumfetta semitriloba	TILIACEAE
		Urena lobata	MALVACEAE
mɨrakɨrawaran	mɨra-kɨrawa-ran	*Sida cordifolia*	MALVACEAE
		Sida santaremensis	MALVACEAE
mɨrameyu	mɨra-meyu	*Orthostichorsis crinite*	PTEROBRYACEAE
		Taxithelium planum	SEMATOPHYLLACEAE
mɨriši'ɨ	mɨriši-'ɨ	*Mauritia flexuosa*	ARECACEAE

Other Unclassified Plants (*Continued*)

Folk Genus	Folk Species	Botanical Species	Family
mukaya	mukaya	*Acrocomia* sp. 1	ARECACEAE
muru'ɨ	muru-'ɨ	*Astrocaryum murumuru*	ARECACEAE
nanaran (= kurupinana)	nana-ran	*Ananas nanas*	BROMELIACEAE
owi	owi	*Geonoma baculifera*	ARECACEAE
owiran	owi-ran	*Geonoma leptospadix*	ARECACEAE
paši'ɨ	paši-'ɨ	*Socratea exorrhiza*	ARECACEAE
pinuwa'ɨ	pinuwa-'ɨ	*Oenocarpus distichus*	ARECACEAE
piri'a	piri'a	*Bactris setosa*	ARECACEAE
	piri'a-hu-'ɨ	*Bactris major*	ARECACEAE
piri'aran	piri'a-ran	*Bactris tomentosa*	ARECACEAE
takwar	takwar	*Guadua glomerata*	POACEAE
takwarɨ	takwar-ɨ	*Lasiacis ligulata*	POACEAE
	takwar-ɨ-pinī	*Piresia goeldii*	POACEAE
		Olyra latifolia	POACEAE
takwarɨran	takwar-ɨ-ran	aff. *Olyra*	POACEAE
		Ichnanthus sp. 1	POACEAE
tɨrɨrɨ	tɨrɨrɨ	*Scleria secans*	CYPERACEAE
tukumã'ɨ	tukumã-'ɨ	*Astrocaryum vulgare*	ARECACEAE
u'ɨwaran	u'ɨwa-ran	*Digitaria insularis*	POACEAE
		Imperata brasiliensis	POACEAE
urupe	urupe-pɨraŋ	Indeterminate mycotica sp. 1	MYCOTA
wasaí'ɨ	wasaí-'ɨ	*Euterpe oleracea*	ARECACEAE
yawar'ɨ	yawar-'ɨ	*Astrocaryum javari*	ARECACEAE
yetahu'ɨ	yetahu-'ɨ	*Orbignya phalerata*	ARECACEAE
yu'ɨ	yu-ɨ-pihun	*Astrocaryum* sp. 1	ARECACEAE
	yu-'ɨ	*Astrocaryum gynacanthum*	ARECACEAE

Outline for a Classification of Ka'apor Generic Plant Names Showing Number of Names for Each Nomenclatural Type

I. Literal names (simple primary lexemes)

'Tree'	68
'Herb'	8
'Vine'	10
Unclassified domesticate	46
Unclassified nondomesticate	23
TOTAL:	155

II. Metaphorical names

  1. Simple primary lexemes

    A. Modeled on a generic name for an undomesticated plant, plant part, or utilitarian condition (taste, edibility) of plants generally

      1. Model is possessed

        a. Model is possessed by an animal

'Tree'	22
'Herb'	0
'Vine'	0
Unclassified domesticate	0
Unclassified nondomesticate	0
SUBTOTAL	22

        b. Model is possessed by a habitat

'Tree'	1
'Herb'	0
'Vine'	0
Unclassified domesticate	0
Unclassified nondomesticate	
SUBTOTAL	1

      2. Model is unpossessed

'Tree'	14
'Herb'	0
'Vine'	0

Unclassified domesticate	0
Unclassified nondomesticate	1
SUBTOTAL	15

B. Modeled by analogy on a generic name for a domesticate (unpossessed)

'Tree'	16
'Herb'	0
'Vine'	0
Unclassified domesticate	0
Unclassified nondomesticate	0
SUBTOTAL	16

C. Modeled on a name for an animal, animal part, or aspect of an animal (color of feathers, texture of hide)

1. Possessed

'Tree'	20
'Herb'	0
'Vine'	2
Unclassified domesticate	0
Unclassified nondomesticate	1
SUBTOTAL	23

2. Unpossessed

'Tree'	15
'Herb'	0
'Vine'	2
Unclassified domesticate	0
Unclassified nondomesticate	0
SUBTOTAL	17

D. Modeled on a mythical beast or divinity (possessed)

'Tree'	2
'Herb'	0
'Vine'	0
Unclassified domesticate	0
Unclassified nondomesticate	0
SUBTOTAL	2

E. Modeled on location of an animal (unpossessed)

'Tree'	2
'Herb'	0
'Vine'	0
Unclassified domesticate	0
Unclassified nondomesticate	0
SUBTOTAL	2

F. Modeled on name for an inanimate object or event (unpossessed)

1. Cultural

'Tree'	34
'Herb'	0
'Vine'	0
Unclassified domesticate	0
Unclassified nondomesticate	0
SUBTOTAL	34

2. Noncultural

'Tree'	4
'Herb'	0

'Vine'	1
Unclassified domesticate	0
Unclassified nondomesticate	0
SUBTOTAL	5

Total Metaphorical Simple Primary Lexemes: 137

2. Productive primary lexemes

A. Modeled on a generic name for undomesticated plant or plant part (unpossessed)

'Tree'	3
'Herb'	1
'Vine'	2
Unclassified domesticate	0
Unclassified nondomesticate	0
SUBTOTAL	6

B. Modeled on a generic or specific name for a domesticated plant (unpossessed)

'Tree'	2
'Herb'	0
'Vine'	0
Unclassified domesticate	0
Unclassified nondomesticate	0
SUBTOTAL	2

C. Modeled on another life form label (unpossessed)

'Tree'	1
'Herb'	0
'Vine'	0
Unclassified domesticate	0
Unclassified nondomesticate	0
SUBTOTAL	1

D. Modeled on a name for an animal (possessed)

'Tree'	27
'Herb'	16
'Vine'	12
Unclassified domesticate	0
Unclassified nondomesticate	0
SUBTOTAL	55

E. Modeled on a name for a divinity (possessed)

'Tree'	0
'Herb'	0
'Vine'	1
Unclassified domesticate	0
Unclassified nondomesticate	0
SUBTOTAL	1

F. Modeled on name for a personal office or kin type (possessed)

'Tree'	3
'Herb'	0
'Vine'	1
Unclassified domesticate	0
Unclassified nondomesticate	0
SUBTOTAL	4

G. Modeled on name for an inanimate object or event (unpossessed and cultural)

'Tree'	1
'Herb'	1

'Vine'   4
Unclassified domesticate   0
Unclassified nondomesticate   0
SUBTOTAL   6

    H. Modeled on a real or perceived quality (relative hardness, color, taste, smell, shape, effect upon touching [e.g., itchiness], relative size, smoothness, genuineness, falseness) [unpossessed]

'Tree'   12
'Herb'   6
'Vine'   14
Unclassified domesticate   0
Unclassified nondomesticate   0
SUBTOTAL   32

Total Productive Primary Lexemes: 107

  3. Unproductive Primary Lexemes

    A. Modeled on generic name for a nondomesticate

      1. possessed

'Tree'   3
'Herb'   2
'Vine'   0
Unclassified domesticate   0
Unclassified nondomesticate   0
SUBTOTAL   5

      2. unpossessed

'Tree'   0
'Herb'   0
'Vine'   0
Unclassified domesticate   8
Unclassified nondomesticate   0
SUBTOTAL   8

    B. Modeled on a generic name for a domesticate

      1. possessed

'Tree'   4
'Herb'   10
'Vine'   1
Unclassified domesticate   0
Unclassified nondomesticate   1
SUBTOTAL   16

      2. unpossessed

'Tree'   0
'Herb'   3
'Vine'   5
Unclassified domesticate   0
Unclassified nondomesticate   4
SUBTOTAL   12

    C. Name incorporates 'mistaken' life-form constituent

      1. possessed

'Tree'   1
'Herb'   0
'Vine'   1
Unclassified domesticate   2
Unclassified nondomesticate   2

SUBTOTAL	6
2. unpossessed	
'Tree'	0
'Herb'	1
'Vine'	0
Unclassified domesticate	4
Unclassified nondomesticate	0
SUBTOTAL	5
D. Obscure Names	
'Tree'	2
'Herb'	10
'Vine'	4
Unclassified domesticate	7
Unclassified nondomesticate	9
SUBTOTAL	32
Total Unproductive Primary Lexemes:	84
Total Primary Lexemes (folk generics):	483

Bibliography

Abreu de Albquerque, C. R. and J. A. Juca Soares. 1968. Malva (*Urena lobata* L.). *Circular IPEAN* 13:3–20. Belém.

Ackerknecht, E. H. 1946. Natural diseases and rational treatment in primitive medicine. *Bulletin of the History of Medicine* 19(5): 467–497.

Aguiar, G. F. Souza and W. A. Neves. 1991. Postmarital residence and within-sex genetic diversity among the Urubu-Ka'apor Indians, Brazilian Amazon. *Human Biology* 63(4): 467–488.

Albuquerque, J. M. 1980. *Plantas Tóxicas no Jardim e no Campo*. Belém: Ministério da Educação e Cultura, Faculdade de Ciências Agrárias do Pará, Serviço de Documentação e Informação.

Albuquerque, M. de and E. M. Ramos Cardoso. 1980. *A Mandioca no Trópico Umido*. Brasília: Editerra.

Alcorn, J. B. 1981. Huastec noncrop resource management: Implications for prehistoric rain forest resource management. *Human Ecology* 9:395–417.

Alcorn, J. B. 1984a. *Huastec-Mayan Ethnobotany*. Austin: University of Texas Press.

Alcorn, J. B. 1984b. Development policy, forests, and peasant farms: Reflections on Huastec-managed forests contributions to commercial production and resource conservation. *Economic Botany* 38(4): 389–406.

Allem, A. C. 1987. *Manihot esculenta* is a native of the Neotropics. *FAO/IBPGR Plant Genetics Resources Newsletter* 71:22–24.

Almanak do Maranhão. 1859. Município de Tury-assú. São Luís, Maranhão.

Almeida, C. M. de. 1851. "O Tury-assu." Rio de Janeiro: Typografia de Agostinho de Freitas Guimarães.

Almeida, C. M. de 1874. *Memórias para a História do Extincto Estado do Maranhão*, vol. 2. Rio de Janeiro.

Anderson, A. B., P. H. May and M. J. Balick. 1991. *The Subsidy from Nature: Palm Forests, Peasantry, and Development on an Amazon Frontier*. New York: Columbia University Press.

Anderson, A. B. and D. A. Posey. 1985. Manejo de cerrado pelos índios Kayapó. *Boletim do Museu Paraense Emílio Goeldi*, sér. Bot. 2(1): 77–98.

Anderson, A. B. and D. A. Posey. 1989. Management of a tropical scrub savanna by the Gorotire Kayapó of Brazil. *Advances in Economic Botany* 7:159–173.

Anderson, E. 1969. *Plants, Man, and Life*. (Revised edition). Berkeley: University of California Press.

Andreazza, M. 1974. Pronunciamento por ocasião da entrega ao tráfego das rodovias Belém-Brasilia e Belém-São Luís, em 13 de fevereiro de 1974. Page 202 in "Perspectivas para os transportes." Brasília: Ministério de Transportes.

Anonymous. 1897. Extracção de borracha. *Diário do Maranhão* (23 September). São Luís, Maranhão.

Anonymous. 1912a. Os índios do Gurupy. *Correio da Tarde* (12 January). Rio de Janeiro.

Anonymous. 1912b. Cruzada Gonçalves Dias: os índios do Gurupy. *Pacotilha da Manhã* (8 January). Rio de Janeiro.

Anonymous. 1915. "Relatório do Ministério de Agricultura." Rio de Janeiro: Ministério de Agricultura.

Anonymous. 1917. "Relatório do Ministério de Agricultura." Rio de Janeiro: Ministério de Agricultura.

Anonymous 1920a. Os selvícolas do Gurupy. *Folha do Norte* (26 January). Belém.

Anonymous. 1920b. Os índios Urubús em Bragança. *Estado do Pará* (22 October). Belém.

Anonymous. 1928. Os índios Urubús: O terror das regiões maranhenses do Alto Gurupy. *O Correio da Manhã* (13 January). Rio de Janeiro.

Anonymous. 1929a. A pacificação dos índios Urubús. *O País* (17 December). Rio de Janeiro.

Anonymous. 1929b. Nos sertões do extremo Norte. *O Globo* (14 January). Rio de Janeiro.

Anonymous. 1970. Considerações sobre o Projeto de Colonização do Alto Turi-PCAT (unpublished report). Recife, Pernambuco, and São Luís, Maranhão: SUDENE (Superintendência de Desenvolvimento do Nordeste).

Anonymous. 1972. *O Projeto de Colonização do Alto Turi*. Recife: Ministério do Interior.

Anonymous. 1975. "Informe sintético sobre o Projeto de Colonização do Alto Turi e a COLONE (Companhia de Colonização do Nordeste)." Unpublished report. São Luís, Maranhão.

Anonymous. 1980a. Os Tembé: Em busca da indianidade perdida? *Porantim* (January–February). São Paulo.

Anonymous. 1980b. Nordeste 80: Cinco projetos prioritários. *Jornal do Brasil*, Suplemento Especial, (28 November). Rio de Janeiro.

Anonymous. 1980c. "Síntese de informações sobre o projeto de colonização do Alto Turi." Unpublished report. São Luís, Maranhão.

Anonymous. 1981. *Sinopse preliminar do censo demográfico*, vol. 1, tomo 1. Brasília: Instituto Brasileiro de Geografia e Estatística.

Anonymous 1992. Extração clandestina de madeira lesa o Pará e ameaça os índios. *O Liberal* (1 February). Belém.

Araújo Brusque, F. C. de. 1862. Relatório apresentado a assemblea legislativa da Província do Pará na primeira sessão da XIII Legislatura. Belém: Typographia Frederico Carlos Rhossard.

APE (Arquivo Público do Estado—São Luís, Maranhão):
  1834. *Magistrados*, vol. 2, document no. 91.
  1839. *Magistrados*, vol. 14, document no. 114.
  1840. *Magistrados*, vol. 14, document no. 145.
  1854. *Magistrados*, vol. 114, document no. 53.
  1874. *Chefe de Polícia*, vol. 24, page 374.
  1878. *Presidente*, vol. 2.
  1880. *Magistrados*, vol. 49, page 241.
  1885. *Diversas Províncias*, pages 105–106.
Arnaud, E. 1978. Notícia sobre os índios Araweté, Rio Xingu, Pará. *Boletim do Museu Paraense Emílio Goeldi, Antropologia*, no. 71.
Atran, S. 1983. Covert fragmenta and the origins of the botanical family. *Man* 18(1): 51–71.
Atran, S. 1985. The nature of folk botanical life-forms. *American Anthropologist* 87(2): 298–315.
Atran, S. 1990. *Cognitive foundations of natural history: Towards an anthropology of science*. New York: Cambridge University Press.
Augé, M. 1985. Introduction. Pages 1–15 in M. Augé, ed., *Interpreting Illness* (History and Anthropology, vol. 2). London: Harwood Academic Publishers.
Azevedo, J. L. de. 1930. Os Jesuítas no Grão Pará, suas missões e a colonização. Coimbra.
Baena, A. M. 1838. "Compêndio das eras da Província do Pará." Belém.
Baena, A. M. 1848. Respostas dadas em 1847 ao Exmo. Presidente da Província do Pará . . . sobre a communicação mercantil entre a dita província e a de Goyaz. *Revista do Instituto Histórico e Geográphico Brasileiro*, tomo X: 80–107.
Bailey, L. H., E. Z. Bailey, and staff of the Liberty Hyde Hortorium. 1976. *Hortus Third: A Concise Dictionary of Plants Cultivated in the United States and Canada*. New York: Mac-Millan.
Balée, W. 1984a. The ecology of ancient Tupi warfare. Pages 241–265 in R. B. Ferguson, ed., *Warfare, Culture, and Environment*. New York: Academic Press.
Balée, W. 1984b. "The Persistence of Ka'apor Culture." Ph.D. dissertation, Columbia University. Ann Arbor: Microfilms International.
Balée, W. 1985. Ka'apor ritual hunting. *Human Ecology* 13(4): 485–510.
Balée, W. 1986. Análise preliminar de inventário florestal e a etnobotânica Ka'apor (Maranhão). *Boletim do Museu Paraense Emilio Goeldi*, ser. Bot. 2(2): 141–167.
Balée, W. 1987. Cultural forests of the Amazon. *Garden* 11(6): 12–14, 32.
Balée, W. 1988a. The Ka'apor Indian wars of lower Amazonia, ca. 1825–1928. Pages 155–169 in R. R. Randolph, D. Schneider, and M. N Diaz (eds.), *Dialectics and Gender*. Boulder: Westview Press.
Balée, W. 1988b. Indigenous adaptation to Amazonian palm forests. *Principes* 32(2): 47–54.
Balée, W. 1989a. The culture of Amazonian forests. *Advances in Economic Botany* 7:1–21.
Balée, W. 1989b. Nomenclatural patterns in Ka'apor ethnobotany. *Journal of Ethnobiology* 9(1):1–24.
Balée, W. 1989c. Cultura na vegetação da Amazônia brasileira. Pages 95–109 in W. A. Neves, ed., Biologia e ecologia humana na Amazônia: Avaliação e perspectivas. *Coleção Eduardo Galvão*. Belém: Museu Paraense Emílio Goeldi, CNPq.
Balée, W. 1990. Ka'apor: Forêt en otage dans l'etat du Maranhão. *Ethnies* 11–12:106–110.
Balée, W. 1992a. People of the fallow: A historical ecology of foraging in Lowland South

America. Pages 35–57 in K. H. Redford and C. Padoch (eds.), *Conservation of Neotropical Forests: Building on Traditional Resource Use*. New York: Columbia University Press.

Balée, W. 1992b. Indigenous history and Amazonian biodiversity. Pages 185–197 in H. K. Steen and R. P. Tucker (eds.), *Changing Tropical Forests: Historical Perspectives on Today's Challenges in Central and South America*. Durham, North Carolina: Forest History Society.

Balée, W. and D. G. Campbell. 1990. Evidence for the successional status of liana forest (Xingu River basin, Amazonian Brazil). *Biotropica* 22(1): 36–47.

Balée, W. and D. Daly. 1990. Resin classification by the Ka'apor Indians. *Advances in Economic Botany* 8:24–34.

Balée, W. and A. Gely. 1989. Managed forest succession in Amazonia: The Ka'apor case. *Advances in Economic Botany* 7:129–158.

Balée, W. and D. Moore. 1991. Similarity and variation in plant names in five Tupí-Guaraní languages (Eastern Amazonia). *Bulletin of the Florida Museum of Natural History, Biological Sciences* 35(4): 209–262.

Barbosa, A. Lemos. 1951. *Pequeno Vocabulário Tupi-Português*. Rio de Janeiro: Livraria São José.

Barbosa Rodrigues, J. 1875. *O Rio Capim*. Rio de Janeiro: Typographia Nacional.

Barbosa Rodrigues, J. 1894. *Vocabulário Indígena*. Rio de Janeiro: Typographia Leuzinger.

Barbosa Rodrigues, J. 1905. Mbaé kaá Tapyeté Enoyndaua ou Botânica da nomenclatura indígena. In *3° Congresso Científico Latinoamericano*, vol. 1. Rio de Janeiro: Impressa Nacional.

Bateson, G. 1972. *Steps to an Ecology of Mind*. New York: Random House.

Bendor-Samuel, D. H. 1966. Hierarchical structures in Guajajara. Unpublished Ph.D. dissertation, University of London.

Bennett, J. W. 1976. *The Ecological Transition: Cultural Anthropology and Human Adaptation*. New York: Pergamon Press.

Bergh, B. O. 1976. Avocado. Pages 148–151 in N. W. Simmonds, ed., *Evolution of Crop Plants*. London: Longman.

Berlin, B. 1973. Folk systematics in relation to biological classification and nomenclature. *Annual Review of Ecology and Systematics* 4:259–271.

Berlin, B. 1976. The concept of rank in ethnobiological classification: Some evidence from Aguaruna folk botany. *American Ethnologist* 3:381–399.

Berlin, B. 1992. *Ethnobiological Classification: Principles of Categorization of Plants and Animals in Traditional Societies*. Princeton: Princeton University Press.

Berlin, B., D. E. Breedlove, and P. H. Raven. 1973. General principles of classification and nomenclature in folk biology. *American Anthropologist* 75(1): 214–242.

Berlin, B., D. E. Breedlove, and P. H. Raven. 1974. *Principles of Tzeltal Plant Classification*. New York: Academic Press.

Berlin, B. and P. Kay. 1991. *Basic Color Terms: Their Universality and Evolution*. Berkeley: University of California Press.

Berredo, B. P. de. 1849. *Annaes Históricos do Estado do Maranhão*. Second edition. São Luís: Typographia Maranhense.

Betendorf, J. F. 1909. Chrónica da missão dos Padres da Companhia de Jesus no Estado do Maranhão. *Revista do Instituto Histórico e Geográphico*, tomo 72, series 1, vol. 119.

Bonavia, D. and A. Grobman. 1989. Andean maize: Its origins and domestication. Pages 456–470 in D. R. Harris and G. C. Hillman (eds.), *Foraging and Farming: The Evolution of Plant Exploitation*. London: Unwin Hyman.

Boom, B. M. 1986. A forest inventory in Amazonian Bolivia. *Biotropica* 18(4): 287–294.

Boom, B. M. 1987. Ethnobotany of the Chácobo Indians, Beni, Bolivia. *Advances in Economic Botany* 4:1–68.

BPB (Biblioteca Pública de Belém):
  1822a. Códice 399, document no. 131.
  1822b. Códice 399, document no. 145.
  1825. Códice 429, document no. 1.
  1829. Códice 458, document no. 40.

Brochado, J. P. 1977. *Alimentação na floresta tropical.* Caderno no. 2. Porto Alegre: Universidade Federal do Rio Grande do Sul.

Brown, C. H. 1977. Folk botanical life-forms: Their universality and growth. *American Anthropologist* 79:317–342.

Brown, C. H. 1984. *Language and Living Things.* New Brunswick, New Jersey: Rutgers University Press.

Brown, C. H. 1985. Mode of subsistence and folk biological taxonomy. *Current Anthropology* 26(1): 43–64.

Brown, M. F. 1986. *Tsewa's Gift: Magic and Meaning in an Amazonian Society.* Washington, D.C.: Smithsonian Institution Press.

Brown, S. and A. E. Lugo. 1990. Tropical secondary forests. *Journal of Tropical Ecology* 6:1–32.

Brücher, H. 1989. *Useful Plants of Neotropical Origin and Their Wild Relatives.* Berlin: Springer-Verlag.

Buhler, A. 1948. Dyeing among primitive people. *Ciba Review* 68:2478–2512.

Bush, M. B., R. P. Dolores and P. A. Colinvaux. 1989. A 6,000 year history of Amazonian maize cultivation. *Nature* 340:303–305.

Calogeras, P. 1938. As minas do Brasil e sua legislação. *Coleção Brasiliana*, vol. 134. São Paulo: Companhia Editora Nacional.

Cameron, J. W. and R. K. Soost. 1976. Citrus. Pages 261–264 in N. W. Simmonds, ed., *Evolution of Crop Plants.* London: Longman.

Campbell, A. T. 1989. *To Square with Genesis: Causal Statements and Shamanic Ideas in Wayãpí.* Iowa City: University of Iowa Press.

Campbell, D. G., D. C. Daly, G. T. Prance and U. N. Maciel. 1986. Quantitative ecological inventory of terra firme and várzea tropical forest on the Rio Xingu, Brazilian Amazon. *Brittonia* 38(4): 369–393.

Carneiro, R. L. 1970. Hunting and hunting magic among the Amahuaca of the Peruvian montaña. *Ethnology* 9(4): 331–341.

Carvalho, J. 1954. Aviso do P. I. A. Pedro Dantas (January–December). *Relatórios das Inspetorias: Inspetoria Regional 2 (Pará).* Document in Museu do Índio, Rio de Janeiro: Serviço de Proteção aos Índios.

Carvalho, J. 1958. Aviso do P. I. A. Pedro Dantas (May, June, August, October–December). In *Relatórios das Inspetorias: Inspetoria Regional 2 (Pará).* Document in Museu do Índio, Rio de Janeiro: Serviço de Proteção aos Índios.

Carvalho, J. 1962. Aviso do P. I. A. Pedro Dantas (April, June). In *Relatórios das Inspetorias: Inspetoria Regional 2 (Pará).* Document in Museu do Índio, Rio de Janeiro: Serviço de Proteção aos Índios.

Cavalcante, P. B. 1976. *Frutas Comestíveis da Amazônia.* Third edition. Manaus and Belém: INPA.

Cavalcante, P. B. 1988. *Frutas Comestíveis da Amazônia*. Fourth edition. Belém: Museu Paraense Emílio Goeldi.

Chan, H. T., Jr. 1983. Guava. Pages 351–359 in H. T. Chan, Jr., ed., *Handbook of Tropical Foods*. New York: Marcel Dekker.

Chernela, J. M. 1982. An indigenous system of forest and fish management in the Uaupés basin of Brazil. *Cultural Survival Quarterly* 6(2): 17–18.

Chmyz, I. and Z. C. Sauner. 1971. Nota prévia sobre as pesquisas arqueológicas no Vale do Rio Piquiri. *Dédalo* 7(13): 7–35.

Clastres, P. 1968. Ethnographie des indiens Guayaki (Paraguay-Brésil). *Journal de la Société des Americanistes* 57:9–61.

Clastres, P. 1972. The Guayaki. Pages 138–174 in M. G. Bicchieri (ed.), *Hunters and Gatherers Today*. New York: Holt, Rinehart, and Winston.

Clastres, P. 1973. Éléments de démographie amérindienne. *L'Homme* 13 (1,2): 23–36.

Cleary, D. 1990. *Anatomy of the Amazon Gold Rush*. London: Macmillan Press with St. Anthony's College.

Clement, C. R. 1989. A center of crop genetic diversity in western Amazonia. *Bioscience* 39(9): 624–631.

Coimbra, C. E. A., Jr. 1985. Estudos de ecologia humana entre os Suruí do Parque Indígena Aripuana, Rondônia: Plantas de importância econômica. *Boletim do Museu Paraense Emílio Goeldi, Antropologia* 2(1): 37–55.

Conklin, H. 1954. The Relation of Hanunóo Culture to the Plant World. Unpublished Ph.D. dissertation, Yale University.

Conklin, H. 1969. Lexicographical treatment of folk taxonomies. Pages 41–59 in S. A. Tyler, ed., *Cognitive Anthropology*. New York: Holt, Rinehart and Winston.

Cooper, J. M. 1949. Fire making. Pages 283–292 in J. H. Steward ed., *Handbook of South American Indians (vol. 5): The Comparative Ethnology of South American Indians*, Bulletin 143, Bureau of American Ethology. Washington, D.C.: Government Printing Office.

Coppens, W. 1975. Contribución al estudio de las actividades de subsistencia de los Hotis del Rio Kaima. *Boletin Indigenista Venezolano* 16(12): 65–78.

Coppens, W. and P. Mitrani. 1974. Les indiens Hoti. *L'Homme* 14(3–4): 131–142.

Corrêa, M. Pio. 1984. *Dicionário das Plantas Úteis do Brasil e das Exóticas Cultivadas*. 6 volumes. Brasilia: Ministério da Agricultura, IBDF.

Coursey, D. G. 1968. The edible aroids. *World Crops* 20(4): 25–30.

Coursey, D. G. 1976. Yams. Pages 70–74 in N. W. Simmonds, ed., *Evolution of Crop Plants*. London: Longman.

Culpepper, N. 1813 (15th edition; orig.1680). *English Physician and Complete Herbal*. London: W. Lewis.

Cunha, P. 1987. Análise fonêmica da língua Guajá. Unpublished master's thesis, UNICAMP, Campinas, São Paulo.

Daly, D. C. and G. T. Prance. 1989. Brazilian Amazon. Pages 401–425 in D. G. Cambell and H. D. Hammond (eds.), *Floristic Inventory of Tropical Countries: The Status of Plant Systematics, Collections, and Vegetation Plus Recommendations for the Future*. Bronx: New York Botanical Garden.

Davis, E. W. and J. A. Yost. 1983a. The ethnomedicine of the Waorani of Amazonian Ecuador. *Journal of Ethnopharmacology* 9(2): 272–297.

Davis, E. W. and J. A. Yost. 1983b. The ethnobotany of the Waorani of eastern Ecuador. *Botanical Museum Leaflets* 3:159–211.

Dean, W. 1984. Indigenous populations of the São Paulo–Rio de Janeiro coast: Trade, aldeamento, slavery and extinction. *Revista de História*, n.s., 117:3–26.

de Certeau, Michel. 1988. *The Writing of History.* Translated by T. Conley. New York: Columbia University Press.

Dempsey, J. M. 1975. *Fiber crops.* Gainesville: The University Presses of Florida.

Denevan, W. M. 1971. Campa subsistence in the Gran Pajonal, eastern Peru. *The Geographical Review* 61(4): 496–518.

Denevan, W. M. 1976. The aboriginal population of Amazonia. Pages 205–234 in W. Denevan, ed., *The Native Population of the Americas in 1492.* Madison: University of Wisconsin Press.

Denevan, W. M. 1992. The pristine myth: The landscape of the Americas in 1492. *Annals of the Association of American Geographers* 82(3): 369–385.

Denevan, W. M. and C. Padoch (eds.). 1988. Swidden-fallow agroforestry in the Peruvian Amazon. *Advances in Economic Botany* 5:1–106.

Denevan, W. M. and J. M. Treacy. 1988. Young managed fallows at Brillo Nuevo. *Advances in Economic Botany* 5:8–46.

Dodt, G. 1939. Descripção dos Rios Paranaíba e Gurupy. *Coleção Brasiliana* vol. 138. São Paulo: Cia. Editora Nacional.

Domingues, V. 1953. *O Turiaçu.* São Luís, Maranhão.

Dooley, R.A. 1982. *Vocabulário do Guarani.* Brasília: Summer Institute of Linguistics.

Ducke, A. 1946. Plantas de cultura precolombiana na Amazônia brasileira: Notas sôbre as espécies ou formas espontâneas que supostamente lhes teriam dado origem. *Boletim Técnico do Instituto Agronômico do Norte* 8:1–24.

Ducke, A. 1949. Notas sobre a flora Neotrópica, II: As leguminosas da Amazônia brasileira. *Boletim Técnico do Instituto Agronômico do Norte* 18:1–248.

Ducke, A. 1953. As espécies brasileiras do gênero *Theobroma* L. *Boletim Técnico do Instituto Agronômico do Norte* 28:1–20.

Ducke, A. and G. A. Black. 1953. Phytogeographical notes on the Brazilian Amazon. *Anais da Academia Brasileira de Ciências* 25(1): 1–46.

Fanshawe, D. B. 1954. Forest types of British Guiana. *The Caribbean Forester* 15(3): 73–111.

FAO (Food and Agriculture Organization of the United Nations). 1986. Food and fruit-bearing forest species 3: Examples from Latin America. *FAO Forestry Paper* 44/3. Rome: FAO.

Ferwerda, F. P. 1976. Coffees. Pages 257–260 in N. W. Simmonds, ed., *Evolution of Crop Plants.* London: Longman.

Ford, R. I., ed. 1978. The nature and status of ethnobotany. *Anthropological Papers* no. 67. Ann Arbor: Museum of Anthropology, University of Michigan.

Foucault, M. 1973. *The Order of Things: An Archaeology of the Human Sciences.* London: Tavistock.

Frake, C. O. 1969. The ethnographic study of cognitive systems. Pages 28–41 in S. A. Tyler, ed., *Cognitive Anthropology.* New York: Holt, Rinehart and Winston.

Frazer, J. G. 1963 (abridged edition; orig. 1922). *The Golden Bough: A Study in Magic and Religion.* New York: Macmillan.

Freiberg, M. 1982. *Snakes of South America.* Neptune, New Jersey: T. F. H. Publications.

Fróes Abreu, S. 1931. *Na terra das palmeiras.* Rio de Janeiro: Oficina Industrial Gráfica.

Fróis, R. L. 1953. Estudo sôbre a Amazônia maranhense e seus limites florísticos. *Revista Brasileira de Geografia* 15(1): 96–100.

Galvão, E. 1979. *Encontro de Sociedades: Índios e Brancos no Brasil.* Rio de Janeiro: Editora Paz e Terra.

Gellner, E. 1988. *Plough, Sword and Book: The Structure of Human History.* London: Collins Harvill.

Gentry, A. H. 1988. Tree species richness of upper Amazonian forests. *Proceedings of the National Academy of Sciences USA (Ecology)* 85:156–159.

Gerber, P. R., ed. 1991. *Ka'apor: Menschen des Waldes und ihre Federkunst, Eine bedrohte Kultur in Brasilien.* Zurich: OZV Offizin Zürich-Verlang, Völkerkundemuseum der Universität Zürich.

Gilmour, J. S. L. 1961. Taxonomy. Pages 27–45 in A. M. Macleod and L. S. Cobley (eds.), *Contemporary Botanical Thought.* Edinburgh: Oliver and Boyd.

Glenboski, L. L. 1983. *The Ethnobotany of the Tukuna Indians, Amazonas, Colombia.* Bogotá: Universidad Nacional de Colombia.

Gomes, M. P. 1977. The ethnic survival of the Tenetehara Indians of Maranhão, Brazil. Ph.D. dissertation, University of Florida. Ann Arbor: Microfilms International.

Gomes, M. P. 1988. *Os Índios e o Brasil.* Petrópolis, Brazil: Editora Vozes.

Goodenough, W. H. 1956. Componential analysis and the study of meaning. *Language* 32(1): 195–216.

Goody, J. 1977. *The Domestication of the Savage Mind.* Cambridge: Cambridge University Press.

Gottsberger, G. 1978. Seed dispersal by fish in the inundated regions of Humaitá, Amazonia. *Biotropica* 10(3): 170–183.

Goulding, M. 1980. *The Fishes and the Forest.* Berkeley: University of California Press.

Greig-Smith, P. 1983. *Quantitative Plant Ecology.* Berkeley: University of California Press.

Grenand, P. 1980. Introduction à l'étude de l'univers Wayãpi: Ethnoécologie des indiens du Haut-Oyapock (Guyane Française). *Langues et Civilisations à Tradition Orale* 40. Paris: SELAF.

Grenand, P. 1982. Ainsi parlaient nos ancêtres. *Travaux et documents de L'ORSTOM,* no. 148. ORSTOM, Paris.

Gross, D. R. 1982a. The Indians and the Brazilian frontier. *Journal of International Affairs* 36(1): 1–14.

Gross, D. R., G. Eiten, N. Flowers, M. F. Leoi, M. Ritter and D. Werner. 1979. Ecology and acculturation among native peoples of central Brazil. *Science* 206:1043–1050.

Guimarães, J. 1887. Relatório da Commissão de reconhecimento do Rio Parauá e suas margens. Pages 62–67 in *Relatório do Presidente da Província do Maranhão (Anexo), José Bento de Araújo.* São Luís, Maranhão.

Hames, R. B. 1980. Game depletion and hunting zone rotation among the Ye'kwana and Yanomamo of Amazonas, Venezuela. Pages 31–66 in R. B. Hames, ed., *Working Papers on South American Indians (No. 2): Studies in Hunting and Fishing in the Neotropics.* Bennington, Vermont: Bennington College.

Hames, R. B. 1983. The settlement pattern of a Yanomamö population bloc: A behavioral ecological interpretation. Pages 393–427 in R. B. Hames and W. T. Vickers (eds.), *Adaptive Responses of Native Amazonians.* New York: Academic Press.

Hames, R. B. 1987. Game conservation or efficient hunting? Pages 92–107 in B. J. McCay and J. M. Acheson (eds.), *The Question of the Commons: The Culture and Ecology of Communal Resources.* Tucson: University of Arizona Press.

Hames, R. B. and W. T. Vickers. 1982. Optimal diet breadth theory as a model to explain variability in Amazonian hunting. *American Ethnologist* 9(2): 358–378.

Harlan, J. R. 1971. Agricultural origins: Centers and non-centers. *Science* 174:468–474.

Harlan, J. R. 1975. *Crops and Man*. Madison, Wisconsin: American Society of Agronomy, Crop Science Society of America.

Harris, D. R. 1989. An evolutionary continuum of people-plant interaction. Pages 11–26 in D. R. Harris and G. C. Hillman (eds.), *Foraging and Farming: The Evolution of Plant Exploitation*. London: Unwin Hyman.

Harris, D. R. and G. C. Hillman, eds. 1989. *Foraging and Farming: The Evolution of Plant Exploitation*. London: Unwin Hyman.

Hawkes, J. G. 1989. The domestication of roots and tubers in the American tropics. Pages 481–503 in D. R. Harris and G. C. Hillman (eds.), *Foraging and Farming: The Evolution of Plant Exploitation*. London, Unwin Hyman.

Hays, T. 1979. Plant classification and nomenclature in Ndumba, Papua New Guinea Highlands. *Ethnology* 18(3): 253–270.

Hays, T. 1983. Ndumba folk biology and general principles of ethnobotanical classification and nomenclature. *American Anthropologist* 85(3): 592–611.

Headland, T. N. 1983. An ethnobotanical anomaly: The dearth of binomial specifics in a folk taxonomy of a Negrito hunter-gatherer society in the Philippines. *Journal of Ethnobiology* 2(3): 109–120.

Heiser, C. B., Jr. 1989. Domestication of Cucurbitaceae: *Cucurbita* and *Lagenaria*. Pages 471–480 in D. R. Harris and G. C.Hillman (eds.), *Foraging and Farming: The Evolution of Plant Exploitation*. London: Unwin Hyman.

Hemming, J. 1978. *Red Gold: The Conquest of the Brazilian Indians*. London: MacMillan.

Hern, W. M. 1975. The illness parameters of pregnancy. *Social Science and Medicine* 9:365–372.

Hoehne, F. C. 1939. *Plantas e Substâncias Vegetais Tóxicas e Medicinais*. São Paulo: Departamento de Botânica do Estado.

Holmberg, A. 1969. *Nomads of the Long Bow*. Garden City, New York: Natural History Press.

Holmes, Walter C. 1990. *Flore Louisiane: An Ethno-Botanical Study of French-speaking Louisiana*. Lafayette: University of Southwestern Louisiana.

Horton, R. 1967. African traditional thought and western science. *Africa* 38:50–71, 155–187.

Horton, R. 1982. Tradition and modernity revisited. Pages 201–260 in M. Hollis and S. Lukes (eds.), *Rationality and Relativism*. Cambridge, Massachusetts: MIT Press.

Huber, J. 1909. Mattas e madeiras amazônicas. *Boletim do Museu Goeldi (Museu Paraense) de Historia Natural e Ethnographia* 6:91–225.

Hunn, E. 1982. The utilitarian factor in folk biological classification. *American Anthropologist* 84(4): 830–837.

Hunn, E. and D. H. French. 1984. Alternatives to taxonomic hierarchy: The Sahaptin case. *Journal of Ethnobiology* 4(1): 73–92.

Hurly, J. 1928. *Nos Sertões do Gurupy*. Belém.

Hurly, J. 1932a. *O Rio Gurupy*. Belém.

Hurly, J. 1932b. *Chorographia do Pará e Maranhão*. Belém.

Hurly, J. 1936. *Traços Cabanos*. Belém: Instituto Lauro Sodré.

Huxley, F. 1957. *Affable Savages: An Anthropologist Among the Urubu Indians of Brazil*. New York: Viking Press.

Irvine, D. 1989. Succession management and resource distribution in an Amazonian rain forest. *Advances in Economic Botany* 7:223–237.

Jackson, J. 1983. *The Fish People: Linguistic Exogamy and Tukanoan Identity in the Northwest Amazon.* Cambridge: Cambridge University Press.

Jennings, D. L. 1976. Cassava. Pages 81–84 in N. W. Simmonds, ed., *Evolution of Crop Plants.* London: Longman.

Johnson, A. 1975. Time allocation in a Machiguenga community. *Ethnology* 14:301–310.

Johnson, A. 1978. In search of the affluent society. *Human Nature* 1(9).

Johnson, A. 1989. How the Machiguenga manage resources: Conservation or exploitation of nature? *Advances in Economic Botany* 7:213–222.

Kakumasu, J. 1986. Urubu-Kaapor. Pages 326–403 in D. C. Derbyshire and G. K. Pullum (eds.), *Handbook of Amazonian Languages.* Amsterdam: Mouton de Gruyter.

Kakumasu, J. 1988. *Dicionário por Tópicos, Urubu-Kaapor-Português.* Brasília: Fundação Nacional do Índio, Summer Institute of Linguistics.

Kaplan, H. and K. Hill. 1985. Food sharing among Aché foragers: Tests of explanatory hypotheses. *Current Anthropology* 26(2): 223–246.

Kaufman, T. 1990. Language history in South America: What we know and how to know more. Pages 13–67 in D. L. Payne, ed., *Amazonian Linguistics: Studies in Lowland South American Languages.* Austin: University of Texas Press.

Kay, P. 1971. Taxonomy and semantic contrast. *Language* 68:866–887.

Kensinger, K. and W. H. Kracke, eds. 1981. Food taboos in lowland South America. *Working Papers on South American Indians*, number 3. Bennington, Vermont: Bennington College.

Kiemen, M. 1954. *The Indian Policy of Portugal in the Amazon Region, 1614–1693.* Washington, D. C.: Catholic University of America Press.

Kingsbury, J. M. 1964. *Poisonous Plants of the United States and Canada.* Englewood Cliffs, New Jersey: Prentice-Hall.

Kipnis, R. 1990. *Comparative Ethnoecology in Eastern Amazonia.* Unpublished supplementary report to the Jessie Smith Noyes Foundation. On file, Comparative ethnoecology project. Belém: Museu Goeldi.

Klein, R. M. 1987. *The Green World.* New York: Harper and Row.

Kozák, V., D. Baxter, L. Williamson and R. L. Carneiro. 1979. The Héta Indians: Fish in a dry pond. *Anthropological Papers of the American Museum of Natural History* 55(6): 353–434.

Kracke, W. 1978. *Force and Persuasion: Leadership in an Amazonian Society.* Chicago: University of Chicago Press.

Kubitzki, K. 1985. The dispersal of forest plants. Pages 192–206 in G. T. Prance and T. Lovejoy (eds.), *Key Environments: Amazonia.* New York: Pergamon Press.

Kuhn, T. S. 1977. Logic of discovery or psychology of research? Pages 266–292 in T. S. Kuhn, *The Essential Tension: Selected Studies in Scientific Tradition and Change.* Chicago: University of Chicago Press. (Reprinted from I. Lakatos and A. Musgrave, eds. 1970. *Criticism and the Growth of Knowledge.* Cambridge: Cambridge University Press.)

Kunkel, G. 1984. *Plants for Human Consumption: An Annotated Checklist of the Edible Phanerograms and Ferns.* Koenigstein: Koeltz Scientific Books.

Laderman, C. 1987. Review of "Primitive polluters: Semang impact on the Malaysian tropical rain forest ecosystem." *Journal of Asian Studies* 46(3): 703–704.

Lange, A. 1914. *The Lower Amazon.* New York: Putnam's.

Larrick, J., W. Yost, J. Kaplan, G. King and J. Mayhall. 1979. Patterns of health and disease among the Waorani Indians of eastern Ecuador. *Medical Anthropology* 3:147–191.

Lathrap, D. 1968. The 'hunting' economies of the tropical forest zone of South America: An attempt at historical perspective. Pages 23–29 in R. B. Lee and I. DeVore (eds.), *Man the Hunter*. Chicago: Aldine.

Lathrap, D. 1970. *The Upper Amazon*. London: Thames and Hudson.

Lathrap, D. 1977. Our father the cayman, our mother the gourd: Spinden revisited, or a unitary model for the emergence of agriculture in the New World. Pages 713–752 in C. A. Reed, ed., *The Origins of Agriculture*. The Hague: Mouton.

LeCointe, P. 1947. Árvores e plantas úteis (indígenas e aclimadas). 2a. edição. *Companhia Editora Brasiliana*, sér. 5a, vol. 251. São Paulo: Companhia Editora Nacional.

Lemle, M. 1971. Internal classification of the Tupí-Guaraní linguistic family. Pages 107–129 in D. Bendor Samuel, ed., *Tupi Studies I*, Publication no. 29. Norman, Oklahoma: Summer Institute of Linguistics Publications in Linguistics and Related Fields.

Leon, J. 1984. The spread of Amazonian crops in Mesoamerica: The botanical evidence. Pages 166–173 in D. Stone, ed., *Pre-Columbian Plant Migration*. Cambridge: Harvard University Press.

Léry, J. de 1960 (orig. French 1586). *Viagem à Terra do Brasil*. Translated by S. Milliet. São Paulo: Livraria Martins Editora.

Lévi-Strauss, C. 1966. *The Savage Mind*. Chicago: University of Chicago Press.

Lévy-Bruhl, L. 1923 (orig. French 1922). *Primitive Mentality*. Translated by L. Clare. Oxford: Clarendon Press.

Lewis, W. H. and M. P. F. Elvin-Lewis. 1977. *Medical Botany: Plants Affecting Man's Health*. New York: Wiley.

Lisboa, C. de. 1967. *História dos Animais e Árvores do Maranhão*. Lisbon: Publicações do Arquivo Histórico Ultramarino e Centro de Estudos Históricos Ultramarinos.

Lisboa, M. A. 1935. A bacia do Gurupy e suas minas de ouro. *Departamento Nacional da Produção Mineral, Boletim* no. 7. Rio de Janeiro.

Lisboa, P. L. B., U. N. Maciel and G. T. Prance. 1987. Perdendo Rondônia. *Ciência Hoje* 6(36): 48–56.

Lisboa, P. L. B., U. N. Maciel and G. T. Prance. 1991. Some effects of colonization on the tropical flora of Amazonia: A case study from Rondônia. *Kew Bulletin* 46(2): 187–204.

Lopes, R. 1916. *O Torrão Maranhense*. Rio de Janeiro: Typographia do Jornal do Commércio.

Lopes, R. 1934. Os Tupis do Gurupy (Ensaio comparativo). *Actas y Trabajos científicos del XXV Congreso Internacional de Americanistas*, tomo I: 139–171. Buenos Aires.

Loukotka, C. 1929. Le Šetá: Un nouveau dialecte Tupi. *Journal de la Société des Americanistes* 21(2): 373–398.

Lounsbury, F. G. 1956. A semantic analysis of the Pawnee kinship usage. *Language* 32:158–194.

Loureiro Fernandes, L. 1959. The Xetá: A dying people of Brazil. *Bulletin of the International Committee on Urgent Anthropological and Ethnological Research* 2:22–26.

Loureiro Fernandes, L. 1964. Les Xetá et les palmiers de la forêt de Dourados: Contribution à l'ethnobotanique du Paraná. Pages 39–43 in *VI Congrès International des Sciences Anthropologiques et Ethnologiques*, tome II. Paris: Musée de L'Homme.

Maack, R. 1968. *Geografia Física do Estado do Paraná*. Curitiba: Banco de Desenvolvimento do Paraná e Instituto de Biologia e Pesquisas Tecnológicas.

Machado, E. O. 1854. Relatório do Presidente da Província do Maranhão, 3 May 1854. São Luís, Maranhão.

Magalhães, A. C. 1985. Parakanã. Pages 19–51 in *Povos Indígenas no Brasil 8: Sudeste do Para (Tocantins)*. São Paulo: CEDI.

Manzatti, L. 1989. *Relatório: Levantamento da Ictiofauna Próximo à Aldeia de Gurupiuna, Reserva Indígena Alto Turiaçu*. Unpublished supplementary report to the Ford Foundation, on file in Comparative Ethnoecology Project. Belém: Museu Goeldi.

Marett, R. R. 1957. Fetishism. Pages 201–202 in E. R. A. Seligman, ed., *Encylopaedia of the Social Sciences*, vol. 5. New York: MacMillan.

Mariz, A. C. 1975. Relatório sobre os trabalhos na área dos índios Kaapor e Guajá. Copy filed with FUNAI, Brasília.

Marques, C. 1870. *Diccionário Histórico-Geográphico da Província do Maranhão*. Rio de Janeiro.

Martin, M. K. 1969. South American foragers: A case study in cultural devolution. *American Anthropologist* 71(2): 243–260.

Martin, P. S. 1984. Prehistoric overkill: The global model. Pages 354–403 in P. S. Martin and R. G. Klein (eds.), *Quarternary Extinctions: A Prehistoric Revolution*. Tucson: University of Arizona Press.

May, P. H., A. B. Anderson, M. J. Balick and J. M. F. Frazão. 1985. Subsistence benefits from the babassu palm (*Orbignya martiana*). *Economic Botany* 39(2): 113–129.

Maybury-Lewis, D. 1967. *Akwĕ-Shavante Society*. Oxford: Clarendon Press.

McKey, D. and S. Beckerman. N. d. Chemical ecology of manioc. Unpublished manuscript.

Medina, J. C. 1959. *Plantas Fibrosas da Flora Mundial*. Campinas, São Paulo: Instituto Agronômico.

Meggers, B. J. 1971. *Amazonia: Man and Culture in a Counterfeit Paradise*. Arlington Heights, Illinois: AHM Publishing.

Meggers, B. J., O. F. Dias, E. T. Miller and C. Perota. 1988. Implications of archeological distributions in Amazonia. Pages 275–294 in W. R. Heyer and P. E. Vanzolini (eds.), *Proceedings of a Workshop on Neotropical Distribution Patterns*. Rio de Janeiro: Academia Brasileira de Ciências.

Métraux, A. 1928. *La Civilisation Matérielle des Tribus Tupí-Guaraní*. Paris: Librairie Orientaliste Paul Geuthner.

Métraux, A. 1948. The Guarani. Pages 69–94 in J. H. Steward, ed., *Handbook of South American Indians (Vol.3): The Tropical Forest Tribes*. Bulletin 143, Bureau of American Ethnology. Washington, D.C.: Government Printing Office.

Métraux, A. and H. Baldus. 1946. The Guayaki. Pages 435–444 in J. H. Steward, ed., *Handbook of South American Indians (Vol. 2): The Marginal Tribes*. Washington D.C.: Government Printing Office.

Migliazza, E. 1982. Linguistic prehistory and the refuge model in Amazonia. Pages 497–519 in G. T. Prance (ed.), *Biological Diversification in the Tropics*. New York: Columbia University Press.

Milliken, R., P. Miller, S. R. Pollard and E. V. Wandelli. 1992. *The Ethnobotany of the Waimiri Atroari Indians of Brazil*. Kew, U.K.: Royal Botanic Gardens.

Milton, K. 1991. Comparative aspects of diet in Amazonian forest dwellers. *Philosophical Transactions of the Royal Society of London*, B 334:253–263.

MINTER (Ministério do Interior). 1984. *Atlas Climatológico da Amazônia Brasileira*. Brasília: SUDAM, PHLA (Projeto de Hidrologia e Climatologia da Amazônia).

Mitchell, J. D. and S. A. Mori. 1987. *The Cashew and its Relatives (Anacardium: Anacardiaceae)*, Memoirs of the New York Botanical Garden. Bronx: New York Botanical Garden.

Monachino, J. 1949. A revision of *Ryania* (Flacourtiaceae). *Lloydia* 12:1–29.

Moran, E. F. 1990. *A Ecologia Humana das Populações da Amazônia*. Petrópolis: Editora Vozes.

Mori, S. A., B. M. Boom, A. M. de Carvalho and T. S. dos Santos. 1983. Southern Bahian moist forests. *Botanical Review* 49:155–232.

Moretti, C. and P. Grenand. 1982. Les nivrées ou plantes ichtyotoxiques de la Guyane Française. *Journal of Ethnopharmacology* 6:139–140.

Mors, W. B. 1991. Plants active against snake bite. *Economic and Medicinal Plant Research (Vol. 5): Plants and Traditional Medicine* :353–372.

Moura, P. de 1936. O Rio Gurupy. *Serviço Geológico e Mineralógico, Boletim* no. 78. Rio de Janeiro.

Muniz, P. 1925. *Município de Ourém*. Belém.

Murphy, R. F. 1960. *Headhunter's Heritage: Social and Economic Change Among the Mundurucu Indians*. Berkeley: University of California Press.

Murphy, R. F. 1979. Lineage and lineality in lowland South America. Pages 217–224 in M. Margolis and W. Carter (eds.), *Brazil: Anthropological Perspectives, Essays in Honor of Charles Wagley*. New York: Columbia University Press.

Murphy, Y. and R. F. Murphy. 1985 (second edition). *Women of the Forest*. New York: Columbia University Press.

Myers, N. 1983. *A Wealth of Wild Species*. Boulder: Westview Press.

Myers, N. 1988. Threatened biotas: "Hot spots" in tropical forests. *The Environmentalist* 8(3): 187–208.

NAS (National Academy of Sciences). 1975. *Underexploited Tropical Plants with Promising Economic Value*. Washington, D.C.: National Academy of Sciences.

Nee, M. 1990. The domestication of *Cucurbita* (Cucurbitaceae). *Economic Botany Supplement* (P. K. Bretting, ed.) 44(3): 56–68.

Newson, L. A. 1992a. Old World epidemics in early colonial Ecuador. Pages 84–112 in N. D. Cook and W. G. Lovell (eds.), *Secret Judgments of God*. Norman: University of Oklahoma Press.

Newson, L. A. 1992b. Cultural influences on the impact of Old World diseases in colonial Latin America. Paper read at the annual meeting of the American Anthropological Association, San Francisco.

Nimer, E. 1972. Climatologia da região Nordeste do Brasil. *Revista Brasileira de Geografia* 34(2): 3–51.

Nimuendaju, C. 1914. Vocabulários da língua geral do Brasil nos dialectos dos Manajé do Rio Ararandéua, Tembé do Rio Acará Pequeno e Turiwara do Rio Acará Grande, Est. do Pará. *Zeitschrift für Ethnologie* 46:615–618.

Nimuendaju, C. 1946. *The Eastern Timbira*. University of California Publications in American Archaeology and Ethnology, vol. 41. Berkeley: University of California.

Nimuendaju, C. 1948a. The Turiwara and Aruã. Pages 193–198 in J. H. Steward, ed., *Handbook of South American Indians (Vol. 3): The Tropical Forest Tribes*, Bulletin 143, Bureau of American Ethnology. Washington, D.C.: Government Printing Office.

Nimuendaju, C. 1948b. The Tucuna. Pages 713–725 in J. H. Steward, ed., *Handbook of South American Indians (Vol. 3): The Tropical Forest Tribes*, Bulletin 143, Bureau of American Ethnology. Washington, D.C.: Government Printing Office.

Nimuendaju, C. 1948c. The Guajá. Pages 135–136 in J. H. Steward, ed., *Handbook of South American Indians (Vol. 3): The Tropical Forest Tribes*, Bulletin 143, Bureau of American Ethnology. Washington, D.C.: Government Printing Office.

Nimuendaju, C. 1983. *Os Apinayé.* Conselho Nacional de Desenvolvimento Cientifico e Tecnologico. Belém: Museu Paraense Emílio Goeldi.

Nimuendaju, C. and A. Métraux. 1948. The Amanayé. Pages 199–202 in *Handbook of South American Indians (Vol. 3): The Tropical Forest Tribes.* Bulletin 143, Bureau of American Ethnology. Washington, D.C.: Government Printing Office.

Noronha, J. M. de. 1856. Roteiro da viagem da cidade do Pará até as últimas colonias dos dominios portuguezes em os Rios Amazonas e Negro. Pages 1–102 in *Notícias para a História e Geografia das Nações Ultramarinas que Vivem nos Dominios Portuguezes ou lhes são Visinhas.* Lisbon: Academia Real das Ciências.

NRC (National Research Council). 1983. *Changing Climate: Report of the Carbon Dioxide Assessment Committee.* Washington, D.C.: National Academy Press.

NSB (National Science Board). 1989. *Loss of Biological Diversity: A Global Crisis Requiring International Solutions.* Washington, D.C.: National Science Foundation.

Oldfield, M. L. and J. B. Alcorn. 1987. Conservation of traditional agroecosystems. *Bioscience* 37(3): 199–208.

Oliveira, J. C. de. 1951. *Folclore Amazônica,* vol. 1. Belém: São José.

Ong, W. J. 1967. *The Presence of the Word: Some Prolegomena for Cultural and Religious History.* New Haven, Connecticut: Yale University Press.

Oren, D. C. 1988. Uma reserva biológica para o Maranhão. *Ciência Hoje* 8(44): 36–45.

Parker, E. 1992. Forest islands and Kayapó resource management in Amazonia: A reappraisal of the *Apêtê. American Anthropologist* 94(2): 406–428.

Pennington, T. D. 1981. *Meliaceae,* monograph no. 28, *Flora Neotropica.* Bronx: New York Botanical Garden.

Pesce, C. 1985. *Oil Palms and Other Oilseeds of the Amazon.* Translated and edited by D. V. Johnson. Algonac, Michigan: Reference Publications.

PIB (*Povos Indígenas no Brasil, 1987/88/89/90*). 1991a. Aconteceu, Especial 18. Colonos podem ser expulsos da Reserva do Alto Turiaçu (page 367). São Paulo: CEDI.

PIB (*Povos Indígenas no Brasil, 1987/88/89/90*). 1991b. Aconteceu, Especial 18. Quadrilha promove invasão das reservas Alto Turiaçu e Caru (page 368). São Paulo: CEDI.

Pickersgill, B. 1976. Pineapple. Pages 15–18 in N. S. Simmonds, ed., *Evolution of Crop Plants.* London: Longman.

Pickersgill, B. 1989. Cytological and genetical evidence on the domestication and diffusion of crops within the Americas. Pages 426–439 in D. R. Harris and G. C. Hillman (eds.), *Foraging and Farming: The Evolution of Plant Exploitation.* London: Unwin Hyman.

Pickersgill, B. and C. B. Heiser, Jr. 1977. Origins and distribution of plants domesticated in the New World tropics. Pages 803–835 in C. A. Reed, ed., *Origins of Agriculture.* The Hague: Mouton.

Pinkley, H. 1973. The ethno-ecology of the Kofán. Unpublished Ph.D. dissertation, Harvard University.

Piperno, D. R. 1989. Non-affluent foragers: Resource availability, seasonal shortages, and the emergence of agriculture in Panamanian tropical forests. Pages 538–554 in D. R. Harris and G. C. Hillman (eds.), *Foraging and Farming: The Evolution of Plant Exploitation.* London: Unwin Hyman.

Piperno, D. R. 1990. Aboriginal agriculture and land usage in the Amazon basin, Ecuador. *Journal of Archaeological Science* 17:665–677.

Pires, J. M. and G. T. Prance. 1985. The vegetation types of the Brazilian Amazon. Pages 109–145 in G. T. Prance and T. Lovejoy (eds.), *Key Environments: Amazonia*. New York: Pergamon Press.

Platzmann, J. 1896. *O Diccionário Anonymo da Língua Geral do Brasil*. Leipzig: B. G. Teubner (orig. Anon., 1795, Lisbon: Officina Patriarcal).

Plucknett, D. L. 1976. Edible aroids. Pages 10–12 in N. W. Simmonds, ed., *Evolution of Crop Plants*. London: Longman.

Plucknett, D. L., N. J. H. Smith, J. T. Williams and N. Murthi Anishetty. 1983. Crop germplasm conservation and developing countries. *Science* 220:163–169.

Popper, K. R. 1959. *The Logic of Scientific Discovery*. New York: Harper Torchbooks.

Posey, D. A. 1983. Indigenous knowledge and development: An ideological bridge to the future. *Ciência e Cultura* 35(3): 877–894.

Posey, D. A. 1984. Indigenous ecological knowledge and development of the Amazon. Pages 135–144 in E. Moran, ed., *The Dilemma of Amazonian Development*. Boulder: Westview Press.

Posey, D. A. 1985. Ethnoecology as applied anthropology in Amazonian development. *Human Organization* 43(2): 95–107.

Posey, D. A. 1986. Topics and issues in ethnoentomology with some suggestions for the development of hypothesis-generation and testing in ethnobiology. *Journal of Ethnobiology* 6(1): 99–120.

Prance, G. T. 1972. *Chrysobalanaceae*, monograph no. 9, *Flora Neotropica*. New York: Hafner Publishing Company.

Prance, G. T. 1973. Phytogeographic support for the theory of Pleistocene forest refuges in the Amazon Basin, based on evidence from distribution patterns in Caryocaraceae, Chrysobalanaceae, Dichapetalaceae and Lecythidaceae. *Acta Amazonica* 3(1): 5–28.

Prance, G. T. 1982. Forest refuges: Evidence from woody angiosperms. Pages 137–158 in G. T. Prance, ed., *Biological Diversification in the Tropics*. New York: Columbia University Press.

Prance, G. T., W. Balée, B. M. Boom and R. L. Carneiro. 1987. Quantitative ethnobotany and the case for conservation in Amazonia. *Conservation Biology* 1(4): 296–310.

Projeto Gurupi. 1975. *Relatório Final de Etapa*, vol. 1. Belém: Companhia de Pesquisa de Recursos Minerais, Superintendência Regional de Belém.

Projeto Radam. 1973. *Levantamento de Recursos Naturais*, vol. 3. Rio de Janeiro: Ministério das Minas e Energia, Departamento Nacional da Produção Mineral.

Purseglove, J. W. 1968. *Tropical Crops: Dicotyledons*, 2 volumes. New York: Wiley.

Queiroz, H. L. 1990. Annual Report to the Ford Foundation. Unpublished manuscript.

Queiroz, H. L. 1991. 'Checklist' preliminar da fauna de mamíferos da Amazonia maranhense. Paper read at Congresso Brasileiro de Zoologia, Salvador, Bahia.

Queiroz, H. L. 1992. A New Species of Capuchin Monkey, Genus *Cebus* Erxleben, 1777 (Cebidae: Primates) from Eastern Brazilian Amazonia. *Goeldiana* (*Zoologia*) 15: 1–13.

Raiol, D. A. 1970. *Motins Políticos*, vol. 3. Belém: Universidade Federal do Pará.

Rambo, A. T. 1985. *Primitive Polluters: Semang Impact on the Malaysian Tropical Rain Forest Ecosystem*. Anthropological Papers no. 76, Museum of Anthropology. Ann Arbor: University of Michigan.

Randall, R. A. and E. S. Hunn. 1984. Do life-forms evolve or do uses for life? Some doubts about Brown's hypotheses. *American Ethnologist* 11:329–349.

Raven, P. 1988. Tropical floristics tomorrow. *Taxon* 37(3): 549- 560.

Redford, K. H. 1991. The ecologically noble savage. *Cultural Survival Quarterly* 15(1): 46–48. (Reprinted from *Orion Nature Quarterly*, 1990, 9(3): 24–29.

Redford, K. H. 1992. The empty forest. *Bioscience* 42(6): 412–422.

Redford, K. H., B. Klein and C. Murcia. 1992. Incorporation of game animals into small-scale agroforestry systems in the neotropics. Pages 333–358 in K. H. Redord and C. Padoch (eds.), *Conservation of Neotropical Forests*. New York: Columbia University Press.

Redford, K. H. and A. M. S. Stearman. 1989. Local peoples and the Beni Biosphere Reserve, Bolivia. *Vida Sylvestre Neotropical* 2(1): 49–56.

Rehm, S. and G. Espig. 1984. *Die Kulturpflanzen der Tropen und Subtropen*. Gottingen: Verlag Engen Ulmen.

Ribeiro, D. 1951. Atividades científicas da secção de estudos do Serviço de Proteção aos Índios. *Sociologia* 8(4): 363–385.

Ribeiro, D. 1955. Os índios Urubus: Ciclo anual das atividades de subsistência de uma tribo da floresta tropical. *Anais do XXXI Congresso Internacional de Americanistas* 1:125–157. São Paulo: Editora Anhembi. (Reprinted in D. Ribeiro, 1976, *Uirá sai à Procura de Deus*. Rio de Janeiro: Paz e Terra.)

Ribeiro, D. 1956. Convívio e contaminação: Efeitos dissociativos da depopulação provocada por epidemias em grupos indígenas. *Sociologia* 18(1): 3–50.

Ribeiro, D. 1970. *Os Índios e a Civilização: A Integração das Populações Indígenas no Brasil Moderno*. Rio de Janeiro: Editora Civilização Brasileira.

Ribeiro, D. and B. Ribeiro. 1957. *Arte Plumária dos Índios Kaapor*. Rio de Janeiro: Editora Civilização Brasileira.

Ribeiro, J. P. 1913. *Exposição Nacional de Borracha de 1913: Maranhão*, Monographia número 4, Indústria e Commércio. Rio de Janeiro: Ministério de Agricultura.

Rivière, P. 1984. *Individual and Society in Guiana: A Comparative Study of Amerindian Social Organization*. New York: Cambridge University Press.

Rice, F. J. D. 1930. A pacificação e identificação das afinidades lingüísticas da tribu Urubu dos Estados do Pará e Maranhão, 1928–1929. *Journal de la Société des Americanistes* 22:311–316.

Ridley, H. N. 1930. *The Dispersal of Plants Throughout the World*. Ashford, Kent, England: L. Reeve and Co.

Rindos, D. 1984. *The Origins of Agriculture: An Evolutionary Perspective*. New York: Academic Press.

Rivet, P. 1924. Les Indiens Canoeiros. *Journal de la Société des Americanistes*, n.s. 16:169–181.

Rizzini, C. T. 1963. Nota prévia sôbre a divisão fitogeográfica do Brasil. *Revista Brasileira de Geografia* 25(1): 3–64.

Rodrigues, A. D. 1988. Proto-Tupi evidence for agriculture. Paper read at First International Ethnobiology Conference, Belém.

Rocha, A. I. de, M. L. H. da Silva, A. P. Mourão and M. P. Cava. 1968. *A Presença de Alcaloides em Espécies Botânicas da Amazônia*. CNPQ/INPA, Publicação no. 12. Manaus: CNPQ/INPA.

Roe, P. G. 1982. *The Cosmic Zygote: Cosmology in the Amazon Basin*. New Brunswick, New Jersey: Rutgers University Press.

Roeder, M. 1967. Recursos naturais especialmente solos no Noroeste do Maranhão, Brasil. Masters thesis, Instituto Interamericano de Ciencias Agrícolas de la OEA, Centro de Enseñanza e Investigación. Turrialba, Costa Rica.

Rogers, D. J. and S. G. Appan, 1973. *Manihot, Manihotoides (Euphorbiaceae)*, monograph no. 13. *Flora Neotropica*. New York: Hafner Press.

Roosevelt, A. 1987. Chiefdoms in the Amazon and Orinoco. Pages 153–185 in R. D. Drennan and C. A. Uribe (eds.), *Chiefdoms in the Americas*. Lanham, Maryland: University Presses of America.

Roosevelt, A. 1989a. Resource management in Amazonia before the Conquest. *Advances in Economic Botany* 7:30–62.

Roosevelt, A. 1989b. Lost civilizations of the lower Amazon. *Natural History*, February: 74–83.

Roosevelt, A. 1991. *Moundbuilders of the Amazon*. San Diego: Academic Press.

Roosmalen, M. G. M. van. 1985. *Fruits of the Guianan Flora*. Institute of Systematic Botany, Utrecht University.

Roth, I. 1981. *Structural Patterns of Tropical Barks*. Berlin: Gebruder Borntraeger.

Ruiz de Montoya, A. 1976 (reprint of 1639 edition). *Tesoro de la Lengua Guarani*, vol. 3. Leipzig: B. G. Teubner.

Saldarriaga, J.G. and D.C. West. 1986. Holocene fires in the northern Amazon basin. *Quaternary Research* 26:358–366.

Salomão, R. P., M. F. F. Silva and N. A. Rosa. 1988. Inventário ecológico em floresta pluvial tropical de terra firme, Serra Norte, Carajás, Pará. *Boletim do Museu Paraense Emílio Goeldi, sér. Bot.* 4(1): 1–46.

Sauer, C. O. 1950. Cultivated plants of South and Central America. Pages 487–543 in *Handbook of South American Indians*, (Vol. 6): Bulletin 143, Bureau of American Ethnology. Washington, D.C.: Government Printing Office.

Scarpa, A. and A. Guerci. 1982. Various uses of the castor oil plant (*Ricinus communis* L.): A review. *Journal of Ethnopharmacology* 5:117–137.

Schultes, R. E. 1977. Diversas plantas comestíveis nativas do Nordeste da Amazônia. *Acta Amazonica* 7(3): 317–327.

Schultes, R. E. 1984. Amazonian cultigens and their northward and westward migration in pre-Columbian times. Pages 19–37 in D. Stone, ed., *Pre-Columbian Plant Migration*. Cambridge: Harvard University Press.

Schulz, J. P. 1960. *Ecological Studies on Rain Forest in Northern Surinam*. Amsterdam: Noord-Hollandsche Uitgevers Maatschappij.

Shulman, S. 1986. Seeds of controversy. *Bioscience* 36(1): 647- 651.

Silva, M. 1943. Rescenciamento dos índios Timbiras, 20 de outubro. Unpaginated in *Documentos dos Postos Luíz Horta, Gorotire, Uaçá, Pedro Dantas, Felipe Cumarão, Maracassumé* (no series or volume number). Serviço de Proteção aos Índios. Document in Museu do Índio, Rio de Janeiro).

Silva, M. and E. R. Maia. 1943. Rescenciamento dos Índios Urubus, 20 de outubro. Unpaginated in *Documentos dos Postos Luíz Horta, Gorotire, Uaçá, Pedro Dantas, Felipe Camarão, Maracassumé* (no series or volume number). Serviço de Proteção aos Índios. Document in Museu do Índio, Rio de Janeiro).

Silva, M. F. da, P. L. B. Lisboa and R. C. L. Lisboa. 1977. *Nomes Vulgares de Plantas Amazônicas*. Belém: INPA.

Singh, D. 1976. Castor. Pages 84–86 in N. W. Simmonds, ed., *Evolution of Crop Plants*. London: Longman.

Singh, L. B. 1976. Mango. Pages 7–9 in N. W. Simmonds, ed., *Evolution of Crop Plants*. London: Longman.

Sleumer, H. O. 1980. *Flacourtiaceae*, Monograph no. 22. *Flora Neotropica*. Bronx: New York Botanical Garden.

Smith, N. J. H. 1980. Anthrosols and human carrying capacity in Amazonia. *Annals of the Association of American Geographers* 70(4): 553–566.

Smith, N. J. H. 1981. *Man, Fishes, and the Amazon*. New York: Columbia University Press.

Smith, P. M. 1976. Minor crops. Pages 301–324 in N. W. Simmonds, ed., *Evolution of Crop Plants*. London: Longman.

Smole, W. 1976. *The Yanoama Indians: A Cultural Geography*. Austin: University of Texas Press.

Sombroek, W. G. 1966. *Amazon Soils: A Reconnaissance of the Soils of the Brazilian Amazon Region*. Wageningen: Centre for Agricultural Publications and Documentation.

Sousa, G. S. de. 1974 (orig. 1587). *Notícia do Brasil*. São Paulo: Departamento de Assuntos Culturais do MEC.

Sousa, M. P., M. E. O. Matos, F. J. A. Matos, M. I. L. Machado and A. A. Craveiro. 1991. *Constituintes Químicos Ativos de Plantas Medicinais Brasileiras*. Fortaleza, Ceará: Laboratório de Produtos Naturais.

Sperber, D. 1985. *On Anthropological Knowledge*. Cambridge: Cambridge University Press.

Sperber, D. and D. Wilson. 1986. *Relevance: Communication and Cognition*. Cambridge: Harvard University Press.

Sponsel, L. E. 1992. The environmental history of Amazonia: Natural and human disturbances, and the ecological transition. Pages 233–251 in H. K. Steen and R. P. Tucker (eds.), *Changing Tropical Forests: Historical Perspectives on Today's Challenges in Central and South America*. Durham, North Carolina: Forest History Society.

Stearman, A. M. 1987. *No Longer Nomads: The Sirionó Revisited*. Lanham, Maryland: Hamilton Press.

Stearman, A. M. and K. H. Redford. 1992. Commercial hunting by subsistence hunters: Sirionó Indians and Paraguayan caiman in lowland Bolivia. *Human Organization* 51(3): 235–244.

Stephens, S. G. 1973. Geographical distribution of cultivated cottons relative to probable centres of domestication in the New World. Pages 239–254 in A. R. Srb, ed., *Genes, Enzymes, and Populations*. New York: Plenum.

Sternberg, H. O. 1975. *The Amazon River of Brazil*. New York: Springer Verlag.

Stocks, A. 1987. Resource management in an Amazon Várzea lake ecosystem. Pages 108–120 in B. J. McCay and J. M. Acheson (eds.), *The Question of the Commons: The Culture and Ecology of Communal Resources*. Tucson: University of Arizona Press.

Stone, E. 1962. *Medicine Among the American Indians*. New York: Hafner Publishing.

SUDAM (Superintendência do Desenvolvimento de Amazônia). 1976. *Polamazônia (Programa de Pólos Agropecuários e Agrominerais da Amazônia): Pré-Amazônia Maranhense*. Belém: SUDAM.

Tambiah, S. J. 1990. *Magic, Science, Religion, and the Scope of Rationality*. Cambridge: Cambridge University Press.

Taylor, C. 1982. Rationality. Pages 87–105 in M. Hollis and S. Lukes (eds.), *Rationality and Relativism*. Cambridge, Mass.: MIT Press.

Ter Welle, B. J. H. 1976. Silica grains in woody plants of the neotropics, especially Surinam. *Leiden Botanical Series* 3:107–142.

Tippo, W. and W. L. Stern. 1977. *Humanistic Botany*. New York: Norton.

Tirpak, D. 1990. The Intergovernmental Panel on Climate Change (IPCC): The U.S. position. Pages 41–43 in *Proceedings of the Conference on Tropical Forestry Response Options to Global Climate Change*. São Paulo: Brazilian Institute for the Environment and Renewable Natural Resources (IBAMA), University of São Paulo (and) Washington, D.C.: Environmental Protection Agency.

Toral, A. 1985. Avá-Canoeiro: Os Índios na clandestinidade. Pages 274–275 in *CEDI, Aconteceu, Especial 15*. São Paulo: Sagarana Editora Ltda.

Toral, A. 1986, Situação e perspectivas de sobrevivência dos Avá-Canoeiro. Unpublished manuscript. São Paulo: Centro Ecumênico de Documentação e Informação.

Turner, N. J. 1974. Plant taxonomic systems and ethnobotany of three contemporary Indian groups of the Pacific Northwest (Haida, Bella Coola, and Lilloet). *Syesis* 7(Suppl. 1): 1–104.

Turner, T. 1992. Things fall apart in the Amazon: Advocacy wars and two cases of Paulinho Payakan. Unpublished manuscript.

Tylor, E. B. 1958 (orig. 1871). *Primitive Culture*. New York: Harper and Brothers.

Ugent, D., S. Pozorski, and T. Pozorski. 1986. Archaeological manioc from coastal Peru. *Economic Botany* 40:78–102.

Van Steenis, C. G. G. J. 1958. Rejuvenation as a factor for judging the status of vegetation types: The biological nomad theory. Pages 212–215 in *Proceedings of a Symposium on Humid Tropics Vegetation*. Paris: UNESCO.

Vasconcellos, S. de. 1865 (orig. 1663). *Chronica da Companhia de Jesus no Estado do Brasil*. Lisbon: A. J. Fernando Lopes.

Vellard, J. 1934. Les indiens Guayaki. *Journal de la Société des Americanistes* 26:223–292.

Vellard, J. 1939. *Une Civilisation du Miel: Les Indiens Guayakis du Paraguay*. Paris: Librairie Gallimard.

Viana, A. 1975. *As Epidemias do Pará*. Belém: Universidade Federal do Pará.

Vickers, W. T. 1980. An analysis of Amazonion hunting yields as a function of settlement age. Pages 7–29 in R. B. Hames, ed., *Working Papers on South American Indians (No. 2): Studies in hunting and fishing in the Neotropics*. Bennington, Vermont: Bennington College.

Vickers, W. T. and T. Plowman. 1984. Useful plants of the Siona and Secoya Indians of eastern Ecuador. *Fieldiana, Botany*, n.s., no. 15. Chicago: Field Museum of Natural History.

Vieira, A. 1925. *Cartas do Padre Antônio Vieira*, vol. 1. Edited by J. L. de Azevedo. Coimbra.

Viveiros de Castro, E. 1986. *Araweté: Os Deuses Canibais*. Rio de Janeiro: Jorge Zahar.

Vogel, V. J. 1970. *American Indian Medicine*. Norman: University of Oklahoma Press.

Wagley, C. 1977. *Welcome of Tears. The Tapirapé Indians of Central Brazil*. New York: Oxford University Press.

Wagley, C. and E. Galvão. 1949. *The Tenetehara Indians of Brazil: A Culture in Transition*. New York: Columbia University Press.

Wallace, A. F. C. 1966. *Religion: An Anthropological View*. New York: Random House.

Watson, J. 1952. Cayuá culture change: A study in acculturation and methodology. *American Anthropologist Memoirs* 73(54), no. 2, part 2.

Wessels-Boer, J. G. 1965. *Palmae, Flora of Suriname*, vol. 5. Leiden: E. J. Brill.

Whitaker, T. W. and W. P. Bemis. 1976. Cucurbits. Pages 64–69 in N. W. Simmonds, ed., *Evolution of Crop Plants*. London: Longman.

Whitehead, R. A. 1976. Coconut. Pages 221–224 in N. W. Simmonds, ed., *Evolution of Crop Plants*. London: Longman.

Whitmore, T. C. 1992. *An Introduction to Tropical Rain Forests.* New York: Oxford University Press.

Yen, D. E. 1976. Sweet potato. Pages 42–44 in N. W. Simmonds, ed., *Evolution of Crop Plants.* London: Longman.

Yost, J. A. 1978. *El Desarrollo Comunitário y la Supervivencia Étnica: El Caso de los Huaorani, Amazonia Ecuatoriana.* Mérida, Mexico.

Yost, J .A. 1981a. People of the forest: The Waorani. Pages 96–115 in *Ecuador: In the Shadow of the Volcanoes.* Quito: Ediciones Libri Mundi.

Yost, J. A. 1981b. Twenty years of contact: The mechanisms of change in Wao ("Auca") culture. Pages 677–704 in N. E. Whitten, Jr., ed., *Cultural Transformations and Ethnicity in Modern Ecuador.* Urbana: University of Illinois Press.

Index

Note: All page numbers in italics refer to tables and figures.

Printed in the USA
CPSIA information can be obtained
at www.ICGtesting.com
JSHW051456221024
72172JS00010B/86

9 780231 074858